MATHS IN ACTION

Higher Mathematics

Second Edition

E Mullan

OXFORD
UNIVERSITY PRESS

Great Clarendon Street, Oxford, OX2 6DP, United Kingdom

Oxford University Press is a department of the University of Oxford. It furthers the University's objective of excellence in research, scholarship, and education by publishing worldwide. Oxford is a registered trade mark of Oxford University Press in the UK and in certain other countries

First published in 2014

British Library Cataloguing in Publication Data
Data available

978-1-4085-2381-0

1 3 5 7 9 10 8 6 4 2

Printed in Great Britain

Acknowledgements
The publishers would like to thank the following for permissions to use their photographs: **Cover**: joanek/iStockphoto; **p1 l**: IgorGolovniov/Shutterstock; **p1 r**: rook76/Shutterstock; **p4**: cafaphotos/iStockphoto; **p5**: maxphotography/iStockphoto; **p27 top**: Hugh Threlfall/Alamy; **p31**: wynnter/iStockphoto; **p45**: Science Photo Library; **p62**: Ivan Vdovin/Alamy; **p72**: Art Directors & TRIP/Alamy; **p87**: GeorgiosArt/iStockphoto; **p108**: sassphotos/iStockphoto; **p115 top**: the_guitar_mann/iStockphoto; **p118**: sassphotos/iStockphoto; **p123**: Miriam and Ira D. Wallach Division of Art, Prints and Photographs/New York Public Library/Science Photo Library; **p127**: jamesbenet/iStockphoto; **p140**: GeorgiosArt/iStockphoto; **p141**: teddy2007b/iStockphoto; **p144 top**: ooyoo/iStockphoto; **p144 bottom**: ewg3D/iStockphoto; **p170**: theasis/iStockphoto; **p17 7**: perkmeup/iStockphoto; **p184 (tablet)**: marlanu/iStockphoto; **p184 (compass)**: DNY59/iStockphoto; **p188**: d-l-b/iStockphoto; **p199**: ZU_09/iStockphoto; **p211**: 4774344sean/iStockphoto; **p217 top**: sorendls/iStockphoto; **p219**: -Oxford-/iStockphoto

All other photographs were provided by the author.

Microsoft product screenshots reprinted with permission from Microsoft Corporation.

We are grateful for permission to reproduce material from the Higher Mathematics Course Assessment Specification:

p212–13: Copyright © Scottish Qualifications Authority

Contents

Introduction

This textbook addresses the needs of students following the Curriculum for Excellence Higher Mathematics course as described in the specifications published by the Scottish Qualifications Authority.

The book has been structured in order that the units – Expressions and functions (Unit 1), Relationships and calculus (Unit 2), and Applications (Unit 3) – can be easily identified.

Where every endeavour has been made to place the three units in distinct sections of the book, it has not been felt appropriate in three places:

- Factorising a cubic expression (Unit 1) has been associated with solving the cubic equation (Unit 2), as the latter provides a raison d'être for the former.
- Simplifying using the laws of logs (Unit 1) is included with the rest of logs and exponential exposition (Unit 2), viz. equations and linearization, to avoid fragmentation of the topic.
- The introduction to integration (Unit 2) is linked to the application of integration (Unit 3).

The chapters have been organised in the general order of the SQA units. (However, this should not inhibit your preferred running order of topics.)

It is assumed that students have covered the content of a course equivalent to that described in the National 5 Maths specifications. In particular, the student should have covered an introduction to vectors and should be able to find the equation of a straight line given a point (a, b) on the line and the gradient, m, using the formula $y - b = m(x - a)$.

All chapters have a similar background structure:

- An introductory section, 'Before we start…', provides an opportunity to learn about the historical context and people who developed the topic – a possible entry into the subject, and a background for discussion, approach and purpose.

  Some people initially feel that they don't have time for any extension to the introduction to a topic. The paradox is that if a student has a firm grip on the subject at the onset, development speeds up and the topic is covered to completion quicker.

  For example, exposing the student to the old-fashioned notion of log tables leads to an acceleration of the understanding of the concept and the laws of logs.
- The 'What you need to know' section is designed to check that the knowledge required for the chapter is in place, that any minor gap in the students' understanding is bridged, and that students are up to speed, ready to commence a new topic.
- Exposition, with worked examples, is followed by exercises containing drill and problem solving situations. In line with the CfE philosophy, much of the contexts are chosen to show application and occurrence in the broad perspective of the everyday world.
- 'Preparation for assessment' provides atomistic, short response, and extended response questions, of the calibre that might be encountered during assessment. Cross-curricular questions are also included.
- The chapter concludes with a summary to supply the student with a quick reference to the requirements of the course covered by the chapter.

Chapter 14 provides an opportunity for the student to prepare for course assessment. The general chapter structure is utilised slightly differently.

- 'Before we start...' becomes a listing of the 'given' formulae in the external exam. Students should get in the habit of using such a reference.
- 'What you need to know' lists the mandatory skills as defined by the SQA[1].
- The first set of questions will act as an auditing exercise for students. Questions wrong or undoable should act as an alarm that some mandatory skill needs to be revised.
- Then follows an exercise on non-calculator questions both short and extended response. Questions of this type could occur in a Paper 1 course assessment. Marks have been associated with each task as a guide to their possible worth in the course assessment. These marks, the author's judgement, may differ slightly from the judgement and opinion of the SQA.
- The last exercise visits the questions that could occur in a Paper 2 course assessment.

Answers to all exercise questions are included at the back of the book.

Throughout the book, various icons have been used to identify particular features.

  indicate the most appropriate mode of tackling a problem. As part of the assessment measures students' numerate abilities without the aid of a calculator, it is essential that the parts marked 'non-calculator' are indeed done without a calculator.

 indicates a question for which some research or investigative work would enrich the student's experience.

 indicates opportunities to explore the power of the spreadsheet. The use of IT for mathematical investigations should be thoroughly explored.

The education of students to Higher Level in mathematics is the principal aim of the book. Passing the exam is a secondary, though essential, target. Many students will go on to further education to use the maths that they have learned.

Proofs, justifications and exposition, many 'non-examinable' materials, are included in the book. Students should be exposed to these for a holistic experience and for the future.

Finally, I would like to thank Robin Howat, an author of the previous Maths in Action Higher series, for his time, support and opinion throughout the project.

E Mullan
November 2013

---

1 SQA state the list is correct as of December 2013.

1 Algebraic functions

Before we start...

In 1673 we find the first documented use of the word 'function' by **Leibniz** to mean a mathematical function.

In 1734 **Euler** invented the notation $f(x)$ to represent a function of x.

Germany and Switzerland are proud of their famous sons.
Each has appeared on his country's postage stamps.

You should already have been introduced to the idea of a function in the National 5 course.

Let us begin with a quick reminder.

A **relationship** between sets of numbers can be described using a rule, e.g. $f: x \rightarrow x + 4$ or $g: x \rightarrow x^2 + 4$

... the relationship called f maps x onto $x + 4$ or the relationship called g maps x onto $x^2 + 4$.

These relationships can be illustrated using arrow diagrams.

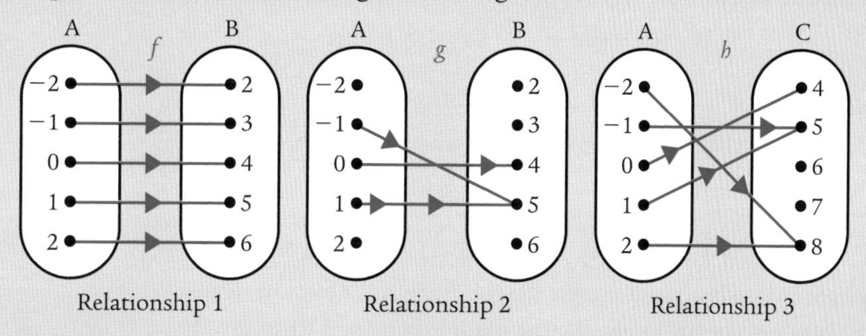

Relationship 1 Relationship 2 Relationship 3

The set from which numbers are taken is called the **domain**. In the above, the domains are A = {−2, −1, 0, 1, 2}.

The target set is called the **codomain**. The codomains in the above are B = {2, 3, 4, 5, 6) and C = {4, 5, 6, 7, 8}.

The set of numbers 'used' in the codomain is called the **range**. In the first case this is B and in the second it is {4, 5}.

In Relationship 1, the value that a number, x, maps onto is called the **image** of x under f and is denoted by $f(x)$.

In Relationship 2, the image of x under g is denoted by $g(x)$.

In this course our interest lies with relationships where each member of the domain has **one and only one** image.

These we call **functions**.

With the given domain and codomain:

Relationship 1 is a function: each member of A has an image ... and only one image.

Relationship 2 is NOT a function: 2 and −2 do not have an image.

Relationship 3 is a function: ... even though some members 'share' an image.

In a relation such as 'is greater than', each member may have multiple images ...
this is NOT a function.

What you need to know

1 Given that $f(x) = 4x - 1$ and that the domain and range are the real numbers,

a evaluate: i $f(3)$ ii $f(-2)$ iii $f(0{\cdot}5)$ iv $f(\pi)$.

b for what value of x is: i $f(x) = 3$ ii $f(x) = 0$ iii $f(x) = x$?

2 $D(x)$ is a function that gives the number of days in week x of the year.

a What is the domain of this function?

b What is the range of the function?

c Evaluate: i $D(5)$ ii $D(20)$.

d Why would this function be referred to as a **constant** function?

3 The function $f(x) = 2x + 1$ is referred to as a **linear** function.

a Why might that be?

b Write down an expression for: i $f(a)$ ii $f(a + 2)$.

c Describe the biggest suitable domain and range of f.

4 a Given that the **reciprocal** function, $g(x) = \frac{36}{x}$ is a function, what number cannot be in the domain?

b Evaluate: i $g(4)$ ii $g(6) - g(9)$ iii $g(2) \times g(3)$.

5 A function $f(x) = x^2 - x - 2$ is defined on all real numbers.

a What would be a suitable name to give such a function?

b Evaluate: i $f(1)$ ii $f(-1)$ iii $f(0)$.

c For what values of x is $f(x) = 0$?

6 As the blade of a wind turbine rotates, the height, h metres, of its tip from the ground is a function of the angle, $\theta°$, the blade has turned through after passing the supporting tower.

It is given by $h(\theta) = 50 - 20 \cos \theta°$ ... a **trigonometric** function.

a Calculate the height of the tip of the blade when θ is:

i 60 ii 90 iii 180.

b For what value of θ will the height *first* be 60 metres?

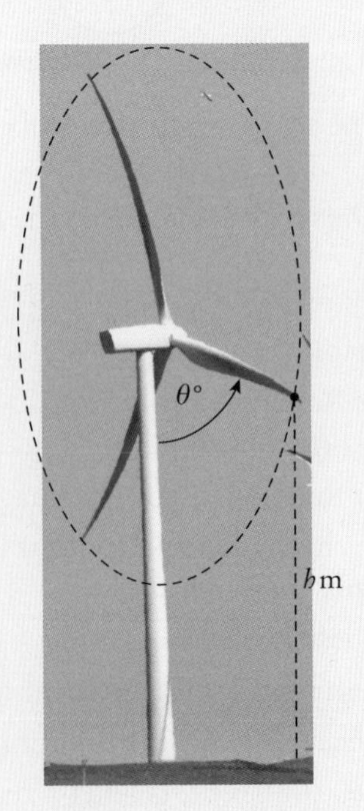

1.1 Graphing functions

A function, $f(x)$, can be described by drawing the graph $y = f(x)$, i.e. plotting the points $(x, f(x))$.

Example 1

A contractor knows the cost of a job, £C, is a function of x, the number of hours he works per day.

$$C(x) = 10x + \frac{90}{x}, \text{ where } 1 \leqslant x \leqslant 6$$

a Draw a graph of the function indicating **i** the domain **ii** the range.

b For how many hours a day must he work to make the job as cheap as possible?

a

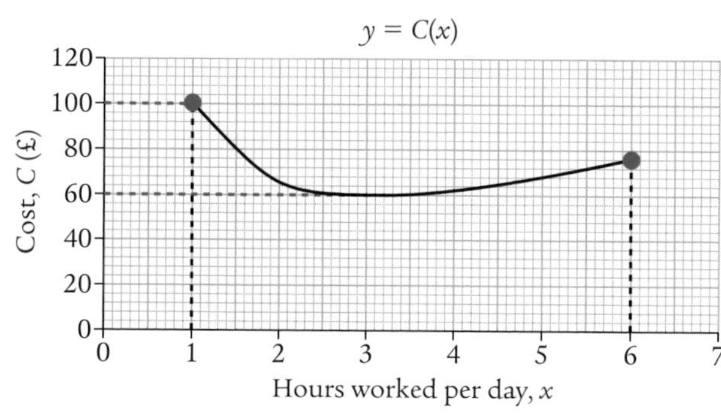

i The black broken lines indicate the domain, $1 \leqslant x \leqslant 6$.
ii The purple broken lines indicate the range, $60 \leqslant C \leqslant 100$.

b The job is at the cheapest £60 when he works 3 hours a day.

[Note: It is common for a practical problem to impose practical limits on the domain and range. The curve is not drawn beyond the domain.]

Exercise 1.1

1 **i** Make a quick sketch of each graph in the domain prescribed.
ii State the range in each case.

a $y = 2x + 5; x \geqslant 2$

b $y = x^2 + 1; -2 \leqslant x \leqslant 2$

c $y = \frac{1}{x}; 1 \leqslant x \leqslant 5$

d $y = \sqrt{x}; 1 \leqslant x \leqslant 49$

e $y = \sin x°; -90 \leqslant x \leqslant 90$

2 Division by zero is undefined, so the largest domain for the function $f(x) = \frac{1}{x}$ is the set of real numbers **except** the number 0. This can be written as $R - \{0\}$.

Write the largest domain of each of the following in a similar way:

a $\frac{1}{x+2}$ **b** $\frac{3}{x-1}$ **c** $\frac{x-2}{x-3}$

d $\frac{x}{(x-1)(x-2)}$ **e** $\frac{x}{x^2-4x+3}$

3 The graphs of three functions with restricted domains are shown.
For each, write down: **i** the domain **ii** the range.

a

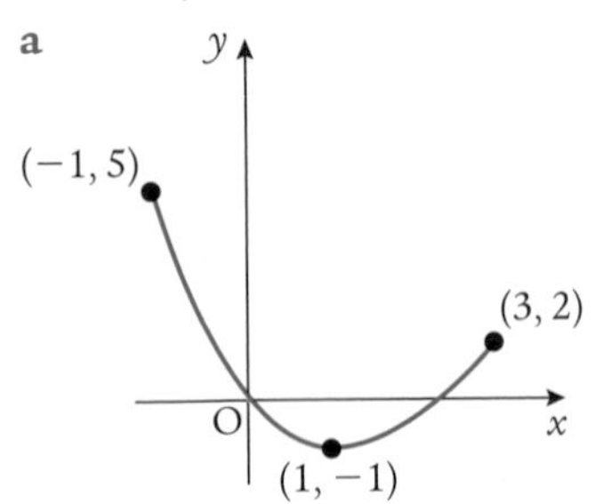

b

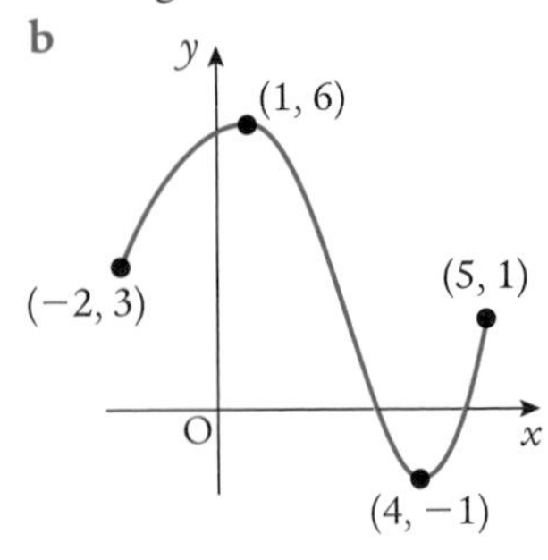

c

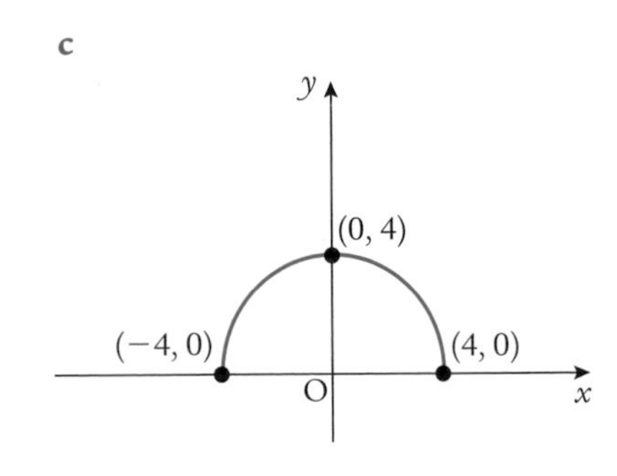

4 Edinburgh to Glasgow on the motorway is a distance of 45 miles.
The average speed, x mph, lies between 20 mph and 60 mph.

a Write down a function for $T(x)$, the time taken for the journey in hours at an average speed of x mph.

b Make a rough sketch of the function indicating the domain and the range.

5 An archaeologist studying in an ancient cemetery estimates the height of a person by measuring the length of the femur[1].
The height of the person, H cm, is a function of the length of the femur, x cm.
The function differs for males and females.

$$H_{\text{male}}(x) = \frac{x + 30{\cdot}3}{0{\cdot}45} \text{ and } H_{\text{female}}(x) = \frac{x + 26{\cdot}0}{0{\cdot}40}$$

The formula gives a good estimate for $30 \leqslant x \leqslant 48$.

a Calculate the range of each function.

b Make a rough sketch of both functions on one chart.

6 As the price of a loaf of bread goes up, the number of people willing to buy it goes down.
If $D(x)$ represents the number of people (counted in thousands) willing to buy the loaf when it costs £x, then

$$D(x) = x^2 - 10x + 25.$$

a How many people does the model predict would want the loaf if it were:
i £0·50 **ii** £5?

b However, the model is only valid for $0{\cdot}40 \leqslant x \leqslant 2{\cdot}00$.
What is the range of the function?

c Sketch the function.

7 Consider $f(x) = \sqrt{x}$.

a Evaluate: **i** $f(4)$ **ii** $f(0{\cdot}49)$ **iii** $f(0)$.

b What happens on your calculator when you try to evaluate $f(-1)$?

c What is the domain of $f(x)$?

d $f(x) = \sqrt{x}$ is accepted as a function with a range of $0 \leqslant y < \infty$.
Why is this restriction imposed on the range?

[Note: This is why you have to give the solution to the equation, say, $x^2 = 4$, as $\sqrt{4}$ or $-\sqrt{4}$ ... $\sqrt{4}$ is NOT equal to 2 or -2.]

---

[1] The bone from the hip to the knee.

8 A jeweller wants a presentation box which is a centimetre high. He begins with a cardboard square of length x cm, which he folds as shown to form the box.
The length of the box is $(x - 2)$ cm and the breadth is $(x - 3)$ cm.
The length of box must lie between 1 cm and 5 cm.

a What is the domain of x?

b Express the volume V cm^3 as a function of x.

c State the range of $V(x)$.

d Sketch $y = V(x)$ and show the domain and range on the graph.

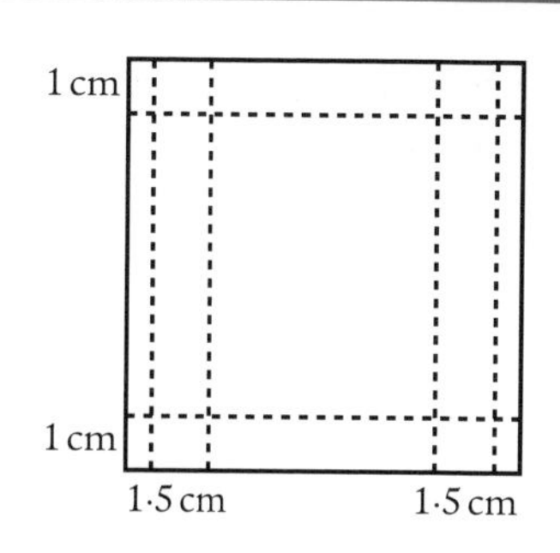

9 A metronome swings back and forth keeping time for a musician. The height, H cm, of the bob above the desk top is a function of $\theta°$, the angle the arm makes with the vertical.

$H(\theta) = 8 + 10\cos\theta°;\ -60 \leqslant \theta \leqslant 60$

a Calculate the range of the function.

b Calculate $H(50) - H(55)$.

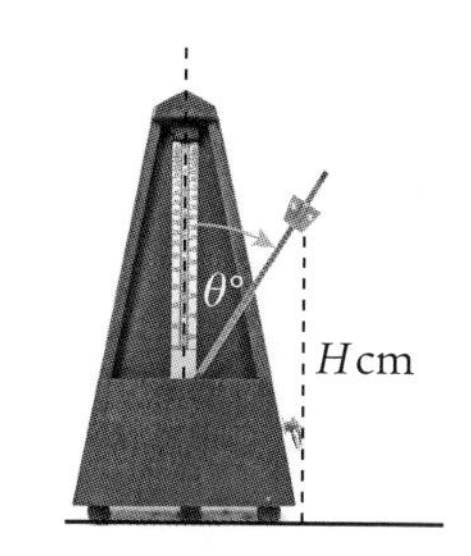

10 In the domain $0 \leqslant \theta \leqslant 180$, for what value is $\tan\theta°$ not defined?

11 List the restrictions on the various functions of a standard scientific calculator.

1.2 Composite functions

When two functions are combined so that the output from the first is used as the input of the second, a new **composite** function is created.

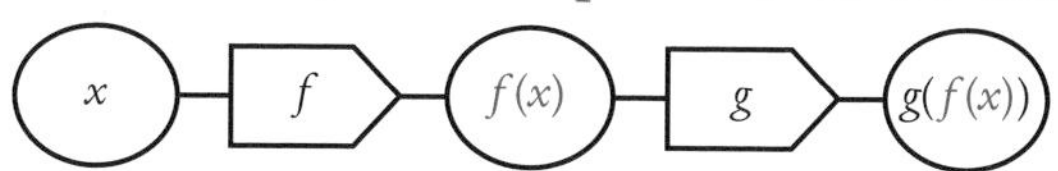

The range of the first function can't include elements that cannot be in the domain of the second.

Remember the definition of a function.

The formulae of the functions can be combined to form the composite function.

For example, let two functions be defined by $f(x) = 3x - 1$ and $g(x) = x^2 + x + 1$, and let a third function be defined by $h(x) = g(f(x))$.

Then $h(x) = g(f(x)) = g(3x - 1)$
$= (3x - 1)^2 + (3x - 1) + 1$
$= 9x^2 - 6x + 1 + 3x - 1 + 1$
$h(x) = 9x^2 - 3x + 1$

Example 1

The functions f and g are defined by $f(x) = 2x + 1$ and $g(x) = 3 - x$.
Find expressions for:

a $f(g(x))$ b $g(f(x))$ c $g(g(x))$ d $f(f(x))$.

a $f(g(x)) = f(3 - x) = 2(3 - x) + 1 = 6 - 2x + 1 = 7 - 2x$

b $g(f(x)) = g(2x + 1) = 3 - (2x + 1) = 3 - 2x - 1 = 2 - 2x$

c $g(g(x)) = g(3 - x) = 3 - (3 - x) = 3 - 3 + x = x$

d $f(f(x)) = f(2x + 1) = 2(2x + 1) + 1 = 4x + 2 + 1 = 4x + 3$

Example 2

Two functions are defined by $f(x) = x + 1$ and $g(x) = \dfrac{1}{x+1}$ on a suitable domain.

a Calculate $f(1)$ and hence $g(f(1))$.

b Calculate $f(-2)$ and $g(-2)$ and comment on the value of $g(f(-2))$.

c **i** Find an expression for $g(f(x))$.

ii Consider this expression at $x = -2$.

iii Comment.

a $f(x) = x + 1 \Rightarrow f(1) = 1 + 1 = 2 \qquad \Rightarrow g(f(1)) = g(2) = \dfrac{1}{2+1} = \dfrac{1}{3}$

b $f(-2) = -2 + 1 = -1 \qquad g(-2) = \dfrac{1}{-2+1} = -1$

$g(f(-2)) = g(-1) = \dfrac{1}{-1+1}$ which is undefined.

c **i** $f(x) = x + 1 \Rightarrow g(f(x)) = g(x + 1) = \dfrac{1}{x+1+1} = \dfrac{1}{x+2}$

ii When $x = -2$, this expression is undefined.

iii Even though $f(-2)$ and $g(-2)$ are defined, $g(f(-2))$ is not defined.

Exercise 1.2

1 For each pair of functions, defined on the largest suitable domain, find a formula for: **i** $f(g(x))$ and **ii** $g(f(x))$.

State the domain of the composite if it differs from the set of real numbers.

a $f(x) = 2x + 3, g(x) = x - 4$

b $f(x) = 4 - 3x, g(x) = 5x - 1$

c $f(x) = 7x + 1, g(x) = 1 - 5x$

d $f(x) = x + 1, g(x) = x^2$

e $f(x) = x^2 + 1, g(x) = 3x - 1$

f $f(x) = 6x - 2, g(x) = x^2 - 3x + 1$

g $f(x) = \dfrac{1}{x}, g(x) = x^2$

h $f(x) = \sin(x°), g(x) = x + 1$

i $f(x) = x + 1, g(x) = \dfrac{1}{x}$

j $f(x) = \sqrt{x}, g(x) = \cos(x°)$

k $f(x) = x^2, g(x) = \sqrt{x}$

l $f(x) = 2x - 1, g(x) = \dfrac{x+1}{2}$

2 For each of the composite functions in question **1**, give $f(g(0))$ and $g(f(0))$ where it exists.

Note that if $g(0)$ is undefined, then so is $f(g(0))$.

3 **a** $f(x) = \dfrac{x}{100}$ is a function that can be used to convert centimetres to metres.

$g(x) = \dfrac{x}{1000}$ is a function that can be used to convert metres to kilometres.

Find $g(f(x))$ and say what it could be used for.

b $f(x) = \dfrac{5(x-32)}{9}$ can be used to convert Fahrenheit temperatures to Celsius.

$g(x) = x - 273$ can be used to convert Celsius temperatures to degrees absolute (kelvin).

Find $g(f(x))$ and say what it could be used for.

c $f(x) = x^2$ gives the area in square metres of a carpet of side x metres.

$g(x) = 10x + 20$ gives the cost of laying a carpet of area $x\ \text{m}^2$.

Find $g(f(x))$ and say what it does in this context.

4 A function is defined by $f(x) = 2x + 1$.

a Find an expression for $f(f(x))$ and state its domain and range.

b $g(x) = \frac{1}{x}, x \neq 0$. Find an expression for $g(g(x))$ giving its domain and range.

c $h(x) = g(f(x))$. Find an expression for it. Give the domain and range of h.

d Find an expression for $h(h(x))$ giving its domain and range.

5 Functions can be used to generate number sequences.

Consider the function $f(x) = \frac{x + 1}{2}$.

a Evaluate: i $f(65)$ ii $f(f(65))$ iii $f(f(f(65)))$ iv $f(f(f(f(65))))$.

b Find formulae in their simplest form for:

i $f(x)$ ii $f(f(x))$ iii $f(f(f(x)))$ iv $f(f(f(f(x))))$.

c Can you spot a general pattern?

6 a In each case, find $f(g(x))$.

i $f(x) = x + 1, g(x) = x - 1$ ii $f(x) = 3x + 1, g(x) = \frac{x - 1}{3}$

iii $f(x) = (x - 1)^2, g(x) = 1 + \sqrt{x}$ iv $f(x) = \frac{x}{3}, g(x) = 3x$

v $f(x) = \frac{x}{x + 1}, g(x) = \frac{x}{1 - x}$

b Examine $g(f(x))$ and comment on your answers.

c Consider the functions $f(x)$ below and find $f(f(x))$. Comment on your answers for each.

i $f(x) = -x + 9$ ii $f(x) = 4 - x$ iii $f(x) = \frac{1 - 3x}{3}$

7 Functions $f(x) = x^2$ and $g(x) = x - 1$, where x is a real number.

a Find: i $f(g(x))$ ii $g(f(x))$.

b In general $f(g(x)) \neq g(f(x))$, but for what value of x is this not true in this case?

1.3 Inverse functions

When $f(g(x)) = x$ for all x in the domain of f, then we call g the inverse of f and denote it by f^{-1}.
Thus, $f^{-1}(f(x)) = x$. It is also true that $f(f^{-1}(x)) = x$.

The inverse function maps $f(x)$ back onto x ... it 'undoes' the function.

This can only happen if the inverse operation is a function ... i.e. every member of its domain, B, has **one and only one** image in its range, A.

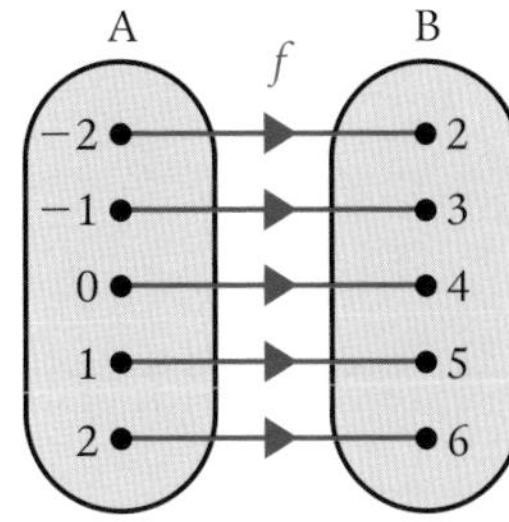

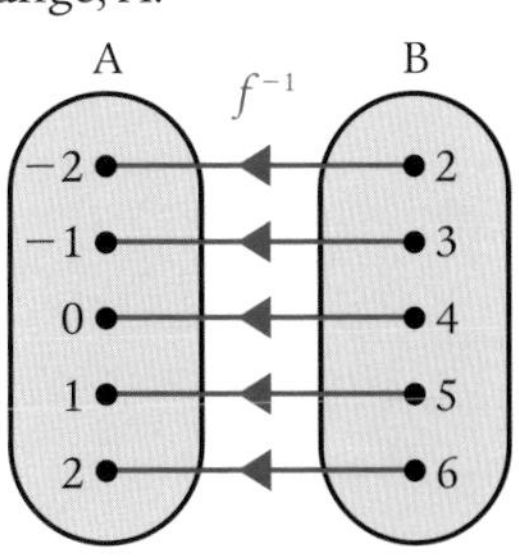

This is already happening in the other direction, i.e. mapping set A onto set B is a function.

Each member of one set has a unique member to which it is 'twinned' in the other set.

This kind of function is referred to as having **one-to-one correspondence**.

(We often have to restrict the domain or range to achieve this one-to-one correspondence if we wish a function to have an inverse.)

If the point (x, y) lies on the curve $y = f(x)$, then (y, x) lies on the curve of $y = f^{-1}(x)$, i.e. if the function exists then: $y = f(x) \Leftrightarrow x = f^{-1}(y)$.

This fact allows us to:

i find a formula for the inverse of a function where it exists

ii sketch $y = f^{-1}(x)$ if we are given a sketch of $y = f(x)$.

[Note: A sketch can be handy as a diagnostic tool ...]

- If a line that runs in the y-direction can be found to cut the graph of a relation at more than one point ... then **the relation is not a function** ... some value of x has more than one image.
  So the relation shown in Figure 1 is not a function.
  We could restrict the **range** and make it a function.

Figure 1

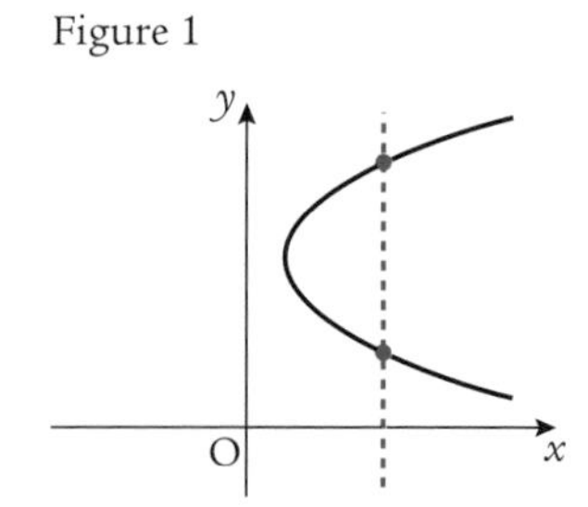

- If a line that runs in the x-direction can be found to cut the graph of a function at more than one point ... then **the function has no inverse** ... some particular value of y has more than one image.
  So the function shown in Figure 2 has no inverse.
  We could restrict the **domain** and make it have an inverse.

Figure 2

Any calculator manual will list the restrictions the makers have used to create functions with inverses.

Example 1

Find the inverse function in each case:

a $f(x) = 3x + 2$ **b** $f(x) = x^2, x \geqslant 0$ **c** $f(x) = \frac{1}{x-1}, x > -1$.

a If $y = 3x + 2$ defines the function then $x = 3y + 2$ will define the inverse function.

$x = 3y + 2 \Rightarrow x - 2 = 3y \Rightarrow y = \frac{x-2}{3}$ ... making y the subject

Thus, $f^{-1}(x) = \frac{x-2}{3}$.

b If $y = x^2$ defines the function then $x = y^2$ will define the inverse function.

$y = \sqrt{x}, x \geqslant 0$ ... making y the subject

Thus, $f^{-1}(x) = \sqrt{x}, x \geqslant 0$.

c If $y = \frac{1}{x+1}$ defines the function then $x = \frac{1}{y+1}$ defines the inverse.

$x = \frac{1}{y+1} \Rightarrow xy + x = 1 \Rightarrow y = \frac{1-x}{x}$

Thus, $f^{-1}(x) = \frac{1-x}{x}, x > 0$.

Example 2

The sketch shows the graph, $y = f(x)$, of a function near the origin and some points that lie on it.

Sketch the graph of the inverse function.

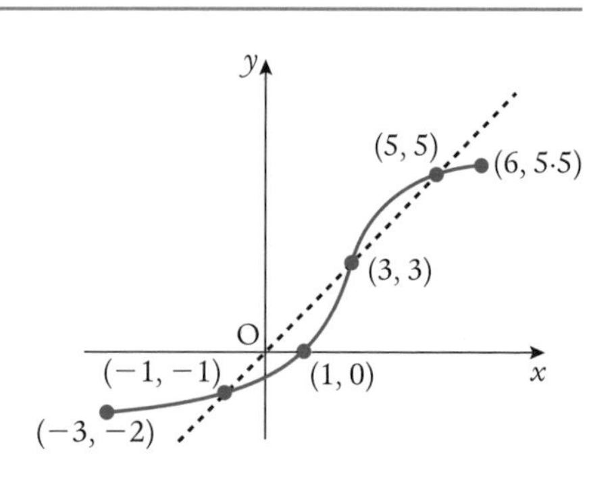

If (x, y) lies on the function, (y, x) lies on the inverse function.

So we have the points $(-2, -3)$, $(-1, -1)$, $(0, 1)$, $(3, 3)$, $(5, 5)$ and $(5{\cdot}5, 6)$ on the graph of the inverse.

[Note that we are wanting the image of the curve under reflection in the line $y = x$.]

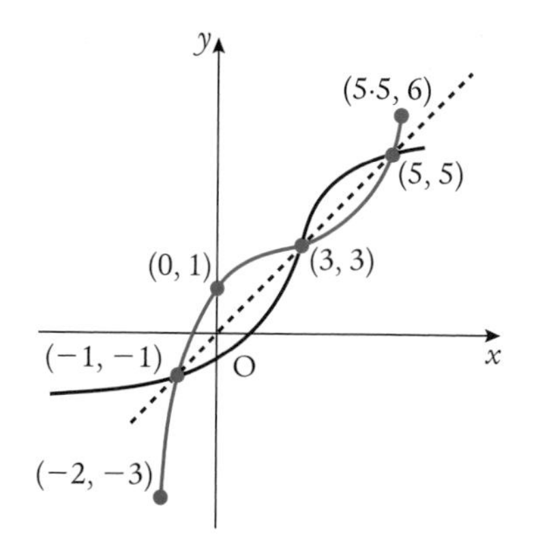

Exercise 1.3

1 Each of the following functions has an inverse. Find its formula.

a $f(x) = 3x + 5$ b $f(x) = 4x$ c $f(x) = 1 - 3x$

d $f(x) = 6 - x$ e $f(x) = \frac{2}{x}; x \neq 0$ f $f(x) = \frac{2x}{x - 1}; x \neq 1$

g $f(x) = \sqrt{x + 2}; x > -2$ h $f(x) = 1 - \sqrt{x}; x > 0$ i $f(x) = \frac{1}{1 - \sqrt{x}}; x > 1$

2 Here is a sketch of $y = 2 - x$.

The line $y = x$ has been marked as a broken line.

a What can you say about the reflection of the line $y = 2 - x$ in $y = x$?

b If $f(x) = 2 - x$, state the inverse function and comment.

c What is special about linear functions which are their own inverses?

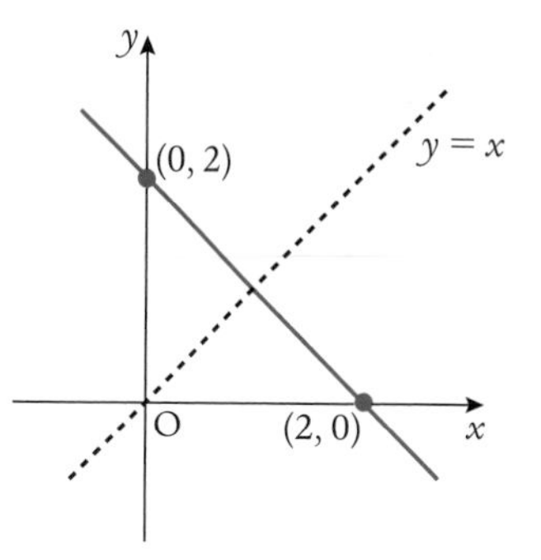

3 Sketch the inverse of each function by considering reflection in the line $y = x$.

a i

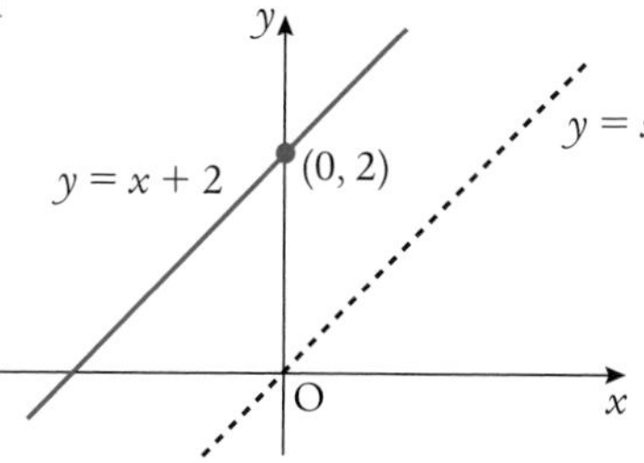

ii

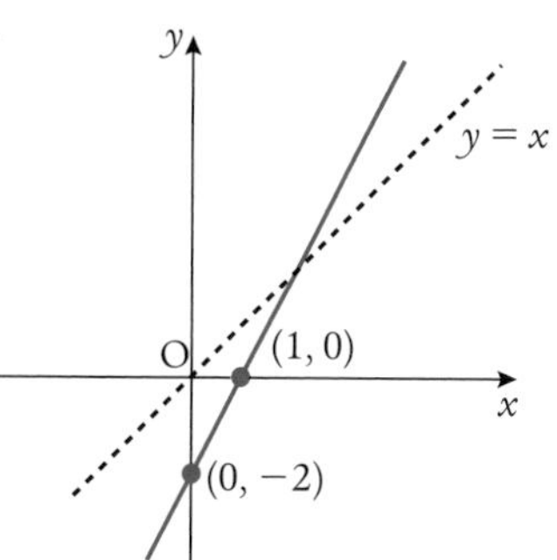

iii

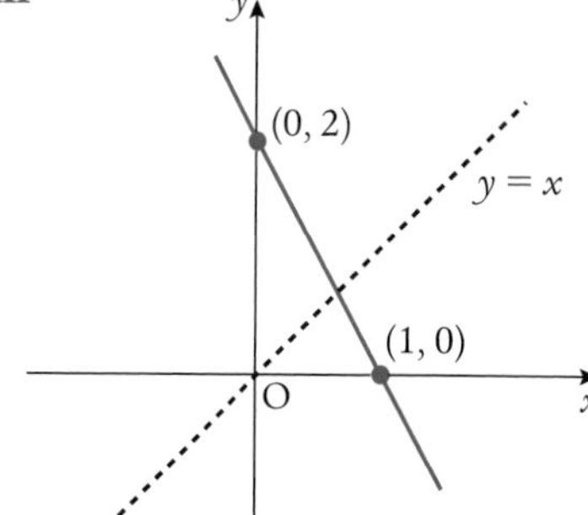

b i

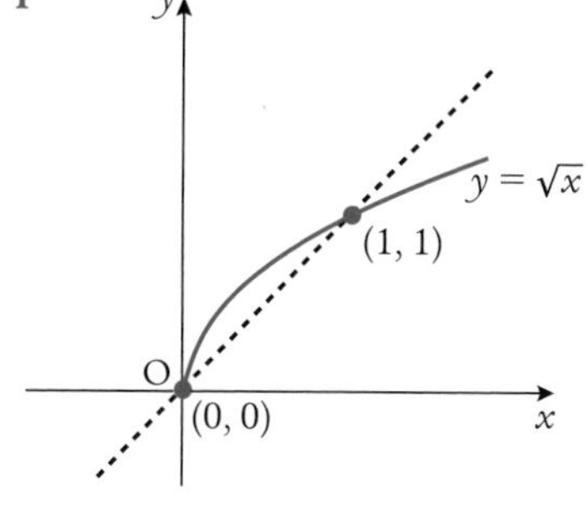

ii

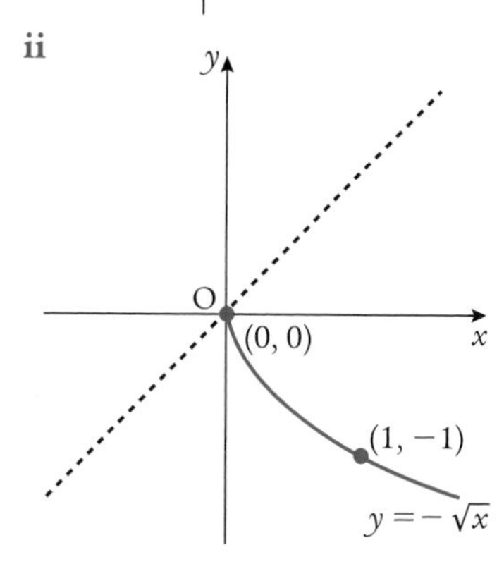

iii

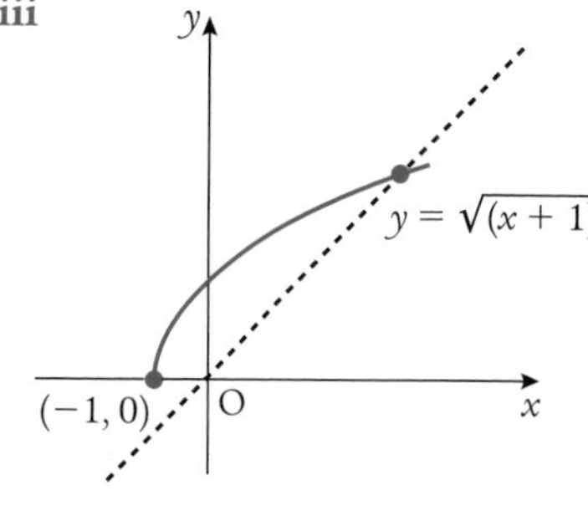

4 Identify which of these functions have inverses.

a $f(x) = x^2 + 1$ **b** $f(x) = x^3$ **c** $f(x) = x^3 - 2x + 5$

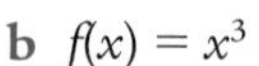

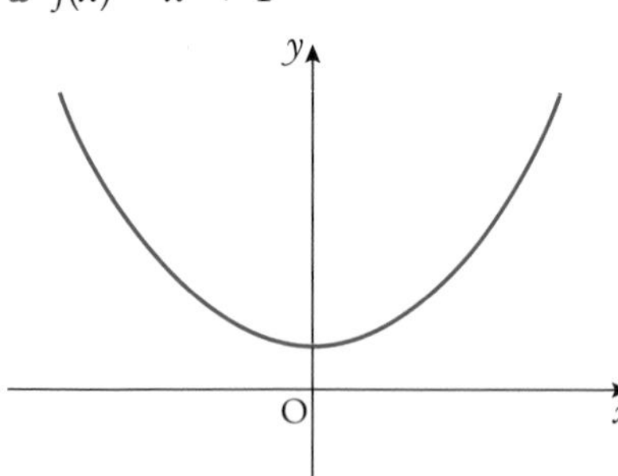

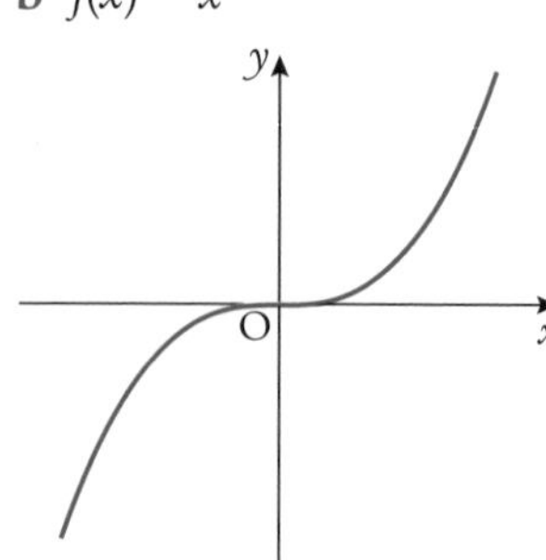

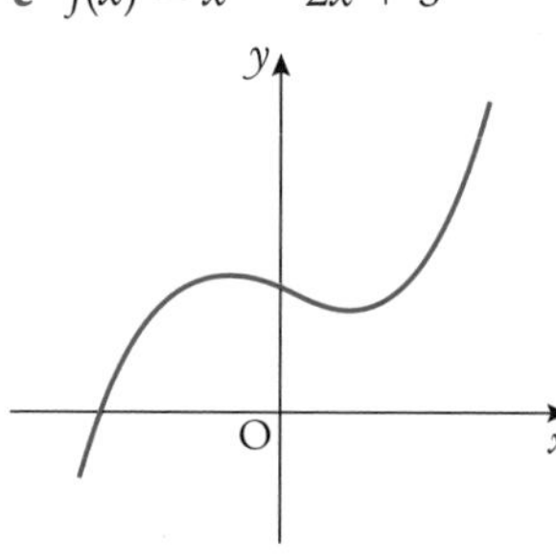

5 The diagram shows a part of the function $f(x) = \cos x°$, where x is a real number ($x \in R$). One section has been highlighted in purple.

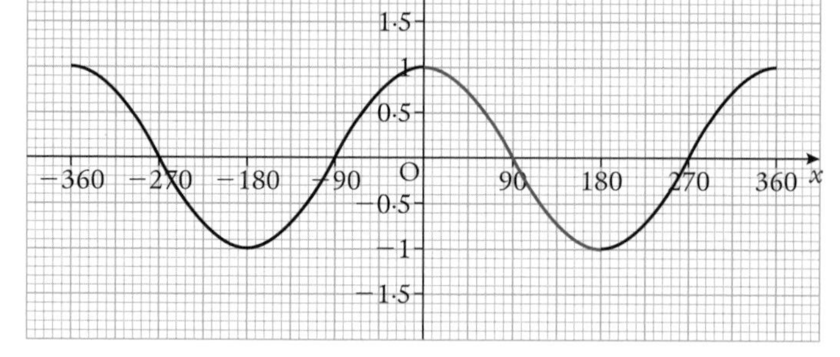

a What is the range of the function?

b How can you tell that this function does not have an inverse?

c The purple section shows the part of the curve used by calculator makers.

i What is the range and domain of this restricted function?

ii How do you know that this restricted function has an inverse?

iii What is the domain and range of this inverse function?

iv Make a sketch of this inverse function ... now called $\sin^{-1} x$ on many calculators.

d This shows part of the function $f(x) = \sin x°, x \in R$.

Can you find the biggest suitable domain that includes $x = 0$ that might allow the calculator makers to define a function $f(x) = \sin x°$ that has an inverse?

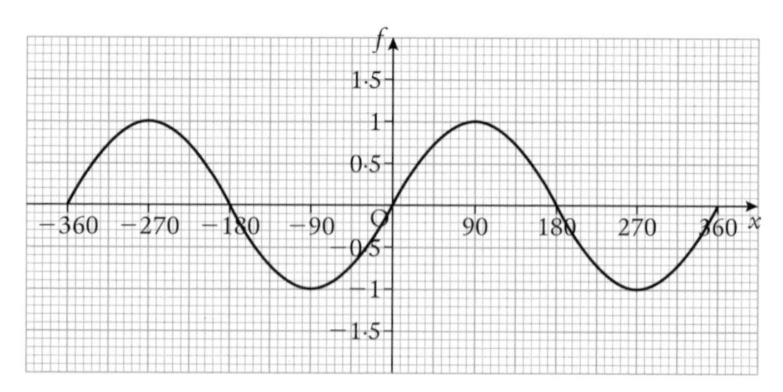

1.4 Completing the square

From National 5 you should already know the algebraic technique called 'completing the square'. It is useful for finding the turning point on a parabola.

Thus it is useful for finding the restrictions on the domain and range of a quadratic function so that it has an inverse.

We have the expansion $(x + a)^2 = x^2 + 2ax + a^2$, where a is a constant.

Usefully re-arranged the expression becomes $x^2 + 2ax = (x + a)^2 - a^2$.

An expression with two varying terms has been modified to become an expression with only one varying term.

Example 1

a Express $x^2 + 6x - 4$ in the form $(x + a)^2 + b$.

b Hence find the minimum value of the function $f(x) = x^2 + 6x - 4$ and the value of x where it occurs.

c What restrictions must be placed on its domain and range for $f(x)$ to have an inverse?

d State the equation of the inverse.

a Comparing $x^2 + 6x$ with $x^2 + 2ax$ we see that $a = 6 \div 2 = 3$.
So $x^2 + 6x = (x + 3)^2 - 3^2$.
$x^2 + 6x - 4 = (x + 3)^2 - 3^2 - 4 = (x + 3)^2 - 13$

b $f(x) = (x + 3)^2 - 13$. The **smallest** value that the term $(x + 3)^2$ can have is zero.
This will occur at $x = -3$. We then have $f(-3) = -13$.

c The graph of this will be a parabola with $(-3, -13)$ as the minimum turning point.
A quick sketch lets you identify a suitable part of the curve
... using a domain $-3 \leqslant x < \infty$ will give a function (shown in purple) with a range of $-13 \leqslant f(x) < \infty$, which has an inverse.

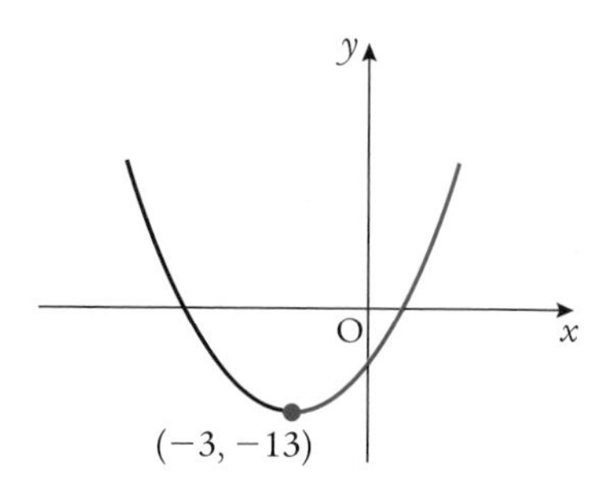

d Given $f(x) = (x + 3)^2 - 13$; $-3 \leqslant x < \infty$.
Letting $y = (x + 3)^2 - 13$ represent the function,
$x = (y + 3)^2 - 13$ represents the inverse.
$\Rightarrow f^{-1}(x) = \sqrt{x + 13} - 3$; $-13 \leqslant x < \infty$.

Example 2

Express $3x^2 + 6x - 2$ in the form $a(x + b)^2 + c$, where a, b and c are constants.

$3x^2 + 6x - 2 = 3(x^2 + 2x) - 2$ ... to get an $x^2 + 2ax$ form
$= 3[(x + 1)^2 - 1^2] - 2$ ... completing the square
$= 3(x + 1)^2 - 3 - 2$
$= 3(x + 1)^2 - 5$ ... the required form

Exercise 1.4

1 Express each function in the form $(x + a)^2 + b$, where a and b are constants.

a $x^2 + 4x - 3$ **b** $x^2 + 10x + 1$ **c** $x^2 - 8x + 4$ **d** $x^2 - 2x - 1$
e $x^2 + 3x + 1$ **f** $x^2 + x - 5$ **g** $x^2 - 3x + 6$ **h** $x^2 - x - 1$
i $x^2 + \frac{1}{2}x - \frac{3}{4}$ **j** $x^2 + 0{\cdot}4x + 0{\cdot}1$

2 **a** Express $x^2 - 4x - 2$ in the form $(x + a)^2 + b$.

b Hence find the minimum value of the function $f(x) = x^2 - 4x - 2$ and the value of x where it occurs.

c What restrictions must be placed on its domain and range for $f(x)$ to have an inverse?

d State the equation of the inverse.

3 Repeat question **2** with the following functions.

a $g(x) = x^2 + 12x - 1$ **b** $h(x) = x^2 - 8x + 12$ **c** $k(x) = x^2 - x + 5$
d $p(x) = x^2 - 2x - 1$ **e** $q(x) = x^2 - 7x + 8$ **f** $t(x) = x^2 - \frac{1}{3}x + \frac{5}{6}$

4 Express each function in the form $f(x) = a(x + b)^2 + c$, where a, b and c are constants.

a $f(x) = 2x^2 + 12x - 5$ b $g(x) = 3x^2 + 18x + 1$ c $h(x) = 4x^2 - 8x + 3$

d $k(x) = 5x^2 - 10x + 2$ e $m(x) = 2x^2 - 6x + 2$ f $t(x) = 3x^2 - 2x + 1$

g $f(x) = 3 + 2x - x^2$ h $g(x) = 1 - 6x - x^2$ i $h(x) = 4 + 3x - x^2$

5 a Find the minimum value of the function $f(x) = 4x^2 + 24x - 3$ and the value of x where it occurs.

b Make a sketch of the function.

c What restrictions must be placed on its domain and range for $f(x)$ to have an inverse?

d State the equation of the inverse.

6 The function $g(x) = \dfrac{12}{x^2 + 2x + 4}$ has the real numbers as its domain.

a Express $x^2 + 2x + 4$ in the form $(x + a)^2 + b$.

b Find the **maximum** value of $g(x)$ and the value of x where it occurs. Explain your strategy.

1.5 Related functions

You have seen how the graph of the inverse function, $y = f^{-1}(x)$, can be attained from $y = f(x)$ by reflection in the line $y = x$.

Other transformations can be used to produce other functions related to $f(x)$.

Ideally you will get a spreadsheet or graphing calculator now, and explore how the graphs of:

$$y = 2x^2, y = (2x)^2, y = x^2 + 1, y = (x + 1)^2, y = -x^2, y = (-x)^2$$

relate to the graph of $y = x^2$.

Example:

In A1, B1 and C1 type Headings: x, x^2 and (2x)^2.

In A2 type −4, in A3 type =A2+1 ... fill down to A10.

In B2 type =A2^2 ... the symbol ^ stands for 'to the power of' ... fill down to B10.

In C2 type =(2*A2)^2 ... the symbol * stands for 'times' ... fill down to C10.

Select A1 to C10 and ask for a chart. Select 'scatter' then select 'Scatter with Smooth Lines'.

You should get this:

	A	B	C
1	x	x^2	(2x)^2
2	−4	16	64
3	−3	9	36
4	−2	4	16
5	−1	1	4
6	0	0	0
7	1	1	4
8	2	4	16
9	3	9	36
10	4	16	64
11			

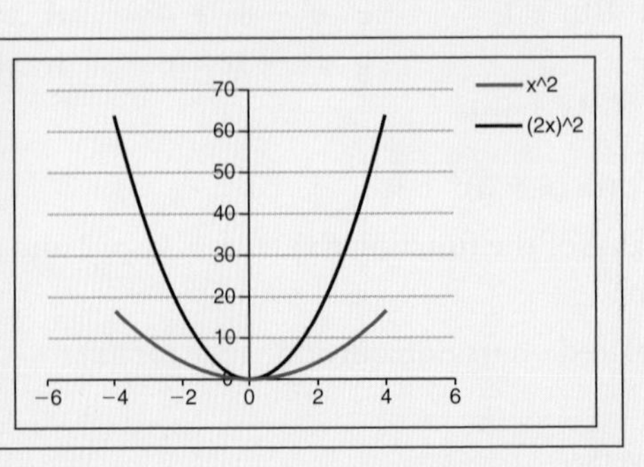

The purple curve outlines $y = x^2$ and the black curve $y = (2x)^2$ on the same domain.

Including row 1 in the 'Select', each series has been named. The comparative graph can be cut and pasted into a Word document for your records.

Given the graph of $y = f(x)$ ... and some of the points on it, say, (0, 0), (2, 4), (8, −1)

$y = 2f(x)$ ... stretches the curve in the y-direction by a factor of 2 ... (0, **0**), (2, **8**), (8, −**2**)
$y = f(2x)$ ... squashes the curve by halving each x-coordinate ... (**0**, 0), (**1**, 4), (**4**, −1)

$y = f(x) + 3$ ... translates the curve in the y-direction upwards by 3 ... (0, **3**), (2, **7**), (8, **2**)
$y = f(x + 3)$ ... translates the curve in the x-direction to the left by 3 ... (−**3**, 0), (−**1**, 4), (**5**, −1)

$y = -f(x)$ ... reflects the curve in the x-axis ... (0, 0), (2, −4), (8, 1)
$y = f(-x)$ ... reflects the curve in the y-axis ... (0, 0), (−2, 4), (−8, −1)

Other functions can be made using a composition of transformations. For example,
$y = 2f(x) + 3$ ... double the y-coordinate and add 3 ... (0, 3), (2, 11), (8, 1)
$y = f(2x - 1)$ ... add 1 then halve each x-coordinate ...(0·5, 0), (1·5, 4), (4·5, −1)
$y = f(2x) + 1$ ... halve the x-coordinate; add 1 to y ... (0, 1), (1, 5), (4, 0)

Example 1

This is part of the graph $y = f(x)$.
Some points that lie on the curve have been given.
Make sketches of the related functions:

a $y = 3f(x)$ **b** $y = -f(x)$
c $y = f(x - 2)$ **d** $y = 3f(x) - 2$

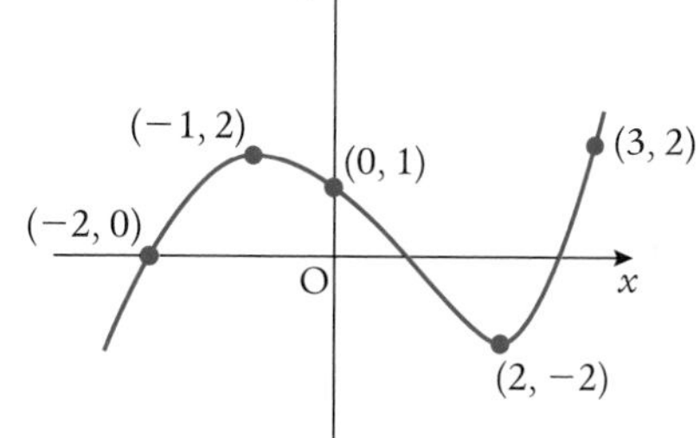

a $y = 3f(x)$
each y-coordinate is trebled
... a stretch in the y-direction.

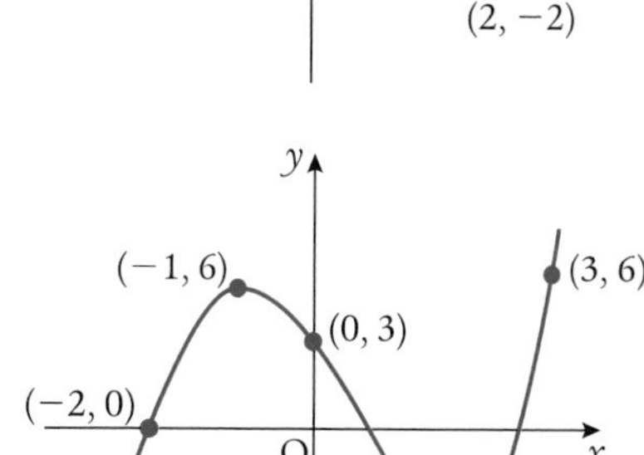

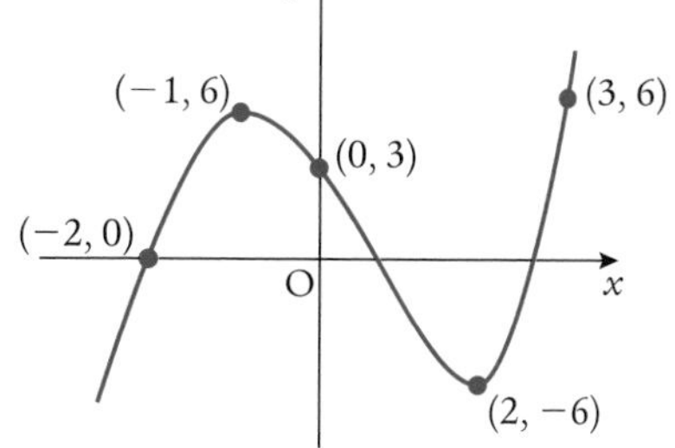

b $y = -f(x)$
multiply each y-coordinate by −1
... reflect in the x-axis.

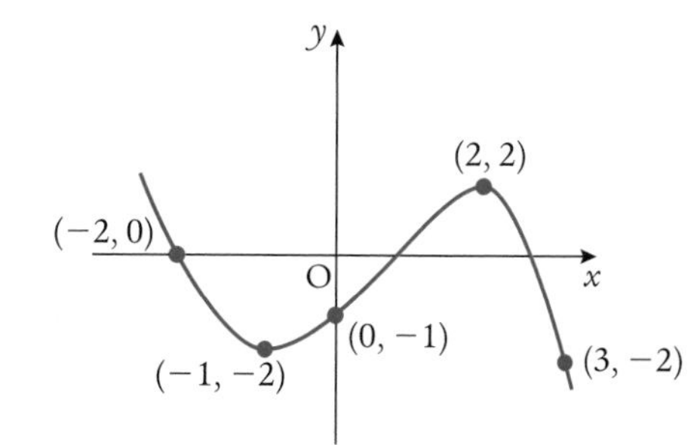

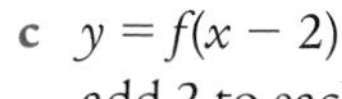

c $y = f(x - 2)$
add 2 to each x-coordinate.
... translating curve 2 to right.

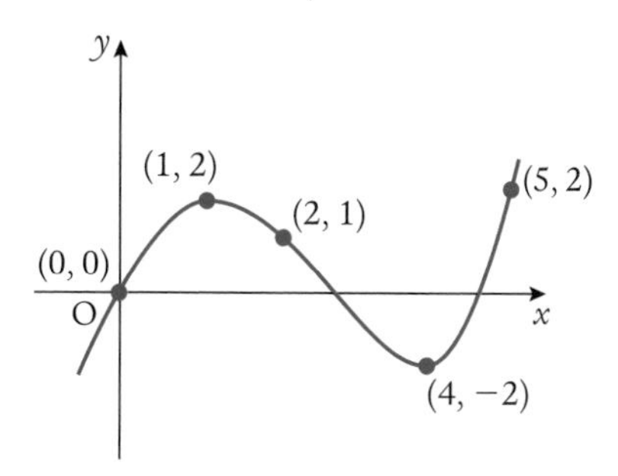

d $y = 3f(x) - 2$
multiply each y-coordinate by 3
then subtract 2 from each y-coordinate
... stretch then translate down.

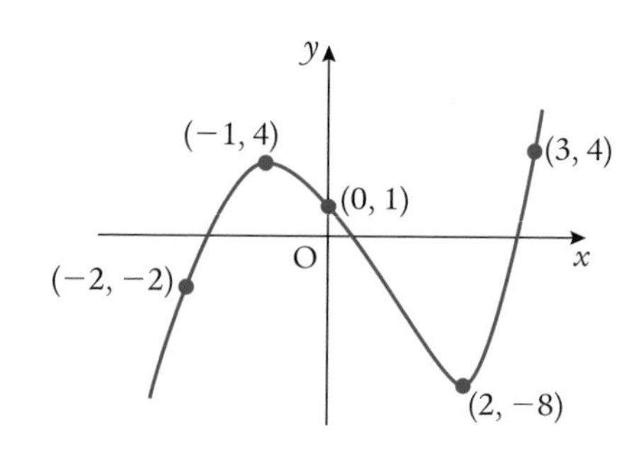

Exercise 1.5

1 This is part of the graph $y = f(x)$. Its domain is $-4 \leqslant x \leqslant 5$. Some points that lie on the curve have been given.

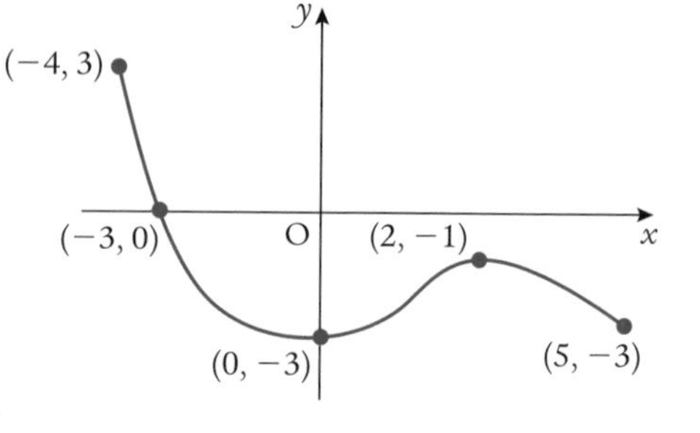

a State the range of $f(x)$.

b For each of these related functions make a sketch and state the domain and range of the function.

i $y = 4f(x)$ **ii** $y = -f(x)$ **iii** $y = -4f(x)$
iv $y = f(-x)$ **v** $y = f(2x)$ **vi** $y = f(x) - 4$
vii $y = 4 - f(x)$ **viii** $y = f(x + 2) + 1$ **ix** $y = 3f(x) - 4$

2 $F(x) = 3x + 1, 0 \leqslant x \leqslant 4$

a Sketch the function and state its range.

b Sketch the graph of each of these related functions that have the same domain. Give also their ranges.

i $2F(x)$ **ii** $F(3x)$ **iii** $-F(x)$ **iv** $F(-x)$ **v** $F(x - 1) + 2$

3 The quadratic function $f(x) = x^2 - 6x + 11$ where x is real, passes through the point A(0, 11).

a By completing the square, identify its minimum value and hence state its range.

b Sketch the curve $y = x^2 - 6x + 11$.

c **i** State the equation of $f(2x)$ and sketch its curve.
ii State the equation of $f(2x + 1)$ and sketch its curve.[2]
iii Sketch the curve of $3 - f(x)$.

d Give the y-intercept (where $x = 0$) and the minimum turning point of the curve:
i $y = f(3x - 3)$ **ii** $y = f(2x + 2)$ **iii** $y = f(ax + b)$.

4 The reciprocal function, $\frac{36}{x}, x \neq 0$, passes through (1, 36), (2, 18), (3, 12) and (4, 9).

a Make a sketch of $f(x)$.

b $h(x) = f(x - 1)$
i State the equation of $h(x)$, give its domain and sketch it.
ii $g(x) = 2f(x) + 1$. Make a sketch and state the domain.
iii Make a sketch of $h(x) = 3 - f(2x)$.

5 Let $g(x) = \sin x°, 0 \leqslant x \leqslant 360$.
Make sketches of the following, stating the range of each:

a $g(x) + 1$ **b** $g(x - 30)$ **c** $-g(x)$
d $g(-x)$ **e** $3g(x)$ **f** $g(3x)$.

[2] The y-intercept is found where $x = 0$. In this case at $f(2 \times 0 + 1) = f(1)$.

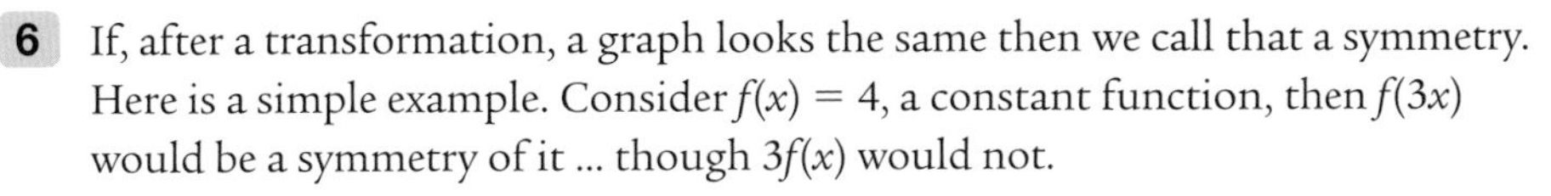

6 If, after a transformation, a graph looks the same then we call that a symmetry. Here is a simple example. Consider $f(x) = 4$, a constant function, then $f(3x)$ would be a symmetry of it ... though $3f(x)$ would not.

a Given $f(x) = x^2$, which of these is a symmetry of it:

i $f(\sqrt{x})$ **ii** $-f(x)$ **iii** $f(-x)$?

b Given $g(x) = 2x + 1$.

i For $x = 1$ to 4, list four points on the line $y = 2x + 1$.

ii List four points on the graph of $h(x) = 2 + g(x - 1)$.

iii Comment.

iv Can you find a symmetry for the function $k(x) = 4x - 3$?

Preparation for assessment

1 A stretched limo taxi driver will only take on journeys from 1 mile to 10 miles long.
His charge, £C, is related to the distance travelled, x miles, by the function $C(x) = 2x + 3$.

a Calculate: **i** C(5) **ii** C(3).

b What is: **i** the domain **ii** the range of the cost function?

c For what value of x is $C(x) = 19$?

2 A builder estimates that the cost of his deliveries from one supplier can be modelled by

$$D(x) = x^2 - 6x + 11$$

where D is the cost in thousands of pounds and x tonnes of material are delivered each time. This weight always lies between 1 and 7 tonnes.

a Find: **i** $D(1)$ **ii** $D(7)$ **iii** $D(4) - D(3)$.

b **i** Express $D(x)$ in the form $a(x + b)^2 + c$.

ii Find the minimum value of D.

c What is the range of the function?

3 A function is defined by $f(x) = \dfrac{60}{x+1}, x \neq -1$.

a Find: **i** $f(29)$ **ii** $f(9)$.

b Find an expression for $f^{-1}(x)$.

c For what value of x is $f(x) = 0{\cdot}5$?

4 Two functions are defined by $f(x) = x^2 + 1$ and $g(x) = 2x + 3$.

a Find: **i** $f(g(x))$ **ii** $g(f(x))$.

b For what value of x does $f(g(x)) = 2g(f(x))$?

5 Functions f and g are defined by $f(x) = \sin x°$ and $g(x) = \sqrt{1 - x^2}$.
Both have the domain $-1 \leqslant x < 1$.

a Find an expression for $f(g(x))$ and the value of $f(g(0{\cdot}5))$ to 2 significant figures.

b **i** Find an expression for $g(f(x))$.

ii Simplify the expression.

6 The functions $f(x) = x^2 + 2x + 2$ and $g(x) = \frac{4}{x}$ are defined on a suitable domain.

a $h(x) = g(f(x))$. Find an expression for $h(x)$.

b Express $f(x)$ in the form $(x + a)^2 + b$.

c Find the maximum value of $h(x)$.

7 The diagram shows the graph $y = f(x)$ close to the origin.
Sketch the graph of:

a $f(x - 1)$

b $f(x - 1) + 2$

c $-f(x) - 1$.

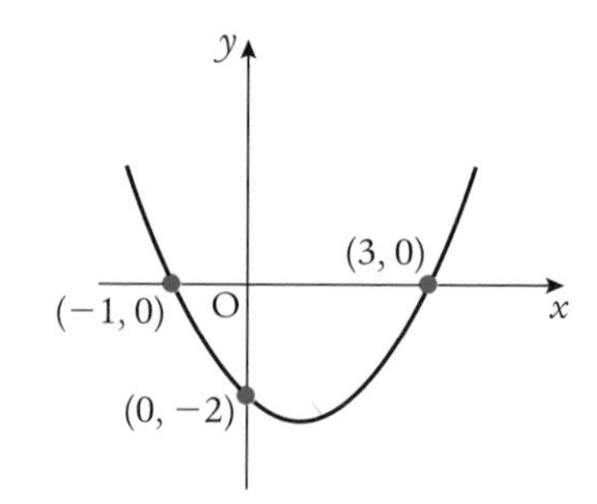

8 The diagram depicts $y = g(x)$ close to the origin.

a State one reason why you know that $g^{-1}(x)$ doesn't exist.

b Sketch: **i** $y = f(-x)$ **ii** $y = f(-x) + 2$.

c Give the largest domain that includes $x = 0$, for which $g(x)$ has an inverse.

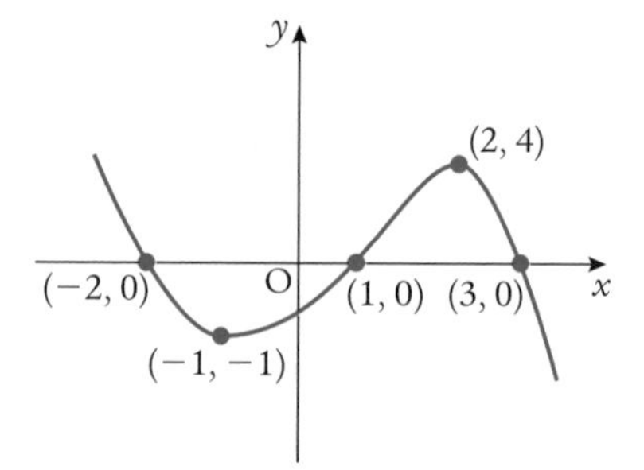

Summary

1 A **function** is a rule that relates each member of one set to exactly one member of another.

2 A **formula** can be used to describe the function
e.g. $f(x) = x^2 + 2x + 1, x \in R$ or $f: x \rightarrow x^2 + 2x + 1, x \in R$.

- f is the letter denoting the function,
- $f(x)$ is the image of x under this function,
- $x \in R$, tells us that, in this example, x is a real number.

3 The **domain** of a function is the set of numbers, x, that can be used in the function.
The **range** of a function is the set of numbers that makes up all images of x under the function.

4 A **graph** can be used to describe the function. The graph of the function f, has equation $y = f(x)$.

5 When **modelling** a situation by a function, the domain can be restricted by practical considerations.
The graph should not be drawn outside the bounds of the domain.
For example, *The Highway Code* gives the breaking distance, d m, for particular speed v mph.
d is a function of v: $d(v) = \frac{v^2}{66}$, $20 \leqslant v \leqslant 70$.
The graph would look like this with the domain indicated by the dashed black lines and the range by the purple dashed lines

- domain: $20 \leqslant v \leqslant 70$
- range: $6 \leqslant d \leqslant 74$.

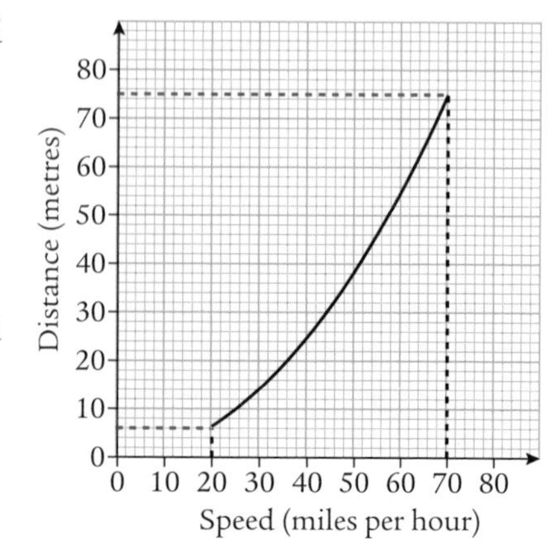

6 **Composite functions** can be created when the output from one function is used as the input for another. Performing function f on x followed by performing g on the image $f(x)$ produces a new function denoted by $g(f(x))$.
In general, the order in which the functions are performed is important: $f(g(x)) \neq g(f(x))$.

7 The **inverse of function** f is denoted by f^{-1}.

$$f(f^{-1}(x)) = f^{-1}(f(x)) = x.$$

The inverse exists only when the function makes a one-to-one correspondence between the domain and range. A formula can be found by considering $y = f(x) \Leftrightarrow x = f^{-1}(y)$.

8 The **graph of the inverse** can be obtained by reflecting the graph of the function in the line $y = x$.

9 **Related functions**

If the graph of a function, $y = f(x)$, is given, then the graphs of related functions can be found by considering transformations.

Function	Point	Transformation
$y = f(x)$	(x, y)	
$y = af(x)$	(x, ay)	Stretch in y-direction
$y = f(x) + b$	$(x, y + b)$	Translation up in y-direction
$y = af(x) + b$	$(x, ay + b)$	Stretch then translate
$y = f(cx)$	$(x/c, y)$	Squash in x-direction
$y = f(x + d)$	$(x - d, y)$	Translate left in x-direction
$y = f(cx + d)$	$([x - d]/c, y)$	Translate then squash
$y = af(cx + d) + b$	$([x - d]/c, ay + b)$	All of the above
$y = -f(x)$	$(x, -y)$	Reflection in x-axis
$y = f(-x)$	$(-x, y)$	Reflection in y-axis
$y = -f(-x)$	$(-x, -y)$	Reflection in origin/half-turn

2 Trigonometry 1 – using radians

Before we start...

In mechanics, optics, cartography, astronomy and maths, problems often crop up involving **very small angles**. The Moon only subtends an angle of 0·5° at your eye!

The unit of angular measure we historically use is the **degree**, defined by $360° = 1$ revolution. Is there a more useful unit than that?

Consider for the moment that we have a new unit with **k units per revolution**.

ABD is the sector of a circle. BD is a chord and BC is a tangent at B cutting AD at C.

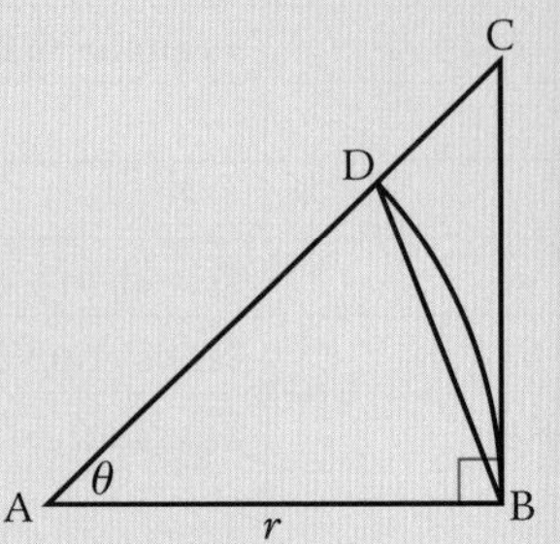

Let the radius be r units long and $\angle CAB$ be θ angular units.

So $AB = AD = r$ and $CB = r\tan\theta$.

From the diagram we see the area of the sector ABD lies between the area of the small triangle ABD and the large triangle ABC.

$$\triangle ABD \leqslant \text{sector } ABD \leqslant \triangle ABC$$

$$\Rightarrow \tfrac{1}{2}r.r.\sin\theta \leqslant \frac{\theta}{k}.\pi r^2 \leqslant \tfrac{1}{2}r.r.\tan\theta$$

$$\Rightarrow \sin\theta \leqslant \frac{2\theta}{k}\pi \leqslant \tan\theta \quad \text{... dividing throughout by } \tfrac{1}{2}r^2$$

$$\Rightarrow 1 \leqslant \frac{2\pi\theta}{k\sin\theta} \leqslant \frac{1}{\cos\theta} \quad \text{... } \sin\theta > 0 \text{ for acute angles, so we can divide by } \sin\theta$$

$$\Rightarrow 1 \geqslant \frac{k\sin\theta}{2\pi\theta} \geqslant \cos\theta \quad \text{... taking the reciprocal and noting that if } A < B \text{ then } \frac{1}{A} > \frac{1}{B}$$

$$\Rightarrow 1 \geqslant \frac{k}{2\pi}.\frac{\sin\theta}{\theta} \geqslant \cos\theta$$

Now, for very small angles, **$\cos\theta \approx 1$** ... e.g. $\cos 0{\cdot}5° = 0{\cdot}99996192... = 1{\cdot}0000$ (to 5 s.f.). So for very small angles,

$$\Rightarrow 1 \geqslant \frac{k}{2\pi}.\frac{\sin\theta}{\theta} \geqslant 1 \text{ ... i.e. } \frac{k}{2\pi}.\frac{\sin\theta}{\theta} \approx 1.$$

We can choose k to be anything we like ... that is the new unit we are looking for.

If we choose k to be 2π then $\dfrac{\sin\theta}{\theta} \approx 1$ and so **$\sin\theta \approx \theta$** for very small angles.

This is very convenient for astronomers and the likes. They can use the new unit for measuring angles. Working with **small angles** and **the new unit**, they can say **$\cos\theta \approx 1$, $\sin\theta \approx \theta$** and also **$\tan\theta \approx \theta$**.

This new unit of angular measure is called the **radian** and is defined by **2π radians = 1 revolution**.

What you need to know

1 a Given that $\sin 30° = 0{\cdot}5$, state the value of:

i $\sin 150°$ ii $\sin 210°$ iii $\sin 330°$.

b Given that $\cos 60° = 0{\cdot}5$, state the value of:

i $\cos 120°$ ii $\cos 240°$ iii $\cos 300°$.

c Given that $\tan 45° = 1$, state the value of:

i $\tan 135°$ ii $\tan 225°$ iii $\tan 315°$.

2 Here is the graph of $y = 3\sin(2x)° + 1, 0 \leq x \leq 360$.

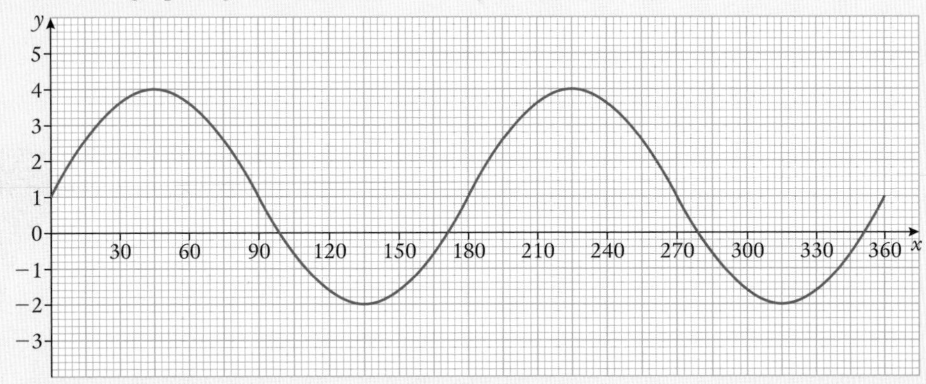

a State the value of: **i** the amplitude of the wave **ii** the period of the function.

b What effect does the '+1' have on the graph?

c What is the maximum value of the function and for what values of x does it occur?

d The curve first cuts the x-axis when $x° = 100°$, to the nearest degree. Where else does it cut the axis in the domain $0 \leq x \leq 360$?

3 The curve shows $y = 2 - \cos(x - 30)°, 0 \leq x \leq 360$.

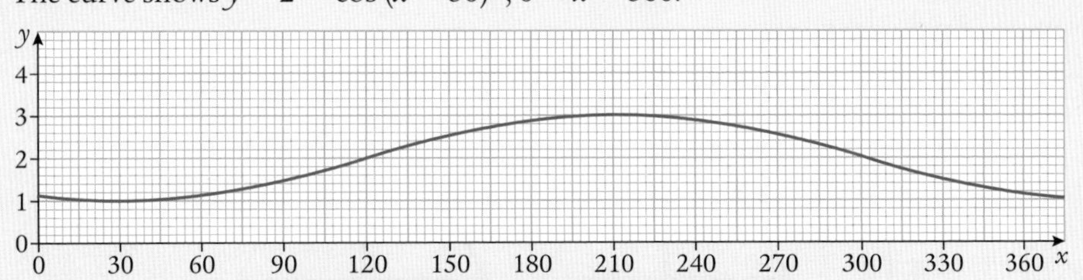

If we consider the graph as a transformation of the graph $y = \cos x°$,

a what effect does adding 2 have on the graph

b what effect does making the cosine negative have

c what effect does subtracting 30 from x have?

4 Make a sketch of $y = \tan x°, 0 \leq x \leq 360$.

5 Solve:

a $\sin x° = 0{\cdot}625, 0 \leq x \leq 360$

b $3\cos x° + 1 = 2, 0 \leq x \leq 360$

c $\cos(2x)° = 0{\cdot}5, 0 \leq x \leq 360$

d $3\sin(2x - 30)° = 1, 0 \leq x \leq 360$.

6 Complete each statement:

a $\sin^2 x° + \cos^2 x° = ?$

b $\dfrac{\sin x}{\cos x} = ?$

2.1 Some additional facts

Using radians

When working in radians, it is best, where possible, to give answers as expressions containing π. Answers are then exact, and rounding errors are avoided.

2π radians = 1 revolution = 360°
π radians = a half-turn = 180°
$\frac{\pi}{2}$ radians = half of a half-turn = $(\frac{1}{2} \times 180)° = 90°$
$\frac{5\pi}{6}$ radians = five-sixths of a half-turn = $(\frac{5}{6} \times 180)° = 150°$
x radians = $\left(\frac{180}{\pi}x\right)$ degrees
x degrees = $\left(\frac{\pi}{180}x\right)$ radians

1 degree = $\frac{\pi}{180}$ radians, i.e. about 0·01745 radians, so work to 3 decimal places at least when working with decimals and radians.

Note that each of the equations above requires that the units are mentioned ... $\pi \neq 180$.

However, when it is clear in the context, we do not use a symbol for radians.

sin x ... means the sine of x radians.

sin $x°$ ... means the sine of x degrees.

Exact values

Using a square of side 1 unit and an equilateral triangle of side 2 units, the exact values of the trig ratios for 30°, 45°, 60° and 90° can be found.

In radians these angles are $\frac{\pi}{6}, \frac{\pi}{4}, \frac{\pi}{3}$ and $\frac{\pi}{2}$ radians.

These can be handy if you don't have a calculator.

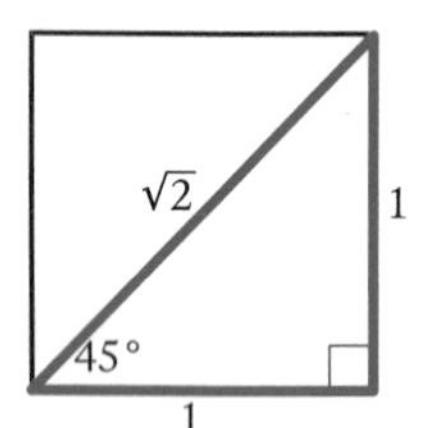

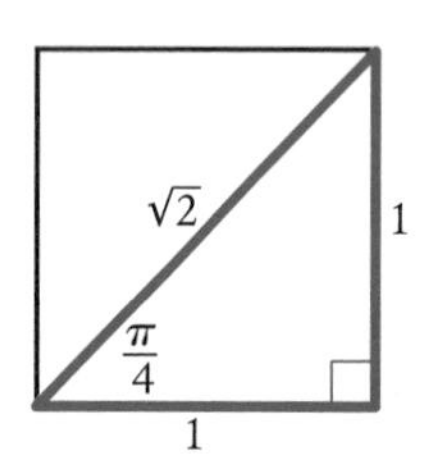

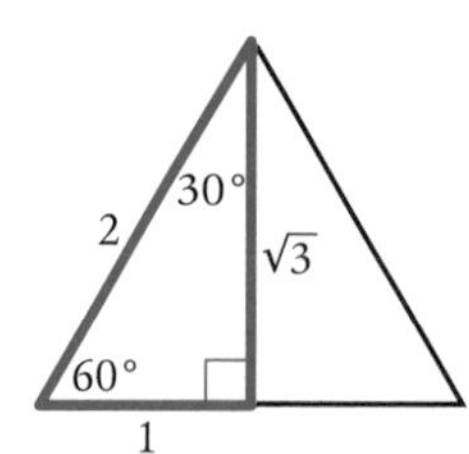

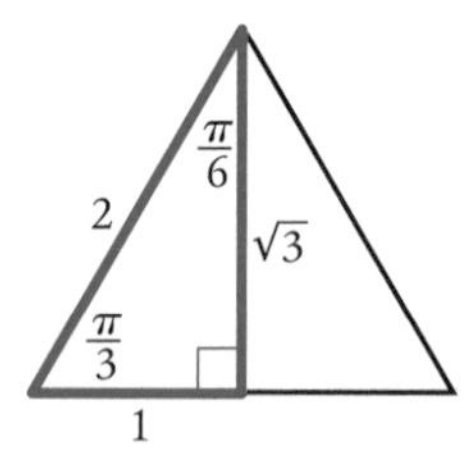

Halving the square: the side is 1 and the diagonal, by Pythagoras' theorem, is $\sqrt{2}$.
Halving the triangle: half the base is 1 and the altitude, by Pythagoras' theorem, is $\sqrt{3}$.

Negative angles

By convention, as we rotate an arm anticlockwise about the origin we open out a positive angle.
Rotating clockwise we open out a **negative** angle.

We already know that:
$\sin a = \frac{y}{r}$, $\cos a = \frac{x}{r}$, $\tan a = \frac{y}{x}$.

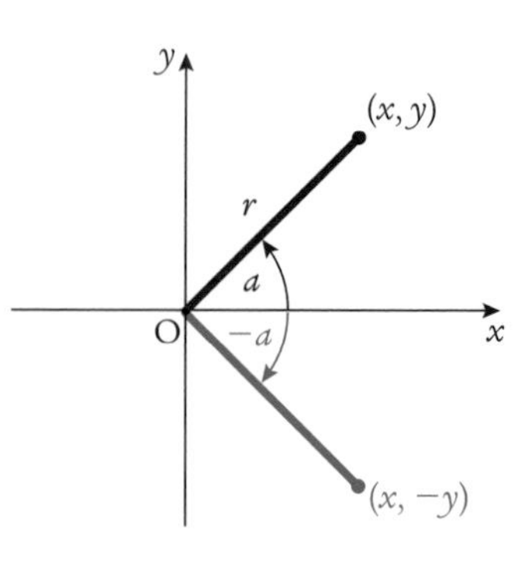

We can see that:

$$\sin(-a) = \frac{-y}{r} = -\sin a$$

$$\cos(-a) = \frac{x}{r} = \cos a$$

$$\tan(-a) = \frac{-y}{x} = -\tan a.$$

Note also that:

$\sin(-a) = \sin(360 - a)$ ... the arm ends at the same point

$\cos(-a) = \cos(360 - a)$

$\tan(-a) = \tan(360 - a)$.

The trig graphs can be extended to the left. Here we see $y = \sin x, -\pi \leqslant x \leqslant \pi$.

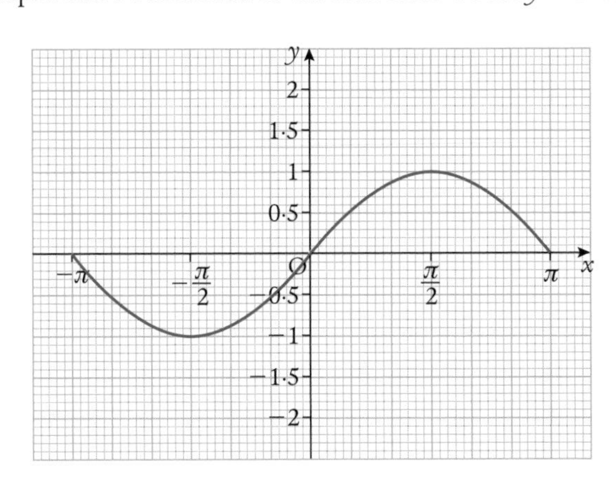

Example 1

Find the value of:

a $\sin\left(-\frac{\pi}{6}\right)$ **b** $\cos\left(\frac{5\pi}{6}\right)$ **c** $\tan\left(\frac{4\pi}{3}\right)$.

a $\sin\left(-\frac{\pi}{6}\right) = -\sin\left(\frac{\pi}{6}\right) = -\frac{1}{2}$ ... using the 'exact values' triangles

b $\cos\left(\frac{5\pi}{6}\right) = \cos\left(\pi - \frac{\pi}{6}\right) = -\cos\left(\frac{\pi}{6}\right) = -\frac{\sqrt{3}}{2}$

c $\tan\left(\frac{4\pi}{3}\right) = \tan\left(\pi + \frac{\pi}{3}\right) = \tan\left(\frac{\pi}{3}\right) = \sqrt{3}$

sin $\pi - \theta$	All θ
tan $\pi + \theta$	cos $2\pi - \theta$

Example 2

Solve:

a $2\sin x° = 1; 0 \leqslant x \leqslant 360$ **b** $2\sin x = 1; 0 \leqslant x \leqslant 2\pi$.

a $2\sin x° = 1$

$\Rightarrow \sin x° = \frac{1}{2}$

$\Rightarrow x = 30 \text{ or } 180 - 30$

$\Rightarrow x = 30 \text{ or } 150$

b $2\sin x = 1$

$\Rightarrow \sin x = \frac{1}{2}$

$\Rightarrow x = \frac{\pi}{6} \text{ or } \pi - \frac{\pi}{6}$ ... from 'exact value' triangle

$\Rightarrow x = \frac{\pi}{6} \text{ or } \frac{6\pi}{6} - \frac{\pi}{6} = \frac{5\pi}{6}$

Exercise 2.1

1 We know $\sin(180 - x)° = \sin x°$.
The equivalent statement using radians is $\sin(\pi - x) = \sin x$.

a Write down the equivalent statement using radians for:

i $\sin(180 + x)° = -\sin x°$ ii $\cos(360 - x)° = \cos x°$
iii $\cos(180 + x)° = -\cos x°$ iv $\tan(180 - x)° = -\tan x°$
v $\sin(360 - x)° = -\sin x°$ vi $\tan(360 - x)° = -\tan x°$.

b We know $\sin(90 - x)° = \cos x°$.
Write down the equivalent statement using radians.

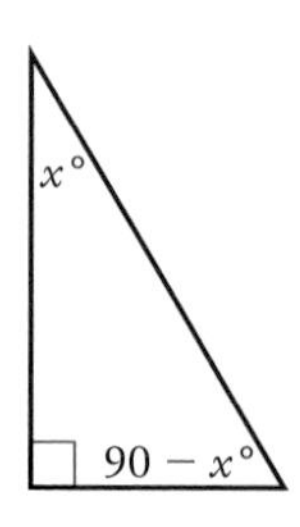

c The first quadrant lies between 0° and 90°.
The second quadrant lies between 90° and 180°.
The third quadrant lies between 180° and 270°.
The fourth quadrant lies between 270° and 360°.
Write down the equivalent statements in radians.

2 Copy and complete each table.

	30°	45°	60°	90°
sin	$\frac{1}{2}$			
cos	$\frac{\sqrt{3}}{2}$			
tan	$\frac{1}{\sqrt{3}}$			Undefined

	$\frac{\pi}{6}$	$\frac{\pi}{4}$	$\frac{\pi}{3}$	$\frac{\pi}{2}$
sin	$\frac{1}{2}$			
cos	$\frac{\sqrt{3}}{2}$			
tan	$\frac{1}{\sqrt{3}}$			Undefined

3 Use the degree table from question **2** to find the value of:

a $\cos 135°$ b $\sin 210°$ c $\tan 240°$ d $\sin(-30)°$
e $\cos(-45)°$ f $\sin(-150)°$ g $\cos(-135)°$ h $\tan(-60)°$
i $\tan(-300)°$ j $\sin(-120)°$.

4 Use the radian table from question **2** to find the value of:

a $\cos\frac{5\pi}{4}$ $\left[\text{Hint: } \frac{5\pi}{4} = \pi + \frac{\pi}{4}\right]$ b $\sin\frac{5\pi}{6}$ c $\tan\frac{5\pi}{3}$
d $\sin\left(-\frac{\pi}{3}\right)$ e $\cos\left(-\frac{\pi}{4}\right)$ f $\sin\left(-\frac{\pi}{2}\right)$ g $\cos\left(-\frac{3\pi}{4}\right)$
h $\tan\left(-\frac{\pi}{3}\right)$ i $\tan\left(\frac{3\pi}{2}\right)$ j $\sin\left(\frac{3\pi}{2}\right)$.

5 Express the following in radians, leaving π in your answer:

a 10° b 36° c 200° d 300°
e 150° f 25° g 19° h 345°.

6 From 200 m up, the angle of depression to the foot of a building was measured as 10°.

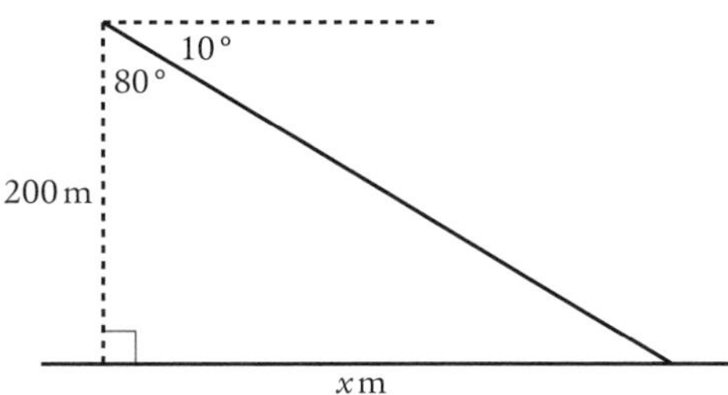

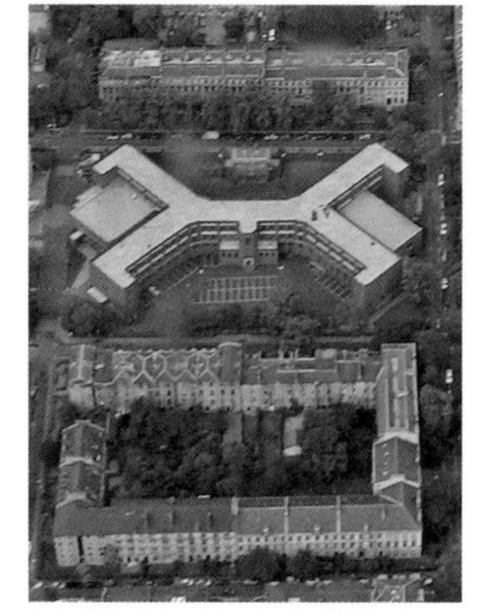

a Calculate the distance, x m, correct to 3 significant figures.

b Convert 80° to radians, correct to 3 significant figures.

c Set your calculator to radian mode and use your answer to part **b** to calculate the value of x again ... to 3 significant figures.

d i Compare your answers to **a** and **c**: what is the difference?

ii To how many figures would you need to work so that the answer using the radian measure was the same as the answer using degrees?

[When working with radian measure, don't switch between units and don't be too severe with your rounding mid-calculation.]

7 To calculate the length of an arc of a sector we use the formula:

$$L = \frac{x}{360} \times 2\pi r$$

where $x°$ is the angle at the centre and r units is the length of the radius.

a What would be the equivalent formula if working in radians?

b State the formula for the area of a sector when the angle at the centre is x radians.

c What is the length of the arc when $x = 1$ radian and the radius is r cm?

d What is the area of a sector where the angle at the centre is 2 radians?

2.2 Sketching graphs

Exercise 2.2

1 The graph shows $y = \sin x$ in the domain $-\frac{\pi}{2} < x < \frac{7\pi}{2}$.

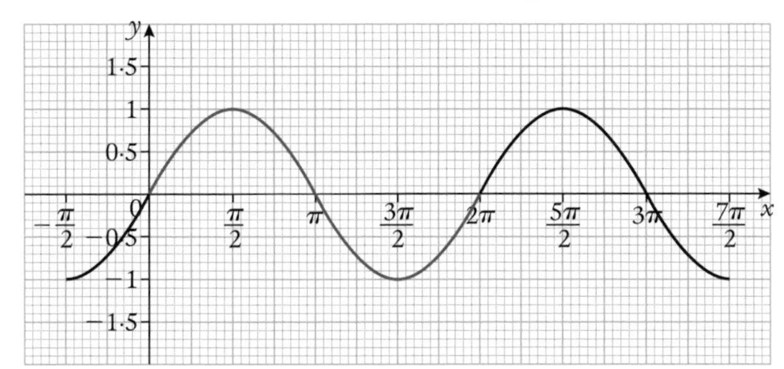

a What is the period of the function?

b State its maximum value and the value of x between 0 and 2π where it is attained.

c When is this maximum next attained?

d Between what values in this domain is the function negative?

e Make a sketch in the same domain of:

i $y = \cos x$ ii $y = \tan x$, $\left(x \neq \frac{n\pi}{2} \text{ where } n \text{ is odd}\right)$.

2 Graphs of the trigonometric functions can be explored with the aid of a spreadsheet.

To get π you need to type PI() ...

if you type =PI() into a cell, you get 3.141592654.

In cell A1 type 0 ... zero.

In cell A2 type =A1+PI()/12 ... $\frac{\pi}{12}$ is added to the contents of the cell above, A1.

Fill this formula down the column to cell A25 ... generating 0 to 2π in steps of $\frac{\pi}{12}$.

In B1 type =SIN(A1)

Fill this formula down to cell B25 ... note that the spreadsheet works in radians as its default.

You are going to explore functions of the form

$y = \text{A} \sin(\text{B}x + \text{C}) + \text{D}$

So type A, B, C, D in cells D1, E1, F1, G1 respectively.

In cells D2, E2, F2, G2 enter some values ... let's say 3, 2, =PI()/4, 1 respectively.

Note that we have given C the value $\frac{\pi}{4}$.

In C1 type =$D$2*SIN($E$2*A1+$F$2)+$G$2

Fill this formula down to cell C25.

Highlight all three columns of figures.

Select 'Charts' and pick 'Scatter'. From the menu choose 'Scatter with Smooth Lines'.

A graph similar to this will appear:

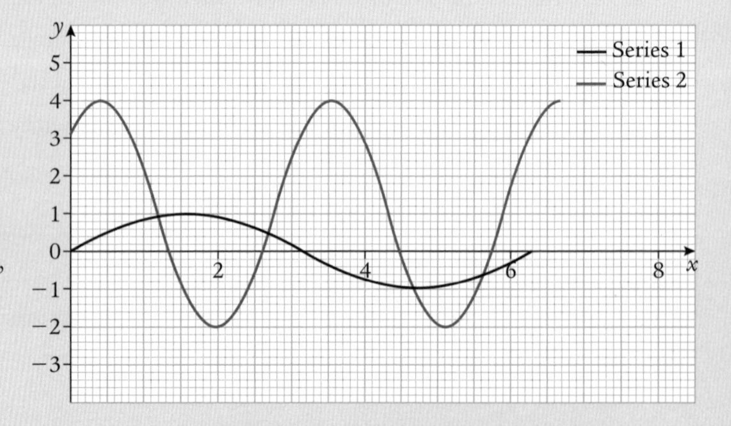

The curve marked 'Series 1' is $y = \sin x$. This will remain fixed so you can compare it with the transformed wave.

The curve marked 'Series 2' is $y = 3\sin(2x + \frac{\pi}{4}) + 1$.

As you alter the values in D2, E2, F2, and G2, the graph will instantly change, letting you see the effect that each constant has on the wave's shape and position.

Make C and D zero to begin with to explore the influence of A and B.

Make A and B both 1 to explore the influence of C and D.

3 Make sketches of the following curves in the domain given.

a $y = 2\sin x + 1; -\frac{\pi}{2} \leqslant x \leqslant \frac{\pi}{2}$

b $y = 3\cos x - 1; -\frac{\pi}{2} \leqslant x \leqslant \frac{\pi}{2}$

c $y = \sin 2x + 1; 0 \leqslant x \leqslant 2\pi$

d $y = 2\cos 3x; 0 \leqslant x \leqslant 2\pi$

e $y = \sin\left(x + \frac{\pi}{3}\right); 0 \leqslant x \leqslant 2\pi$

f $y = 2\cos\left(x - \frac{\pi}{6}\right); 0 \leqslant x \leqslant 2\pi$

g $y = 2\sin\left(x - \frac{\pi}{4}\right); 0 \leqslant x \leqslant 2\pi$

h $y = 3\cos\left(2x - \frac{\pi}{6}\right) + 1; 0 \leqslant x \leqslant 2\pi$

i $y = \sin\left(x - \frac{\pi}{2}\right); 0 \leqslant x \leqslant 2\pi$ ... and comment

j $y = 2\tan x + 1; -\frac{\pi}{2} < x < \frac{\pi}{2}$

4 Suggest suitable equations for each of the waves.

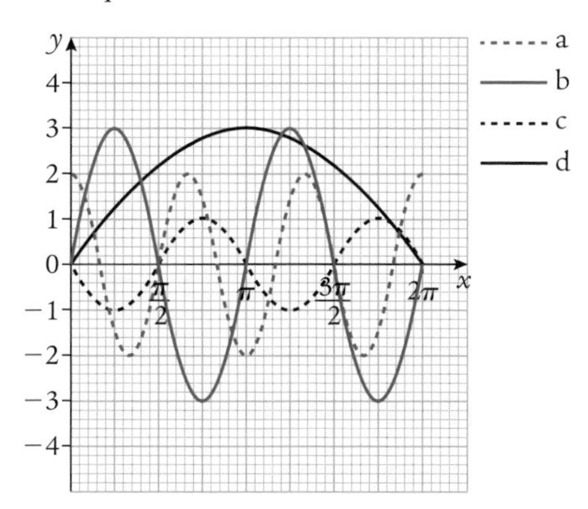

5 The Moon has a cycle of 28 days. Astronomers model the cycle using the formula

$$P(x) = 50 - 50 \cos\left(\frac{\pi}{14}x\right)$$

where x is the number of days since the new Moon and P is the percentage of the disc that is lit.

a What percentage of the Moon is lit 7 days after the new Moon?

b Calculate: **i** $P(14)$ **ii** $P(21)$ **iii** $P(28)$.

c Sketch the function in the domain $0 \leqslant x \leqslant 56$.

2.3 Trigonometric equations

When using a calculator to solve the equation $\sin x = 0{\cdot}4$, having checked that the calculator is in RADIAN mode, we would use the button sequence

[$\sin^{-1}$] [( ] [0] [.] [4] [ )] [=]

to give the answer 0·41151685.

In this book we will use the notation $\sin^{-1}(x)$ to represent **the value the calculator returns** when you input x.

Example 1

Solve: $2 \sin x + 1 = 1{\cdot}8$; $0 \leqslant x \leqslant 4\pi$.

$2 \sin x + 1 = 1{\cdot}8$

$\Rightarrow \quad 2 \sin x = 0{\cdot}8$

$\Rightarrow \quad \sin x = 0{\cdot}4$

$\Rightarrow \quad x = \sin^{-1}(0{\cdot}4)$ or $\pi - \sin^{-1}(0{\cdot}4)$ or $2\pi + \sin^{-1}(0{\cdot}4)$ or $2\pi + [\pi - \sin^{-1}(0{\cdot}4)]$

$\Rightarrow \quad x = 0{\cdot}412$ or $2{\cdot}730$ or $6{\cdot}695$ or $9{\cdot}013$ radians (to 3 d.p.).

[Note: The first answer came from the calculator, the second from the symmetry of the sine wave, $\sin x = \sin(\pi - x)$, the third and fourth from the period of the wave, $\sin x = \sin(2\pi + x)$.
Remember:

- the symmetry of the cosine gives us ... $\cos x = \cos(2\pi - x)$
- the symmetry of the tangent gives us ... $\tan x = \tan(\pi + x)$.]

Example 2

Solve: **a** $\sqrt{3}\tan x = 1; 0 \leqslant x \leqslant 2\pi$ **b** $\sqrt{3}\tan x = -1; 0 \leqslant x \leqslant 2\pi$.

a $\sqrt{3}\tan x = 1$

$\Rightarrow \tan x = \dfrac{1}{\sqrt{3}}$

$\Rightarrow x = \tan^{-1}\left(\dfrac{1}{\sqrt{3}}\right)$ or $\pi + \tan^{-1}\left(\dfrac{1}{\sqrt{3}}\right)$

$\Rightarrow x = \dfrac{\pi}{6}$ or $\pi + \dfrac{\pi}{6} = \dfrac{7\pi}{6}$ ... using the 'exact values' triangles

b $\sqrt{3}\tan x = -1$

$\Rightarrow \tan x = -\dfrac{1}{\sqrt{3}}$

When dealing with the 'exact values' triangles and negative values for $\sin x$, $\cos x$ or $\tan x$, we are better referring to the quadrant diagram:

sin $\pi - \theta$	All θ
tan $\pi + \theta$	cos $2\pi - \theta$

The tangent is **negative** in the 2nd and 4th quadrant, so we calculate

$x = \pi - \tan^{-1}\left(\dfrac{1}{\sqrt{3}}\right)$ or $2\pi - \tan^{-1}\left(\dfrac{1}{\sqrt{3}}\right)$

$\Rightarrow x = \pi - \dfrac{\pi}{6}$ or $2\pi - \dfrac{\pi}{6}$

$\Rightarrow x = \dfrac{5\pi}{6}$ or $\dfrac{11\pi}{6}$

Example 3

Solve: $2 + 6\cos\left(2x + \dfrac{\pi}{3}\right) = 5; 0 \leqslant x \leqslant 2\pi$.

$2 + 6\cos\left(2x + \dfrac{\pi}{3}\right) = 5$

$\Rightarrow 6\cos\left(2x + \dfrac{\pi}{3}\right) = 3$

$\Rightarrow \cos\left(2x + \dfrac{\pi}{3}\right) = \dfrac{1}{2}$

$\Rightarrow 2x + \dfrac{\pi}{3} = \cos^{-1}\left(\dfrac{1}{2}\right)$ or $2\pi - \cos^{-1}\left(\dfrac{1}{2}\right)$ or $2\pi + \cos^{-1}\left(\dfrac{1}{2}\right)$ or $2\pi + \left(2\pi - \cos^{-1}\left(\dfrac{1}{2}\right)\right)$

$\Rightarrow 2x + \dfrac{\pi}{3} = \dfrac{\pi}{3}$ or $2\pi - \dfrac{\pi}{3}$ or $2\pi + \dfrac{\pi}{3}$ or $4\pi - \dfrac{\pi}{3}$

$\Rightarrow 2x + \dfrac{\pi}{3} = \dfrac{\pi}{3}$ or $\dfrac{5\pi}{3}$ or $\dfrac{7\pi}{3}$ or $\dfrac{11\pi}{3}$

$\Rightarrow 2x = 0$ or $\dfrac{4\pi}{3}$ or $\dfrac{6\pi}{3}$ or $\dfrac{10\pi}{3}$

$\Rightarrow x = 0$ or $\dfrac{2\pi}{3}$ or $\dfrac{3\pi}{3}$ or $\dfrac{5\pi}{3}$

$\Rightarrow x = 0$ or $\dfrac{2\pi}{3}$ or π or $\dfrac{5\pi}{3}$

Example 4

Solve: $3\sin^2 x - 5\sin x - 2 = 0; 0 \leqslant x \leqslant 2\pi$.

$$3\sin^2 x - 5\sin x - 2 = 0$$

$\Rightarrow \quad (3\sin x + 1)(\sin x - 2) = 0$

$\Rightarrow 3\sin x + 1 = 0$ or $\sin x - 2 = 0$

> Compare it with
> $3a^2 - 5a - 2 = 0$
> $\Rightarrow (3a + 1)(a - 2) = 0$

$\Rightarrow \sin x = -\frac{1}{3}$ or $\sin x = 2$ (no solutions)

$\Rightarrow \quad x = \sin^{-1}(-\frac{1}{3})$ or $\pi - \sin^{-1}(-\frac{1}{3})$ or $2\pi + \sin^{-1}(-\frac{1}{3})$

$\Rightarrow \quad x = -0{\cdot}340$ or $3{\cdot}481$ or $5{\cdot}943$

Since $-0{\cdot}340$ is not in the domain $0 \leqslant x \leqslant 2\pi$, the solutions are $x = 3{\cdot}481$ or $5{\cdot}943$.

Exercise 2.3

1 Solve each equation in the domain $0 \leqslant x \leqslant 2\pi$.

a $2\sin x = \sqrt{3}$ **b** $2\cos x - \sqrt{3} = 0$ **c** $4\sin x + 1 = 3$ **d** $\tan x = \sqrt{3}$

e $\sqrt{2}\sin x = 1$ **f** $\sqrt{2}\cos x + 1 = 0$ **g** $6\sin x + 5 = 2$ **h** $3 - 4\cos x = 5$

2 At a level crossing, the height of the end of the barrier as it is being raised can be modelled by:

$$h(x) = 3\sin x + 1; -\frac{\pi}{2} \leqslant x \leqslant \frac{\pi}{2}.$$

The positive angles model the barrier on the left side of the road; the negative angles model the one on the right-hand side.

For what values of x is $h(x) = 2{\cdot}5$?

3 Solve, being mindful of the domain:

a $3\sin x + 1 = 2; 0 \leqslant x \leqslant \pi$ **b** $2\cos x + 5 = 7; 0 \leqslant x \leqslant 2\pi$

c $5\tan x + 2 = 3; 0 \leqslant x \leqslant 4\pi$ **d** $5 - 3\cos x = 4; 0 \leqslant x \leqslant 2\pi$

e $4 - 2\tan x = 5; 0 \leqslant x \leqslant 2\pi$ **f** $3 + 2\cos x = 1; 0 \leqslant x \leqslant 3\pi$

g $3\cos x + 7 = 6; -\pi \leqslant x \leqslant \pi$ **h** $6 + 5\cos x = 3, -\pi \leqslant x \leqslant 2\pi$.

4 The circumference of the round window in the church is modelled by:

$$h(x) = 2\sin x + 10; 0 \leqslant x \leqslant 2\pi$$

where h m is the height from the ground to a point x radians round from the vertical.

a Calculate: **i** $h(1)$ **ii** $h(2)$ **iii** $h\left(\frac{\pi}{3}\right)$.

b Calculate the values of x when $h(x) = 11$.

5 Solve:

a $2\cos\left(2x - \frac{\pi}{6}\right) - \sqrt{3} = 0; 0 \leqslant x \leqslant 2\pi$ **b** $2\sin 3x - \sqrt{3} = 0; 0 \leqslant x \leqslant 2\pi$

c $10\sin\left(2x - \frac{\pi}{4}\right) + 1 = 6; 0 \leqslant x \leqslant 2\pi$ **d** $\tan\left(2x + \frac{\pi}{6}\right) = \sqrt{3}; 0 \leqslant x \leqslant 2\pi$

e $\sqrt{2}\cos\left(3x - \frac{\pi}{2}\right) = 1; 0 \leqslant x \leqslant 2\pi$ **f** $5\sin(2x - 1) = 2; 0 \leqslant x \leqslant 2\pi$.

6 The depth of water in the harbour, d metres, can be modelled by:

$$d(x) = 3\sin\left(\frac{\pi}{6}x + 5\right) + 4;\ 0 \leqslant x \leqslant 24$$

where x is the number of hours since midnight.

a i What is the deepest the water gets?
  ii At what times does it occur in a day?

b When is low tide?

c At what time is the depth 2 metres?

7 Solve each equation in the domain $0 \leqslant x \leqslant 2\pi$.

a $2\sin^2 x - 3\sin x + 1 = 0$ b $3\cos^2 x + 11\cos x + 6 = 0$

c $6\tan^2 x - \tan x - 1 = 0$ d $4\sin^2 x + 5\sin x + 1 = 0$

e $3\cos^2 x + 10\cos x + 8 = 0$ f $5\tan^2 x + 23\tan x + 12 = 0$

8 Astronomers studying a binary system of stars to measure the combined brightness over several months.

They model it by:

$$B(x) = 8 - \sin x - 6\sin^2 x$$

where x is the number of months since the start of observations and B is the brightness in suitable units.
When was $B = 7$ in the 10 months of observations?

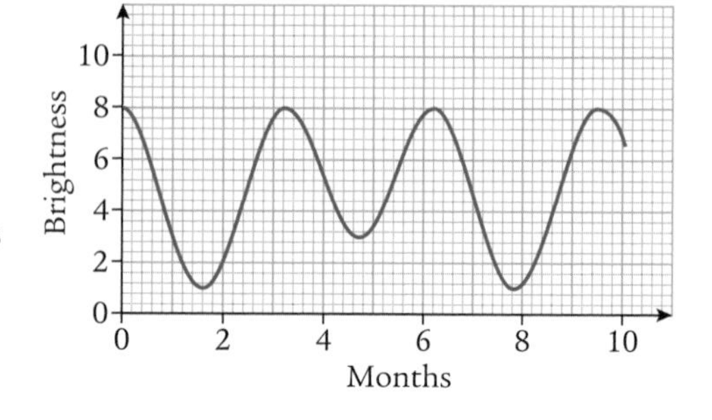

Preparation for assessment

1 Convert 1·25 radians to degrees.

2 Evaluate $\sin\left(-\frac{1}{2}\right) + \cos\left(\frac{\sqrt{3}}{2}\right)$.

3 ABC is a triangle with AB = 2 units, AC = 1 unit and angle BAC = $\frac{\pi}{3}$ radians.
Calculate, without the aid of a calculator, the exact value of:

a the area of ABC b the length of BC.

4 Solve the equation $3\sin x + 4 = 5;\ 0 \leqslant x \leqslant 2\pi$.

5 The head of a piston, rising and falling in a cylinder, can be modelled by the function

$$h(x) = 2\cos 3x + 10$$

where h is the height of the head of the piston x seconds after measurements were started.

a What is the height of the piston head after 3 seconds?

b Calculate the values of x in the interval $0 \leqslant x \leqslant 2\pi$ when the height is 9 units.

6 The height of a chair in a Ferris wheel can be modelled by

$$h(x) = 5\sqrt{3} - 10\cos\left(2x - \frac{\pi}{6}\right);\ 0 \leqslant x \leqslant 2\pi$$

where h is the height in metres x minutes into the ride.

a Calculate the times during the ride when $h = 0$.

b When is the chair below the embarkation point?

7 Solve: $2 - \sin x - 3\sin^2 x = 0;\ 0 \leqslant x \leqslant 2\pi$.

Summary

1 Radian measure

2π radians $= 360°$

$\Rightarrow 1$ radian $= \dfrac{360}{2\pi} = \dfrac{180}{\pi}$ degrees

- x radians $= \dfrac{180}{\pi}x$ degrees
- x degrees $= \dfrac{\pi}{180}x$ radians

2 Exact values for the trigonometric ratios for $30°$, $45°$, $60°$, $90°$ and their equivalent in radians can be read from these triangles:

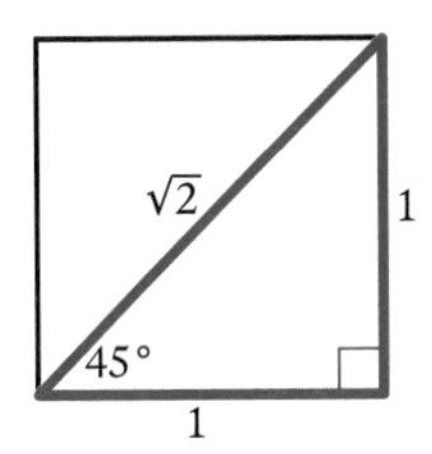

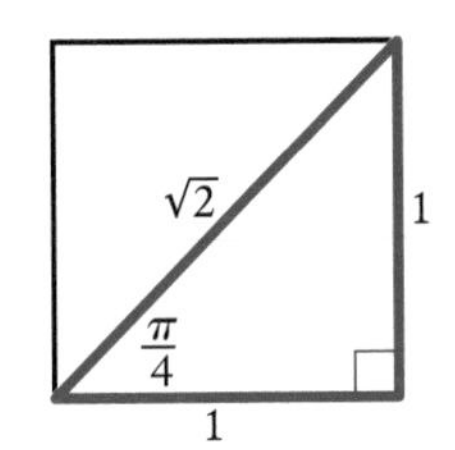

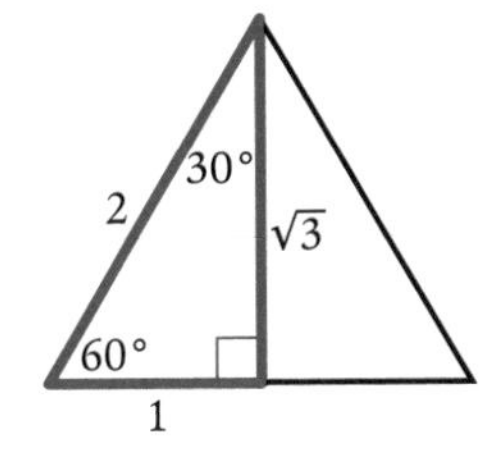

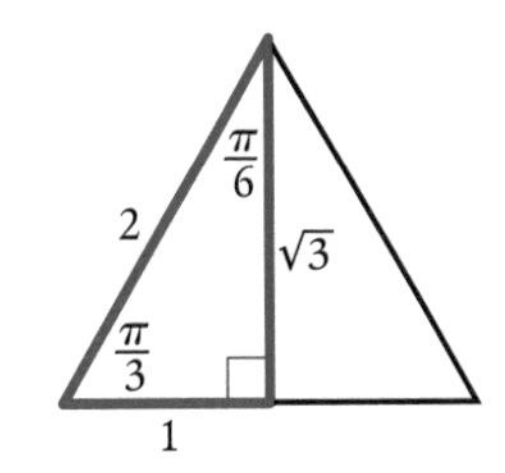

3 By convention,

- angles opened from the x-axis, **anticlockwise** are considered **positive**
- angles opened from the x-axis, **clockwise** are considered **negative**.

$\sin(-x) = -\sin(x)$ $\quad\cos(-x) = \cos(x)$ $\quad\tan(-x) = -\tan(x)$

4 Graphs

A quick sketch of $y = \sin x$ and $y = \cos x$ often helps in problems.

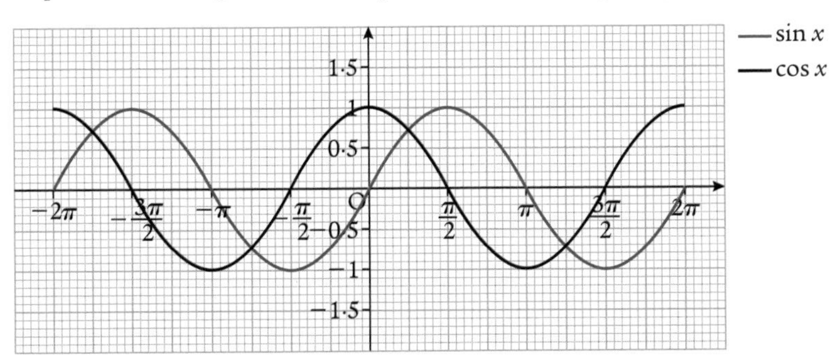

Note from the graph that $\sin x = \cos\left(x - \dfrac{\pi}{2}\right)$ ... a translation of $\dfrac{\pi}{2}$ units to the right.

The curve $y = \tan x$

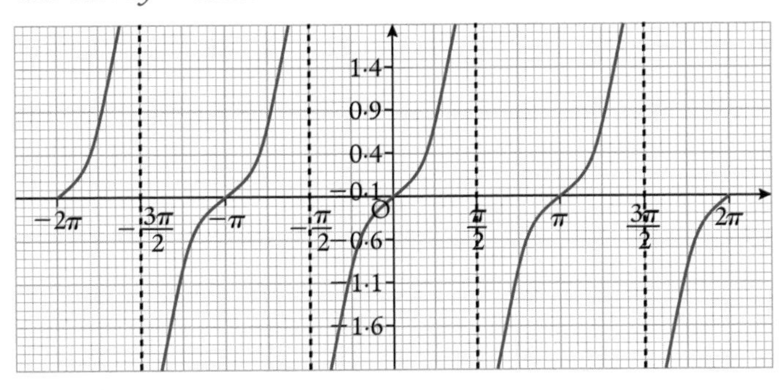

5 The **symmetries** of the curves reveal relationships ... summarised by

$\sin(\pi - x) = \sin x$ $\cos(\pi - x) = -\cos x$ $\tan(\pi - x) = -\tan x$	$\sin x$ $\cos x$ $\tan x$
$\sin(\pi + x) = -\sin x$ $\cos(\pi + x) = -\cos x$ $\tan(\pi + x) = \tan x$	$\sin(2\pi - x) = -\sin x = \sin(-x)$ $\cos(2\pi - x) = \cos x = \cos(-x)$ $\tan(2\pi - x) = -\tan x = \tan(-x)$

sin	All
tan	cos

6 Transformations

For example, to sketch $y = a\sin(bx + c) + d$ ... start with the known points on $y = \sin x$.

i $y_{\text{new}} = ay_{\text{old}} + d$

ii $x_{\text{new}} = \dfrac{x_{\text{old}} - c}{b}$

All functions can be transformed like this.

7 Equations

- reduce the equation to the form $\sin x = B$, $\cos x = B$ or $\tan x = B$
- the 1st solution, obtained from the calculator, is $\sin^{-1} B$, $\cos^{-1} B$ or $\tan^{-1} B$
- the 2nd solution, from the symmetries, is $\pi - \sin^{-1} B$, $2\pi - \cos^{-1} B$ or $\pi + \tan^{-1} B$.

Further solutions are found by adding (or subtracting) any multiple of 2π to both the 1st and 2nd solutions.

Check that you only have solutions that are in the prescribed domain.

If working without a calculator using exact values, symmetries should be used to deduce the value of, say, $\sin^{-1}(-\frac{1}{2})$.

3 Trigonometry 2 – compound angle formulae

Before we start...

Around AD 150, the Greek astronomer Ptolemy, living in Alexandria, produced a book, now called the *Almagest*, which contained among other things a basic relationship between $\sin x$, $\cos x$, $\sin y$, $\cos y$ and $\sin(x + y)$.

Ptolemy did not use the notation nor units we use today, and it is only through mathematical historians' efforts that the connection to today's work has been established.

By considering any one counter-example we easily establish that in general

$$\sin(x + y) \neq \sin x + \sin y.$$

Try $x = 30$ and $y = 60$.

So what **does** $\sin(x + y)$ equal?

A **basic fact** you learn when beginning trigonometry is that given a right-angled triangle, with hypotenuse h and angle $x°$, the side opposite $x°$ has length $h \sin x°$, and the side adjacent to $x°$ has length $h \cos x°$.

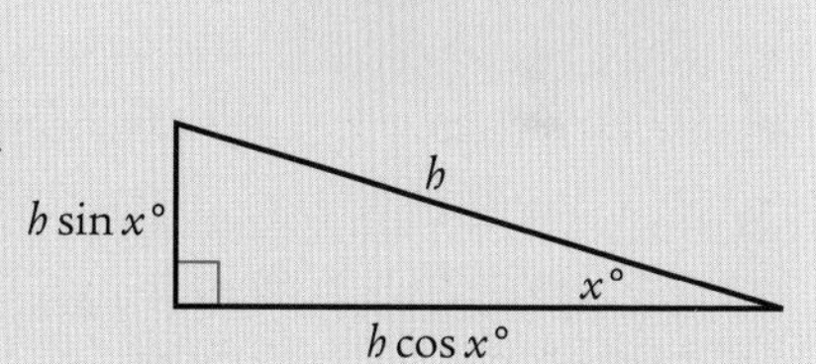

[The units of length and angular measure used don't affect the argument but for the sake of clarity the argument below is pursued in degrees. Units have been omitted from the diagrams.]

Consider the right-angled triangle PQR with hypotenuse 1 and angle RPQ $= A°$.

Using the **basic fact** above, we have RQ $= \sin A°$ and PQ $= \cos A°$.

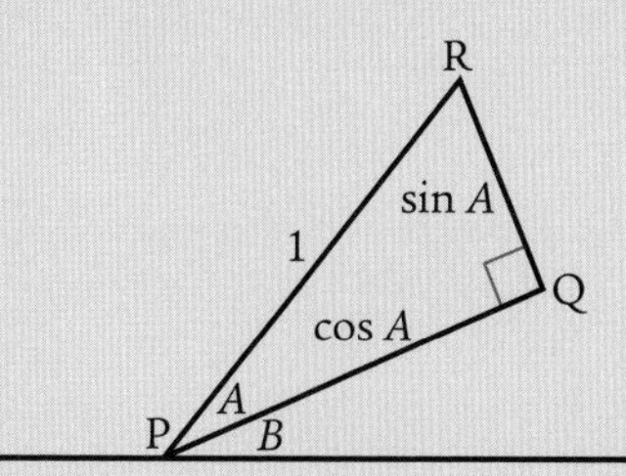

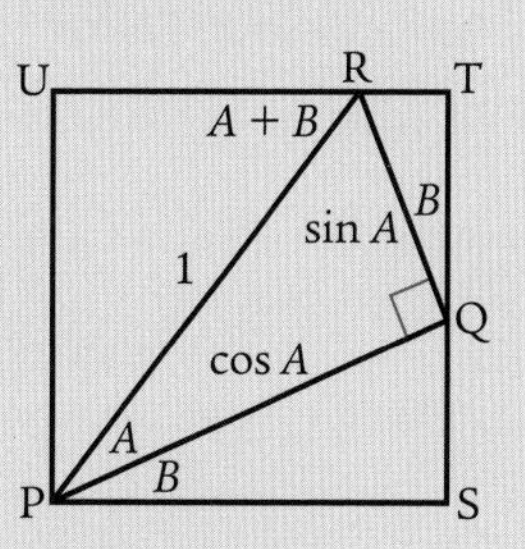

A line passes through P at an angle of $B°$ to PQ.

Lines are drawn through P, R and Q to complete a rectangle PSTU.

- $\angle QPS = B° \Rightarrow \angle PQS = (90 - B)° \Rightarrow \angle RQT = B°$
- $\angle URP = \angle RPS$ (alternate angles) $\Rightarrow \angle URP = (A + B)°$

Now, using the **basic fact** in the three right-angled triangles formed, we get the lengths as marked in purple in the diagram:

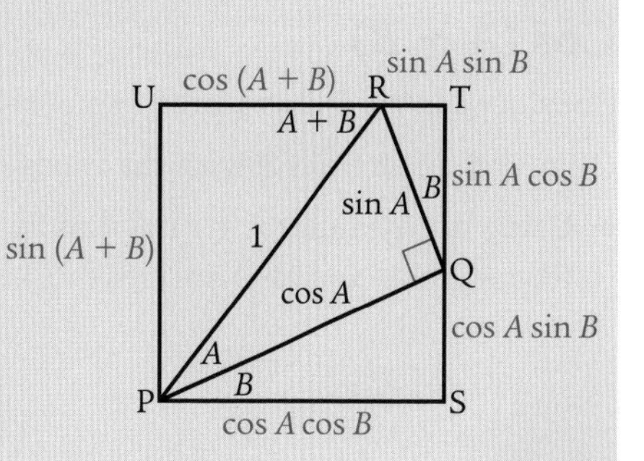

Opposite sides of the rectangle are equal:

i PU = QT + QS $\Rightarrow \sin(A + B)° = \sin A° \cos B° + \cos A° \sin B°$

ii UR = PS − RT $\Rightarrow \cos(A + B)° = \cos A° \cos B° - \sin A° \sin B°$.

These two relationships form the basis of this chapter.

What you need to know

1 In the diagram, the point P has coordinates (x, y).

The line OP makes an angle of $\theta°$ with the x-axis.

Let OP $= r$ units.

Thus $x = r\cos\theta°$ and $y = r\sin\theta°$.

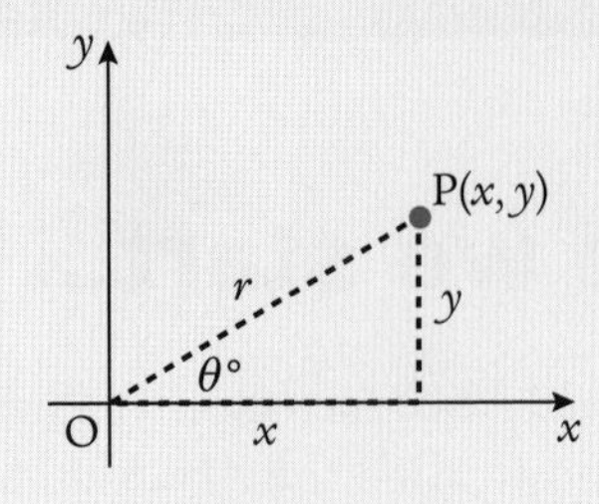

Find the coordinates of Q, R, S and T.

a

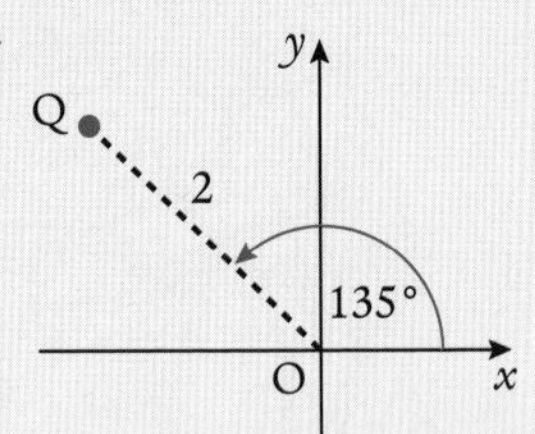

b

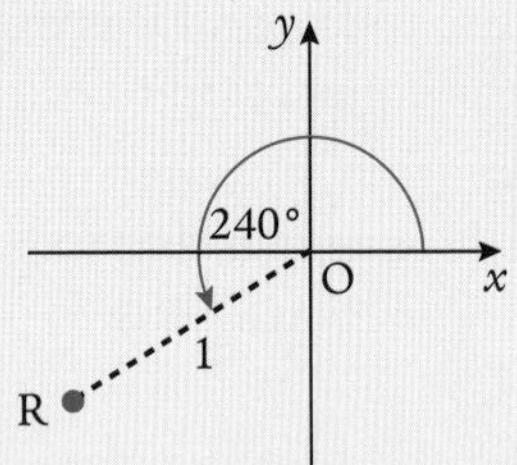

c

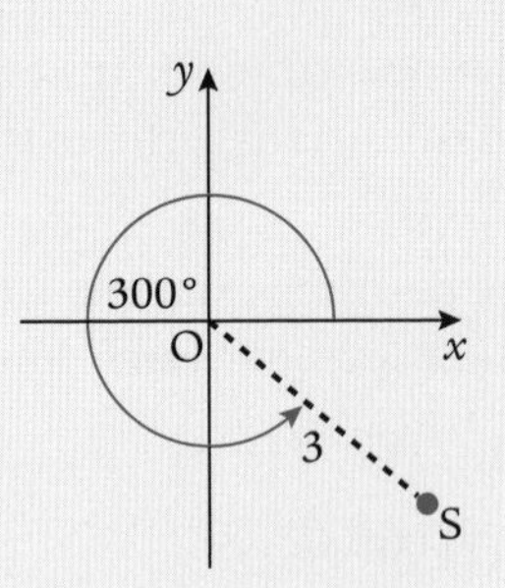

d

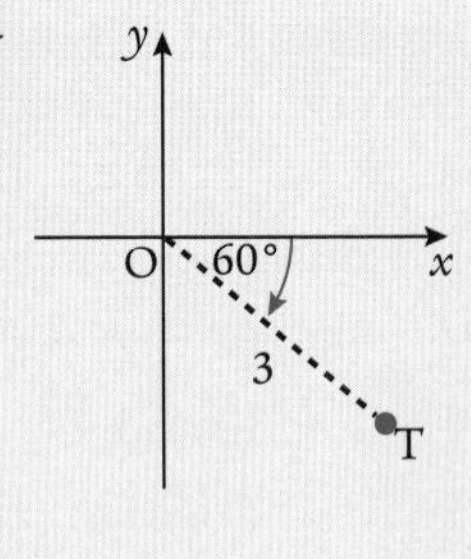

2 Verify that $\sin(30 + 60)°$ does not equal $\sin 30° + \sin 60°$.

3 $\sin(-x)° = -\sin x°$ and $\cos(-x)° = \cos x°$

Use this to help you find the **exact** value of:

a $\sin(-60)°$ **b** $\cos(-30)°$ **c** $\sin(-30)°$ **d** $\cos(-45)°$ **e** $\sin(-45)°$.

4 $\cos(180 - x)° = -\cos x°$. In a similar fashion, simplify:

a $\sin(180 - x)°$ **b** $\cos(360 - x)°$ **c** $\sin(90 - x)°$

d $\cos(90 - x)°$ **e** $\cos(180 + x)°$.

5 a If $\sin^2 x° = 0{\cdot}36$, what is the value of $\cos^2 x°$?

b If $\sin x° = \frac{4}{5}$; $0 \leq x \leq 90$, what is the value of $\cos x°$?

6 Calculate the distance between P(1, 5) and Q(4, 9) using the formula $PQ^2 = (x_P - x_Q)^2 + (y_P - y_Q)^2$.

3.1 Cos $(A \pm B)$

Cos $(A + B)$

Consider a circle radius 1 unit, centre the origin with two radii OP and OQ drawn.

OP makes an angle of A with the x-axis and OQ makes an angle of $-B$ with the x-axis.

- Using the formulae $x = r\cos\theta$ and $x = r\sin\theta$, the coordinates of P are $(\cos A, \sin A)$ and of Q are $(\cos(-B), \sin(-B))$.

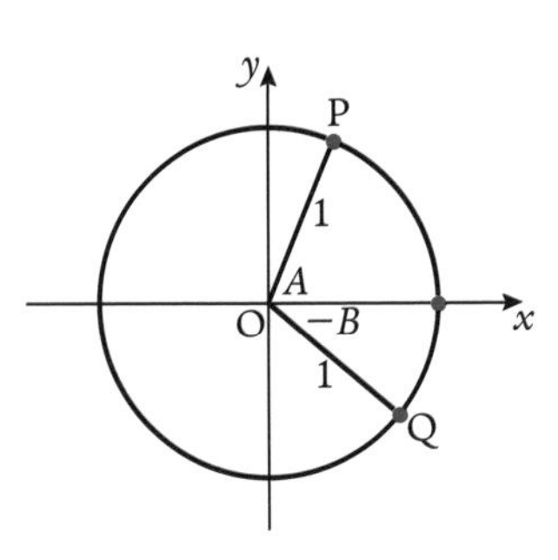

- Using the distance formula:

  $PQ^2 = (x_P - x_Q)^2 + (y_P - y_Q)^2$

  $\Rightarrow PQ^2 = (\cos A - \cos(-B))^2 + (\sin A - \sin(-B))^2$

  $\Rightarrow PQ^2 = (\cos A - \cos B)^2 + (\sin A + \sin B)^2$ ... using $\cos(-x) = \cos x$ and $\sin(-x) = -\sin x$

  $\Rightarrow PQ^2 = \cos^2 A - 2\cos A\cos B + \cos^2 B + \sin^2 A + 2\sin A\sin B + \sin^2 B$

  $\Rightarrow PQ^2 = (\cos^2 A + \sin^2 A) + (\cos^2 B + \sin^2 B) + 2\sin A\sin B - 2\cos A\cos B$

  $\Rightarrow PQ^2 = 2 + 2\sin A\sin B - 2\cos A\cos B$

- If we rotate the whole circle so that Q lands on the x-axis.

  P moves to $P'(\cos(A+B), \sin(A+B))$ and Q to $Q'(1, 0)$.

  $P'Q'^2 = (x_{P'} - x_{Q'})^2 + (y_{P'} - y_{Q'})^2$

  $\Rightarrow P'Q'^2 = (\cos(A+B) - 1)^2 + (\sin(A+B) - 0)^2$

  $\Rightarrow P'Q'^2 = \cos^2(A+B) - 2\cos(A+B) + 1 + \sin^2(A+B)$

  $\Rightarrow P'Q'^2 = 2 - 2\cos(A+B)$

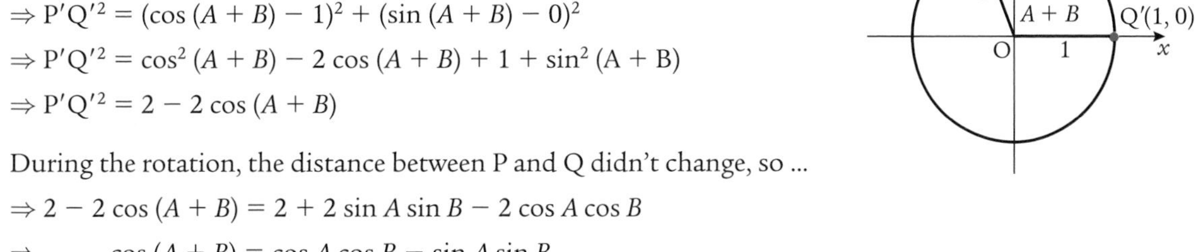

- During the rotation, the distance between P and Q didn't change, so ...

  $\Rightarrow 2 - 2\cos(A+B) = 2 + 2\sin A\sin B - 2\cos A\cos B$

  $\Rightarrow \quad \cos(A+B) = \cos A\cos B - \sin A\sin B$

  The same result as in the introduction, but now proved for all sizes of angles A and B ... in degrees or radians.

Cos $(A - B)$

Since the formula works for all angles, it will work when B is replaced by $(-B)$.

So, $\cos(A + (-B)) = \cos A\cos(-B) - \sin A\sin(-B)$

$\Rightarrow \quad \cos(A - B) = \cos A\cos B + \sin A\sin B$.

Example 1

The angles A and B are acute. Cos $A = \frac{3}{5}$ and $\sin B = \frac{5}{13}$.

Find the exact value of $\cos(A + B)$.

Since both are acute, we can sketch right-angled triangles using the data given and compute the missing side using Pythagoras' theorem.

$\cos(A+B) = \cos A\cos B - \sin A\sin B$

$\Rightarrow \cos(A+B) = \frac{3}{5}\cdot\frac{12}{13} - \frac{4}{5}\cdot\frac{5}{13} = \frac{36}{65} - \frac{20}{65} = \frac{16}{65}$

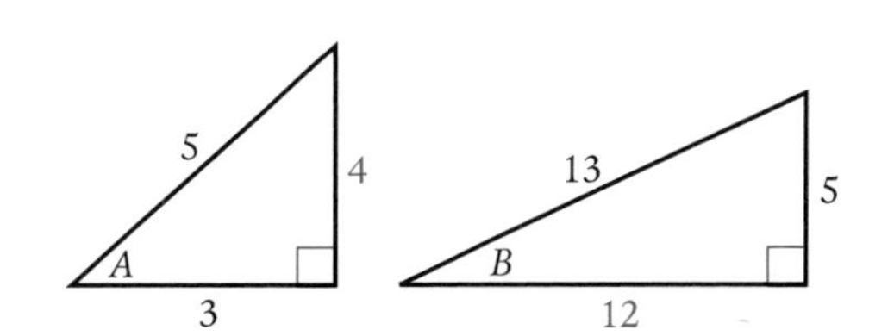

Example 2

Expand $4\cos(x + 30)°$.

$$
\begin{aligned}
4\cos(x+30)° &= 4(\cos x°\cos 30° - \sin x°\sin 30°) \\
&= 4\cos x°\cos 30° - 4\sin x°\sin 30° \\
&= 4.\frac{\sqrt{3}}{2}\cos x° - 4.\tfrac{1}{2}\sin x° \\
&= 2\sqrt{3}\cos x° - 2\sin x°
\end{aligned}
$$

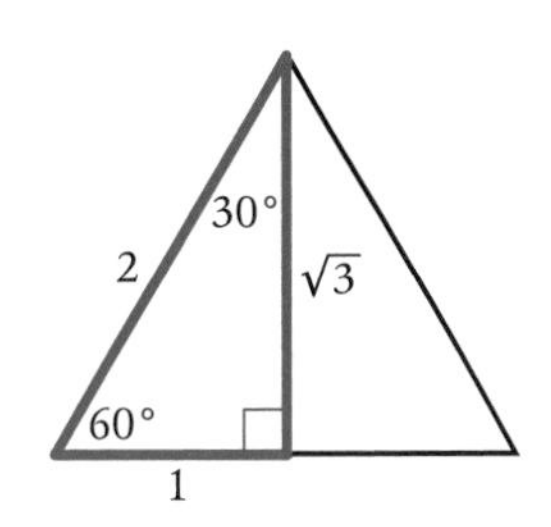

Exercise 3.1

1 Expand the following:

a $\cos(x + y)$ b $\cos(p - q)$ c $\cos(2p + q)$ d $\cos(3a + 2b)$.

2 a Verify $\cos(A + B) = \cos A \cos B - \sin A \sin B$ using $A = \frac{\pi}{3}$ and $B = \frac{\pi}{6}$.

b Verify $\cos(A - B)° = \cos A° \cos B° + \sin A° \sin B°$ using $A = 120$ and $B = 30$.

c Verify that $\cos(180 - x)° = -\cos x°$ by expanding and simplifying the left-hand side.

3 a Simplify: i $\cos(90 + x)°$ ii $\cos(90 - x)°$.

b i $270 = 180 + 90$. Expand $\cos(180 + 90)°$ to verify that $\cos 270° = 0$.

ii Simplify $\cos(270 + x)°$.

c Express $\cos 30°$ in terms of the sines and cosines of $10°$ and $20°$.

4 a Find the **exact** value of:

i $\cos 75°$ [Hint: $75 = 45 + 30$] ii $\cos 105°$ iii $\cos 15°$.

b Find the **exact** value of: i $\cos\frac{5\pi}{12}$ ii $\cos\frac{\pi}{12}$.

5 a A and B are acute angles. $\sin A° = \frac{3}{5}$ and $\sin B° = \frac{5}{13}$.
Find the exact value of $\cos(A + B)°$.

b P and Q are acute angles. $\cos P = \frac{12}{13}$ and $\sin Q = \frac{7}{25}$.
Find the exact value of $\cos(P - Q)$.

c C and D are acute angles. $\cos C = \frac{8}{17}$ and $\cos D = \frac{20}{29}$.
Find the exact value of $\cos(C + D)$.

6 Expand and simplify each expression where possible without using a calculator.

a $3\cos(x - 30)°$ b $4\cos(x + 60)°$ c $-2\cos(x - 45)°$

d $5\cos\left(x + \frac{\pi}{4}\right)$ e $-\cos\left(x + \frac{\pi}{6}\right)$ f $-3\cos\left(x - \frac{\pi}{3}\right)$

7 a How do you know that $\cos 20° \cos 70° - \sin 20° \sin 70° = 0$?

b What is the **exact** value of:

i $\cos 123° \cos 57° - \sin 123° \sin 57°$

ii $\cos 240° \cos 120° - \sin 240° \sin 120°$

iii $\cos 200° \cos 20° + \sin 200° \sin 20°$?

c Simplify each expression as far as possible without a calculator.

i $\cos 62° \cos 49° - \sin 62° \sin 49°$

ii $\cos 310° \cos 30° + \sin 310° \sin 30°$

iii $\sin 27° \sin 17° - \cos 27° \cos 17°$

8 A quadrilateral ABCD is right-angled at A and C.
Its sides are AB = 24 cm, BC = 15 cm, CD = 20 cm and DA = 7 cm.

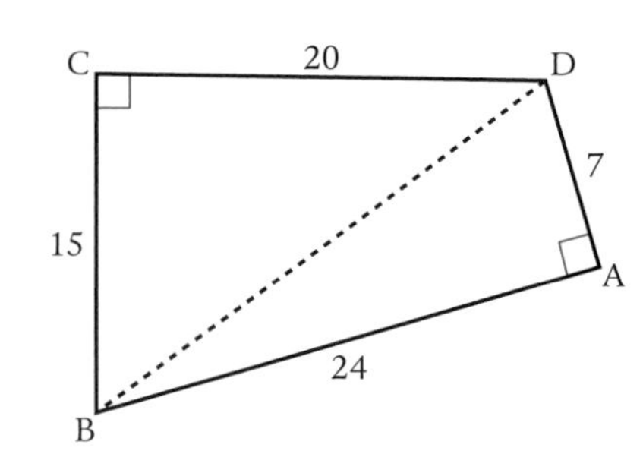

a Calculate the length of the diagonal BD.

b Hence calculate the **exact** size of $\cos\angle ABC$ without a calculator.

c Calculate the exact size of $\cos\angle ADC$ and comment.

9 A photographer on a boat at P takes a shot of the Forth Bridge.

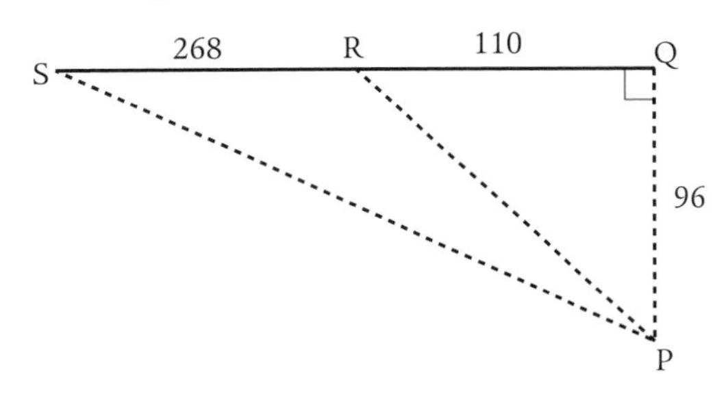

The diagram gives some of the distances in metres.

a Calculate, using Pythagoras' theorem, the distance: **i** PR **ii** PS.

b Using the formula for cos $(A - B)$, calculate the value of cos ∠SPR.

c Hence, calculate the size of ∠SPR to the nearest degree.

10 a **i** Expand and simplify: $\cos(A + B) + \cos(A - B)$.

ii Hence simplify $\cos(40 + 20)^\circ + \cos(40 - 20)^\circ$.

b Given that $67 = 42 + 25$ and $17 = 42 - 25$, express $\cos 67^\circ + \cos 17^\circ$ as the product of cosines.

3.2 Sin($A \pm B$)

$$\cos(A - B) = \cos A \cos B + \sin A \sin B$$

$\Rightarrow \cos((90^\circ - A) - B) = \cos(90^\circ - A)\cos B + \sin(90^\circ - A)\sin B$ ... replacing A by $(90^\circ - A)$

$\Rightarrow \cos(90^\circ - (A + B)) = \cos(90^\circ - A)\cos B + \sin(90^\circ - A)\sin B$ ...re-arranging $\cos((90^\circ - A) - B)$

$\Rightarrow \sin(A + B) = \sin A \cos B + \cos A \sin B$ ... $\cos(90 - x)^\circ = \sin x^\circ$ and $\sin(90 - x)^\circ = \cos x^\circ$

Replace B by $-B$,

$\sin(A + (-B)) = \sin A \cos(-B) + \cos A \sin(-B)$

$\Rightarrow \sin(A - B) = \sin A \cos B - \cos A \sin B$

Example 1

Find the exact value of sin 75°.

$$\begin{aligned}\sin 75^\circ &= \sin(30 + 45)^\circ \\ &= \sin 30^\circ \cos 45^\circ + \cos 30^\circ \sin 45^\circ \\ &= \frac{1}{2}\cdot\frac{1}{\sqrt{2}} + \frac{\sqrt{3}}{2}\cdot\frac{1}{\sqrt{2}} \quad \text{... using 'exact value' triangles} \\ &= \frac{1 + \sqrt{3}}{2\sqrt{2}}\end{aligned}$$

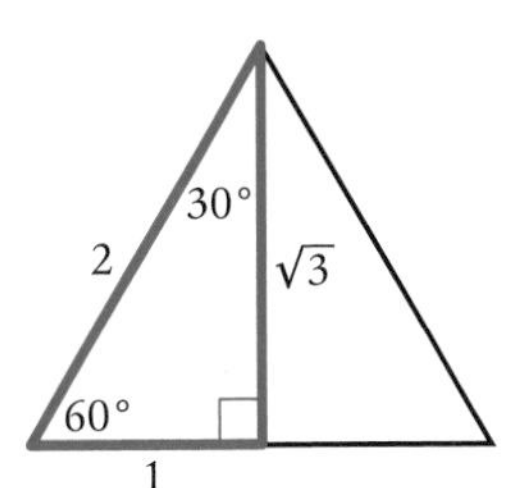

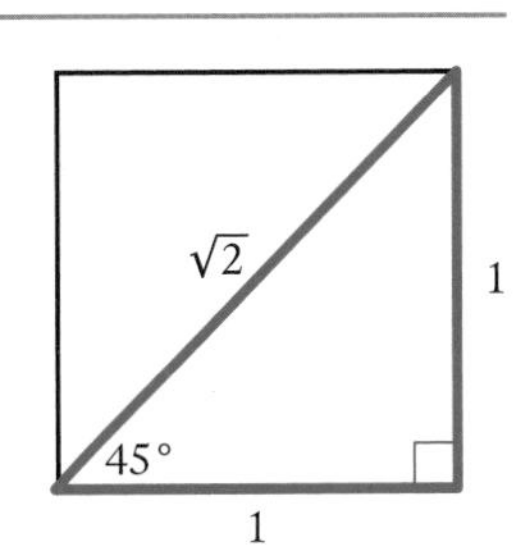

Example 2

Triangle ABC has sides, AB = 25 cm, BC = 28 cm, AC = 17 cm.
The altitude from A meets BC at D, which is 8 cm from C.

a Calculate AD, the length of the altitude.

b Calculate the **exact** value of sin ∠BAC without the aid of a calculator.

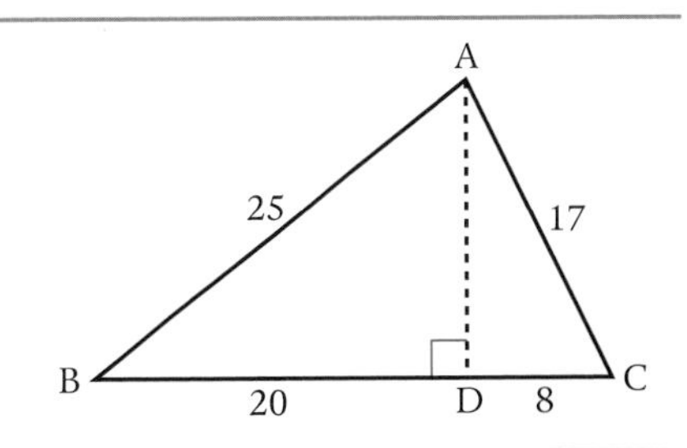

a By Pythagoras' theorem: $AD = \sqrt{17^2 - 8^2} = 15$ cm.

b $\sin \angle BAC = \sin (\angle BAD + \angle CAD)$

$$= \sin \angle BAD \cos \angle CAD + \cos \angle BAD \sin \angle CAD$$

$$= \frac{20}{25} \cdot \frac{15}{17} + \frac{15}{25} \cdot \frac{8}{17} = \frac{300}{425} + \frac{120}{425}$$

$$= \frac{420}{425} = \frac{84}{85}$$

Exercise 3.2

1 Use the expansion of $\sin (A - B)$ to:

a verify $\sin (180 - A)° = \sin A°$

b simplify $\sin (90 + A)°$

c find the exact value of: **i** $\sin\left(\frac{\pi}{4} - \frac{\pi}{6}\right)$ **ii** $\sin\left(\frac{\pi}{3} + \frac{\pi}{4}\right)$.

2 **a** Find the **exact** value of:

i $\sin 75°$ **ii** $\sin 105°$ **iii** $\sin 15°$.

b Find the **exact** value of: **i** $\sin \frac{5\pi}{12}$, given $\frac{5\pi}{12} = \frac{3\pi}{12} + \frac{2\pi}{12}$ **ii** $\sin \frac{\pi}{12}$.

3 Expand and simplify where possible without using a calculator.

a $2 \sin (x - 60)°$ **b** $3 \sin (x + 45)°$ **c** $-2 \sin (x + 45)°$

d $3 \sin\left(x - \frac{\pi}{6}\right)$ **e** $-\sin\left(x + \frac{\pi}{4}\right)$ **f** $1 - 2 \sin\left(x - \frac{\pi}{6}\right)$

4 Express the following as simply as possible without a calculator.

a $\sin 25° \cos 45° + \cos 25° \sin 45°$

b $\sin 70° \cos 21° - \cos 70° \sin 21°$

c $\sin 94° \cos 6° + \cos 94° \sin 6°$

d $2 \sin 100° \cos 65° - 2 \cos 100° \sin 65°$

e $2 \sin 20° \sin 31° + 2 \cos 20° \cos 31°$

f $3 \cos 70° \sin 21° - 3 \sin 70° \cos 21°$

5 ABCD is a parallelogram with a base, BC, of length 52 mm.
Adjacent to BC is CD = 29 mm.
The altitude DE = 21 mm, meeting BC extended by 20 mm at E.

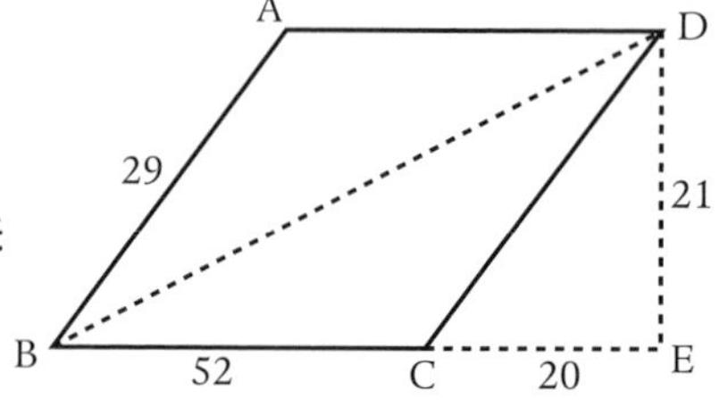

a Calculate the length of the diagonal BD.

b Use the fact that $\angle ABD = \angle ABC - \angle DBC$ to find the **exact** value of:

i $\sin \angle ABD$ **ii** $\sin \angle BDC$.

c Find $\sin \angle BAD$.

6 The three angles of a triangle are A, B, and C. We know $C = 180 - (A + B)$.
Given that $\sin A = \frac{3}{5}$ and $\sin B = \frac{12}{13}$, calculate the value of $\sin C$.

7 ABC is a right angled triangle with base AB = 3 and height BC = 2.
M is the midpoint of BC.

a Calculate, leaving surds in your answer, the length of: **i** AM **ii** AC.

b Hence calculate the value of: **i** $\sin \angle CAM$ **ii** $\cos \angle CAM$ **iii** $\tan \angle CAM$.

8 A and B are acute angles such that $\sin A = \frac{2}{3}$ and $\cos B = \frac{2}{7}$.

a Without a calculator, find the exact value of: **i** $\sin(A + B)$ **ii** $\sin(A - B)$.

b Find the exact value of: **i** $\cos(A + B)$ **ii** $\cos(A - B)$.

c Find the value of $\tan(A + B)$, rationalising the denominator.

9 A small crane for town repairs has an extended arm of length 8 m. At the end of the arm is an elbow with an extension which holds a cage. Several measurements (in metres) were made while the arm was in the position shown.

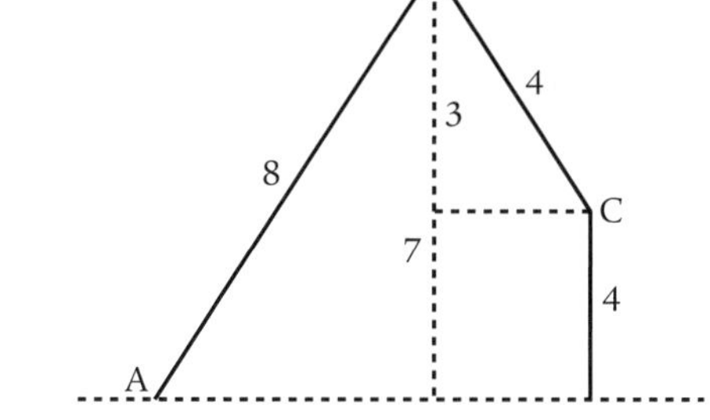
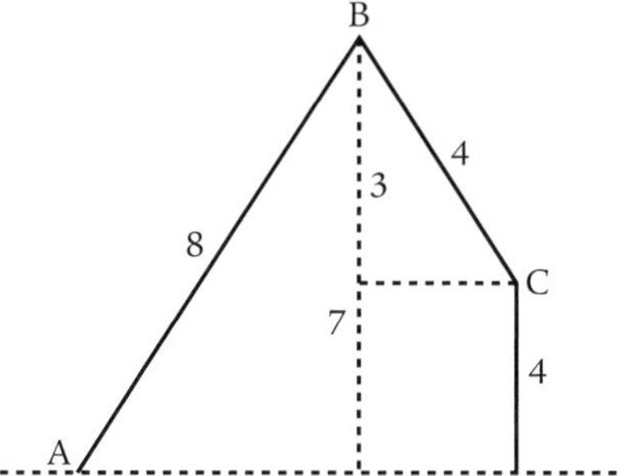

a Keeping surds in your answer, calculate the **exact** size of $\sin \angle ABC$.

b Hence with the aid of a calculator find the size of $\angle ABC$.

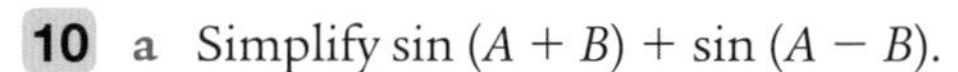

10 **a** Simplify $\sin(A + B) + \sin(A - B)$.

b Two angles add to make 50° and subtract to give 10°.
What are the sizes of the two angles?

c Without the aid of a calculator show that $\sin 50° + \sin 10° = \cos 20°$.

d Use a similar strategy to simplify $\sin 70° + \sin 50°$.

e Simplify $\sin 105° + \sin 15°$.

11 Open a spreadsheet and in column A generate a list of angles from 0 to 2π.
(You can get instructions for how to do this in Chapter 2, Exercise 2.2.)
You are going to explore the possibility that $\sin(A + B) = \sin A + \sin B$.

In B1 type =SIN(A1) ... we are going to let A vary from 0 to 2π.

In G1 type a number between 0 and 2π ... remember that π = PI().

This will hold the value of B which will be a fixed value as A varies.

In C1 type =B1+SIN($G$1) ... $\sin(A) + \sin(B)$.

In D1 type =SIN(A1+$G$1) ... $\sin(A + B)$.

Fill down B1, C1, D1 to row 25.

Highlight all four columns of figures.

Select ‘Charts’ and pick ‘Scatter’. From the menu choose ‘Scatter with Smooth Lines’.

A graph similar to this will appear:
we have put =3*PI()/4 into G1.

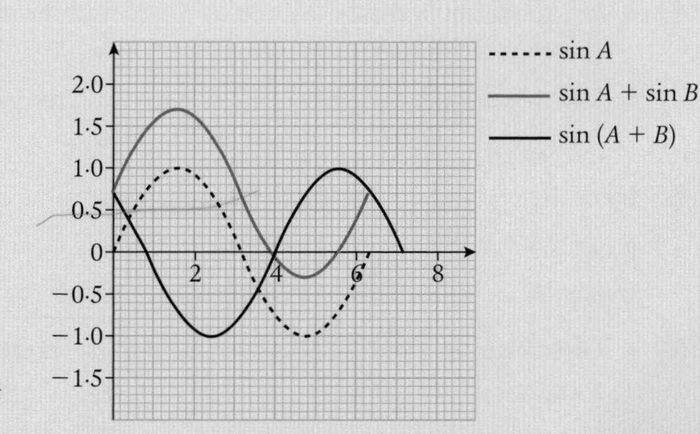

With different values of B, you will notice:

a the $y = \sin A$ curve is parallel to the $\sin A + \sin B$ curve – can you explain why?

b the curve $y = \sin A + \sin B$ intersects the curve $y = \sin(A + B)$ at three places – can you explain the three intersections?

3.3 Double angle formulae

We have now devised four formulae:

i $\cos(A + B) = \cos A \cos B - \sin A \sin B$
ii $\cos(A - B) = \cos A \cos B + \sin A \sin B$
iii $\sin(A + B) = \sin A \cos B + \cos A \sin B$
iv $\sin(A - B) = \sin A \cos B - \cos A \sin B$.

If we replace B by A in each formula we generate four other facts (two of which you should already know).

i $\cos(A + A) = \cos A \cos A - \sin A \sin A$ $\Rightarrow \cos 2A = \cos^2 A - \sin^2 A$ ... ①
ii $\cos(A - A) = \cos A \cos A + \sin A \sin A$ $\Rightarrow 1 = \cos^2 A + \sin^2 A$ ... ②
iii $\sin(A + A) = \sin A \cos A + \cos A \sin A$ $\Rightarrow \sin 2A = 2 \sin A \cos A$ ... ③
iv $\sin(A - A) = \sin A \cos A - \cos A \sin A$ $\Rightarrow \sin 0 = 0$ ... ④

Adding ① and ② gives: $\cos 2A + 1 = 2\cos^2 A$ $\Rightarrow \cos 2A = 2\cos^2 A - 1$ ... ⑤

Subtracting ② from ① gives $\cos 2A - 1 = -2\sin^2 A$ $\Rightarrow \cos 2A = 1 - 2\sin^2 A$ ... ⑥

So, we have four new identities we can work with:

$$\cos 2A = \cos^2 A - \sin^2 A$$
$$= 2\cos^2 A - 1$$
$$= 1 - 2\sin^2 A$$
$$\sin 2A = 2 \sin A \cos A.$$

Example 1

In an isosceles triangle ABC with altitude AD, $\angle BAD = x°$. It is given that $\sin x° = \frac{3}{5}$.

Calculate the size of:

a $\sin 2x°$ b $\cos 2x°$.

a Since $x°$ is in a right-angled triangle, it is acute.
We can place $x°$ in a right-angled triangle of our own so that $\sin x° = \frac{3}{5}$.
We use Pythagoras to get the third side, 4 ... and can read off the sine and cosine ratios.

$$\sin 2x = 2 \sin x \cos x = 2 \times \frac{3}{5} \times \frac{4}{5} = \frac{24}{25}$$

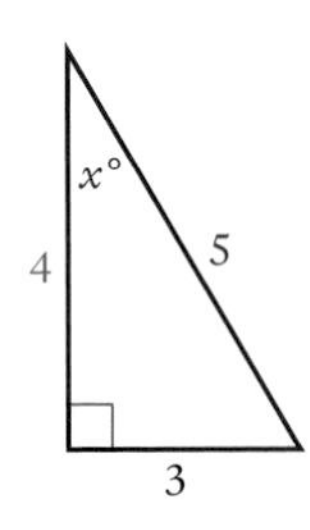

b We have the choice of three identities. We'll use $\cos 2x = \cos^2 x - \sin^2 x$.

$$\cos 2x = \left(\frac{4}{5}\right)^2 - \left(\frac{3}{5}\right)^2 = \frac{16}{25} - \frac{9}{25} = \frac{7}{25}$$

In the following exercise **exact** values are expected but you may use a calculator.

Exercise 3.3

1 a Use the formula $\cos 2x = 2\cos^2 x - 1$ to find the value of $\cos 2x$ when $\cos x$ equals:
i 0·4 ii 0·6 iii 1 iv 0.

b Use the formula $\cos 2x = 1 - 2\sin^2 x$ to find the value of $\cos 2x$ when $\sin x$ equals:
i 0·5 ii 0·1 iii 1 iv 0.

2 If $\cos A = \frac{5}{13}$ and $\sin B = \frac{4}{5}$, calculate the exact value of:

a $\sin 2A$ b $\cos 2A$ c $\sin 2B$
d $\cos 2B$ e $\sin 2(A + B)$ f $\cos(2A + B)$.

3 a Given that $\sin x° = \frac{21}{29}$ and x is acute, find the exact value of:
i $\cos x°$ ii $\sin 2x°$ iii $\cos 2x°$.

b Use your results from **a** to get the exact value of: **i** $\sin 4x°$ **ii** $\cos 4x°$.

4 The base angle of an isosceles triangle is $x°$.

a Find an expression in x for the apex angle, $a°$.

b If $\sin x° = \frac{7}{25}$, find the exact value of: i $\sin a°$ ii $\cos a°$.

c Calculate the exact value of: i $\sin 2a°$ ii $\cos 2a°$.

5 An obtuse angle of size $x°$ is bisected.

The sine of $\left(\frac{x}{2}\right)° = \frac{84}{85}$.

a Rewrite $\cos 2A = \cos^2 A - \sin^2 A$, using $A = \frac{x}{2}$.

b Find the exact value of: i $\sin x°$ ii $\cos x°$.

6 $y = \cos 2x - \cos x$

a Substitute for $\cos 2x$ using the appropriate formula so that y is expressed in terms of $\cos x$ only.

b The expression is in quadratic form. Factorise it.

c Similarly, convert the following to quadratic form and factorise.

i $\cos 2x + 3\cos x - 1$ ii $\cos 2x + \sin x + 2$

iii $\cos 2x + \sin x$ iv $2\cos 2x + 7\cos x + 5$

v $2\cos 2x - 7\sin x$ vi $3\cos 2x + \cos x + 1$

7 $y = \sin 2x + 2\sin x$

a i Substitute for $\sin 2x$.

ii Factorise the resultant expression.

b Similarly, convert and factorise the following.

i $3\sin 2x + \sin x$ ii $2\sin 2x - \cos x$

iii $\sin 2x + 4\sin x$ iv $3\cos x - 2\sin 2x$

8 a $3x = 2x + x$

Use this fact to express $\sin 3x$ in terms of $2x$ and x.

b Expand the expressions in $2x$ to express $\sin 3x$ in terms of $\sin x$ only.

c Find an expression for $\cos 3x$.

9 Simplify:

a $2\sin\frac{\pi}{3}\cos\frac{\pi}{3}$ b $\cos^2\frac{\pi}{4} - \sin^2\frac{\pi}{4}$ c $2\cos^2\frac{\pi}{12} - 1$

d $1 - 2\sin^2\frac{\pi}{8}$ e $\sin^2\frac{5\pi}{12} - \cos^2\frac{5\pi}{12}$ f $2\sin^2\frac{2\pi}{3} + 1$.

10 The London Eye has a radius of r m.

As it rotates anticlockwise the point B moves away from the bottom of the wheel, A, heading for the top of the wheel at C.

The angle ADB opens out, as does angle ACB.

A theorem in mathematics states that the angle subtended by a chord at the centre of a circle is twice the angle subtended at the circumference.

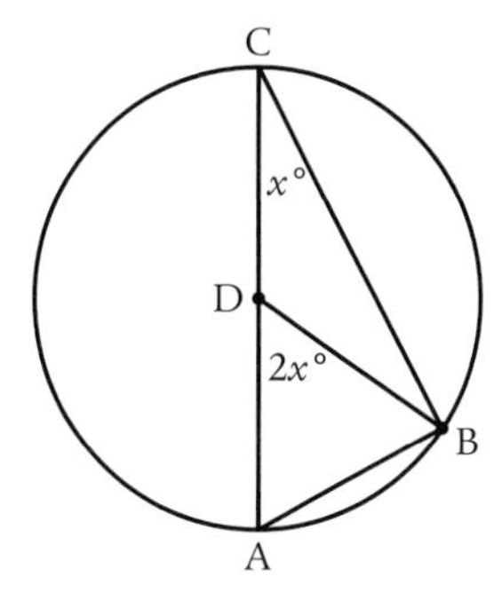

a Use the triangle ABD and the cosine rule to find an expression for the length of the chord AB in terms of r and x.

b Use the triangle ABC to find another expression for AB.

c By equating these expressions, verify one of the expansions for $\cos 2x$.

3.4 Adding waves

We have looked at how expressions like $4 \sin (x + 30)°$ can be expanded, viz.

$$4 \sin x° \cos 30° + 4 \cos x° \sin 30° = 2\sqrt{3} \sin x° + 2 \cos x°.$$

The process can be reversed.

We can take an expression of the form $a \sin x° + b \cos x°$ and express it in the form $p \sin (x + q)°$.

Example 1

Express $3 \sin x° + 4 \cos x°$ in the form $p \sin (x + q)°$.

Expand $p \sin (x + q)°$ ... $p \sin x° \cos q° + p \cos x° \sin q°$

Compare with the original expression ... $3 \sin x° + 4 \cos x°$

Equating the coefficients of $\sin x°$ ... $p \cos q° = 3$ ... ①

Equating the coefficients of $\cos x°$ ... $p \sin q° = 4$ ... ②

Square ① and ② and add ... $p^2 \cos^2 q° + p^2 \sin^2 q° = 3^2 + 4^2$

$\Rightarrow p^2 (\cos^2 q° + \sin^2 q°) = 25$

$\Rightarrow p^2 = 25$

$\Rightarrow p = 5$

Divide ② by ① $\quad \dfrac{p \sin q°}{p \cos q°} = \dfrac{4}{3} \Rightarrow \tan q° = \dfrac{4}{3}$

$\Rightarrow q = \tan^{-1}\left(\dfrac{4}{3}\right)$ or $180 + \tan^{-1}\left(\dfrac{4}{3}\right) = 53{\cdot}1$ or $233{\cdot}1$ (to 1 d.p.)

From ① and ② we see that both $\sin q°$ and $\cos q°$ are positive ... so q is in the 1st quadrant.

So $q = 53{\cdot}1$.

Thus, $3 \sin x° + 4 \cos x° = 5 \sin (x + 53{\cdot}1)°$.

Waves

Both $f(x) = a \sin x$ and $g(x) = b \cos x$ when graphed produce waves.

The fact that $f(x) + g(x)$ can be expressed as, say, $p \cos (x + q)°$ means it is a wave also.

The amplitude of the resultant wave is p.

Thus the maximum and minimum values of $f(x) + g(x)$ can be quickly found by expressing it in the form $p \cos (x + q)°$.

Example 2

a Express $8 \sin x° + 15 \cos x°$ in the form $p \cos(x + q)°$.

b Sketch $y = 8 \sin x°$, $y = 15 \cos x°$, and $y = 8 \sin x° + 15 \cos x°$.

a Expand $p \cos (x + q)°$ ... $p \cos x° \cos q° - p \sin x° \sin q°$

Compare with the original expression ... $8 \sin x° + 15 \cos x°$

Equating the coefficients of $\sin x°$ ... $-p \sin q° = 8$ ... ①

Equating the coefficients of $\cos x°$ ... $p \cos q° = 15$ ... ②

Square ① and ② and add ... $p^2 \sin^2 q° + p^2 \cos^2 q° = 8^2 + 15^2$

$\Rightarrow p^2 (\sin^2 q° + \cos^2 q°) = 289$

$\Rightarrow p^2 = 289$

$\Rightarrow p = 17$

Divide ① by ② $\qquad \dfrac{-p\sin q^\circ}{p\cos q^\circ} = \dfrac{8}{15} \Rightarrow \tan q^\circ = -\dfrac{8}{15}$

$$\Rightarrow q = \tan^{-1}\left(-\frac{8}{15}\right) \text{ or } 180 + \tan^{-1}\left(-\frac{8}{15}\right) \text{ or } 360 + \tan^{-1}\left(\frac{-8}{15}\right)$$

$$= -28{\cdot}1 \text{ or } 151{\cdot}9 \text{ or } 331{\cdot}9 \text{ (to 1 d.p.)}$$

From ① the sine is negative [3rd or 4th quadrant]

from ② the cosine is positive [1st or 4th quadrant]

... so q is in the 4th quadrant.

So $q = 331{\cdot}9$

b Thus, $8\sin x^\circ + 15\cos x^\circ = 17\cos(x + 331{\cdot}9)^\circ$

... a cosine wave with an amplitude of 17, shifted 331·9 to the left.

This angle, 331·9°, is known as the **phase angle**.

[So, for example, the point (360, 1) on $y = \cos x^\circ$ becomes (28·1, 17).]

The curve cuts the y-axis when $x = 0$, i.e. $8\sin 0^\circ + 15\cos 0^\circ = 15$.

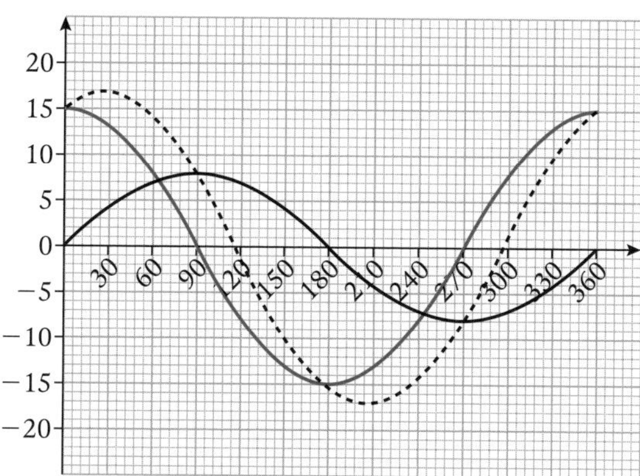

Exercise 3.4

1 For the function $y = a\sin(x + b)$; $0 \leqslant b \leqslant \frac{\pi}{2}$, state, in terms of a and/or b,

a the maximum value of the function

b the minimum value of the function

c the coordinates to the first turning point to the right of the origin.

2 Express each function in the requested form, then state:

i the maximum value of the function **ii** the minimum value of the function

iii the value of the function when $x = 0$

iv the shift in the x-direction when compared to the basic wave $y = \sin x^\circ$ or $y = \cos x^\circ$.

a $12\sin x^\circ + 5\cos x^\circ$; $0 \leqslant x \leqslant 360$ in form $a\sin(x + b)^\circ$

b $24\sin x^\circ - 7\cos x^\circ$; $0 \leqslant x \leqslant 360$ in form $a\sin(x - b)^\circ$

c $20\cos x^\circ + 21\sin x^\circ$; $0 \leqslant x \leqslant 360$ in form $a\cos(x - b)^\circ$

d $\sqrt{3}\cos x^\circ - \sin x^\circ$; $0 \leqslant x \leqslant 360$ in form $a\cos(x + b)^\circ$

3 Express each function in the requested form being careful when you equate coefficients.

a $3\sin x^\circ + 4\cos x^\circ$; $0 \leqslant x \leqslant 360$ in form $a\sin(x - b)^\circ$

b $60\sin x^\circ - 11\cos x^\circ$; $0 \leqslant x \leqslant 360$ in form $a\sin(x + b)^\circ$

c $1{\cdot}6\sin x^\circ + 6{\cdot}3\cos x^\circ$; $0 \leqslant x \leqslant 360$ in form $a\cos(x + b)^\circ$

d $-\sqrt{2}\cos x^\circ - \sin x^\circ$; $0 \leqslant x \leqslant 360$ in form $a\cos(x - b)^\circ$

4 Express each function in the requested form.

a $\sqrt{3}\sin x + \cos x;\ 0 \leqslant x \leqslant 2\pi$ in form $a\sin(x + b)$

b $\sin x - \sqrt{3}\cos x;\ 0 \leqslant x \leqslant 2\pi$ in form $a\sin(x - b)$

c $\sin x + \cos x;\ 0 \leqslant x \leqslant 2\pi$ in form $a\cos(x + b)$

d $3\sin x - 4\cos x;\ 0 \leqslant x \leqslant 2\pi$ in form $a\cos(x - b)$

5 $f(x) = 8\sin x° + 15\cos x°;\ 0 \leqslant x \leqslant 360$, has been expressed in the form $a\cos(x + b)°$ in Example 2 above.

a Express it also in the form:

i $a\sin(x + b)°$ ii $a\sin(x - b)°$ iii $a\cos(x - b)°$.

b Confirm, by considering a sketch in the domain $0 \leqslant x \leqslant 360$, that these expressions indeed all produce the same graph.

6 a Sketch the curve $y = \sin x + \cos x + 3;\ 0 \leqslant x \leqslant 2\pi$, annotating the maxima, minima and y-intercept. [Hint: Express $\sin x + \cos x$ in a suitable form.]

b Sketch $y = \sin x° - \sqrt{3}\cos x° + 2;\ 0 \leqslant x \leqslant 360$.

7 Find the maximum and minimum value of the function $y = \dfrac{30}{24\cos x + 7\sin x + 35}$.

8 Two wheels revolve independently. The height of a marked point on wheel A can be modelled by $y = 4\sin x + 6$, where x is the number of seconds since the wheel was set in motion. The height of a point on wheel B can be modelled by $y = 5\cos x + 4$.

a Form an equation which describes the time when the two points are the same height.

b Express $4\sin x - 5\cos x$ in the form $a\sin(x + b)$.

c Hence solve the equation to find when the points are the same height in the domain $0 \leqslant x \leqslant 2\pi$.[1]

9 Using suitable units, the rotation of two radar antennae on a ship can be modelled by:

$y = 5\cos x + 20$ and $y = 3\sin x + 20$

where y metres is the distance from the point to the observer and x seconds is the time since observations began.

a Find the times when the distances are equal over the period $0 \leqslant x \leqslant 2\pi$.

b What is this distance?

10 To get more value from spreadsheet investigations you should customise the ribbon so that the 'Developer' tab is visible. [In 'Help', type 'Customize Ribbon' for instructions. You only have to do this once.]

Once you have this tab, you can add controls to your spreadsheet.

In A1 Type 0 (zero); In A2 type =A1+10, (10 more than the cell above); fill down to row 37.

You now have the angles from 0 to 360 in steps of 10.

In B1 type =3*SIN(PI()/180*A1). This is the wave $y = 3\sin x°$. The wave can be changed to suit the question. Note that, since we're working in degrees, we must multiply the angle by $\pi/180$ to make it radians. Spreadsheet functions use radians.

In C1 type =4*COS(PI()/180*A1). This is the wave $y = 4\cos x°$. This wave can also be changed.

[1] Full coverage of the solution to this type of equation will come up in Chapter 8.

In D1 type =B1+C1, adding the two waves.

In E1 type =$F$2*SIN(PI()/180*(A1+$G$2)), this is a wave of form $a \sin(x + b)$ where you have control of the values of a and b.

Fill B1, C1, D1 and E1 down to row 37.

In F1 and G1, type 'a' and 'b' respectively.

In F2 and G2 type =F3-10 and =G3-90 respectively.

Now use the developer tab and select a 'spin button' ... draw it in cells F4 to F6 by dragging.

Right-click the button and select 'Format Control' from the pop-up menu.

Make minimum value 0; maximum value 20; incremental change 1 and Cell link F3.

You can't make the minimum negative.

Now by clicking the spinner you adjust the value in F3 so that it goes from 0 to 20 in steps of 1.

Right-click the button, copy and paste and place it in G4 to G6.

Right-click the button and select 'Format Control' from the pop-up menu.

Make minimum value 0; maximum value 180; incremental change 10 and Cell link G3.

Note that what you typed in F2 and G2 changes the ranges from 0 to 20 and 0 to 180 to -10 to 10 and -90 to 90.

Select the first 5 columns of data and draw a **smooth line scatter** graph.

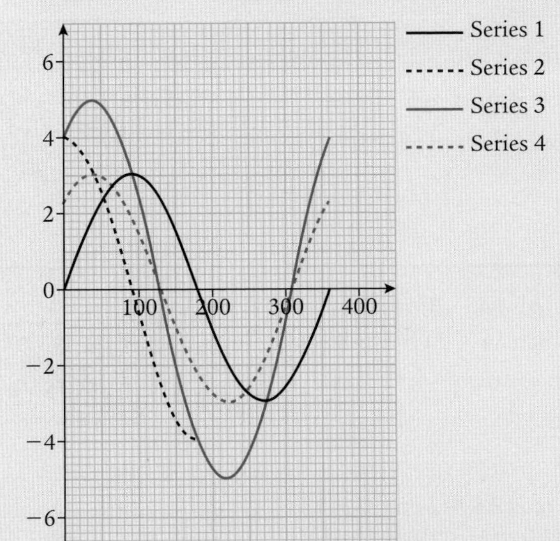

F	G
a	b
3	50
13	140

Columns F and G and the spinners in place when the purple curves have the same phase but not yet the same amplitude.

The black curves are the original waves $y = 3 \sin x°$ and $y = 4 \cos x°$.

The solid purple curve is their sum.

The broken purple curve is the test curve of the form $y = a \sin(x + b)°$ which you control using the spinners. Use the 1st spinner to get the amplitudes to match; use the 2nd spinner to get the phases to match. You can read off the values of a and b.

The original waves can be altered in B1 and C1 and filled down to create a new problem.

Investigate the link between a and b and the position of the 'sum' wave.

Preparation for assessment

1 Expand and simplify:

a $3\sin(x + 30)°$ b $4\cos\left(x - \frac{\pi}{3}\right)$.

2 A and B are acute angles such that $\sin A = \frac{3}{7}$ and $\cos B = \frac{9}{11}$.
Calculate the exact value of:

a $\sin(A + B)$ b $\cos(A - B)$ c $\cos 2A$

d $\sin 2A$ e $\sin 4A$.

3 Two lines L_1 and L_2 cut the x-axis at $(1, 0)$.
The gradient of L_1 is 0·75; the gradient of L_2 is 0·5.
By considering where the lines cut the line $x = 5$, find the exact value of:

a the sine of the acute angle between the lines

b the cosine of the acute angle between the lines.

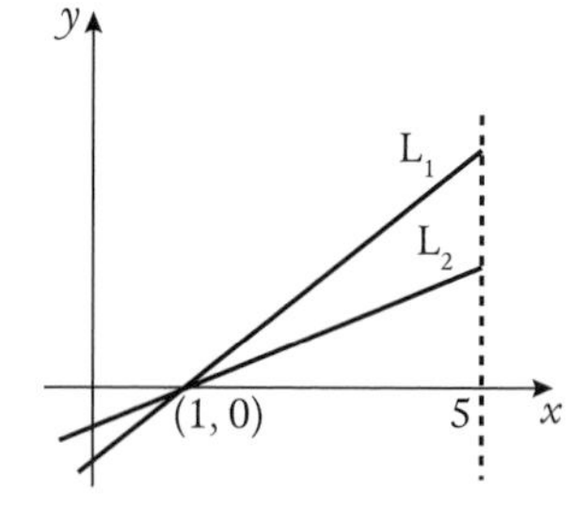

4 a Show that $\frac{\cos 2A}{\sin A + \cos A} = \cos A - \sin A$
when $\sin A + \cos A \neq 0$.

b For what values of A in the interval $0 \leq A \leq 2\pi$ is $\sin A + \cos A = 0$?

5 Calculate the maximum value of:

a $\cos x - \sin x$ b $\cos^2 x - \sin^2 x$.

6 a Express $3\cos x + 4\sin x$ in the form $k\cos(x + a)$ where $k, a > 0$.

b Hence find the minimum value of the function $3\cos x + 4\sin x + 6$.

c Would it be valid to divide by $3\cos x + 4\sin x + 6$ for all values of x?

Summary

1 i $\cos(A + B) = \cos A\cos B - \sin A\sin B$
ii $\cos(A - B) = \cos A\cos B + \sin A\sin B$
iii $\sin(A + B) = \sin A\cos B + \cos A\sin B$
iv $\sin(A - B) = \sin A\cos B - \cos A\sin B$

This is often given as:

$\cos(A \pm B) = \cos A\cos B \mp \sin A\sin B$
$\sin(A \pm B) = \sin A\cos B \pm \cos A\sin B$.

2 i $\cos 2A = \cos^2 A - \sin^2 A$
ii $\cos 2A = 2\cos^2 A - 1$
iii $\cos 2A = 1 - 2\sin^2 A$
iv $\sin 2A = 2\sin A\cos A$

3 $a\sin x + b\cos x$, where a and b real number constants can be expressed in the form:

i $p\sin(x \pm q)$ ii $p\cos(x \pm q)$.

- Expand the desired form.
- Equate coefficients to get equations of the form:
  $p\sin q° = a$ ... ①
  $p\cos q° = b$ ... ②
- Square and add the equations to find p.
- Divide to get $\tan q = \frac{a}{b}$ and hence q.

[By examining the signs of $\sin q$ and $\cos q$ we can decide in which quadrant q lies.]

In this 'tidier' form we can easily sketch the curve and identify maxima and minima.

4 Vectors

Before we start...

In 1843, William Rowan Hamilton a mathematician from Dublin developed the branch of maths known as Quaternions.

The algebra involved was very useful for modelling 3D space and in others' hands it was re-shaped into the study of **vectors**.

It was predicted to become 'a vast influence upon the future of mathematical science' ... and it was.

You have been introduced to the concept and notation for vectors during the National 5 course but it will be useful to begin with a few reminders.

A vector is a quantity with magnitude and direction.

A scalar has magnitude only.

A vector can be represented by a line whose length is proportional to the magnitude of the vector.

Once expressed in geometric terms, the vector problem becomes a problem in geometry.

What is found true about the representative will be true about the vector.

A vector can be resolved into two components at right angles to each other using simple trigonometry.

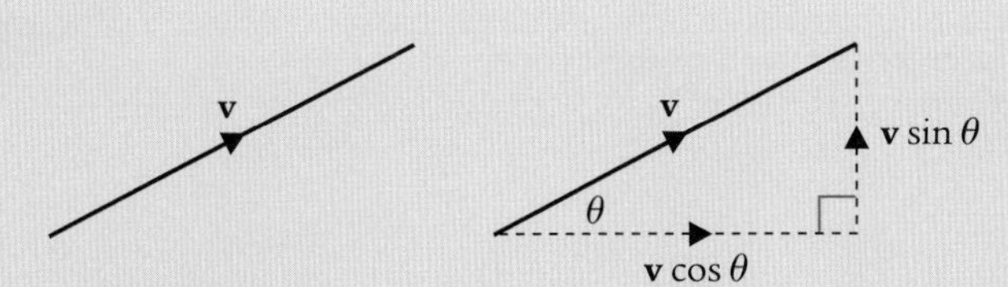

Note that under vector addition $\mathbf{v} = \mathbf{v}\cos\theta + \mathbf{v}\sin\theta$

There are obviously an infinite number of ways to perform this resolution.

Directions are usually chosen to suit the context.

It may be that you are working with an inclined plane in physics ... along the plane and at right angles to the plane would be your most convenient directions to use.

When working on the coordinate plane ... the x-direction and y-direction would be best.

What you need to know

1 Representatives of the vectors **u**, **v**, and **w** have been drawn on the grid.

a On squared paper draw representatives of

- **i** $\mathbf{u} + \mathbf{v}$
- **ii** $\mathbf{v} + \mathbf{w}$
- **iii** $\mathbf{u} + (\mathbf{v} + \mathbf{w})$
- **iv** $(\mathbf{u} + \mathbf{v}) + \mathbf{w}$.

b How do answers **iii** and **iv** in **a** relate to the vector sum $\mathbf{u} + \mathbf{v} + \mathbf{w}$?

2 Three vectors are defined by $\mathbf{p} = \begin{pmatrix}2\\4\end{pmatrix}$, $\mathbf{q} = \begin{pmatrix}5\\-1\end{pmatrix}$, and $\mathbf{r} = \begin{pmatrix}-7\\-3\end{pmatrix}$.

a Express in component form:

i $\mathbf{p} + \mathbf{q}$ ii $\mathbf{q} + \mathbf{r}$ iii $\mathbf{p} + \mathbf{q} + \mathbf{r}$.

b The result of part **a iii** is known as a zero vector.
State the components of the vector that would need to be added to each of the following to produce the zero vector.

i $\begin{pmatrix}1\\6\end{pmatrix}$ ii $\begin{pmatrix}-3\\2\end{pmatrix}$ iii $\begin{pmatrix}5\\-7\end{pmatrix}$ iv $\begin{pmatrix}-2\\-8\end{pmatrix}$

c Simplify:

i $\begin{pmatrix}3\\5\end{pmatrix} - \begin{pmatrix}1\\2\end{pmatrix}$ ii $\begin{pmatrix}-1\\4\end{pmatrix} - \begin{pmatrix}2\\1\end{pmatrix}$ iii $\begin{pmatrix}5\\-6\end{pmatrix} - \begin{pmatrix}-3\\-4\end{pmatrix}$.

d Simplify:

i $\begin{pmatrix}2\\7\end{pmatrix} + \begin{pmatrix}2\\7\end{pmatrix}$ ii $2\begin{pmatrix}2\\7\end{pmatrix}$ iii $5\begin{pmatrix}2\\7\end{pmatrix}$ iv $-3\begin{pmatrix}2\\7\end{pmatrix}$.

e Given that $\mathbf{m} = \begin{pmatrix}1\\3\end{pmatrix}$ and $\mathbf{n} = \begin{pmatrix}-2\\5\end{pmatrix}$, simplify:

i $4\mathbf{m}$ ii $-3\mathbf{n}$ iii $2\mathbf{m} + 3\mathbf{n}$ iv $3\mathbf{m} - 4\mathbf{n}$.

3 The coordinate system can be extended to three dimensions.
B has coordinates (0, 1, 2).

a State the coordinates of the 8 vertices of the cube.

b Going from B to C, we move 2 in the x-direction, -1 in the y-direction and 0 in the z-direction. We write:

$$\overrightarrow{BC} = \begin{pmatrix}2\\-1\\0\end{pmatrix}.$$

Express the following displacements in component form:

i $\overrightarrow{BA}$ ii $\overrightarrow{AC}$.

c Verify that $\overrightarrow{BA} + \overrightarrow{AC} = \overrightarrow{BC}$.

4 Three vectors are defined by $\mathbf{u} = \begin{pmatrix}1\\-2\\3\end{pmatrix}$, $\mathbf{v} = \begin{pmatrix}3\\-1\\0\end{pmatrix}$, $\mathbf{w} = \begin{pmatrix}-4\\5\\1\end{pmatrix}$.
Simplify:

a $\mathbf{u} + \mathbf{v}$ b $\mathbf{v} - \mathbf{w}$ c $2\mathbf{v} + 3\mathbf{u}$ d $4\mathbf{u} - \mathbf{v} + 2\mathbf{w}$.

5 Which vector, $\mathbf{p} = \begin{pmatrix}2\\-4\\5\end{pmatrix}$ or $\mathbf{q} = \begin{pmatrix}1\\6\\-3\end{pmatrix}$, has the greater magnitude?

4.1 Some basics

Position vectors

If, with reference to an origin O, P is a point with coordinates (x_p, y_p, z_p) then the vector $\overrightarrow{OP} = \begin{pmatrix}x_p\\y_p\\z_p\end{pmatrix}$ is called the **position vector** of P and is usually denoted by **p**.[1]

Equal vectors

Two vectors are equal only if their corresponding components are equal.

[1] We will use this convention without further explanation unless such an explanation is felt necessary.

Directed line segment

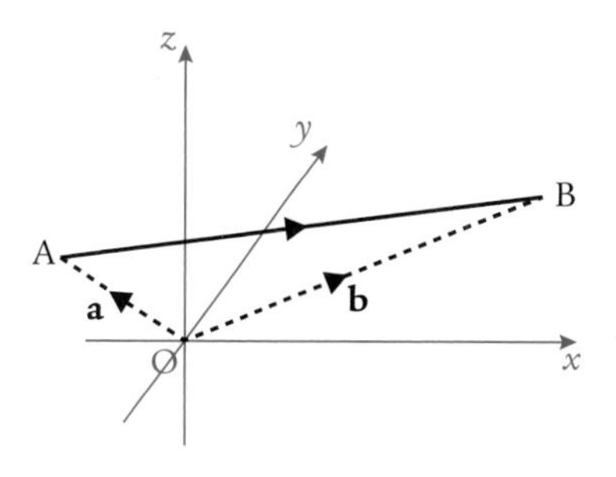

A directed line segment can be expressed simply in terms of the position vectors of its endpoints.

By vector addition, we can see that:

$\mathbf{a} + \overrightarrow{AB} = \mathbf{b}$

$\Rightarrow \overrightarrow{AB} = \mathbf{b} - \mathbf{a}$ ... an important result which we will use often

In terms of components this becomes:

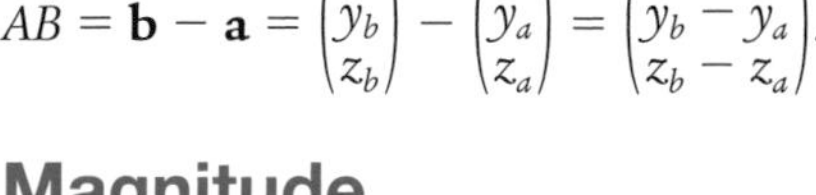

$$\overrightarrow{AB} = \mathbf{b} - \mathbf{a} = \begin{pmatrix} x_b \\ y_b \\ z_b \end{pmatrix} - \begin{pmatrix} x_a \\ y_a \\ z_a \end{pmatrix} = \begin{pmatrix} x_b - x_a \\ y_b - y_a \\ z_b - z_a \end{pmatrix}.$$

Magnitude

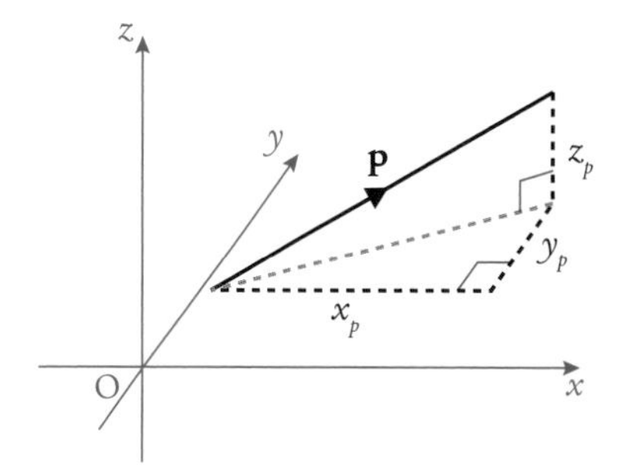

The magnitude of the vector $\mathbf{p} = \begin{pmatrix} x_p \\ y_p \\ z_p \end{pmatrix}$ is denoted by $|\mathbf{p}|$ and can be calculated using the formula $|\mathbf{p}| = \sqrt{x_p^2 + y_p^2 + z_p^2}$.

Similarly, $|\overrightarrow{AB}| = \sqrt{(x_b - x_a)^2 + (y_b - y_a)^2 + (z_b - z_a)^2}$.

Example 1

A (10, 5, 3), B(2, 1, 2), C(5, 2, 3) and D(1, −2, −4) are four points in space.

Show that $|\overrightarrow{AB}| = |\overrightarrow{CD}|$ but $\overrightarrow{AB} \neq \overrightarrow{CD}$.

$$\overrightarrow{AB} = \mathbf{b} - \mathbf{a} = \begin{pmatrix} 2 \\ 1 \\ 2 \end{pmatrix} - \begin{pmatrix} 10 \\ 5 \\ 3 \end{pmatrix} = \begin{pmatrix} -8 \\ -4 \\ -1 \end{pmatrix} \Rightarrow |\overrightarrow{AB}| = \sqrt{(-8)^2 + (-4)^2 + (-1)^2} = 9$$

$$\overrightarrow{CD} = \mathbf{d} - \mathbf{c} = \begin{pmatrix} 1 \\ -2 \\ -4 \end{pmatrix} - \begin{pmatrix} 5 \\ 2 \\ 3 \end{pmatrix} = \begin{pmatrix} -4 \\ -4 \\ -7 \end{pmatrix} \Rightarrow |\overrightarrow{CD}| = \sqrt{(-4)^2 + (-4)^2 + (-7)^2} = 9$$

So $|\overrightarrow{AB}| = |\overrightarrow{CD}| = 9$ but $\overrightarrow{AB} \neq \overrightarrow{CD}$ since corresponding components are not equal.

There are some useful notions associated with the idea of magnitude.

(i) Triangular inequality

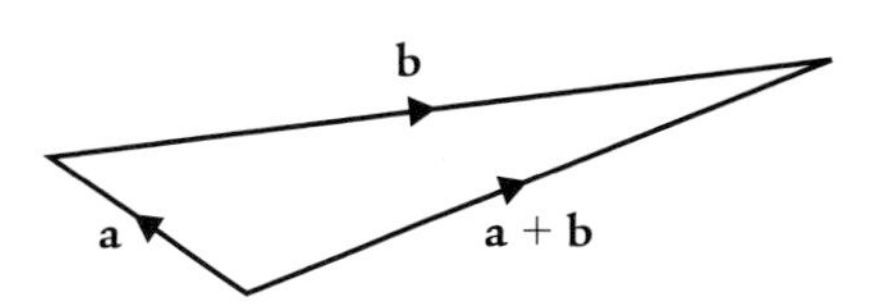

Given two vectors **a** and **b**, in general the representatives of the two vectors and their sum, **a** + **b**, will form a triangle.

We know that any one side will be less than or equal to the sum of the other two sides.

Thus, $|\mathbf{a} + \mathbf{b}| \leqslant |\mathbf{a}| + |\mathbf{b}|$ ... the triangular inequality.

In the case where $|\mathbf{a} + \mathbf{b}| = |\mathbf{a}| + |\mathbf{b}|$, **a**, **b** and **a** + **b** must be parallel.

(ii) Parallel test

Consider when $\mathbf{b} = k\mathbf{a}$ where k is a scalar ...

$|\mathbf{a} + \mathbf{b}| = |\mathbf{a} + k\mathbf{a}| = |\mathbf{a}(1 + k)| = |\mathbf{a}||(1 + k)| = |\mathbf{a}|(1 + k)$ ... $1 + k$ being a scalar

Also, $|\mathbf{a}| + |\mathbf{b}| = |\mathbf{a}| + |k\mathbf{a}| = |\mathbf{a}| + k|\mathbf{a}| = |\mathbf{a}|(1 + k)$.

Thus $|\mathbf{a} + k\mathbf{a}| = |\mathbf{a}| + |k\mathbf{a}|$ ... so $k\mathbf{a}$ is parallel to **a**.

So if $\mathbf{b} = k\mathbf{a}$ then **b** is parallel to **a**. ... another useful result

(iii) Unit vectors

A vector whose magnitude is 1 is called a unit vector.

If $\mathbf{p} = \begin{pmatrix} x_p \\ y_p \\ z_p \end{pmatrix}$, then $u_p = \dfrac{1}{|\mathbf{p}|}\begin{pmatrix} x_p \\ y_p \\ z_p \end{pmatrix}$ is a unit vector parallel to **p**.

Example 2

P (2, 4, 6), Q(8, 7, 8) , R(12, 4, 5) and S(30, 13, 11) are the vertices of a quadrilateral.

a Show that PQ is parallel to RS.

b Find a unit vector parallel to PQ.

a $\overrightarrow{PQ} = \mathbf{q} - \mathbf{p} = \begin{pmatrix} 8 \\ 7 \\ 8 \end{pmatrix} - \begin{pmatrix} 2 \\ 4 \\ 6 \end{pmatrix} = \begin{pmatrix} 6 \\ 3 \\ 2 \end{pmatrix}$

$\overrightarrow{RS} = \mathbf{s} - \mathbf{r} = \begin{pmatrix} 30 \\ 13 \\ 11 \end{pmatrix} - \begin{pmatrix} 12 \\ 4 \\ 5 \end{pmatrix} = \begin{pmatrix} 18 \\ 9 \\ 6 \end{pmatrix} = 3\begin{pmatrix} 6 \\ 3 \\ 2 \end{pmatrix} = 3\overrightarrow{PQ}$

$\overrightarrow{RS} = k\overrightarrow{PQ} \Rightarrow$ RS is parallel to PQ.

b $|\overrightarrow{PQ}| = \sqrt{6^2 + 3^2 + 2^2} = \sqrt{49} = 7$

A unit vector parallel to $\overrightarrow{PQ} = \frac{1}{7}\overrightarrow{PQ} = \frac{1}{7}\begin{pmatrix} 6 \\ 3 \\ 2 \end{pmatrix} = \begin{pmatrix} \frac{6}{7} \\ \frac{3}{7} \\ \frac{2}{7} \end{pmatrix}$.

Collinearity

Four points, A, B, C, and D are collinear if we can show that **i** AB is parallel to CD and **ii** the two lines share a common point.

If asked to prove three points A, B and C are collinear, we should show that AB is parallel to BC, say, and then point out that B is a common point.

Example 3

Relative to a suitable set of axes, a mine shaft is drilled through the points A(1, 5, 2), B(3, 6, −2) and C(6, 7·5, −8).

Show that these three points all lie in a straight line.

$\overrightarrow{AB} = \mathbf{b} - \mathbf{a} = \begin{pmatrix} 3 \\ 6 \\ -2 \end{pmatrix} - \begin{pmatrix} 1 \\ 5 \\ 2 \end{pmatrix} = \begin{pmatrix} 2 \\ 1 \\ -4 \end{pmatrix}$

$\overrightarrow{AC} = \mathbf{c} - \mathbf{a} = \begin{pmatrix} 6 \\ 7{\cdot}5 \\ -8 \end{pmatrix} - \begin{pmatrix} 1 \\ 5 \\ 2 \end{pmatrix} = \begin{pmatrix} 5 \\ 2{\cdot}5 \\ -10 \end{pmatrix}$

Considering the x-components, we find the factor ... 5 ÷ 2 = 2·5.

$\overrightarrow{AC} = \begin{pmatrix} 5 \\ 2{\cdot}5 \\ -10 \end{pmatrix} = 2{\cdot}5\begin{pmatrix} 2 \\ 1 \\ -4 \end{pmatrix} = 2{\cdot}5\overrightarrow{AB}$

which implies AC is parallel to AB

... of course, if the other components don't match then the converse is true. AC would not be parallel to AB

AND ... the point A lies on both AB and AC.

... a common point ... so AB and AC lie on the same straight line.

So A, B, and C lie on the same straight line ... A, B, and C are collinear

Exercise 4.1

1 Give the position vectors of the following points in component form:

a A(1, 2, 7) **b** B(−1, 0, −3) **c** C(−1, −5, 4).

2 P(1, 2, 8), Q(5, −1, 4), R(−3, 0, 1) and S(5, −3, −2) are four points in space.

a Express, in component form, the vectors:

i $\overrightarrow{PQ}$ **ii** $\overrightarrow{PR}$ **iii** $\overrightarrow{RP}$ **iv** $\overrightarrow{RS}$ **v** $\overrightarrow{SR}$.

b If $\overrightarrow{AB}$ has components $\begin{pmatrix} x \\ y \\ z \end{pmatrix}$, state the component of:

i $\overrightarrow{BA}$ **ii** $\overrightarrow{AB} + \overrightarrow{BA}$.

3 Calculate the magnitude of the following vectors:

a $\overrightarrow{AB} = \begin{pmatrix} 2 \\ 2 \\ 1 \end{pmatrix}$ **b** $\overrightarrow{CD} = \begin{pmatrix} 12 \\ 9 \\ 8 \end{pmatrix}$ **c** $\overrightarrow{EF} = \begin{pmatrix} -9 \\ 2 \\ -6 \end{pmatrix}$ **d** $GH = \begin{pmatrix} 8 \\ -19 \\ 4 \end{pmatrix}$.

4 **a** Calculate the magnitude of each vector given the end-points of its representative:

i $\overrightarrow{AB}$ given A(3, 0, 5) and B(14, 10, 7)

ii $\overrightarrow{PQ}$ given P(4, −1, −3) and Q(12, 0, 1)

iii $\overrightarrow{MN}$ given M(2, −3, −2) and Q(4, 3, 1).

b E (−8, −4, −5), F(10, 5, 1) , G(4, −8, 3) and H(12, 11, 7) are four points in space. Show that $|\overrightarrow{EF}| = |\overrightarrow{GH}|$ but $|\overrightarrow{EF}| \neq |\overrightarrow{GH}|$.

5 Verify the triangular inequality using the vectors $\mathbf{a} = \begin{pmatrix} 3 \\ 6 \\ 2 \end{pmatrix}$ and $\mathbf{b} = \begin{pmatrix} 2 \\ 1 \\ 2 \end{pmatrix}$.

6 In 3D, tests for parallel lines are essential.
Arrange the following vectors into parallel pairs.

$\overrightarrow{AB} = \begin{pmatrix} 1 \\ 7 \\ 3 \end{pmatrix}$ $\overrightarrow{CD} = \begin{pmatrix} -2 \\ 6 \\ 10 \end{pmatrix}$ $\overrightarrow{EF} = \begin{pmatrix} 5 \\ -15 \\ -25 \end{pmatrix}$ $\overrightarrow{GH} = \begin{pmatrix} 3 \\ -2 \\ 4 \end{pmatrix}$

$\overrightarrow{IJ} = \begin{pmatrix} -2 \\ -8 \\ 2 \end{pmatrix}$ $\overrightarrow{KL} = \begin{pmatrix} 2 \\ 14 \\ 6 \end{pmatrix}$ $\overrightarrow{MN} = \begin{pmatrix} -12 \\ 8 \\ -16 \end{pmatrix}$ $\overrightarrow{OP} = \begin{pmatrix} 7 \\ 28 \\ -7 \end{pmatrix}$

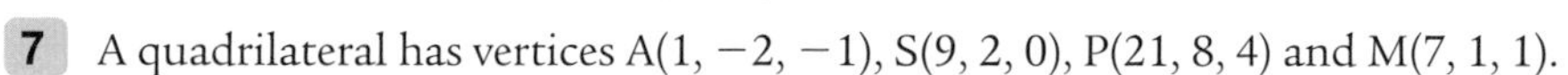

7 A quadrilateral has vertices A(1, −2, −1), S(9, 2, 0), P(21, 8, 4) and M(7, 1, 1).

a Calculate $|\overrightarrow{AS}|$.

b Verify that AMPS is a trapezium.

8 Show that, in each case, the set of three points are collinear.

a A(1, 4, 2), B(2, 3, 4), C(5, 0, 10)

b P(−5, −10, −1), Q(−3, −2, 3), R(0, 10, 9)

c S(3, −1, 4), T(−1, 1, −2), V(9, −4, 13)

d W(3, 7, −9), X(5, 8, −6), Y(6, 8·5, −4·5)

9 Relative to a convenient set of axes and units, an observer on the Earth is fixed at E(1, 3, 6).
A probe is heading for a soft landing on the Moon at a site with coordinates fixed at M(33·5, 43, 16).

a Show that when the probe is at $P_1(11, 18, 12)$ it is not on the line EM.

b The probe is programmed to pass through $P_2(14, 19, 10)$ and $P_3(14, 21, 12)$.
At which of these two points will the Moon, Earth and probe be collinear?

c What is the difference in magnitude between $\overrightarrow{EP_2}$ and $\overrightarrow{EP_3}$?

4.2 Vector paths and geometry

Some rules

1 The representatives of vectors obey the laws of geometry.

2 If $\overrightarrow{AB}$ represents a vector **u** then $\overrightarrow{BA}$ represents its negative, $-\mathbf{u}$.

3 If $\overrightarrow{AB}$ is a representative of the vector **u**, and if $\overrightarrow{CD}$ is parallel to $\overrightarrow{AB}$ and has the same magnitude and sign, then $\overrightarrow{CD}$ is also a representative of **u**.

4 When representatives are drawn nose-to-tail, a path is formed.
If the start of that path is joined to the end of the path, a representative of the sum of all the vectors forming the path is obtained.
This vector sum is often called the **resultant**.

5 If a path ends back where it started, we say the path is closed and the resultant equals the zero vector.

6 Given the representatives of two vectors, we can always find a plane that they both lie on.

7 There is no linear combination of the two vectors that can lift you off the plane.
Thus, if $\mathbf{w} = a\mathbf{v} + b\mathbf{u}$, where a and b are scalars, then **w** must also lie on the plane defined by **u** and **v** ... i.e. **w**, **u** and **v** are co-planar.

Example 1

The vector **v**, lying on the x-y plane makes an angle of $\theta°$ with the x-axis.

a Express **v** in component form, viz. $\begin{pmatrix} x \\ y \\ z \end{pmatrix}$.

b Verify that the components do produce a vector with magnitude $|\mathbf{v}|$.

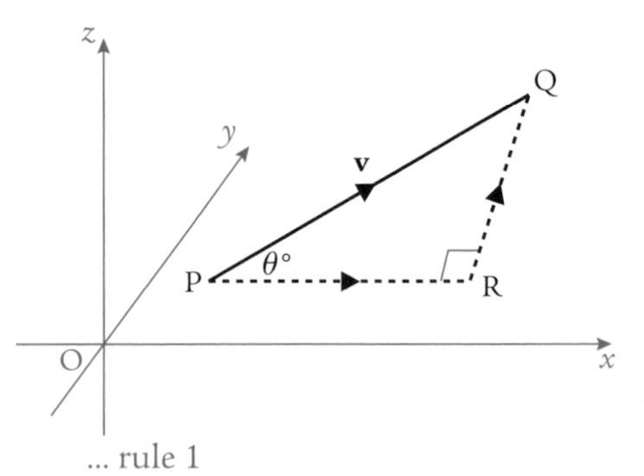

a Let $\overrightarrow{PQ}$ represent **v** with PR parallel to the x-axis and RQ parallel to the y-axis.
Using simple trigonometry we get $\overrightarrow{PR} = \mathbf{v} \cos \theta°$ and $\overrightarrow{RQ} = \mathbf{v} \sin \theta°$. ... rule 1
Giving us the vector addition, $\mathbf{v} = \mathbf{v} \cos \theta° + \mathbf{v} \sin \theta°$.
Note that the vector lies on the x-y plane so its z-component is zero.

$$|\mathbf{v}| = \begin{pmatrix} \mathbf{v} \cos \theta° \\ \mathbf{v} \sin \theta° \\ 0 \end{pmatrix}$$

b $|\mathbf{v}| = \sqrt{x^2 + y^2 + z^2} = \sqrt{\mathbf{v}^2 \cos^2 \theta° + \mathbf{v}^2 \sin^2 \theta° + 0^2}$

$|\mathbf{v}| = \sqrt{\mathbf{v}^2 (\cos^2 \theta° + \sin^2 \theta°)} = \sqrt{\mathbf{v}^2} = |\mathbf{v}|$

Example 2

Three forces act on a kite. Its weight, **d** units, the force of the wind **w** units and the force of the string **u** units holding it back.

When the photo was taken, the kite was stationary.

The diagram shows sketches of two of the forces in space.

Copy the sketch and add the third force.

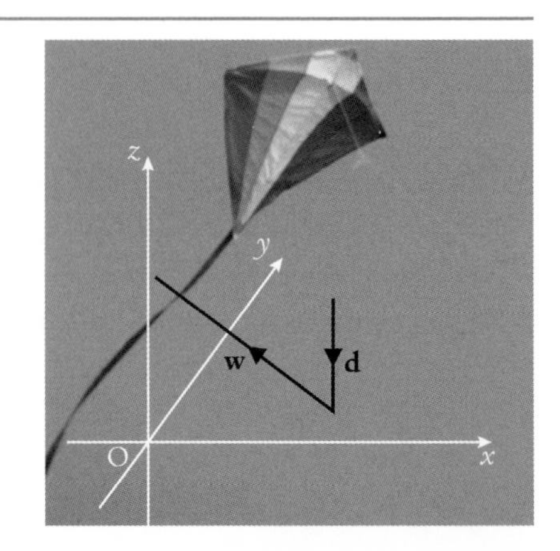

The key feature here is that ... the kite was **stationary**.

This means the resultant of the three forces is the zero vector.

The vector walk created by the representatives of the forces is a closed path. This allows us to draw the third representative, from the *nose* of **w** to the *tail* of **d**.

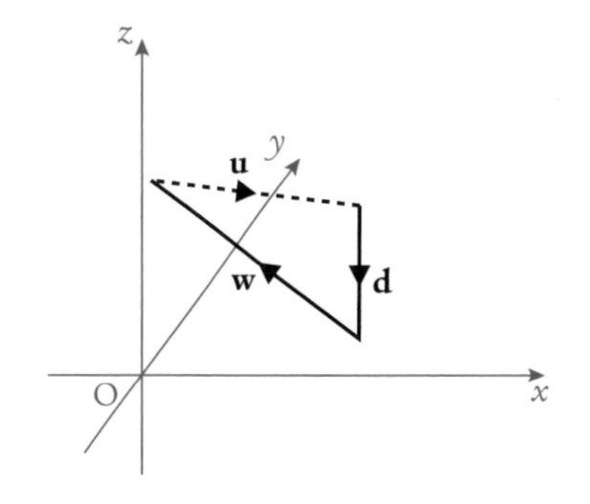

Example 3

If M is the midpoint of AB, express **m** in terms of **a** and **b**.

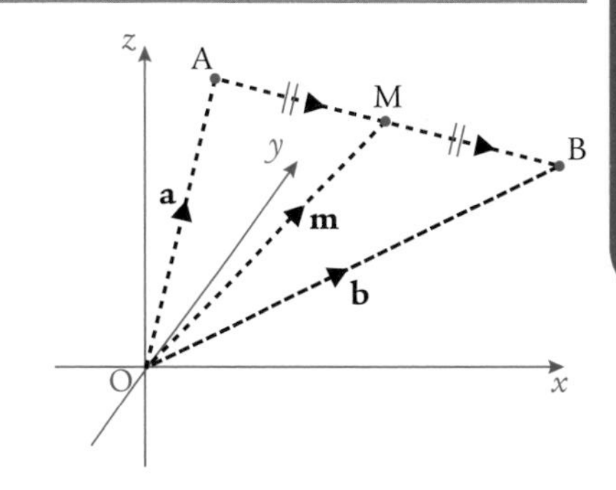

Since M is the midpoint of AB then

$\overrightarrow{AM} = \overrightarrow{MB}$ ... using position vectors

$\Rightarrow \mathbf{m} - \mathbf{a} = \mathbf{b} - \mathbf{m}$

$\Rightarrow 2\mathbf{m} = \mathbf{b} + \mathbf{a}$

$\Rightarrow \mathbf{m} = \frac{1}{2}(\mathbf{b} + \mathbf{a})$ ... this result is worth remembering

Example 4

Show that the vectors $\mathbf{u} = \begin{pmatrix} 3 \\ 1 \\ 7 \end{pmatrix}$, $\mathbf{v} = \begin{pmatrix} 2 \\ -2 \\ 3 \end{pmatrix}$, and $\mathbf{w} = \begin{pmatrix} 5 \\ 7 \\ 15 \end{pmatrix}$, are co-planar.

If the three vectors are co-planar then there exists constants a and b such that $a\mathbf{u} + b\mathbf{v} = \mathbf{w}$.

Equating x-components: $3a + 2b = 5$ ... ①

Equating y-components: $a - 2b = 7$ ... ②

Adding ① + ②: $4a = 12 \Rightarrow a = 3$

Substitute in ②: $3 - 2b = 7 \Rightarrow b = -2$

Equating z-components: $7a + 3b = 15$ ... ③

Check to see if $a = 3$, $b = -2$ is a solution to ③: $7 \times 3 + 3 \times (-2) = 21 - 6 = 15$ ✓

So we find that $3\mathbf{u} - 2\mathbf{v} = \mathbf{w}$.

So **u**, **v** and **w** are co-planar.

Exercise 4.2

1 ABCD is a parallelogram. $\overrightarrow{AB}$ represents the vector **u** and $\overrightarrow{AD}$ represents the vector **v**.

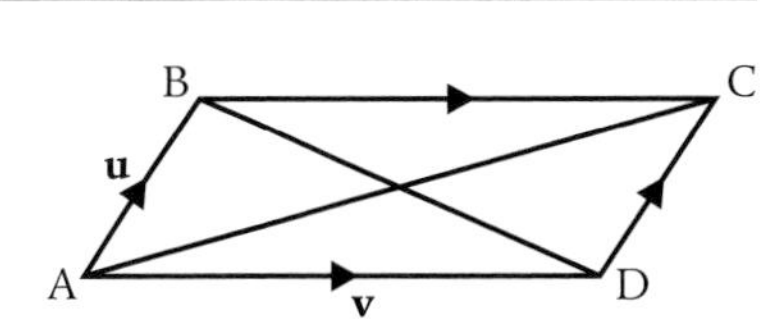

a Express, in terms of **u**, **v**, or both:

i $\overrightarrow{BC}$ **ii** $\overrightarrow{CD}$ **iii** $\overrightarrow{AC}$ **iv** $\overrightarrow{BD}$.

b In terms of $\overrightarrow{AB}$ and $\overrightarrow{AD}$, what is represented by:

i $\overrightarrow{AC}$ **ii** $\overrightarrow{BD}$?

c The position vectors of A, B, C and D are **a**, **b**, **c** and **d** respectively.

i Express **u** in terms of **a** and **b**.

ii By considering that $\overrightarrow{AB} = \overrightarrow{DC}$, prove $\mathbf{a} + \mathbf{c} = \mathbf{b} + \mathbf{d}$.

iii M is the midpoint of AC. Express **m** in terms of **a** and **c**.

iv N is the midpoint of BD. Express **n** in terms of **b** and **d**.

v Use part **cii** to prove M and N are coincident.

2 PQRSTU is a regular hexagon.
$\overrightarrow{PQ}$ and $\overrightarrow{PC}$ represent the vectors **u** and **v** respectively.

a Express, in terms of **u**, **v**, or both:

i $\overrightarrow{ST}$ ii $\overrightarrow{QC}$ iii $\overrightarrow{PU}$ iv $\overrightarrow{UT}$
v $\overrightarrow{QT}$ vi $\overrightarrow{QU}$ vii $\overrightarrow{QS}$ viii $\overrightarrow{RU}$.

b Verify that $\overrightarrow{QS} + \overrightarrow{SU} + \overrightarrow{UQ} = \mathbf{0}$, the zero vector, by using **u** and **v**.

3 The old castle tower is modelled by a cuboid.
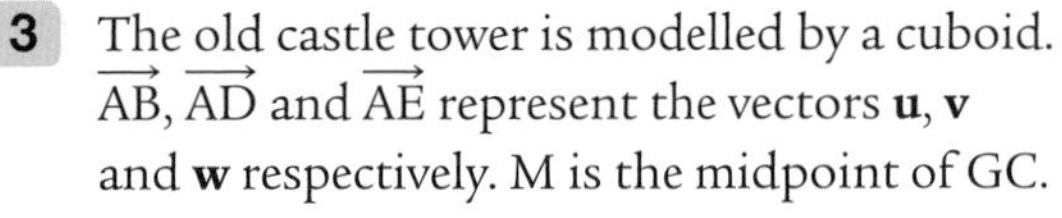
$\overrightarrow{AB}$, $\overrightarrow{AD}$ and $\overrightarrow{AE}$ represent the vectors **u**, **v** and **w** respectively. M is the midpoint of GC.

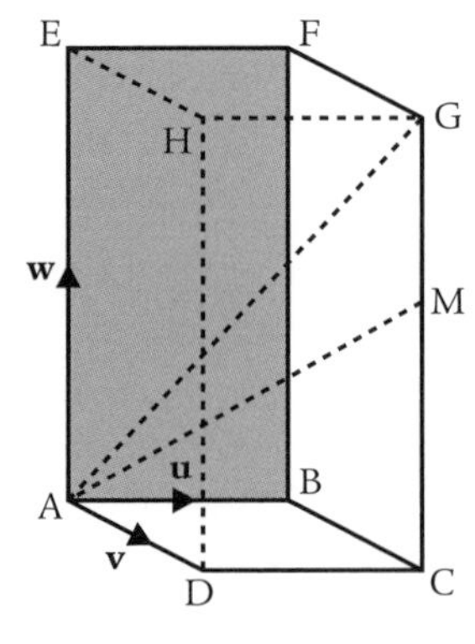

a Express, in terms of **u**, **v** and/or **w**:

i $\overrightarrow{AC}$ ii $\overrightarrow{AH}$
iii $\overrightarrow{AG}$ iv $\overrightarrow{AM}$.

b P is the midpoint of the face BCGF.
Express $\overrightarrow{AP}$ in terms of **u**, **v** and **w**.

c Q is the midpoint of ABFE.
Express $\overrightarrow{QP}$ in terms of **u**, **v** and **w**.
[Hint: $\overrightarrow{AQ} + \overrightarrow{QP} = \overrightarrow{AP}$]

4 The skylight on the roof of a shop is a square based pyramid. The apex of the pyramid, A, is vertically above the centre F, of its base, BCDE.
$\overrightarrow{AB}$, $\overrightarrow{BC}$ and $\overrightarrow{CD}$ represent the vectors **u**, **v** and **w** respectively.

a Find, in terms of **u**, **v** and/or **w**,

i $\overrightarrow{AC}$ ii $\overrightarrow{DA}$
iii $\overrightarrow{EC}$ iv $\overrightarrow{BF}$.

b AF is the vertical height of the pyramid.
Express $\overrightarrow{AF}$ in terms of **u**, **v** and **w**.

c M is the midpoint of AD and N is the midpoint of AB.

i Express $\overrightarrow{MN}$ in terms of **u**, **v** and **w**.
ii Express $\overrightarrow{DB}$ in terms of **u**, **v** and **w**, and comment.

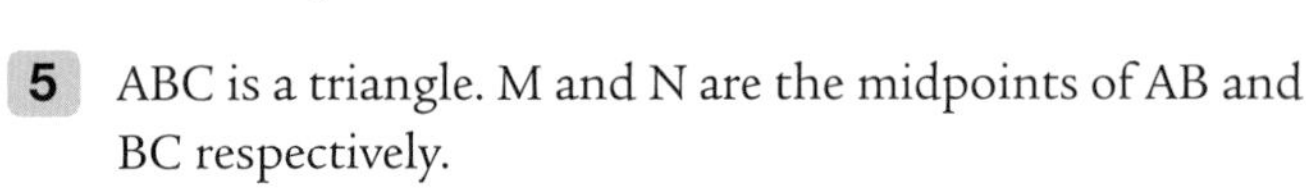

5 ABC is a triangle. M and N are the midpoints of AB and BC respectively.

a Using position vectors,

i express **m** and **n** in terms of **a**, **b** and **c**
ii express $\overrightarrow{MN}$ in terms of **a**, **b** and **c**
iii hence prove that MN is parallel to AC
iv similarly prove that MP is parallel to BC, where P is the midpoint of AC.

b A **median** of a triangle joins a vertex to the midpoint of the opposite side.

i Find the median $\overrightarrow{AN}$ in terms of **a**, **b** and **c**.
ii Similarly express $\overrightarrow{BP}$ and $\overrightarrow{CM}$ in terms of **a**, **b** and **c**.
iii Prove that $\overrightarrow{AN} + \overrightarrow{BP} + \overrightarrow{CM} = \mathbf{0}$.

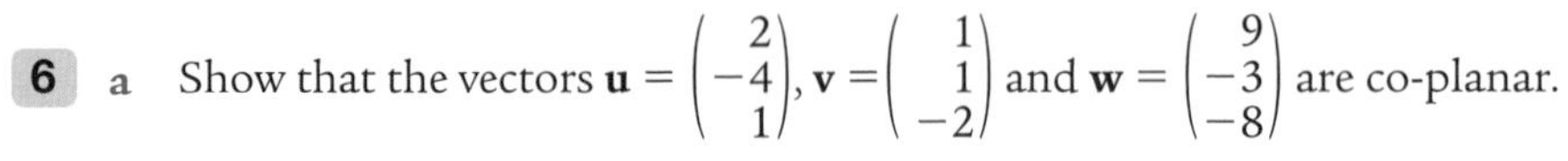

6 a Show that the vectors $\mathbf{u} = \begin{pmatrix} 2 \\ -4 \\ 1 \end{pmatrix}$, $\mathbf{v} = \begin{pmatrix} 1 \\ 1 \\ -2 \end{pmatrix}$ and $\mathbf{w} = \begin{pmatrix} 9 \\ -3 \\ -8 \end{pmatrix}$ are co-planar.

b Show that the vectors $\mathbf{p} = \begin{pmatrix} -3 \\ 1 \\ 2 \end{pmatrix}$, $\mathbf{q} = \begin{pmatrix} 1 \\ 5 \\ 1 \end{pmatrix}$ and $\mathbf{r} = \begin{pmatrix} -10 \\ 14 \\ 9 \end{pmatrix}$ are **not** co-planar.

4.3 Dividing lines

The diagram shows a point P dividing a line AB internally in the ratio $m : n$.

By this we mean, P lies between A and B and $\frac{AP}{PB} = \frac{m}{n}$.

[Note: This does NOT mean AP = m nor PB = n.]

Vectors can be used to analyse such situations.

Given that **a**, **b** and **p** are the position vectors of A, B and P respectively,

$\frac{AP}{PB} = \frac{m}{n} \Rightarrow nAP = mPB.$

Using vectors, $n\overrightarrow{AP} = m\overrightarrow{PB}$

$\Rightarrow n(\mathbf{p} - \mathbf{a}) = m(\mathbf{b} - \mathbf{p})$

$\Rightarrow n\mathbf{p} - n\mathbf{a} = m\mathbf{b} - m\mathbf{p}$

$\Rightarrow n\mathbf{p} + m\mathbf{p} = m\mathbf{b} + n\mathbf{a}$

$\Rightarrow \mathbf{p}(m + n) = m\mathbf{b} + n\mathbf{a}$

$\Rightarrow \qquad \mathbf{p} = \frac{m\mathbf{b} + n\mathbf{a}}{m + n}.$

This formula is commonly referred to as the **section formula**.

You may work afresh at each instance of this type of problem or use the formula.

Example 1

Find the point P that divides the line AB in the ratio 2 : 3, where A is (3, −1, 5) and B is (13, 4, −10).

From the information given, $\mathbf{a} = \begin{pmatrix} 3 \\ -1 \\ 5 \end{pmatrix}$ and $\mathbf{b} = \begin{pmatrix} 13 \\ 4 \\ -10 \end{pmatrix}$.

$\frac{AP}{PB} = \frac{2}{3} \Rightarrow 3AP = 2PB$

$\Rightarrow 3(\mathbf{p} - \mathbf{a}) = 2(\mathbf{b} - \mathbf{p})$

$\Rightarrow 3\mathbf{p} - 3\mathbf{a} = 2\mathbf{b} - 2\mathbf{p}$

$\Rightarrow \qquad 5\mathbf{p} = 2\mathbf{b} + 3\mathbf{a}$

$\Rightarrow \qquad \mathbf{p} = \frac{1}{5}(2\mathbf{b} + 3\mathbf{a})$

$\Rightarrow \qquad \mathbf{p} = \frac{1}{5}\left(2\begin{pmatrix} 13 \\ 4 \\ -10 \end{pmatrix} + 3\begin{pmatrix} 3 \\ -1 \\ 5 \end{pmatrix}\right)$

$\Rightarrow \qquad \mathbf{p} = \frac{1}{5}\begin{pmatrix} 26 + 9 \\ 8 - 3 \\ -20 + 15 \end{pmatrix} = \frac{1}{5}\begin{pmatrix} 35 \\ 5 \\ -5 \end{pmatrix} = \begin{pmatrix} 7 \\ 1 \\ -1 \end{pmatrix}$

So P is the point P(7, 1, −1).

Example 2

Triangle ABC has vertices A(6, 2, 5), B(2, 7, 9) and C(4, 3, −5).

a The median AP cuts BC at P. Find the coordinates of P using the section formula.

b Find the point E which divides the median AP internally in the ratio 2 : 1.

a Being a midpoint, CP : PB = 1 : 1.

$\Rightarrow \mathbf{p} = \dfrac{\mathbf{b} + \mathbf{c}}{1 + 1} = \tfrac{1}{2}(\mathbf{b} + \mathbf{c})$ ... a formula already known

$\Rightarrow \mathbf{p} = \tfrac{1}{2}\left(\begin{pmatrix} 2 \\ 7 \\ 9 \end{pmatrix} + \begin{pmatrix} 4 \\ 3 \\ -5 \end{pmatrix}\right) = \tfrac{1}{2}\begin{pmatrix} 6 \\ 10 \\ 4 \end{pmatrix} = \begin{pmatrix} 3 \\ 5 \\ 2 \end{pmatrix}$

So P is the point P(3, 5, 2).

b E is the point where AE : EP = 2 : 1.

$\Rightarrow \mathbf{e} = \dfrac{2 . \mathbf{p} + 1 . \mathbf{a}}{2 + 1} = \tfrac{1}{3}(2\mathbf{p} + \mathbf{a})$

$\Rightarrow \mathbf{e} = \tfrac{1}{3}\left(2\begin{pmatrix} 3 \\ 5 \\ 2 \end{pmatrix} + \begin{pmatrix} 6 \\ 2 \\ 5 \end{pmatrix}\right) = \tfrac{1}{3}\begin{pmatrix} 12 \\ 12 \\ 9 \end{pmatrix} = \begin{pmatrix} 4 \\ 4 \\ 3 \end{pmatrix}$

E is the point E(4, 4, 3).

Example 3

In what ratio does P(11, 1, 8) divide the line joining A(2, 1, 5) to B(17, 1, 10)?

Let AP : PB = $m : n$.

Consider the x-coordinates only ... and using $\mathbf{p} = \dfrac{m\mathbf{b} + n\mathbf{a}}{m + n}$.

$11 = \dfrac{17m + 2n}{m + n}$

$\Rightarrow 11m + 11n = 17m + 2n$

$\Rightarrow \quad 6m = 9n$

$\Rightarrow \quad \dfrac{m}{n} = \dfrac{9}{6} = \dfrac{3}{2}$

P divides AB in the ratio 3 : 2.

Exercise 4.3

1 Find the point P which divides AB in the given ratio in each case.

a A(1, 2, 4), B(7, 14, 22) in the ratio 5 : 1.

b A(−1, 2, 3), B(4, −3, 13) in the ratio 3 : 2.

c A(4, −3, 1), B(9, −8, 11) in the ratio 1 : 4.

d A(−5, 2, −3), B(−1, −2, 5) in the ratio 3 : 1.

e A(0, 4, 7), B(3, 1, 13) in the ratio 1 : 2.

f A(−1, 0, −3), B(6, −7, 11) in the ratio 3 : 4.

g A(4, 1, 1), B(7, −2, 7) in the ratio 1 : 1.

2 PQR is a triangle with vertices P(4, − 1, 5), Q(8, −5, 13) and R(12, 3, −7).

a Find the point M which divides $\overrightarrow{PQ}$ in the ratio 1 : 3.

b Similarly find the point N which divides PR in the ratio 1 : 3.

c Express $\overrightarrow{MN}$ in component form.

d Prove that MN and QR are parallel.

e Prove that if a line cuts two sides of a triangle in the same ratio, say, $a : b$, then it is parallel to the third side of the triangle.

3 Triangle ABC has vertices A(3, −2, 4), B(−1, 4, −2) and C(11, −2, −2).

a The median AP cuts BC at P. Find the coordinates of P.

b Find the point E which divides the median AP internally in the ratio 2 : 1.

c i The median BQ cuts AC at Q. Find the coordinates of Q.
 ii Find the point F which divides the median BQ internally in the ratio 2 : 1.
 iii Comment.

4 P, Q and R are points on AB, BC and CA respectively so that CP, AQ and BR are the **medians** of the triangle ABC.
Prove that these medians meet at the point[2] whose position vector is $\frac{1}{3}(\mathbf{a} + \mathbf{b} + \mathbf{c})$.

5 In what ratio does:

a P(6, 0, 7) divide the line joining A(3, 0, 6) to B(15, 0, 10)

b P(7, 3, 4) divide the line joining A(1, 3, 2) to B(22, 3, 9)?

c In what ratio does $P(x, y, z)$ divide the line joining $A(a_1, a_2, a_3)$ to $B(b_1, b_2, b_3)$?
 i Express it in terms of a_1, b_1 and x.
 ii Express it in terms of position vectors.

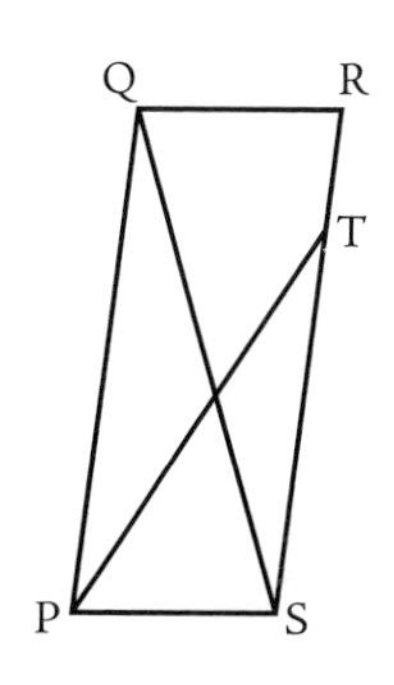

6 PQRS is a parallelogram.
For the following, express your answers using position vectors.

a Find **t** in terms of **r** and **s**, where T divides $\overrightarrow{SR}$ in the ratio 3 : 1.

b Find **v**, where V divides $\overrightarrow{PT}$ in the ratio 4 : 3 in terms of **p**, **r** and **s**.

c Express **q** in terms of **p**, **r** and **s**.

d Find **u** where U divides $\overrightarrow{QS}$ in the ratio 4 : 3 in terms of:
 i **s** and **q** ii **p**, **r** and **s**.
 Comment.

7 A yacht keeps its mast secure by using a guy attached at a point A on the bow and a point B on the mast.
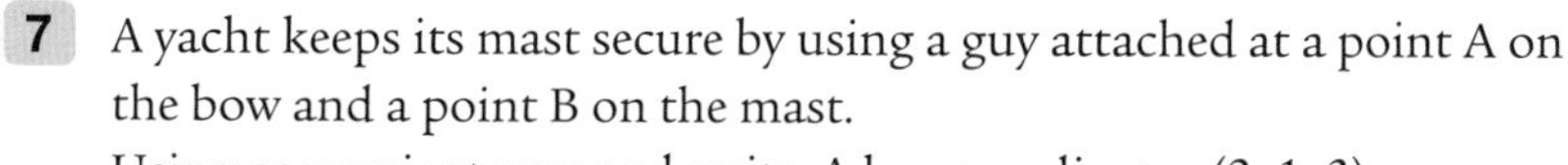
Using convenient axes and units, A has coordinates (2, 1, 3).
The mast runs from C(38, 5, 7) to D(38, 10, 17).
The point B cuts $\overrightarrow{CD}$ in the ratio 4 : 1.

a Calculate the coordinates of the point B.

b Find the magnitude of $\overrightarrow{AB}$ correct to 1 decimal place.

8 A regular tetrahedron has vertices A, B, C and D.
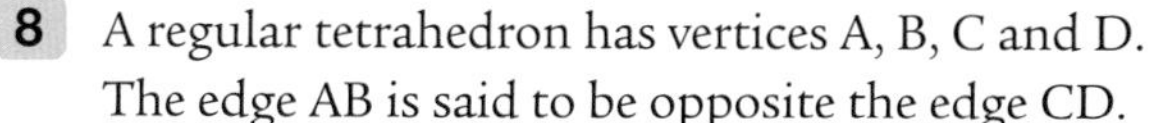
The edge AB is said to be opposite the edge CD.

a In terms of the position vectors of A, B, C and D, find:
 i E, the midpoint of AB ii F, the midpoint of CD.

b Find the midpoint of EF in terms of **a**, **b**, **c**, **d**.

c Repeat this process for opposite edges AC and DB.
Comment.

4.4 Basis vectors and scalar product

Basis vectors

You have already met the concept of the unit vector.
Three very useful unit vectors are the ones parallel to the x, y and z-directions.

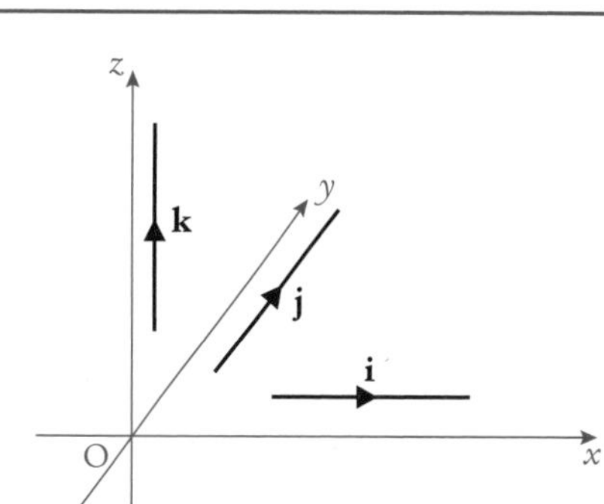

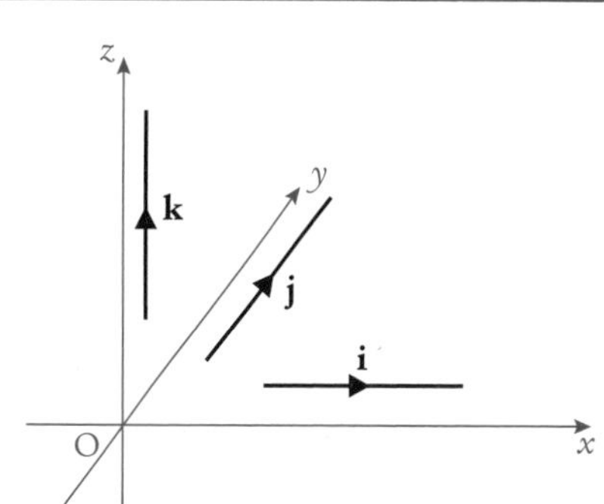

[2] This point is called the centroid of the triangle.

They are defined by

$\mathbf{i} = \begin{pmatrix}1\\0\\0\end{pmatrix}$, $\mathbf{j} = \begin{pmatrix}0\\1\\0\end{pmatrix}$, $\mathbf{k} = \begin{pmatrix}0\\0\\1\end{pmatrix}$, ... often referred to as the **basis vectors**.

Any vector can be expressed in terms of the basis vectors. For example,

$$\begin{pmatrix}3\\1\\-5\end{pmatrix} = \begin{pmatrix}3\\0\\0\end{pmatrix} + \begin{pmatrix}0\\1\\0\end{pmatrix} + \begin{pmatrix}0\\0\\-5\end{pmatrix} = 3\begin{pmatrix}1\\0\\0\end{pmatrix} + 1\begin{pmatrix}0\\1\\0\end{pmatrix} - 5\begin{pmatrix}0\\0\\1\end{pmatrix} = 3\mathbf{i} + \mathbf{j} - 5\mathbf{k}.$$

In general, $\mathbf{a} = \begin{pmatrix}a_1\\a_2\\a_3\end{pmatrix} = a_1\mathbf{i} + a_2\mathbf{j} + a_3\mathbf{k}$.

A sum such as $\begin{pmatrix}2\\1\\3\end{pmatrix} + \begin{pmatrix}1\\0\\4\end{pmatrix} = \begin{pmatrix}3\\1\\7\end{pmatrix}$ would look like $2\mathbf{i} + \mathbf{j} + 3\mathbf{k} + \mathbf{i} + 4\mathbf{k} = 3\mathbf{i} + \mathbf{j} + 7\mathbf{k}$.

In many branches of science, this notation is preferred.

Scalar product

We know how to multiply a vector by a scalar but if we wish to multiply a vector by a vector we have to define what we mean.

In this course we will look at just one such definition.

Remember if a vector, $\mathbf{v}$, is resolved into components, the component $\theta°$ round from $\mathbf{v}$ is $\mathbf{v}\cos\theta°$.

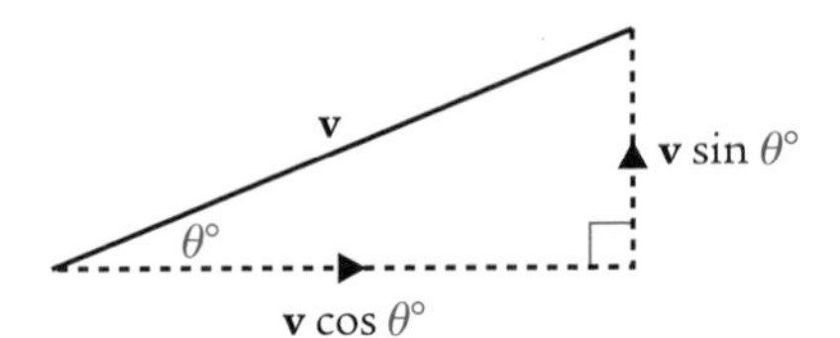

The magnitude of this component is $|\mathbf{v}\cos\theta°| = |\mathbf{v}|\cos\theta°$.

Consider two non-zero vectors, tail-to-tail, with the angle between them $\theta°$ ... and find the component of $\mathbf{b}$ that runs in the same direction as $\mathbf{a}$.

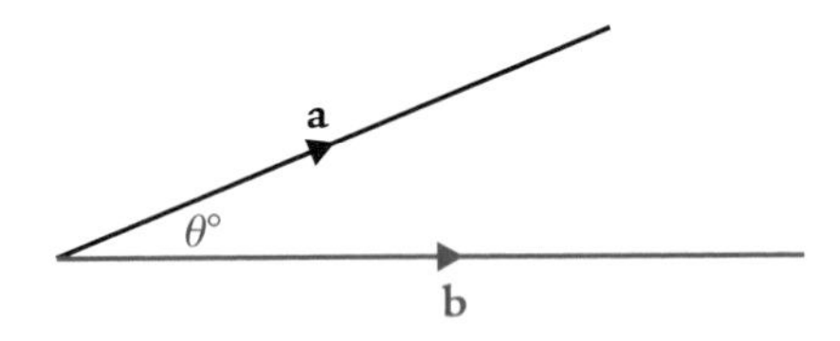

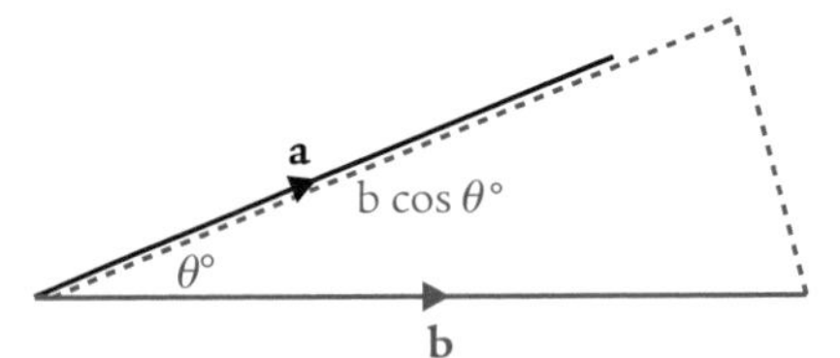

We define the scalar product[3] $\mathbf{a}.\mathbf{b}$ as the product of the two magnitudes $|\mathbf{a}|$ and $|\mathbf{b}|\cos\theta°$

i.e. $\mathbf{a}.\mathbf{b} = |\mathbf{a}||\mathbf{b}|\cos\theta°$, $\mathbf{a}, \mathbf{b} \neq \mathbf{0}$.

Some properties of the scalar product

(1) $\mathbf{a}.\mathbf{b} = \mathbf{b}.\mathbf{a}$ ... since $\mathbf{a}.\mathbf{b} = |\mathbf{a}||\mathbf{b}|\cos\theta° = |\mathbf{b}||\mathbf{a}|\cos\theta° = \mathbf{b}.\mathbf{a}$

(2) $\mathbf{a}.(\mathbf{b} + \mathbf{c}) = \mathbf{a}.\mathbf{b} + \mathbf{a}.\mathbf{c}$ ... the proof is beyond the scope of the course

(3) $\mathbf{a}.\mathbf{a} = |\mathbf{a}|^2$ ... $\mathbf{a}.\mathbf{a} = |\mathbf{a}||\mathbf{a}|\cos 0° = |\mathbf{a}||\mathbf{a}|.1 = |\mathbf{a}|^2$

(4) $\mathbf{i}.\mathbf{i} = \mathbf{j}.\mathbf{j} = \mathbf{k}.\mathbf{k} = 1$ ... e.g. $\mathbf{i}.\mathbf{i} = |\mathbf{i}||\mathbf{i}|\cos 0° = 1.1.1 = 1$

(5) $\mathbf{i}.\mathbf{j} = \mathbf{j}.\mathbf{k} = \mathbf{k}.\mathbf{i} = 0$ ... e.g. $\mathbf{i}.\mathbf{j} = |\mathbf{i}||\mathbf{j}|\cos 90° = 1.1.0 = 0$

(6) If $\mathbf{a}.\mathbf{b} = 0$, then $\mathbf{a}$ is perpendicular to $\mathbf{b}$. Since $\mathbf{a}, \mathbf{b} \neq \mathbf{0}$, then $\cos\theta° = 0 \Rightarrow \theta = 90$.

Scalar product using components

Consider vectors $\mathbf{a}$ and $\mathbf{b}$ such that $\mathbf{a} = \begin{pmatrix}a_1\\a_2\\a_3\end{pmatrix} = a_1\mathbf{i} + a_2\mathbf{j} + a_3\mathbf{k}$ and $\mathbf{b} = \begin{pmatrix}b_1\\b_2\\b_3\end{pmatrix} = b_1\mathbf{i} + b_2\mathbf{j} + b_3\mathbf{k}$

$\mathbf{a}.\mathbf{b} = (a_1\mathbf{i} + a_2\mathbf{j} + a_3\mathbf{k}).(b_1\mathbf{i} + b_2\mathbf{j} + b_3\mathbf{k})$ ... by definition

$\Rightarrow \mathbf{a}.\mathbf{b} = a_1b_1\mathbf{i}.\mathbf{i} + a_1b_2\mathbf{i}.\mathbf{j} + a_1b_3\mathbf{i}.\mathbf{k} + a_2b_1\mathbf{j}.\mathbf{i} + a_2b_2\mathbf{j}.\mathbf{j} + a_2b_3\mathbf{j}.\mathbf{k} + a_3b_1\mathbf{k}.\mathbf{i} + a_3b_2\mathbf{k}.\mathbf{j} + a_3b_3\mathbf{k}.\mathbf{k}$ ... by (2)

$\Rightarrow \mathbf{a}.\mathbf{b} = a_1b_1 + a_2b_2 + a_3b_3$ ... by (4) and (5)

[3] It is often referred to as the dot product.

Example 1

Find a number a such that the vectors $\mathbf{p} = \begin{pmatrix} 3 \\ 1 \\ 4 \end{pmatrix}$ and $\mathbf{q} = \begin{pmatrix} -2 \\ a \\ 3 \end{pmatrix}$ are perpendicular.

If $\mathbf{p}$ and $\mathbf{q}$ are perpendicular then $\mathbf{p} \, . \, \mathbf{q} = 0$.

$$\Rightarrow \begin{pmatrix} 3 \\ 1 \\ 4 \end{pmatrix} . \begin{pmatrix} -2 \\ a \\ 3 \end{pmatrix} = 0$$

$\Rightarrow 3 \, . (-2) + 1 \, . \, a + 4 \, . \, 3 = -6 + a + 12 = 0$

$\Rightarrow a = -6$

The angle between vectors

By definition $\mathbf{a} \, . \, \mathbf{b} = |\mathbf{a}| \, |\mathbf{b}| \cos \theta°$

$$\Rightarrow \cos \theta° = \frac{\mathbf{a} \, . \, \mathbf{b}}{|\mathbf{a}| \, |\mathbf{b}|}.$$

Example 2

Find the acute angle between the vectors $\mathbf{u} = \begin{pmatrix} 4 \\ 1 \\ 8 \end{pmatrix}$ and $\mathbf{v} = \begin{pmatrix} 1 \\ 2 \\ 3 \end{pmatrix}$.

Find the scalar product: $\mathbf{u} \, . \, \mathbf{v} = \begin{pmatrix} 4 \\ 1 \\ 8 \end{pmatrix} . \begin{pmatrix} 1 \\ 2 \\ 3 \end{pmatrix} = 4 \, . \, 1 + 1 \, . \, 2 + 8 \, . \, 3 = 4 + 2 + 24 = 30.$

Find the magnitudes: $|\mathbf{u}| = \sqrt{4^2 + 1^2 + 8^2} = \sqrt{81} = 9$; $|\mathbf{v}| = \sqrt{1^2 + 2^2 + 3^2} = \sqrt{14}$.

Substitute into $\cos \theta° = \dfrac{\mathbf{a} \, . \, \mathbf{b}}{|\mathbf{a}| \, |\mathbf{b}|}$ to get $\cos \theta° = \dfrac{30}{9\sqrt{14}} = 0{\cdot}89087...$

$\Rightarrow \theta° = \cos^{-1}(0{\cdot}89087) = 27{\cdot}0°$ (to 3 s.f.).

Example 3

A helicopter is being used to transport poles.

An observer at A(3, 1, 2) photographs the helicopter at H(4, 5, 50) when the end of the pole is at P(−3, 4, 25).

Calculate the size of the angle HAP.

Plan your strategy by sketching the vectors making sure that they are tail-to-tail. It doesn't have to be realistic.

H (4, 5, 50)

A (3, 1, 2) $\theta°$

P (−3, 4, 25)

$\overrightarrow{AH} = \mathbf{h} - \mathbf{a} = \begin{pmatrix} 4 \\ 5 \\ 50 \end{pmatrix} - \begin{pmatrix} 3 \\ 1 \\ 2 \end{pmatrix} = \begin{pmatrix} 1 \\ 4 \\ 48 \end{pmatrix} \Rightarrow |\overrightarrow{AH}| = \sqrt{1^2 + 4^2 + 48^2} = 48{\cdot}176...$

$\overrightarrow{AP} = \mathbf{p} - \mathbf{a} = \begin{pmatrix} -3 \\ 4 \\ 25 \end{pmatrix} - \begin{pmatrix} 3 \\ 1 \\ 2 \end{pmatrix} = \begin{pmatrix} -6 \\ 3 \\ 23 \end{pmatrix} \Rightarrow |\overrightarrow{AP}| = \sqrt{(-6^2) + 3^2 + 23^2} = 23{\cdot}958...$

$\overrightarrow{AH} . \overrightarrow{AP} = 1 \, . (-6) + 4 \, . \, 3 + 48 \, . \, 23 = 1110$

Substitute into $\cos \theta° = \dfrac{\mathbf{a} \, . \, \mathbf{b}}{|\mathbf{a}| \, |\mathbf{b}|}$ to get $\cos \theta° = \dfrac{1110}{48{\cdot}18 \times 23{\cdot}96} = 0{\cdot}9615$ (to 4 s.f.)

$\Rightarrow \theta° = \cos^{-1}(0{\cdot}9615) = 15{\cdot}9°$ (to 3 s.f.).

Exercise 4.4

1 Express each vector in terms of the basis vectors **i**, **j** and **k**.

a $\begin{pmatrix}3\\7\\9\end{pmatrix}$ b $\begin{pmatrix}-1\\1\\-1\end{pmatrix}$ c $\begin{pmatrix}1\\0\\-2\end{pmatrix}$ d $\begin{pmatrix}0\\0\\6\end{pmatrix}$

2 Calculate the magnitude of each vector.

a $15\mathbf{i} + 10\mathbf{j} + 6\mathbf{k}$ b $\mathbf{i} + 4\mathbf{j} - 8\mathbf{k}$

c $-2\mathbf{i} + 9\mathbf{j} - 6\mathbf{k}$ d $29\mathbf{i} + 22\mathbf{j} + 14\mathbf{k}$

3 Given that $\mathbf{u} = 4\mathbf{i} + 6\mathbf{j} - 2\mathbf{k}$, $\mathbf{v} = -2\mathbf{i} + \mathbf{j} + 3\mathbf{k}$ and $\mathbf{w} = -\mathbf{i} + 6\mathbf{k}$, calculate:

a $7\mathbf{u}$ b $3\mathbf{v} + 2\mathbf{w}$ c $3\mathbf{u} + \mathbf{v}$

d $\mathbf{u} + \mathbf{v} - \mathbf{w}$ e $3(2\mathbf{v} - \mathbf{u})$ f $|2\mathbf{u} + \mathbf{w}|$.

4 a Find a unit vector parallel to $14\mathbf{i} - 5\mathbf{j} + 2\mathbf{k}$, leaving the coefficients as proper fractions.

b For what value of a is $12\mathbf{i} + 3\mathbf{j} - 18\mathbf{k}$ parallel to $a\mathbf{i} + 2\mathbf{j} - 12\mathbf{k}$?

5 a Show that $3\mathbf{i} + 5\mathbf{j} - 2\mathbf{k}$ is perpendicular to $-2\mathbf{i} + 4\mathbf{j} + 7\mathbf{k}$.

b Find a so that $3\mathbf{i} + 5\mathbf{j} + 9\mathbf{k}$ is perpendicular to $4\mathbf{i} + 3a\mathbf{j} - a\mathbf{k}$.

6 AB is a line segment.
The position vectors $\mathbf{a} = 5\mathbf{i} + \mathbf{j} + 2\mathbf{k}$ and $\mathbf{b} = 10\mathbf{i} + 16\mathbf{j} + 12\mathbf{k}$ define its endpoints.
The point P divides $\overrightarrow{AB}$ in the ratio 3 : 2. Express **p** in the form $x\mathbf{i} + y\mathbf{j} + z\mathbf{k}$.

7 Calculate the scalar product, $\mathbf{u} \cdot \mathbf{v}$ for each diagram.

a
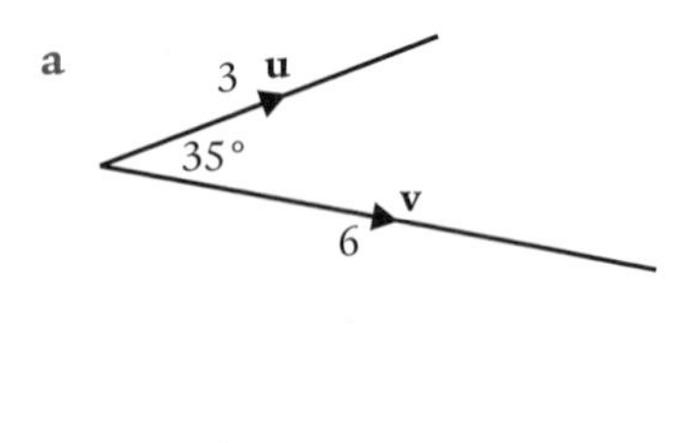

b
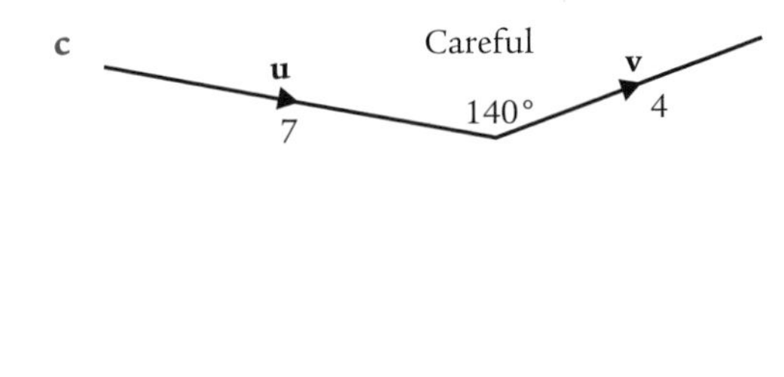

8 The diagram shows an equilateral triangle, ABC of side 2 units.
An altitude AM is drawn.
Calculate the **exact** value of the following scalar products:

a $\overrightarrow{BA} \cdot \overrightarrow{BC}$ b $\overrightarrow{AB} \cdot \overrightarrow{AM}$ c $\overrightarrow{AM} \cdot \overrightarrow{BC}$ d $\overrightarrow{BA} \cdot \overrightarrow{CB}$.

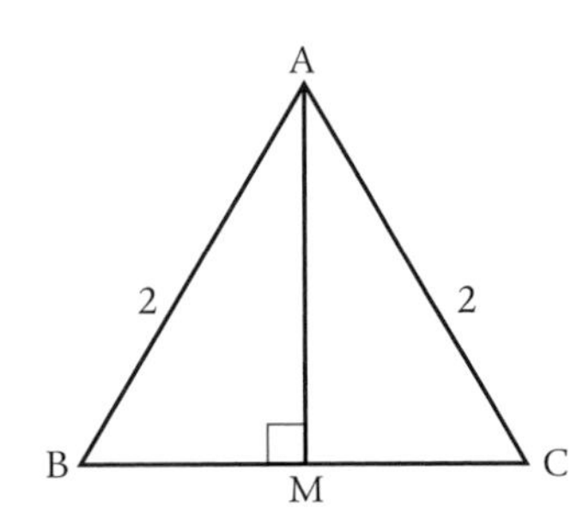

9 The diagonals of the square PQRS intersect at C.
The square has a side of 2 units.
Calculate the **exact** value of:

a $\overrightarrow{SR} \cdot \overrightarrow{SP}$ b $\overrightarrow{SP} \cdot \overrightarrow{SQ}$ c $\overrightarrow{PR} \cdot \overrightarrow{SQ}$ d $\overrightarrow{QP} \cdot \overrightarrow{QS}$.

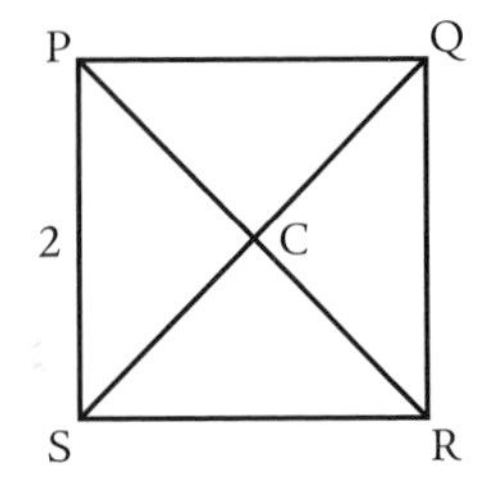

10 Relative to convenient axes and units, the displacements of two vessels are given by $\mathbf{u} = 6\mathbf{i} - 3\mathbf{j} - 2\mathbf{k}$ and $\mathbf{v} = 21\mathbf{i} + 42\mathbf{j} - 2\mathbf{k}$.

a Calculate: i the scalar product $\mathbf{u}.\mathbf{v}$ ii $|\mathbf{u}|$ iii $|\mathbf{v}|$.

b Hence calculate the acute angle between the displacements correct to 1 decimal place.

11 Calculate the acute angle between each pair of vectors:

a $\mathbf{u} = \begin{pmatrix} 2 \\ 1 \\ 2 \end{pmatrix}; \mathbf{v} = \begin{pmatrix} 6 \\ 17 \\ 6 \end{pmatrix}$ b $\mathbf{u} = \begin{pmatrix} -18 \\ 6 \\ 13 \end{pmatrix}; \mathbf{v} = \begin{pmatrix} 15 \\ -12 \\ 16 \end{pmatrix}$

c $\mathbf{u} = \begin{pmatrix} 6 \\ 6 \\ -7 \end{pmatrix}; \mathbf{v} = \begin{pmatrix} 5 \\ 14 \\ 2 \end{pmatrix}$.

12 Three standing stones stand on a moor marking a paleolithic burial site.
Measurements taken by an archaeologist give the tops of the three stones as being at:
A(1, 2, 3·7), B(8, 6, 4·6) and C(5, 12, 5·2).
The axes were set by the archaeologist and the units are metres.

a Calculate the distance AB to 3 significant figures.

b Calculate the size of the angle BAC.

13 Triangle ABC has vertices A(1, −2, 5), B(3, 5, −1) and C(−2, 6, 4).
Calculate the size of each angle in the triangle.

14 The control tower keeps track of flights in its airspace.
Using a convenient coordinate system an observer at the top of the tower has coordinates (15, 25, 4).
A jet approaches and has coordinates (25, 70, 120).
A helicopter hovers nearby with coordinates (20, −10, 100).
Through what angle does the observer turn as he shifts his gaze from the helicopter to the approaching jet?

Preparation for assessment

1 The point A(3, 6, −1) is joined to B(5, −4, 3) to form $\overrightarrow{AB}$, a representative of vector **u**.

a Find the components of **u**.

b Find the magnitude of **u**, leaving your answer as a surd.

2 A vector is represented by $\overrightarrow{AB} = \begin{pmatrix} 3 \\ 12 \\ 4 \end{pmatrix}$.

a Find the value of a such that $\overrightarrow{CD} = \begin{pmatrix} 2a \\ 6 \\ 2 \end{pmatrix}$ is parallel to $\overrightarrow{AB}$.

b Find the value of b such that $\overrightarrow{EF} = \begin{pmatrix} -20b \\ b \\ -12 \end{pmatrix}$ is perpendicular to $\overrightarrow{AB}$.

c Find a unit vector parallel to AB.

3 During a survey of woodlands, three trees are recorded as being at A(5, 1, 2), B(7, 5, −4) and C(10, 11, −13).
Prove that the three trees lie in a straight line.

4 PQRS is a parallelogram. T is the mid-point of QR.
$\overrightarrow{PQ} = \mathbf{u}$ and $\overrightarrow{PS} = \mathbf{v}$.
Express, in terms of **u** and/or **v**

a $\overrightarrow{PR}$ b $\overrightarrow{QT}$ c $\overrightarrow{PT}$

d $\overrightarrow{TS}$ e $\overrightarrow{PR} - \overrightarrow{PT}$.

5 Trying to define a bedding plane of a rock outcrop, a geologist identified two vectors on the plane, viz. $\mathbf{u} = 2\mathbf{i} + \mathbf{j} + 7\mathbf{k}$ and $\mathbf{v} = 3\mathbf{i} - 2\mathbf{j} + 8\mathbf{k}$.
Show that $\mathbf{w} = -\mathbf{i} + 10\mathbf{j} + 4\mathbf{k}$ is also on the plane.

6 Find the point that divides the vector $\overrightarrow{AB}$ in the ratio 4 : 1 where A is (−3, 4, −2) and B is (7, −1, 3).

7 An equilateral triangle and a square share a common base of length 2 units.
Vectors **a**, **b** and **c** are as shown in the diagram.
Evaluate the scalar product $\mathbf{a} . (\mathbf{b} + \mathbf{c})$.

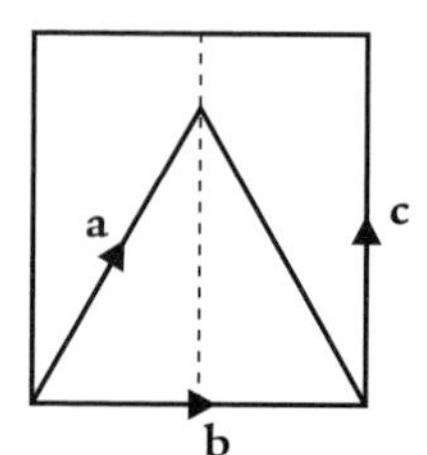

8 Using the coordinate axes given,

a state the coordinates of the points A, B and C on the puzzle cube

b calculate the size of the angle ABC.

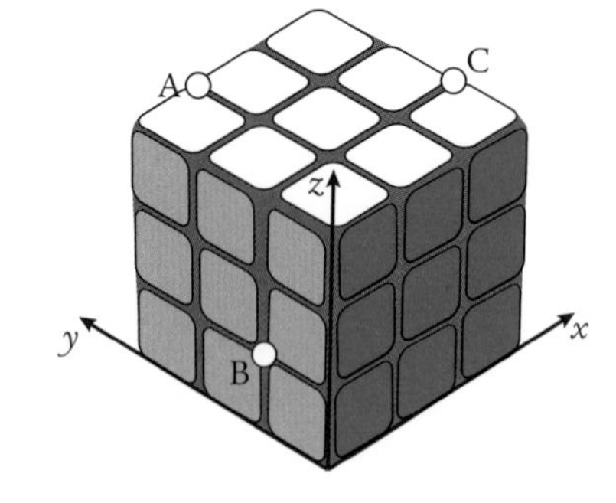

Summary

1 A vector is a quantity with magnitude and direction.
A scalar has magnitude only.

2 If, with reference to an origin O, P is a point with coordinates (x_p, y_p, z_p) then the vector $\overrightarrow{OP} = \begin{pmatrix} x_p \\ y_p \\ z_p \end{pmatrix}$ is called the **position vector** of P and is usually denoted by **p**.

3 $\overrightarrow{AB} = \mathbf{b} - \mathbf{a} = \begin{pmatrix} x_b \\ y_b \\ z_b \end{pmatrix} - \begin{pmatrix} x_a \\ y_a \\ z_a \end{pmatrix} = \begin{pmatrix} x_b - x_a \\ y_b - y_a \\ z_b - z_a \end{pmatrix}$

4 $|\overrightarrow{AB}| = \sqrt{(x_b - x_a)^2 + (y_b - y_a)^2 + (z_b - z_a)^2}$

5 If $\mathbf{b} = k\mathbf{a}$ then **b** is parallel to **a**.

6 If $\mathbf{p} = \begin{pmatrix} x_p \\ y_p \\ z_p \end{pmatrix}$ then $u_p = \dfrac{1}{|\mathbf{p}|}\begin{pmatrix} x_p \\ y_p \\ z_p \end{pmatrix}$ is a unit vector parallel to $\mathbf{p}$.

7 If asked to prove three points A, B and C are collinear, we should show AB is parallel to BC, say, and then point out that B is a common point.

8 If $\mathbf{w} = a\mathbf{v} + b\mathbf{u}$ where a and b are scalars, then $\mathbf{w}$ must also lie on the plane defined by $\mathbf{u}$ and $\mathbf{v}$, ... i.e. $\mathbf{w}$, $\mathbf{u}$ and $\mathbf{v}$ are co-planar.

9 If P divides AB in the ratio $m:n$ then $\mathbf{p} = \dfrac{m\mathbf{b} + n\mathbf{a}}{m + n}$.

10 $\mathbf{i} = \begin{pmatrix} 1 \\ 0 \\ 0 \end{pmatrix}, \mathbf{j} = \begin{pmatrix} 0 \\ 1 \\ 0 \end{pmatrix}, \mathbf{k} = \begin{pmatrix} 0 \\ 0 \\ 1 \end{pmatrix}$ ... are often referred to as the **basis vectors**.

11 The scalar product is defined by $\mathbf{a}.\mathbf{b} = |\mathbf{a}||\mathbf{b}| \cos\theta°$, $\mathbf{a}, \mathbf{b} \neq \mathbf{0}$.

12 In terms of components: $\mathbf{a}.\mathbf{b} = a_1b_1 + a_2b_2 + a_3b_3$.

13 $\cos\theta° = \dfrac{\mathbf{a}.\mathbf{b}}{|\mathbf{a}||\mathbf{b}|}$

5 Quadratic theory

Before we start...

The history of the quadratic equation can be traced back to the times of Ancient Egypt.

However, it was **Brahmagupta** (c 660 AD in India) who published a partial solution.

Al-Khwarismi (c 820 AD in Baghdad ... see stamp) developed this solution, but since he didn't work with negative numbers, he considered there to be 6 different types of quadratic equation.

A complete solution was published by **Bhaskara** around 1100 AD in India.

Each of these solutions depended heavily on verbal description and it wasn't until **Cardano** (c 1545) and **Descartes** (1637) that the algebraic formula we know now emerged.

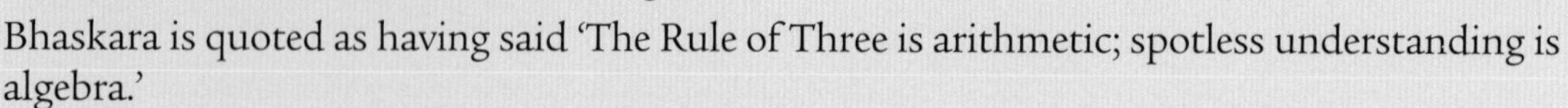

Bhaskara is quoted as having said 'The Rule of Three is arithmetic; spotless understanding is algebra.'

The Rule of Three is an old technique for solving direct proportion problems, learned by students without an understanding of why it worked.

The algebra behind the modern formula depends on the use of the technique 'completing the square'. Read the proof and go for 'spotless understanding'.

$$ax^2 + bx + c = 0, a \neq 0$$

$$\Rightarrow \quad a\left(x^2 + \frac{b}{a}x\right) = -c$$ ... we need the coefficient of x^2 to be 1

$$\Rightarrow \quad a\left(\left(x + \frac{b}{2a}\right)^2 - \left(\frac{b}{2a}\right)^2\right) = -c$$ ... completing the square

$$\Rightarrow \quad a\left(x + \frac{b}{2a}\right)^2 - \frac{b^2}{4a} = -c$$ ... first steps towards making x the subject of the formula

$$\Rightarrow \quad a\left(x + \frac{b}{2a}\right)^2 = \frac{b^2}{4a} - c = \frac{b^2 - 4ac}{4a}$$

$$\Rightarrow \quad \left(x + \frac{b}{2a}\right)^2 = \frac{b^2 - 4ac}{4a^2}$$

$$\Rightarrow \quad x + \frac{b}{2a} = \pm\sqrt{\frac{b^2 - 4ac}{4a^2}} = \pm\frac{\sqrt{b^2 - 4ac}}{2a}$$ ... the square root function only produces a positive answer

$$\Rightarrow \quad x = -\frac{b}{2a} \pm \frac{\sqrt{b^2 - 4ac}}{2a}$$

$$\Rightarrow \quad x = \frac{-b \pm \sqrt{b^2 - 4ac}}{2a}, a \neq 0$$

Consider this: The formula doesn't hold when $a = 0$. Under what other conditions might there be a problem applying the formula?

What you need to know

1 Solve by factorising the quadratic expression.

a $x^2 + x - 12 = 0$ **b** $2x^2 - 3x - 2 = 0$

2 Find the solution to each equation by first completing the square.

a $x^2 + 2x - 35 = 0$ **b** $6x^2 + 7x - 3 = 0$

3 Find the solutions to each equation correct to 1 decimal place.

a $x^2 + 2x - 7 = 0$ **b** $3x^2 + x - 7 = 0$

4 This is a sketch of a quadratic function of the form $y = (x + a)^2 + b$.

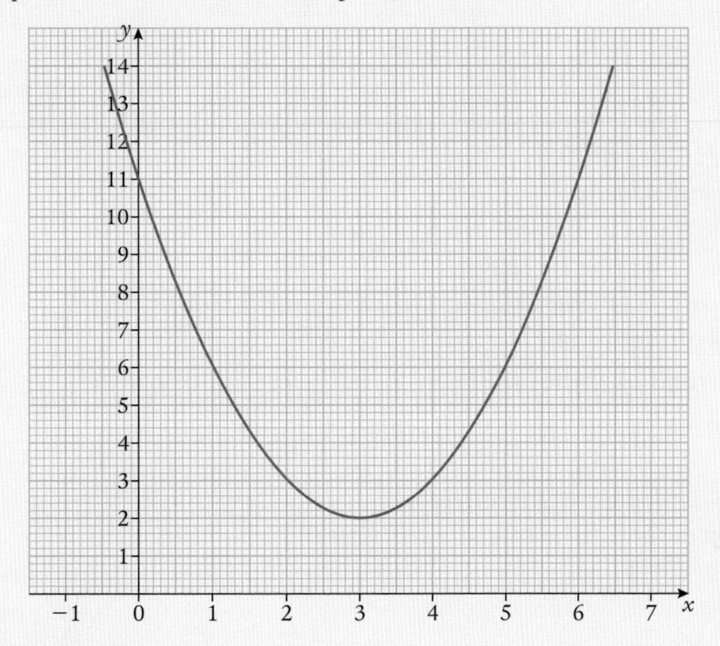

What are the values of a and b?

5 By first completing the square make sketches of the following curves, indicating the coordinates of the turning point and any intersections with the axes.

a $y = x^2 - 6x + 8$ **b** $y = 15 - 2x - x^2$

6 Using suitable units and axes, the inner arch of the Tyne Bridge can be modelled by

$$y = \tfrac{3}{2}(32 - 4x - x^2)$$

where y units is the height above the road x units from the origin.
Find the values of x and y that correspond to:

a the highest point on the arch

b the points where the arch meets the road.

5.1 Inequations

Although not required by the course, we will use a couple of terms for the sake of convenience.

A curve shaped like this: we will call '**concave up**'.

Any curve with equation $y = ax^2 + bx + c$, $a > 0$ is concave up.

It will be a parabola with a minimum turning point.

A curve shaped like this: we will call '**concave down**'.

Any curve with equation $y = ax^2 + bx + c$, $a < 0$ is concave down.

It will be a parabola with a maximum turning point.

Example 1

For what values of x is $x^2 - x - 12 \geqslant 0$?

- Solve the corresponding equation:
  $x^2 - x - 12 = 0 \Rightarrow (x - 4)(x + 3) = 0 \Rightarrow x = 4$ or -3.
- Determine the nature of the turning point:
  Since the coefficient of x^2 is **greater** than zero, the curve is concave **up**.
  We have a minimum turning point.
- Determine the y-intercept ... i.e. when $x = 0$:
  $y = 0^2 - 0 - 12 = -12$.
- Sketch the curve.
- Identify the parts of the domain which correspond to the curve being on or above the x-axis. [$x^2 - x - 12 \geqslant 0$]
  In this case $x \leqslant -3$ or $x \geqslant 4$.

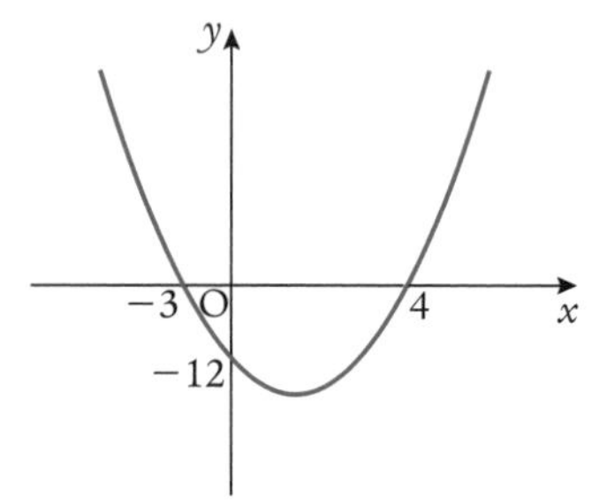

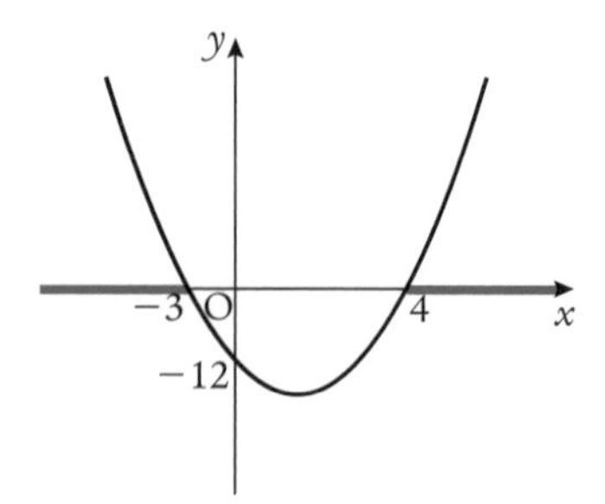

So $x^2 - x - 12 \geqslant 0$ when $x \leqslant -3$ or $x \geqslant 4$.

Example 2

Solve the inequation $5 - 9x - 2x^2 > 0$.

- $(1 - 2x)(5 + x) = 0 \Rightarrow x = \frac{1}{2}$ or -5
- Coefficient of $x^2 < 0$ ... so the curve is concave down and has a maximum turning point.
- Cuts y-axis at $y = 5$.
- A quick sketch identifies the part of the domain where $y > 0$.

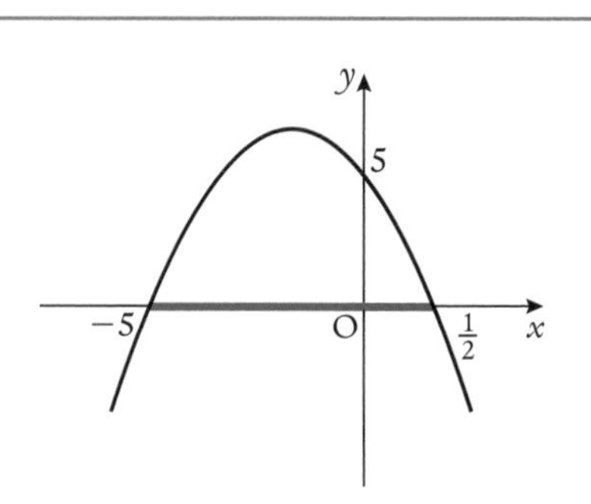

So $5 - 9x - 2x^2 > 0$ when $-5 < x < \frac{1}{2}$.

Exercise 5.1

1 Solve each inequation.

a $x^2 - 5x + 6 \geqslant 0$ b $x^2 - 3x - 4 < 0$ c $1 - x^2 > 0$

d $10 - 3x - x^2 \leqslant 0$ e $x^2 + 11x + 30 \leqslant 0$ f $x^2 + 6x - 7 > 0$

g $8 + 2x - x^2 < 0$ h $-9 - 6x - x^2 \geqslant 0$ i $2x^2 + x - 1 \leqslant 0$

j $6x^2 - 7x + 2 < 0$ k $-2 - 7x - 3x^2 > 0$ l $2 + x - 10x^2 < 0$

m $20x^2 + 12x > 0$ n $-1 + 8x - 7x^2 \leqslant 0$ o $(x + 4)^2 \leqslant 0$

p $x^2 < 4$

2 Re-arrange into standard quadratic form and then solve.

a $x^2 < x + 6$ b $x^2 \geqslant 2x + 24$ c $8 - x^2 \leqslant -7x$

d $5x - 2x^2 > 3$ e $2x^2 - x \leqslant 21$ f $3x^2 > 8x + 16$

3 I'm looking for two numbers whose sum is 30 and whose product is less than 200.

a If one of the numbers is represented by x, find an expression in x for the other.

b Find an expression in x for the product.

c Form an inequation and solve it to find the possible values of x.

d Repeat this process when the sum of the two numbers is 17 and the product must be greater than 60.

4 A gardener has been given 68 m of flexible fencing. He has been told to cordon off a rectangular piece of the garden that is at least 240 m^2 in area.

a Let x m represent the length of the rectangle. Find an expression in x for the breadth of the rectangle.

b Form an inequation and solve it to find all possible suitable values for the length.

5 The lengths of the sides of two rectangles depend on the value of some number x as shown.

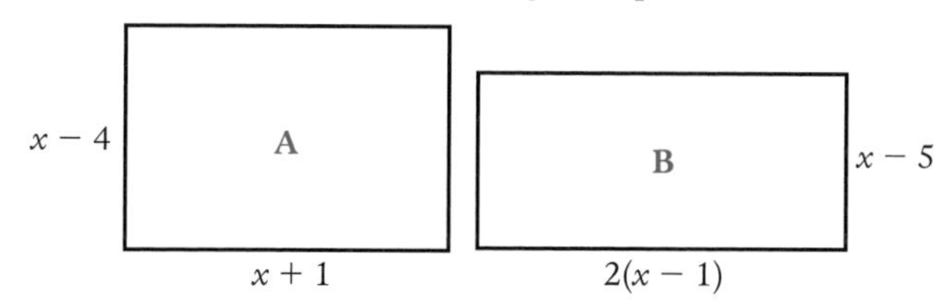

a Write down an expression in terms of x for each area.

b What size restriction must be placed on x if the rectangles are to exist?

c Rectangle A has a greater area than rectangle B.

i Write down an inequation in x.

ii Solve it to find all possible values for x.

6 The side of a bridge is based on two-and-a-half congruent rhombuses as shown. One diagonal of the rhombus is 6 metres longer than the other.

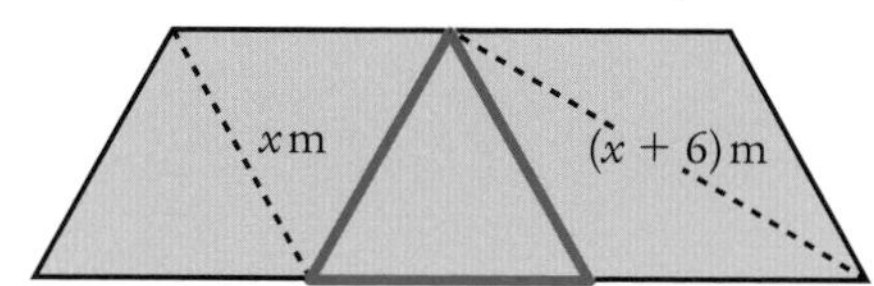

What restriction must be placed on the length of the shorter diagonal if the area of the side of the bridge has to be no more than 2750 m^2?

[Area of a rhombus $= \frac{1}{2}(d_1 \times d_2)$, where d_1 and d_2 are the diagonals.]

5.2 Nature of the roots of a quadratic equation

If $ax^2 + bx + c = 0$,

then $x = \dfrac{-b + \sqrt{b^2 - 4ac}}{2a}$ and $x = \dfrac{-b - \sqrt{b^2 - 4ac}}{2a}$ are the solutions or roots of the equation.

Note that $a \neq 0$ as division by zero is undefined.

Note also that the solutions depend on the expression within the square root:

- if $b^2 - 4ac < 0$, we can't take the square root and so **no real roots exist**
- if $b^2 - 4ac = 0$, both roots are the same, viz. $x = \dfrac{-b}{2a}$ ... we call this **two coincident roots**
- if $b^2 - 4ac > 0$, we get **two distinct real roots**.

It is of interest that if $b^2 - 4ac$ is a **perfect square** then the two distinct roots will be **rational**.

Because this expression can be used to quickly discriminate between these different cases, it is given the name 'discriminant' and is often denoted by Δ (**delta**).

$$\Delta = b^2 - 4ac$$

We use the discriminant to determine the nature of the roots of an equation.

When $a > 0$

$\Delta < 0$ ... no real roots

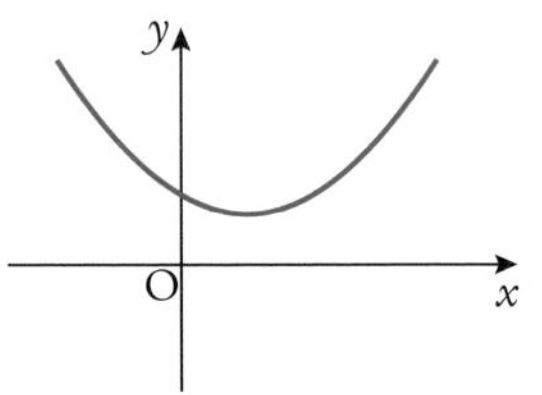

$\Delta = 0$ ... coincident roots

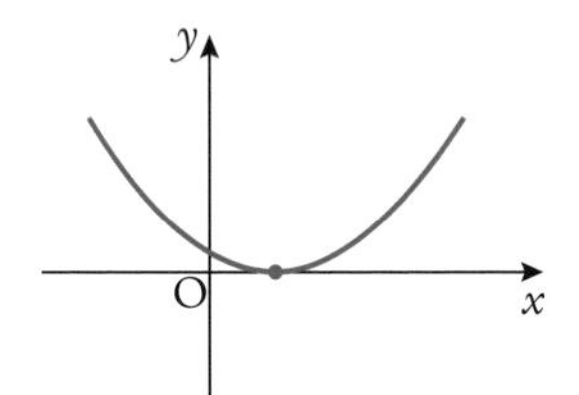

$\Delta > 0$ ... two distinct real roots

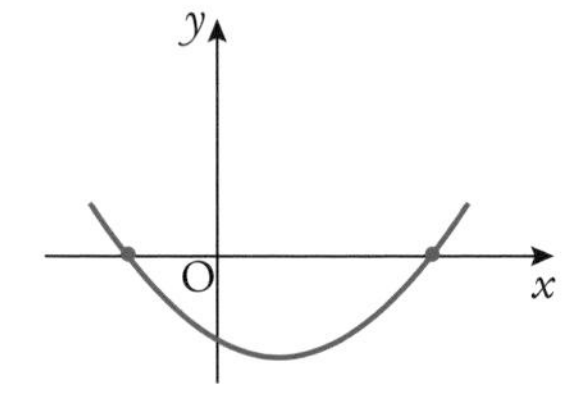

When $a < 0$

$\Delta < 0$ ... no real roots

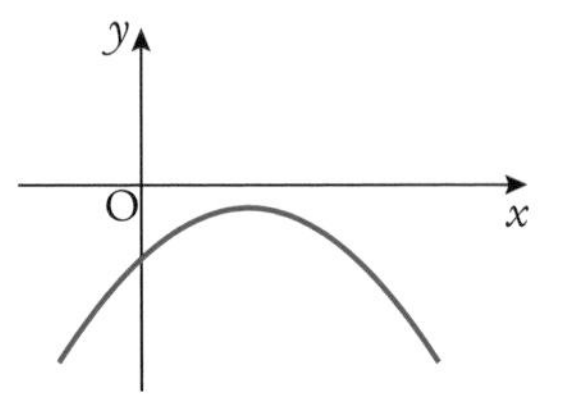

$\Delta = 0$ ... coincident roots

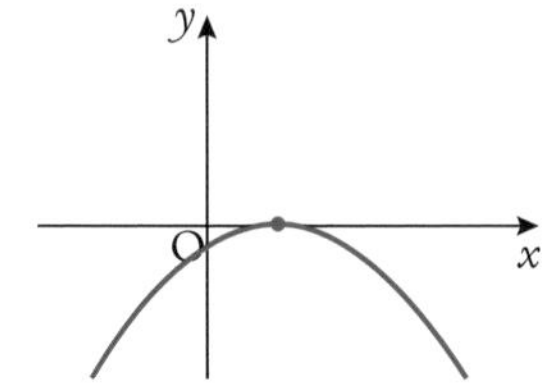

$\Delta > 0$ ... two distinct real roots

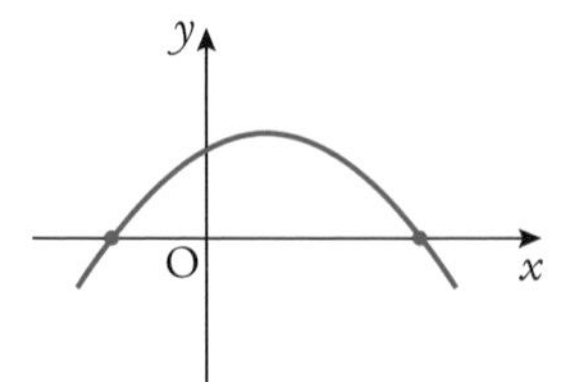

Example 1

Determine the nature of the roots of each equation.

a $3x^2 + 2x - 1 = 0$ **b** $4x^2 - 12x + 9 = 0$ **c** $2x^2 + x + 5 = 0$

a Comparing the equation with the standard quadratic form
$a = 3, b = 2, c = -1 \Rightarrow b^2 - 4ac = 2^2 - 4.3.(-1) = 16$
$b^2 - 4ac > 0 \Rightarrow$ two distinct real roots.

b Comparing the equation with the standard quadratic form
$a = 4, b = -12, c = 9 \Rightarrow b^2 - 4ac = (-12)^2 - 4.4.9 = 0$
$b^2 - 4ac = 0 \Rightarrow$ two coincident roots.

c Comparing the equation with the standard quadratic form
$a = 2, b = 1, c = 5 \Rightarrow b^2 - 4ac = 1^2 - 4.2.5 = -39$
$b^2 - 4ac < 0 \Rightarrow$ no real roots.

Example 2

For what value of k does the equation $kx^2 + 2x + 3 = 0$ have no real roots?

Comparing the equation with $ax^2 + bx + c = 0$,

$a = k, b = 2, c = 3 \Rightarrow b^2 - 4ac = 2^2 - 4\,.\,k\,.\,3 = 4 - 12k.$

For no real roots,

$$b^2 - 4ac < 0 \Rightarrow 4 - 12k < 0$$
$$\Rightarrow k > \tfrac{1}{3}.$$

Example 3

The line $y = mx + 1$ and the curve $y = x^2 + 5x + 2$ will intersect when $x^2 + 5x + 2 = mx + 1$.
For what values of m is the line a tangent to the curve?

The line will be a tangent when there is only one point of intersection,
i.e. when $x^2 + 5x + 2 = mx + 1$ has coincident roots.

$$x^2 + 5x + 2 = mx + 1$$
$$\Rightarrow x^2 + (5 - m)x + 1 = 0$$
$$\Rightarrow \quad b^2 - 4ac = (5 - m)^2 - 4\,.\,1\,.\,1 = (5 - m)^2 - 4$$

For coincident roots, $b^2 - 4ac = 0 \Rightarrow (5 - m)^2 - 4 = 0$

$$\Rightarrow (5 - m)^2 = 4$$
$$\Rightarrow \quad 5 - m = \pm 2$$
$$\Rightarrow \quad m = 5 \pm 2$$
$$\Rightarrow \quad m = 7 \text{ or } 3.$$

So $y = 3x + 1$ and $y = 7x + 1$ are tangents to $y = x^2 + 5x + 2$.

Exercise 5.2

1 Determine the nature of the roots of each equation.

a $2x^2 + 3x - 1 = 0$ **b** $3x^2 + 6x + 3 = 0$ **c** $5x^2 - 2x + 2 = 0$

d $4x^2 + 12x + 9 = 0$ **e** $3x^2 + 1 = 0$ **f** $4 - 5x - 2x^2 = 0$

g $-1 - 3x - x^2 = 0$ **h** $-1 + 6x - 9x^2 = 0$ **i** $-4 - 3x - 2x^2 = 0$

2 Express each equation in the form $ax^2 + bx + c = 0$, then determine the nature of the roots.

a $5x^2 = 1 - 7x$ **b** $3x^2 = 2x - 4$ **c** $x(x + 8) = -5$

d $7x(7x - 2) = -1$ **e** $x^2 + 2x + 4 = 1 + x - x^2$ **f** $x = \dfrac{20x - 4}{25x}$

g $x = \sqrt{\dfrac{7 - x}{2}}$ **h** $\dfrac{4x^2}{2x + 3} = 1$ **i** $x(x + 1) = -1$

3 **a** Prove that the equation $3x^2 + bx - 1 = 0$ has distinct real roots for all values of b.

b For what values of k will the equation $x^2 + (2k - 1)x + k^2 + 1 = 0$ have no real roots?

c For what values of m does the equation $3x^2 + mx + 3 = 0$ have coincident roots?

4 A rectangle has a perimeter of 40 units.

a If the length of the rectangle is x units, find an expression in x for:
i the breadth **ii** the area of the rectangle ... call it $f(x)$.

b If the value of the area, A units^2, is given, x can be calculated.
For what values of A will the equation $f(x) = A$ have distinct real roots?

c For which whole number values of A will the roots be rational?

5 The arch on the underside of the Forth Bridge has been modelled by

$$y = \frac{x}{900}(400 - x).$$

The rail track is modelled by $y = k$ and x and y are measured in metres.

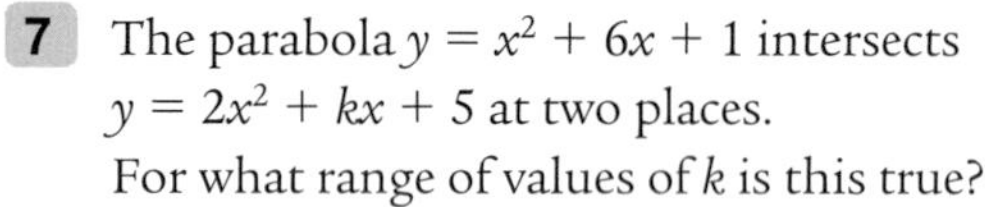

a Form an equation to find where the curve and line intersect.

b Express it in the form $ax^2 + bx + c = 0$.

c For what values of k will this equation have two distinct roots?

d The rail track is a tangent to the curve.

i What is the value of k? ii What is the value of x at this point?

6 a A line of the form $y = mx + 5$ is a tangent to the curve $y = x^2 + 4x + 6$.
For what values of m is this true?

b For what value of c is the line $y = 7x + c$ a tangent to $y = x^2 + x + 12$?

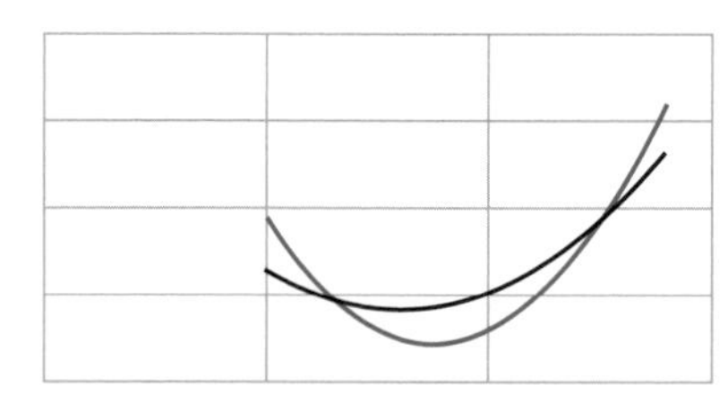

7 The parabola $y = x^2 + 6x + 1$ intersects $y = 2x^2 + kx + 5$ at two places.
For what range of values of k is this true?

8 Using suitable axes and units the dancing waters have been modelled by the two parabola
$y = 10 - 4x - x^2$ and $y = 9 - kx - 2x^2, k \geqslant 0$.
For what values of k will the jets of water intersect at two places?

5.3 Modelling

Consider reversing the process of solving a quadratic equation.

If the roots exist and are $x = p$ and $x = q$,

then $x - p = 0$ or $x - q = 0$

and $(x - p)(x - q) = 0$

and before that ... $r(x - p)(x - q) = 0$ ... to cover all possibilities.

So, if we know the roots of the equation and one other piece of information that lets us work out r, then we can work out the equation of the parabola, viz. $y = r(x - p)(x - q)$.

Example 1

The roots of a quadratic function are -2 and 3. The curve cuts the y-axis at (0, 12).

What is the equation of the parabola expressed in the form $y = ax^2 + bx + c$?

Since the roots are -2 and 3, the equation takes the form $y = r(x + 2)(x - 3)$.

Substituting $x = 0$ and $y = 12$ into the equation:

$12 = r(0 + 2)(0 - 3)$

$\Rightarrow 12 = -6r$

$\Rightarrow \quad r = -2$

$\Rightarrow y = -2(x + 2)(x - 3) = -2(x^2 - x - 6)$

$\Rightarrow y = 12 + 2x - 2x^2$.

Example 2

The parabola is a common feature in bridges.

In this example the arch goes between two piers which are 160 m apart.

The arch rises to a height of 55 m.

Using a set of axes where the origin is set as shown, 20 m from the first pier, model the arch by finding the parabola which fits the data.

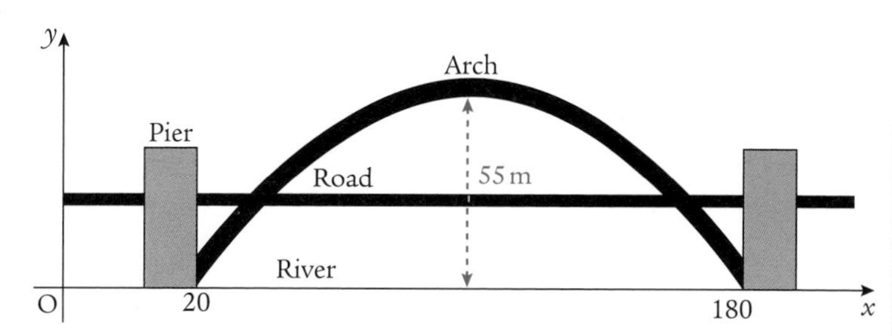

The roots of the equation will be 20 and $160 + 20 = 180$.

So the equation takes the form $y = r(x - 20)(x - 180)$, where x is the distance in metres from the point of origin and y metres is the height of the arch at that point.

Halfway between the roots, at $x = 100$, we know that $y = 55$.

Substituting into the equation:

$55 = r(100 - 20)(100 - 180) = -6400r$

$\Rightarrow r = -\frac{55}{6400} = -\frac{11}{1280}$.

The arch can be modelled by $y = -\frac{11}{1280}(x - 20)(x - 180) = -\frac{11}{1280}(x^2 - 200x + 3600)$.

Exercise 5.3

1 Find the equation of the parabola that fits each set of data.

a The roots are -1 and 5 and it cuts the y-axis at $(0, -4)$.

b The roots are 2 and 7 and it passes through (1, 4).

c The roots are -4 and -2 and its maximum value is 8.

d The roots are 0 and 8 and its minimum value is -5.

2 The roots of a quadratic equation are a and b. The coefficient of x^2 is 1.

a What is the relation between the roots and

i the coefficient of the x term

ii the constant?

b Write down a quadratic equation whose roots sum to 7 and multiply to make 6.

c Write down a quadratic equation whose roots sum to 16, multiply to make 48 and whose maximum value is 32.

3 The Clyde Arc, also known as the squinty bridge, can be modelled by a parabola. Using data found on the web we can sketch the bridge.

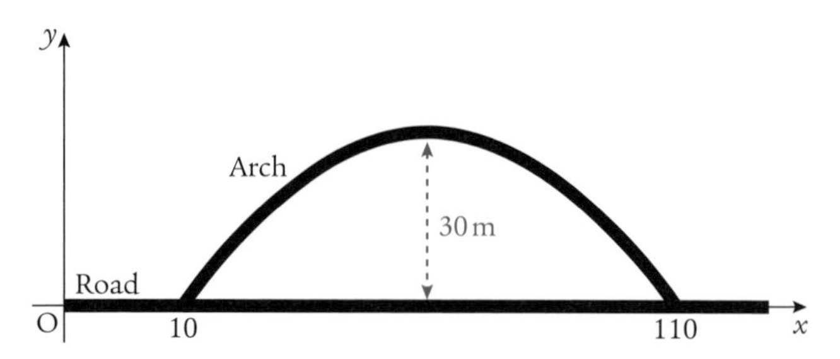

a Find a parabola to model the arc.

b If the arc were continued, where would it cut the y-axis?

4 At the golf school Mary practised her chipping as she was being videoed.
Two seconds after the filming started, she struck the ball.
It remained in the air for 4 seconds before it came to earth.
It reached a maximum height of 20 m during its flight.

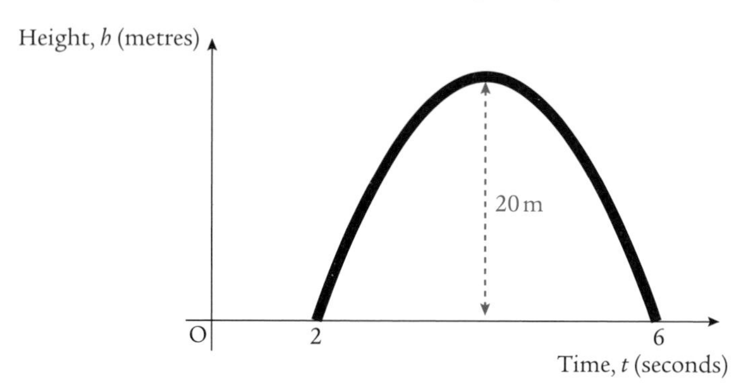

a Find a quadratic model to describe the flight.

b At what times was the ball more than 15 m high?

Preparation for assessment

1 Solve:

a $2x^2 - 17x + 35 \leqslant 0$ b $3 - 14x - 5x^2 < 0$.

2 A rectangle has a perimeter of 40 m.

a If the length is x m, find an expression for the area of the rectangle in terms of x.

b For what values of x is the area greater than 96 m^2?

3 Determine the nature of the roots of each quadratic equation.

a $5x^2 + 17x + 12 = 0$ b $2x^2 + 3x + 5 = 0$ c $9x^2 - 12x + 4 = 0$

4 For what values of n does the quadratic equation $2(n + 1)x^2 + 24x + n$ have no real roots?

5 a Find a parabola whose roots are 7 and -3 and which passes through $(1, 2)$.

b Find the turning point of the parabola and state whether it is a maximum or minimum.

Summary

1 $ax^2 + bx + c = 0 \Rightarrow x = \dfrac{-b \pm \sqrt{b^2 - 4ac}}{2a}, a \neq 0$

2 Any curve with equation $y = ax^2 + bx + c$, $a > 0$ is concave up.
It will be a parabola with a minimum turning point.
Any curve with equation $y = ax^2 + bx + c$, $a < 0$ is concave down.
It will be a parabola with a maximum turning point.

3
- Making use of points **1** and **2** we can quickly sketch the parabola.
- Using the sketch we can identify the parts of the domain which correspond to:
  i the curve below the x-axis ... for $y < 0$
  ii the curve above the x-axis ... for $y > 0$.

4 The expression $\Delta = b^2 - 4ac$ is called the discriminant:
- if $b^2 - 4ac < 0$, the equation has **no real roots**
- if $b^2 - 4ac = 0$, both roots are the same, viz. $x = \dfrac{-b}{2a}$ ... **coincident roots**
- if $b^2 - 4ac > 0$, the equation has **two distinct real roots**
- if $b^2 - 4ac$ is a **perfect square** then the two distinct roots will be **rational**.

5 If the roots of a quadratic equation are $x = p$ and $x = q$, the equation is of the form $y = r(x - p)(x - q)$.
We find r by finding one other point, (x_1, y_1), on the curve and substituting to get:
$r = \dfrac{y_1}{(x_1 - p)(x_1 - q)}$.

6 Polynomials

Before we start...

'Awake! For Morning in the Bowl of Night
Has flung the Stone that puts the Stars to Flight.'

The Rubaiyat of Omar Khayyam

Omar Khayyam (1048–1131 AD) was a mathematician, astronomer and poet in Persia[1].

He is remembered more for his famous poem, **The Rubaiyat**, than for his mathematics.

Importantly, he was one of the first mathematicians to describe a solution to the cubic equation[2].

His methods were completely geometric and based on identifying where a particular parabola and circle intersected.

Algebraic solutions to this problem were not forthcoming till the 16th century with **Tartaglia** and **Cardan**.

The dispute between these two about who had the rights to publish a general solution has become notorious.

Cubic functions can be used to model a variety of situations. We are just going to use one to illustrate the dangers of making assumptions or drawing conclusions based on too little data.

You have probably seen puzzles in magazines giving you three terms of a sequence and asking you to supply the fourth, e.g. 'A sequence starts 5, 9, 13, ..., what is the next term?'

Most people would reply 17, assuming that the nth term was $u_n = 4n + 1$.

However, equally valid would be the cubic expression $u_n = n^3 - 6n^2 + 15n - 5$, which, if you let $n = 1, 2, 3, 4$ produces the sequence 5, 9, 13, 23, ...

How are these two expressions, the linear and the cubic, related?

If we separate out the linear expression, which was proposed first, from the cubic one we get:

$$u_n = (n^3 - 6n^2 + 11n - 6) + (4n + 1).$$

When you know how to, you can factorise the cubic term to get

$$u_n = (n - 1)(n - 2)(n - 3) + (4n + 1).$$

Now letting $n = 1, 2, 3$, we see that the cubic term $= 0$ in each case, leaving the linear expression to generate the expected terms. When $n = 4$ ... it's a different story.

If, given the first three terms of a sequence, you think the nth term is generated by $u_n = f(n)$, then it could also be generated by $u_n = (n - 1)(n - 2)(n - 3) + f(n)$.

1, 1, 1, ... What is the next term?

Douglas Adams[3] would have been pleased to know that using the formula $u_n = \frac{41}{6}(n - 1)(n - 2)(n - 3) + 1$ the answer is 42!

Modelling is powerful.

Spotting patterns can be dangerous.

Making assumptions can be fatal.

Handling polynomials is important.

---

[1] Modern-day Iran.

[2] An equation of the form $ax^3 + bx^2 + cx + d = 0$, where a, b, c, d are constants; $a \neq 0$.

[3] Author of *The Hitchhiker's Guide to the Galaxy*.

What you need to know

1 Completely factorise:

a $x^2 - 2x - 15$ b $2x^2 + 5x - 12$

c $6x^2 - 23x + 20$ d $4x^2 - 18x - 10$.

2 The graph shows $y = f(x)$ near the origin.

$f(x)$ is a quadratic function.

Find the equation of the function.

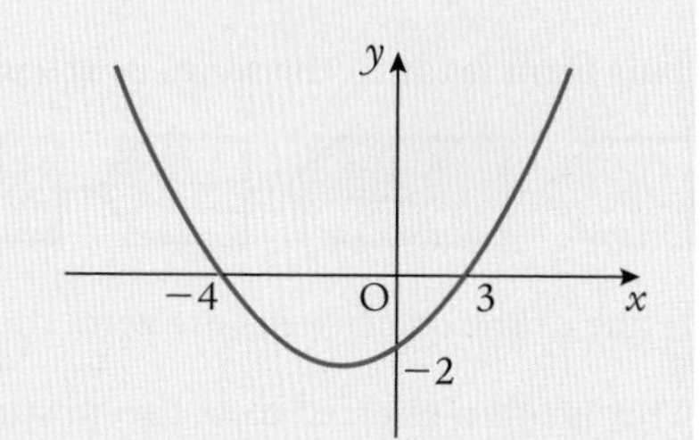

3 Expand:

a $(2x + 3)(3x - 1)$ b $2x(x + 1)(x + 2)$

c $(x + 1)(x^2 + 2x - 3)$ d $(x + 1)(x + 2)(x + 3)$.

4 The function $f(x) = 5x^2 + 15x + 10$ is given.

a Factorise $f(x)$ completely.

b Hence, evaluate: i $f(3)$ ii $f(4)$ iii $f(7)$ iv $f(-1)$.

c By studying the factors say how you know $f(x)$ will be even if x is an integer.

5 Here are the graphs of three functions close to the origin.

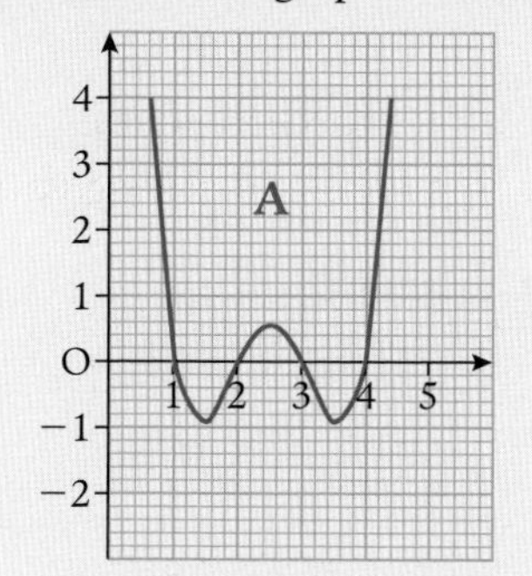

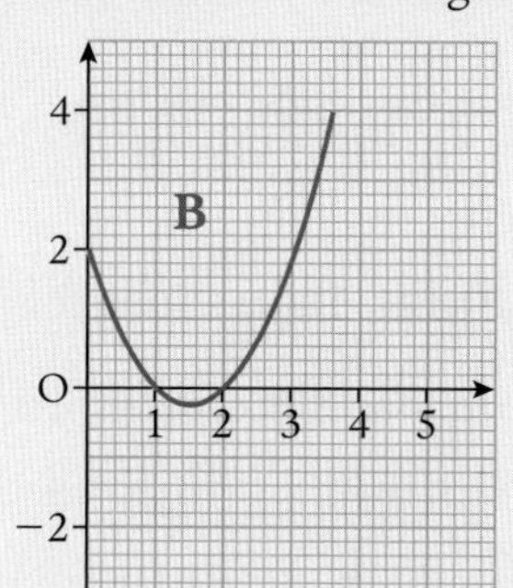

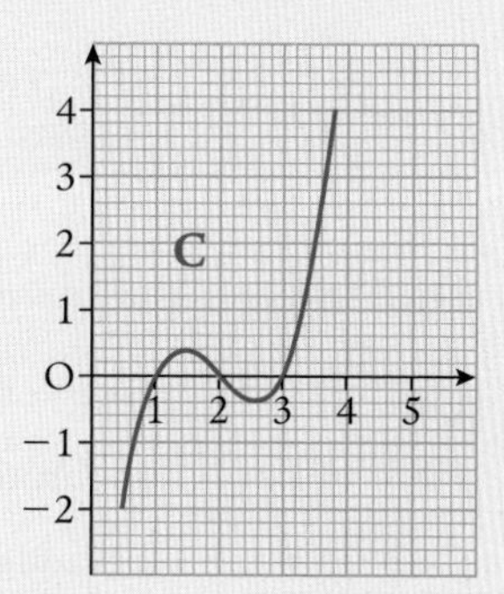

$f(x) = (x - 1)(x - 2)$, $g(x) = (x - 1)(x - 2)(x - 3)$ and $h(x) = (x - 1)(x - 2)(x - 3)(x - 4)$

Which graph goes with which function? Give a reason for your decision.

6.1 Polynomial expressions

Definition

Any expression that takes the form $a_nx^n + a_{n-1}x^{n-1} + a_{n-2}x^{n-2} + \dots + a_2x^2 + a_1x + a_0$, where a_n to a_0 are constants and $a_n \neq 0$ is called a polynomial expression of degree n.

n is a whole number. Any, or all, of a_{n-1} to a_0 could be zero.

3	is a polynomial of degree zero [a constant]
$2x + 1$	is a polynomial of degree one [a linear expression]
$x^2 + 3x - 4$	is a polynomial of degree two [a quadratic expression]
$2x^3 + x^2 - x + 5$	is a polynomial of degree three [a cubic expression]
$x^4 - 3x^3 + 2x^2 + x + 1$	is a polynomial of degree four [a quartic expression].

Evaluation

Finding the value of a cubic at a particular value of x can be quite laborious even with a calculator.

We can create a relatively simple **algorithm**[4] to streamline this work.

The cubic expression $a_3x^3 + a_2x^2 + a_1x + a_0$ can be re-written as $((a_3x + a_2)x + a_1)x + a_0$.

This form leads to the series of steps:

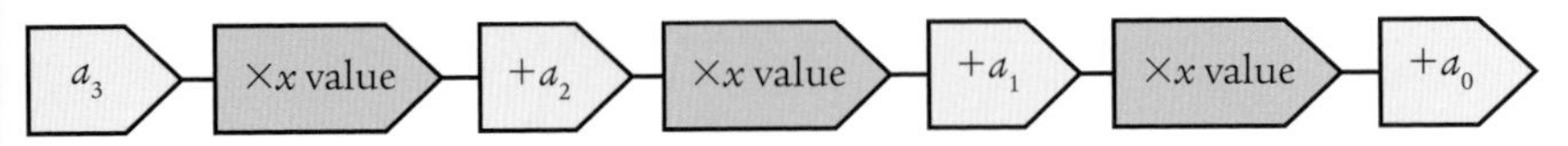

If any of the coefficients are zero, they must still be used.

This is often referred to as the 'nesting' algorithm.

Example 1

Given that $f(x) = 3x^3 + 2x^2 - 4x + 1$, calculate:

a $f(5)$ **b** $f(-1)$.

a Because of the programming of most scientific calculators, '=' must be pushed after each input:

The sequence of buttons to push is: $3 = \times 5 = + 2 = \times 5 = - 4 = \times 5 = + 1 = 406$.

So $f(5) = 406$.

b The sequence of buttons to push is: $3 = \times (-1) = + 2 = \times (-1) = -4 = \times (-1) = + 1 = 4$.

So $f(-1) = 4$.

For future work it is useful to record the intermediate steps. The working for part **a** becoming:

5	3	2	−4	1	... list the x-value followed by the coefficients
		15	85	405	... 3 is brought down then ×5 to give 15
	3	17	81	406	... 2 is added to give 17 then ×5 to give 85
					... −4 is added to give 81 then ×5 to give 405
					... 1 is added to give 406 = $f(5)$

Example 2

Evaluate $2x^3 - 3x^2 - 1$ when $x = -2$.

−2	2	−3	0	−1	... note that even the 0 coefficient is listed
		−4	14	−28	... 2 is brought down then ×(−2) to give −4
	2	−7	14	−29	... −3 is added to give −7 then ×(−2) to give 14
					... 0 is added to give 14 then ×(−2) to give −28
					... −1 is added to give −29 ... **required value**

When $x = -2$, $2x^3 - 3x^2 - 1 = -29$.

---

[4] Algorithm: a step-by-step method of achieving a result.

Exercise 6.1

1 Evaluate each of the functions, $f(x)$, at the values shown using the nesting algorithm.

a	$x^3 + 5x^2 + 2x - 8$;	i $f(0)$	ii $f(3)$
b	$x^3 - 9x^2 + 24x - 16$;	i $f(0)$	ii $f(4)$
c	$x^3 - 2x^2 - 5x + 6$;	i $f(2)$	ii $f(4)$
d	$x^3 - 2x^2 - x + 2$;	i $f(1)$	ii $f(-2)$
e	$x^3 - 7x + 6$;	i $f(3)$	ii $f(-4)$
f	$x^3 - 3x^2 + 4$;	i $f(-1)$	ii $f(2)$
g	$2x^3 + 7x^2 + 6x$;	i $f(-1)$	ii $f(-4)$
h	$3x^3 - 11x^2 + 5x + 3$;	i $f(1)$	ii $f(2)$
i	$4x^3 + 3x^2 - 25x + 6$;	i $f(-3)$	ii $f(1)$
j	$2x^3 - 2x + 1$;	i $f(-1)$	ii $f(1)$

2 Use the structured layout to establish the value of the function, $f(x)$, while showing intermediate working.

a	$x^3 - 4x^2 + x + 6$;	i $f(1)$	ii $f(-2)$
b	$x^3 - 2x^2 - 16x + 32$;	i $f(3)$	ii $f(-1)$
c	$x^3 - 6x^2 + 5x + 12$;	i $f(-1)$	ii $f(3)$
d	$2x^3 - x^2 - 8x + 4$;	i $f(-1)$	ii $f(1)$
e	$4x^3 + x^2 - 36x - 9$;	i $f(-2)$	ii $f(\frac{1}{2})$

3 $f(x) = x^3 + 3x^2 - x - 2$ and $g(x) = 2x^3 + x^2 - 3x + 10$.

a Show that the curves intersect when $x = -2$.

b Which is greater, $f(3)$ or $g(3)$?

4 $f(x) = 3x^3 - ax^2 - 3ax + 4a$, where a is a constant.

a Use the structured layout to find an expression in a for $f(2)$.

b For what value of a is $f(2) = 0$?

5 A box is made from a rectangle by cutting out squares from each corner and folding.
The squares have a side of x units.
The volume of the box is a function of x:
$V(x) = 4x^3 - 30x^2 + 68x - 42$, where $1 < x < 3$.

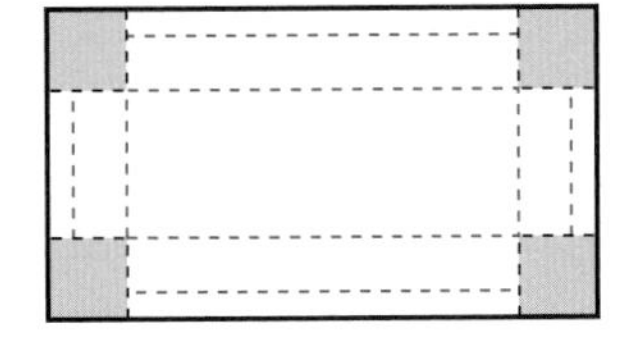

a Evaluate:
i $V(1)$ ii $V(3)$ and comment.

b Which is greater $V(1{\cdot}5)$ or $V(2{\cdot}5)$?

6.2 Algebraic division

Consider the division $27 \div 4 = 6$ remainder 3. Each part has a name.

27 is the dividend; 4 is the divisor; 6 is the quotient; 3 is the remainder.

With a larger dividend we work from left to right recording remainders (in purple) as we go, e.g.

$3458 \div 4$

4	3	4	$^{2}5$	$^{1}8$	r 2
		8	6	4	

For our purposes this is better laid out in a 'long division' format.

$$
\begin{array}{r|rrrrl}
\multicolumn{1}{r}{} & & 8 & 6 & 4 & \text{r}\,2 \\ \cline{2-5}
4 & 3 & 4 & 5 & 8 & \\
 & -3 & 2 & & & \\ \cline{2-3}
 & & 2 & 5 & 8 & \\
 & & -2 & 4 & & \\ \cline{3-5}
 & & & 1 & 8 & \\
 & & & -1 & 6 & \\ \cline{4-5}
 & & & & 2 &
\end{array}
$$

... ① we need to multiply 4 by 8 to get close to 34

... ② subtract $4 \times 8 = 32$ to find remainder

... ③ bring down the rest of the dividend

... repeat steps ① to ③ till you exhaust the dividend

We can interpret the result as: $4 \times 864 + 2 = 3458$.

A similar approach can be taken with algebraic expressions.

The method is best expressed by example.

Example 1

Divide $2x^3 - 9x^2 + 10x - 5$ by $x - 1$.

$$
\begin{array}{r|l}
\multicolumn{1}{r}{} & 2x^2 - 7x + 3 \quad \text{r} -2 \\ \cline{2-2}
x - 1 & 2x^3 - 9x^2 + 10x - 5 \\
 & \underline{2x^3 - 2x^2} \\
 & \quad\; -7x^2 + 10x - 5 \\
 & \quad\; \underline{-7x^2 + 7x} \\
 & \qquad\qquad\; 3x - 5 \\
 & \qquad\qquad\; \underline{3x - 3} \\
 & \qquad\qquad\quad\;\; -2
\end{array}
$$

... ① we need to multiply $x - 1$ by $2x^2$ to get something that starts with $2x^3$ namely $2x^3 - 2x^2$

... ② **subtract** $2x^3 - 2x^2$ to find remainder of $-7x^2$

... ③ bring down the rest of the dividend

... repeat steps ① to ③ multiplying by whatever will lead to the elimination of the leading term till you exhaust the dividend

We can interpret the result as: $(x - 1)(2x^2 - 7x + 3) - 2 = 2x^3 - 9x^2 + 10x - 5$.

An alternative method

Consider the technique of evaluating $2x^3 - 9x^2 + 10x - 5$ when $x = 1$.

$$
\begin{array}{r|rrrr}
1 & 2 & -9 & 10 & -5 \\
 & & 2 & -7 & 3 \\ \hline
 & 2 & -7 & 3 & -2
\end{array}
$$

Note that the bottom line gives the coefficients of the quotient and the remainder.

If we wish to divide a polynomial by $x - b$, then we evaluate the polynomial at $x = b$ using the above algorithm and interpret the intermediate working as described.

For this reason, the algorithm is known as 'synthetic division' ... and is useful when the divisor is of the form $x - b$.

For divisors of the form $ax - b$, where $a \neq 1$, 'algebraic' division may be easier.

Example 2

Use synthetic division to divide $3x^3 - 2x^2 + 4$ by $x + 2$.

$$
\begin{array}{r|rrrr}
-2 & 3 & -2 & 0 & 4 \\
 & & -6 & 16 & -32 \\ \hline
 & 3 & -8 & 16 & -28
\end{array}
$$

... consider $x + 2$ as $x - (-2)$

... note the zero coefficient

$(3x^3 - 2x^2 + 4) \div (x + 2) = 3x^2 - 8x + 16$, remainder -28.

Example 3

Express $\dfrac{2x^3 - 3x^2 + x - 5}{2x - 1}$ in the form $(cx^2 + dx + e) + \dfrac{r}{2x - 1}, x \neq \dfrac{1}{2}$.

$$\frac{2x^3 - 3x^2 + x - 5}{2x - 1} = (2x^3 - 3x^2 + x - 5) \div (2x - 1)$$

$$\begin{array}{r|l} & \quad x^2 - x + 0 \quad \text{r} -5 \\ \hline 2x - 1 & 2x^3 - 3x^2 + x - 5 \\ & 2x^3 - x^2 \\ \hline & \quad -2x^2 + x - 5 \\ & \quad -2x^2 + x \\ \hline & \qquad\quad 0 - 5 \\ & \qquad\quad 0 - 0 \\ \hline & \qquad\quad\ -5 \end{array}$$

$$\frac{2x^3 - 3x^2 + x - 5}{2x - 1} = (x^2 - x) - \frac{5}{2x - 1}, x \neq \frac{1}{2}$$

Note: If you wish to use synthetic division here you must consider $2x - 1$ as $2(x - \frac{1}{2})$.
... use synthetic division with $(x - \frac{1}{2})$ ... then divide the quotient by 2.

$\frac{1}{2}$	2	−3	1	−5
		1	−1	0
	2	−2	0	−5
÷2	1	−1	0	−5

... the remainder is unaffected by the division

... giving the same result.

Exercise 6.2

1 Perform each calculation using the 'long division' format, recording the remainder.

a $(x^3 + 9x^2 + 24x + 16) \div (x + 4)$
b $(x^3 + 4x^2 + x - 5) \div (x + 2)$
c $(x^3 - 4x^2 - 9x + 38) \div (x - 4)$
d $(x^3 - x^2 - x - 1) \div (x + 1)$
e $(x^3 - x^2 - 8x + 15) \div (x - 2)$
f $(x^3 - 7x^2 + 16x - 12) \div (x - 3)$
g $(4x^3 - 10x^2 + 8x + 1) \div (x - 1)$
h $(2x^3 + 7x^2 + 8x + 7) \div (x + 1)$
i $(3x^3 - 2x^2 - 17x - 14) \div (x + 1)$
j $(2x^3 - 13x^2 + 27x - 21) \div (x - 2)$

2 Find the quotient and remainder in each case.

a $(3x^3 - 10x^2 + 11x - 6) \div (3x - 4)$
b $(2x^3 - 9x^2 + 10x - 3) \div (2x - 1)$
c $(3x^3 + 20x^2 + 39x + 20) \div (3x + 2)$
d $(2x^3 + 5x^2 - 6x - 10) \div (2x - 3)$
e $(6x^3 + 13x^2 + 9x + 3) \div (2x + 1)$
f $(15x^3 - 43x^2 + 28x - 4) \div (3x - 2)$

3 Use synthetic division to find the quotient and remainder in each case.

a $(x^3 - 3x^2 - 4x + 16) \div (x - 2)$
b $(x^3 - 7x^2 + 8x + 10) \div (x - 4)$
c $(2x^3 + 5x^2 - 6x - 9) \div (x + 3)$
d $(3x^3 + 11x^2 - 2x - 19) \div (x + 3)$
e $(2x^3 + 2x^2 - 20x + 20) \div (x + 4)$
f $(5x^3 - 23x^2 + 20x + 22) \div (x - 2)$

4 a Use synthetic division to divide $(2x^3 + 5x^2 - 6x - 8)$ by $(2x - 3)$ by considering the divisor as $2(x - \frac{3}{2})$.
[Hint: use $\frac{3}{2}$ in the synthetic division then divide quotient by 2.]

b Use synthetic division to divide:
i $3x^3 + 17x^2 + 22x + 8$ by $3x + 2$
ii $2x^3 - x^2 - 25x - 8$ by $2x + 1$
iii $5x^3 + 21x^2 + 24x + 10$ by $5x + 1$.

5 a Express $\frac{x^3 + 10x^2 + 33x + 40}{x + 3}$, $x \neq -3$ in the form $(cx^2 + dx + e) + \frac{r}{x + 3}$.

b Express $\frac{4x^3 - 7x^2 - 34x - 3}{4x + 1}$, $x \neq -\frac{1}{4}$ in the form $(cx^2 + dx + e) + \frac{r}{4x + 1}$.

6 Divide $ax^3 + bx^2 + cx + d$ by $x - e$

a using 'long division'

b using synthetic division

to prove the algorithm works.

7 $24 \div 6 = 4$ remainder 0.

The fact that the remainder is zero means 6 is a factor of 24.

The fact that the quotient is 4 when the remainder is zero means 4 is also a factor of 24.

a Show that:
i $(x - 4)$ is a factor of $x^3 + 3x^2 - 16x - 48$ and state another factor
ii $(x + 3)$ is a factor of $x^3 + 8x^2 + 21x + 18$ and state another factor
iii $(3x - 1)$ is a factor of $3x^3 - 7x^2 - 22x + 8$ and state another factor.

b For what value of k is $(x + 3)$ a factor of $x^3 + 4x^2 - 3x + k$?

c For what value of k is $(x - 2)$ a factor of $x^3 - 3x^2 - 10x + k$?

6.3 Factorisation of a polynomial

There are two important results that come directly from the above work.

The remainder theorem

If a polynomial expression $f(x)$ is divided by $x - b$ then the remainder is $f(b)$.

The factor theorem

If $f(b) = 0$, then $x - b$ is a factor of the polynomial $f(x)$.

The converse is also true. If $x - b$ is a factor of the polynomial $f(x)$ then $f(b) = 0$.

Definition

If the equation $ax^2 + bx + c = 0$ has no real roots then $ax^2 + bx + c$ is an **irreducible quadratic** ... it cannot be factorised.

Every factor of a polynomial is either a linear or an irreducible quadratic expression.

We will only consider factors with rational coefficients.

Example 1

Factorise $x^3 + 9x^2 + 26x + 24$ completely.

We consider the factors of 24 ... $\pm1, \pm2, \pm3, \pm4, \pm6$, ... using the **nesting** algorithm till we find one that makes $f(x) = 0$.

$f(1)$:	$1 = \times 1 = +9 = \times 1 = +26 = \times 1 = +24$	gives 60 ... not zero
$f(-1)$:	$1 = \times(-1) = +9 = \times(-1) = +26 = \times(-1) = +24$	gives 6 ... not zero
$f(2)$:	$1 = \times 2 = +9 = \times 2 = +26 = \times 2 = +24$	gives 120 ... not zero
$f(-2)$:	$1 = \times(-2) = +9 = \times(-2) = +26 = \times(-2) = +24$	gives 0

So, by the **factor theorem** ... $x - (-2) = x + 2$ is a factor.

Now divide $x^3 + 9x^2 + 26x + 24$ by $x + 2$.

-2	1	9	26	24
		-2	-14	-24
	1	7	12	0

Thus $(x^3 + 9x^2 + 26x + 24) \div (x + 2) = \mathbf{1}x^2 + \mathbf{7}x + \mathbf{12}$.

So $x^2 + 7x + 12$ is a quadratic factor.

This quadratic also factorises: $x^2 + 7x + 12 = (x + 3)(x + 4)$.

Thus, in completely factorised form, $x^3 + 9x^2 + 26x + 24 = (x + 2)(x + 3)(x + 4)$.

Example 2

Factorise completely $2x^3 + 6x^2 + 10x + 6$.

Remove the common factor of 2: $2(x^3 + 3x^2 + 5x + 3)$.

Check $x = \pm1, \pm2, \pm3$ for a value of x which makes the cubic expression zero.

$f(1)$:	$1 = \times 1 = +3 = \times 1 = +5 = \times 1 = +3 =$	gives 12 ... not zero
$f(-1)$:	$1 = \times(-1) = +3 = \times(-1) = +5 = \times(-1) = +3 =$	gives 0 ... $x + 1$ is a factor

Divide the cubic by $(x + 1)$ to find a quadratic factor.

-1	1	3	5	3
		-1	-2	-3
	1	2	3	0

So $x^2 + 2x + 3$ is a quadratic factor. However, note that its discriminant is less than zero.

This quadratic is irreducible. It has no factors.

Thus, in completely factorised form, $2x^3 + 6x^2 + 10x + 6 = 2(x + 1)(x^2 + 2x + 3)$.

Example 3

Factorise the quartic expression $x^4 + 4x^3 - 7x^2 - 22x + 24$ completely.

$f(1)$: $1 = \times 1 = +4 = \times 1 = -7 = \times 1 = -22 = \times 1 = +24 =$ gives 0 ... $x - 1$ is a factor

Divide the quartic by $(x - 1)$ to find a cubic factor.

1	1	4	-7	-22	24
		1	5	-2	-24
	1	5	-2	-24	0

So $x^3 + 5x^2 - 2x - 24$ is a cubic factor.

$f(1) \neq 0$

$f(-1) \neq 0$

$f(2)$: 1 = × 2 = + 5 = × 2 = − 2 = × 2 = − 24 = gives 0 … $x - 2$ is a factor

Divide the cubic by $(x - 2)$ to find a quadratic factor.

$$\begin{array}{r|rrrr} 2 & 1 & 5 & -2 & -24 \\ & & 2 & 14 & 24 \\ \hline & 1 & 7 & 12 & 0 \end{array}$$

So $x^2 + 7x + 12$ is a quadratic factor. Factorising $x^2 + 7x + 12 = (x + 3)(x + 4)$.

So $x^4 + 4x^3 - 7x^2 - 22x + 24 = (x - 1)(x - 2)(x + 3)(x + 4)$.

Exercise 6.3

1 Use the **remainder theorem** to find the remainder when:

a $x^3 - 9x^2 + 20x - 15$ is divided by $x - 6$

b $x^3 + 5x^2 - 9x - 40$ is divided by $x + 3$

c $x^3 + 4x^2 - 7x - 20$ is divided by $x - 2$

d $3x^3 + 19x^2 + 30x + 8$ is divided by $x + 4$.

2 a Show that $x + 2$ is a factor of $x^3 - 6x^2 - x + 30$.

b Factorise $x^3 - 6x^2 - x + 30$ completely.

3 Factorise each expression completely.

a $x^3 + x^2 - 17x + 15$

b $x^3 - x^2 - 14x + 24$

c $3x^3 + 6x^2 - 15x - 18$

d $x^3 - 2x^2 + 2x - 15$

e $x^3 - 21x - 20$

f $x^3 + 8x^2 + 11x - 20$

g $3x^3 - 4x^2 - 3x - 2$

h $5x^3 + 9x^2 - 17x + 3$

i $6x^3 + 13x^2 + x - 2$

j $12x^3 + 41x^2 + 13x - 6$

k $x^4 - x^3 - 21x^2 + x + 20$

l $x^4 + x^3 - 14x^2 - 24x$

m $x^4 - 12x^3 + 47x^2 - 72x + 36$

n $x^4 + 2x^3 - 11x^2 + 16x - 20$

4 a Use the **factor theorem** to show that $x - 3$ is a factor of $x^2 - 9$.

b Show that:

i $x - a$ is a factor of $x^2 - a^2$ and find another factor by division

ii $x - a$ is a factor of $x^3 - a^3$ and find another factor by division

iii $x - a$ is a factor of $x^4 - a^4$ and find another factor by division.

c i How can you tell that $x - a$ is a factor of $x^n - a^n$?

ii Find another factor by division.[5]

5 a For what value of k is $x^3 - 9x^2 + 8x + k$ divisible by $(x - 6)$?

b Completely factorise the expression for this value of k.

6 a For what value of k is $x^3 - 3x^2 - kx + k + 2$ divisible by $(x + 3)$?

b Completely factorise the expression for this value of k.

7 When $f(x) = x^3 - 4x^2 + ax + b$ is divided by $x + 3$, the remainder is zero; when it is divided by $x + 1$ the remainder is 28.

a Find a and b.

b Fully factorise $f(x)$.

[5] It will be useful in later work to know that, for example, $a^4 - b^4 = (a - b)(a^3 + a^2b + ab^2 + b^3)$.

6.4 Polynomial equations

In this course only equations which can be readily factorised will be examined.
The general method developed by Cardan and Tartaglia is beyond the scope of the course.
The method used in this course depends on the truth that if $a \,.\, b \,.\, c = 0$ then $a = 0$ or $b = 0$ or $c = 0$.

Example 1

Solve $2x^3 + 13x^2 + 13x - 10 = 0$.

First look for a factor: a quick search gives $f(-2) = 0 \Rightarrow x + 2$ is a factor.

Divide the cubic by $(x + 2)$ to find a quadratic factor.

−2	2	13	13	−10
		−4	−18	10
	2	9	−5	0

So $2x^2 + 9x - 5$ is a quadratic factor. Factorising $2x^2 + 9x - 5 = (2x - 1)(x + 5)$.

$2x^3 + 13x^2 + 13x - 10 = 0$

$\Rightarrow (x + 2)(2x - 1)(x + 5) = 0$

$\Rightarrow x + 2 = 0$ or $2x - 1 = 0$ or $x + 5 = 0$

$\Rightarrow x = -2$ or $x = \frac{1}{2}$ or $x = -5$

Example 2

Find where the curve $y = x^3 - 4x^2 - 7x + 10$ cuts the x-axis.

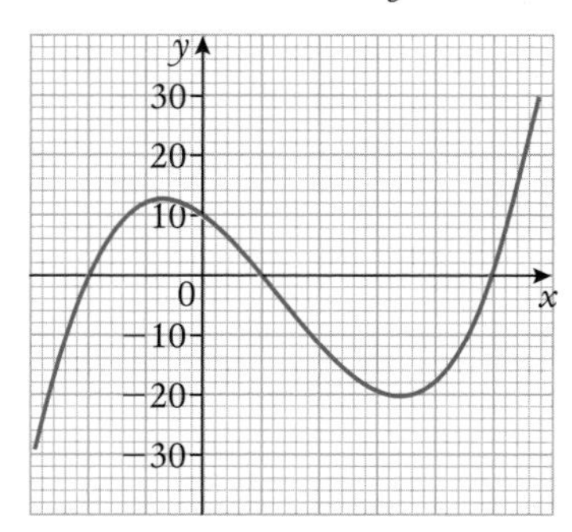

Curve cuts x-axis when $y = 0$, i.e. $x^3 - 4x^2 - 7x + 10 = 0$.

Search the factors of 10 using the nesting algorithm for a value of x that makes $y = 0$.

$x = 1$ is such a value ... so $x - 1$ is a factor.

Divide the cubic by $(x - 1)$ to find a quadratic factor.

1	1	−4	−7	−10
		1	−3	10
	1	−3	−10	0

So $x^2 - 3x - 10$ is a quadratic factor. So $(x - 5)(x + 2)$ are factors.

$x^3 - 4x^2 - 7x + 10 = 0$

$\Rightarrow (x - 1)(x - 5)(x + 2) = 0$

$\Rightarrow x - 1 = 0$ or $x - 5 = 0$ or $x + 2 = 0$

$\Rightarrow x = 1$ or $x = 5$ or $x = -2$

Curve cuts x-axis at $(-2, 0)$, $(1, 0)$ and $(5, 0)$.

Exercise 6.4

1 Solve for x.

a $x^3 - 2x^2 - x + 2 = 0$

b $x^3 + 10x^2 + 19x - 30 = 0$

c $x^3 - 4x^2 + x + 6 = 0$

d $x^3 - 5x^2 + 2x + 8 = 0$

e $x^3 + 10x^2 + 27x + 18 = 0$

f $x^3 - 21x - 20 = 0$

g $6x^3 - 11x^2 - 92x - 15 = 0$

h $2x^3 - 3x^2 + 1 = 0$

i $4x^3 + 25x^2 + 49x + 30 = 0$

j $6x^3 + 5x^2 - 2x - 1 = 0$

k $10x^3 - x^2 - 7x - 2 = 0$

l $12x^3 + 25x^2 - 4x - 12 = 0$

2 Each equation has an associated sketch.

Solve the equation and match it up with its sketch.

Equation 1: $x^3 + 11x^2 + 36x + 26 = 0$

Equation 2: $x^3 - 10x^2 + 27x - 18 = 0$

Equation 3: $x^3 + 5x^2 + 3x - 9 = 0$

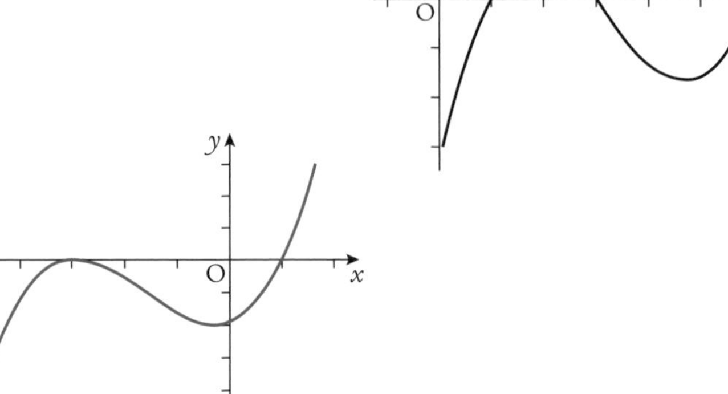

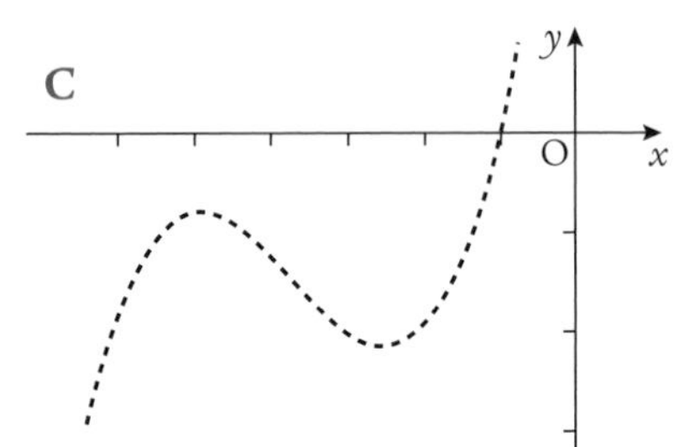

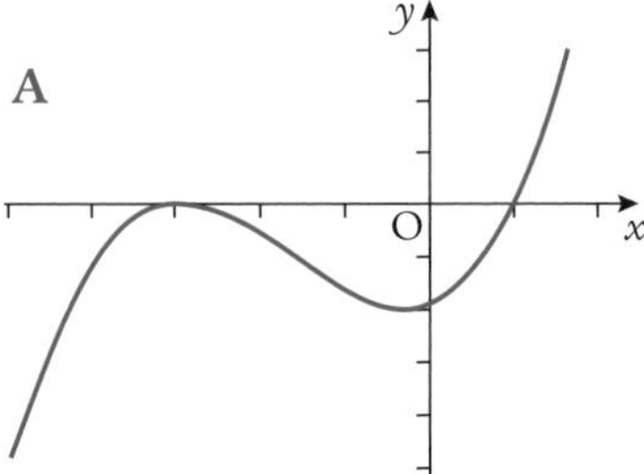

3 Solve the following quartic equations. In each case one solution has been given.

a $x^4 + 6x^3 + 7x^2 - 6x - 8 = 0$; $x = -2$ is one solution.

b $x^4 - 11x^3 + 28x + 36x - 144 = 0$; $x = 6$ is one solution.

c $4x^4 + 11x^3 - 3x^2 - 16x + 4 = 0$; $x = -2$ is one solution.

d $2x^4 + 5x^3 - 3x^2 - x - 3 = 0$; $x = 1$ is one solution.

4 A box is constructed so that its length, breadth and height are $(x - 2)$ units, $(x + 3)$ units and $(x + 9)$ units respectively.

a Write an expression in x for the volume, V units³, of the box.

b i For what values of x is the volume zero?

ii What practical restriction must be placed on x?

c The volume is actually 72 units³.

i Write down a cubic equation to represent the situation.

ii Solve the equation to find the value of x in the context.

5 The depth of water in a harbour is modelled by:

$d = 1 + 3x + 3x^2 - x^3$

where d is the depth in metres and x is the time since low tide, measured in units of 6-hours.

The model is only useful in the domain $0 \leqslant x < 2$.

a Solve the equation $1 + 3x + 3x^2 - x^3 = 6$, giving the roots correct to 1 decimal place where appropriate.

b At what time after low tide was the depth of water equal to 6 metres?

6 The area under a section of the archway of a bridge can be calculated using $A = 9x + 6x^2 - x^3$, where A is the area in square units and x units is the distance from the left-hand side of the bridge. $0 \leqslant x \leqslant 4$.

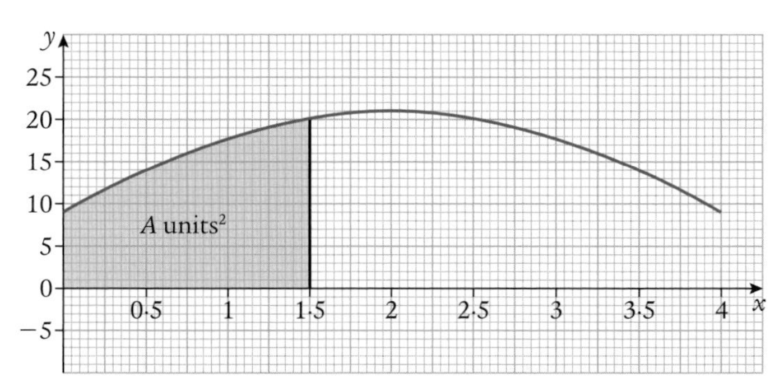

a Solve $9x + 6x^2 - x^3 = 14$.

b For what value of x will the area under the arch be 14 units²?

7 a Show that the line $y = 2x - 5$ is a tangent to the cubic $y = 2x^3 - 17x^2 + 50x - 50$ when $x = 3$.

b Where else does the line cut the curve?

8 Two scientists watched the temperature rise in an experiment.

One thought the rise could be modelled by the cubic $y = 3x^3 - 5x^2 + 3x + 1$, where y °C is the rise in temperature and x is the time in minutes since the experiment started. The other felt that x and y were related by $y = 5x^2 - 8x + 5$.

For what values of x and y were the two models in agreement?

9 The equation $x^3 + 2x^2 - 13x + m$ has a root $x = 1$.

a Find the value of m.

b Calculate the other roots of the equation.

10 $2x^3 - 3x^2 - mx + n$ has roots $x = 3$ and $x = -2$.

a Find the values of m and n.

b Find the third root of the equation.

6.5 Non-rational solutions

When solutions are non-rational, we will not find them by factorising.

If $f(x) = 0$ is a cubic equation then we should try rational values for x and look for $f(x)$ changing sign.

If $f(x_1) < 0$ and $f(x_2) > 0$ then for some value of x between x_1 and x_2, $f(x) = 0$.

For example, if $x_1 < x_2$ and $f(x_1)$ is negative and $f(x_2)$ is positive, the curve must cross the axis between $(x_1, f(x_1))$ and $(x_2, f(x_2))$.

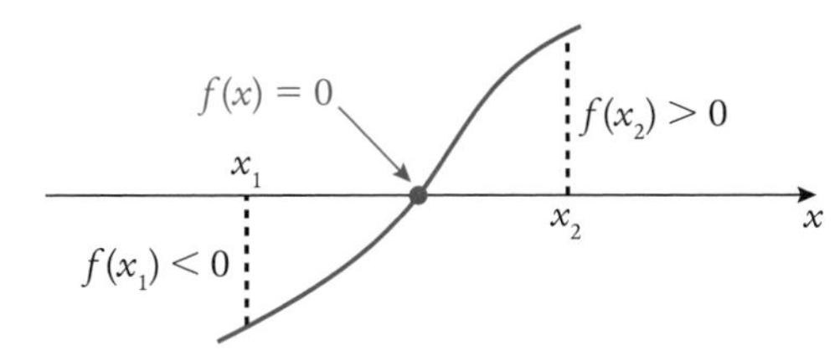

We can then test a value of x between x_1 and x_2 to get closer to the solution.

If this new value, x_3, is positive ... the solution lies between x_1 and x_3.

If this new value is negative ... the solution lies between x_2 and x_3.

This can be repeated as often as we like to refine our solution.

Example 1

a Show that a solution to the equation $x^3 + 2x^2 + x - 1 = 0$ lies between 0 and 1.

b Find the solution correct to 2 decimal places.

a Using the nesting method on $f(x) = x^3 + 2x^2 + x - 1$, $f(0) = -1$ and $f(1) = 3$.
The switch of sign tells us a solution lies in range $0 < x < 1$.

b Try $x = 0{\cdot}5$ ... $f(0{\cdot}5) = 0{\cdot}125$ ... positive.

Solution lies between this and the previous negative answer ... in range $0 < x < 0{\cdot}5$.

Since 0·125 is closer to zero than -1, we might try a value for x closer to 0·5 than 0.

Try $x = 0{\cdot}4$ ... $f(0{\cdot}4) = -0{\cdot}216$ ... negative.

Solution lies between this and the previous positive answer ... in range $0{\cdot}4 < x < 0{\cdot}5$.

Try $x = 0{\cdot}45$ ... $f(0{\cdot}45) = -0{\cdot}053875$ ... negative.

Solution lies between this and the previous positive answer ... in range $0{\cdot}45 < x < 0{\cdot}5$.

Try $x = 0{\cdot}47$ ... $f(0{\cdot}47) = 0{\cdot}015623$ ... positive.

Solution lies between this and the previous negative answer ... in range $0{\cdot}45 < x < 0{\cdot}47$.

Try $x = 0{\cdot}46$ ... $f(0{\cdot}46) = -0{\cdot}01946$ ... negative.

Solution lies between this and the previous positive answer ... in range $0{\cdot}46 < x < 0{\cdot}47$.

Try $x = 0{\cdot}465$ ... $f(0{\cdot}465) = -0{\cdot}002005$ ... negative.

Solution lies between this and the previous positive answer ... in range $0{\cdot}465 < x < 0{\cdot}47$.

To 2 decimal places both 0·465 and 0·47 round to 0·47.

Solution is $x = 0{\cdot}47$ correct to 2 decimal places.

Your working need only show the x-values as they 'close in' on the solution ... which always lies between the bottom values in the two columns.

Negative	Positive
0	1
0·4	0·5
0·45	0·47
0·46	
0·465	

Exercise 6.5

1 $x^3 - 3x^2 - x + 5 = 0$

a Show that this equation has roots between:
i -2 and -1 **ii** 1 and 2 **iii** 2 and 3.

b Find the root that lies between 1 and 2 correct to 2 decimal places.

c Find the root that lies between 2 and 3 correct to 1 decimal place.

2 The equation $x^3 - x - 9 = 0$ has only one root.

a Between what two consecutive whole numbers does the root lie?

b Find the root correct to 1 decimal place.

3 The size of a debt is modelled by the cubic equation $D(x) = 6 + 6x + 3x^2 + x^3$, where x years is the term of the debt.
The model works in the domain $0 \leqslant x < 1$, i.e. it only works when the debt is taken over a period less than a year.
D is the debt in hundreds of pounds.

a What is the size of the debt at the beginning, i.e. $x = 0$?

b i Show that the debt will be £1000 sometime between 0·5 of a year and 0·6 of a year.
ii Solve $6 + 6x + 3x^2 + x^3 = 10$ to find this value of x correct to 2 decimal places.

4 A spherical tank of radius 10 m holds some water.

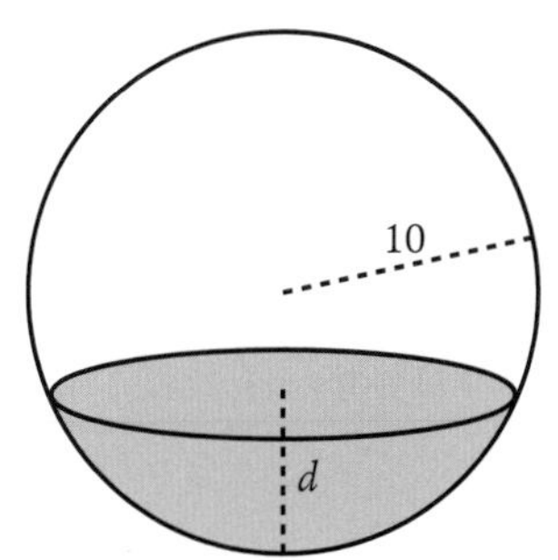

The volume of water can be calculated from the formula $V(d) = 10\pi d^2 - \frac{\pi}{3}d^3$, where d m is the depth of the water at the centre.
The volume of the sphere is $V_{\text{sphere}} = \frac{4000}{3}\pi$.

a Show that when the sphere is one-quarter full $d^3 - 30d^2 + 1000 = 0$.

b Show that this has a solution in the range $6 < d < 7$.

c Find this value correct to 2 decimal places.

Preparation for assessment

1 a $f(x) = x^3 + 2x^2 + 3x - 4$. Calculate: i $f(-1)$ ii $f(1)$.

b $g(x) = x^4 - 2x^3 + x^2 + 5x - 1$. Calculate: i $g(0)$ ii $g(4)$.

2 a Divide $2x^3 + x^2 + 5x - 2$ by $x - 1$ using the 'long division' algorithm.

b Divide $3x^3 + 2x^2 - 4x - 5$ by $x + 2$ using the synthetic division algorithm.

c Divide $x^4 - 2x^3 + 2x^2 - 10x - 1$ by $x - 3$.

3 Factorise:

a $x^3 + 6x^2 + 5x - 12$

b $6x^3 - 25x^2 + 28x - 4$

c $4x^3 - 2x^2 + 4x - 6$

d $x^4 + 4x^3 - 17x^2 - 24x + 36$.

4 Solve:

a $x^3 - 3x^2 - 22x + 24 = 0$

b $x^4 + 7x^3 + 2x^2 - 64x - 96 = 0$.

5 a i Show that the line $y = 5x + 2$ is a tangent to the curve $y = x^3 - 3x^2 - 4x - 3$.
ii State the point of tangency.

b Find the point where the line cuts the curve again.

c i Show that cubic curve $y = x^3 - 3x^2 - 4x - 3$ cuts the x-axis in the region $4 < x < 4{\cdot}5$.
ii Find this point correct to 2 decimal places.

Summary

1 Any expression that takes the form $a_nx^n + a_{n-1}x^{n-1} + a_{n-2}x^{n-2} + \dots + a_2x^2 + a_1x + a_0$, where a_n to a_0 are constants and $a_n \neq 0$ is called a polynomial expression of degree n.

2 The cubic expression $a_3x^3 + a_2x^2 + a_1x + a_0$ can be re-written as $((a_3x + a_2)x + a_1)x + a_0$.

This form leads to the 'nesting' algorithm for evaluating a cubic at a particular value of x.

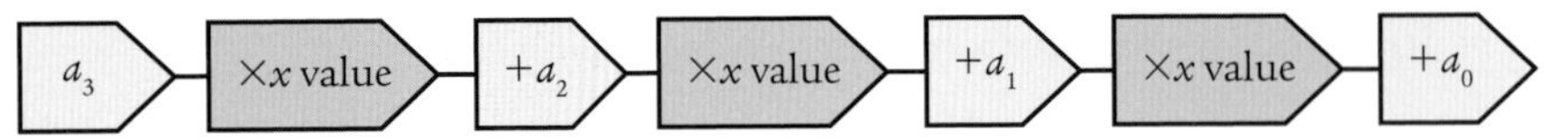

If any of the coefficients are zero, they must still be used.

3 We can divide a polynomial by a linear expression using a long division algorithm or by synthetic division.

4 **The remainder theorem**
If a polynomial expression, $f(x)$, is divided by $x - b$ then the remainder is $f(b)$.

5 **The factor theorem**
If $f(b) = 0$, then $x - b$ is a factor of the polynomial $f(x)$.

The converse is also true. If $x - b$ is a factor of the polynomial $f(x)$ then $f(b) = 0$.

6 **Irreducible quadratic**
If the equation $ax^2 + bx + c = 0$ has no real roots then $ax^2 + bx + c$ is an irreducible quadratic ... it cannot be factorised.

Every factor of a polynomial is either a linear or an irreducible quadratic expression.

7 We can solve polynomial equations of the form $f(x) = 0$ by factorising and setting each factor to zero.

8 When solutions are non-rational, we will not find them by factorising.

If $f(x) = 0$ is a cubic equation then we should try rational values for x and look for $f(x)$ changing sign.

If $f(x_1) < 0$ and $f(x_2) > 0$ then for some value of x between x_1 and x_2, $f(x) = 0$.

7 Differential calculus 1

Before we start...

In 1687, Sir Isaac Newton outlined the theory of gravity in a book we now know as *The Principia*.

We all know how to find the gradient between two points.

His concern was to find the gradient **at** a point.

To develop his theories he invented a new field of mathematics known as Infinitesimal Calculus, where he learned to handle very small amounts. Independently a German mathematician by the name of Leibniz developed basically the same material but used a more 'user-friendly' notation. However, the right to claim priority as the inventor led to protracted and bitter rows between mathematicians and scientists across the whole of Europe.

A committee eventually came down on the side of Newton.

We have looked at one of the arguments in Chapter 2 where, **as long as we work in radians**, when θ is very small,

$\sin\theta \approx \theta$, $\cos\theta \approx 1$ and $\tan\theta \approx \theta$.

We shall develop this kind of argument in this chapter.

To do this we shall look again at expressions of the type $a^3 - b^3$, $a^4 - b^4$, $a^5 - b^5$, etc.

Consider $a^4 - b^4$. We learned in Chapter 6, Exercise 6.3 that $(a - b)$ is a factor and, using synthetic division,

$$\begin{array}{c|ccccc} b & 1 & 0 & 0 & 0 & -b^4 \\ & & b & b^2 & b^3 & b^4 \\ \hline & 1 & b & b^2 & b^3 & 0 \end{array}$$

the other factor is $(a^3 + a^2b + ab^2 + b^3)$ ... i.e. $\boldsymbol{a^4 - b^4 = (a - b)(a^3 + a^2b + ab^2 + b^3)}$. ... ①

Suppose we have the curve $y = x^4$ and a line cutting it at two places, $P(x, x^4)$ and $Q(x + h, (x + h)^4)$.

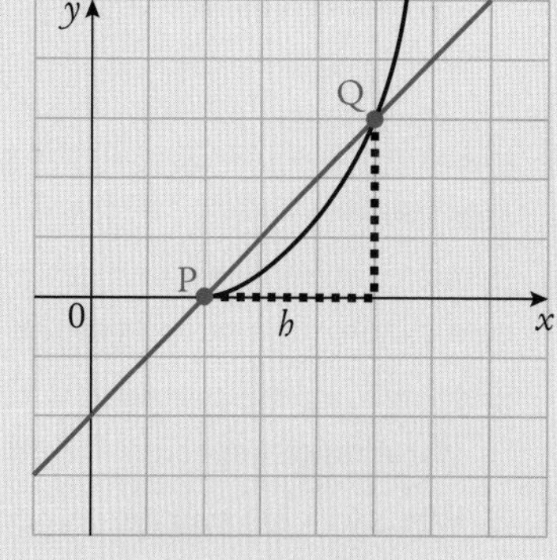

We can find an expression for the gradient of PQ,

$$m_{PQ} = \frac{y_Q - y_P}{x_Q - x_P} = \frac{(x + h)^4 - x^4}{(x + h) - x}.$$

Substituting $x + h$ for a and x for b in the identity labelled ① gives

$$m_{PQ} = \frac{(x + h - x)((x + h)^3 + (x + h)^2x + (x + h)x^2 + x^3)}{h} = (x + h)^3 + (x + h)^2x + (x + h)x^2 + x^3.$$

In this simplified form, we can consider the value of m_{PQ} when h is very small.

[Why could we not do it before the simplification?]

For very small values of h, $m_{PQ} \approx x^3 + x^2 . x + x . x^2 + x^3$

i.e. $m_{PQ} \approx 4x^3$.

What meaning does this have in geometric terms? Remember, h is the x-direction distance between P and Q.

What you need to know

1 Factorise:

a $a^2 - b^2$ b $a^3 - b^3$ c $a^5 - b^5$.

2 a Find an expanded expression for $f(x + h)$, given that:

i $f(x) = x^2 + 4x - 3$ ii $f(x) = x^3$.

b Use the compound angle formulae to help you find an expanded expression for $f(x + h)$, given that:

i $f(x) = \sin x$ ii $f(x) = \cos x$.

3 Find the gradient of the line joining:

a A(3, 4) to B(8, 14) b P(1, 5) to Q(x, y) c M(x, y) to N($x + 3$, $y + 9$).

4 a Given that $f(x) = x^2 + 4$, find the point on the curve with x-coordinate **i** 4 **ii** 6.

b Find the gradient of the line passing through these two points.

c Evaluate the function at the point: i P where $x = 1$ ii Q where $x = 1 + h$.

d i Find an expression for the gradient of the line passing through P and Q.

ii Estimate its value when h is very small.

5 For this question, you need recall the laws of indices.

$x^m \times x^n = x^{m+n}$; $x^m \div x^n = \dfrac{x^m}{x^n} = x^{m-n}$; $(x^m)^n = x^{mn}$; $x^0 = 1$; $\dfrac{1}{x^m} = x^{-m}$; $\sqrt[n]{x^m} = x^{\frac{m}{n}}$

a Express each in the form ax^n, where a is a constant.

i $\dfrac{1}{x}$ ii $\dfrac{2}{x^2}$ iii $\dfrac{4}{x^{-2}}$ iv $\dfrac{1}{2x^3}$ v $\dfrac{2}{3x^{-1}}$

vi $\sqrt{x}$ vii $4\sqrt[3]{x}$ viii $\dfrac{1}{\sqrt{x}}$ ix $\dfrac{2}{\sqrt[3]{x}}$ x $\dfrac{\sqrt{x}}{x}$

b Simplify:

i $x^5 \times x^{-2}$ ii $x^{\frac{1}{2}} \times x^{\frac{2}{3}}$ iii $x^{\frac{3}{4}} \times x^{-\frac{1}{3}}$ iv $x^3 \div x^4$ v $x^{\frac{1}{3}} \div x^{-\frac{2}{3}}$ vi $\sqrt{x} \div \sqrt[3]{x}$.

c Simplify, expressing your answer as the sum of terms of the form ax^n:

i $\dfrac{1 + \sqrt{x}}{x}$ ii $\dfrac{1 + x}{\sqrt{x}}$ iii $\dfrac{x^{\frac{1}{2}}(1 + x^{\frac{1}{2}})}{x^{\frac{2}{3}}}$.

6 In a time–distance graph, the gradient represents velocity.

A salesman makes a journey of 100 miles to a client, then returns to the depot.

a How long did he spend with the client? [Gradient = speed = 0 mph.]

b What was his average velocity in:

i the first hour

ii the last hour of his journey?

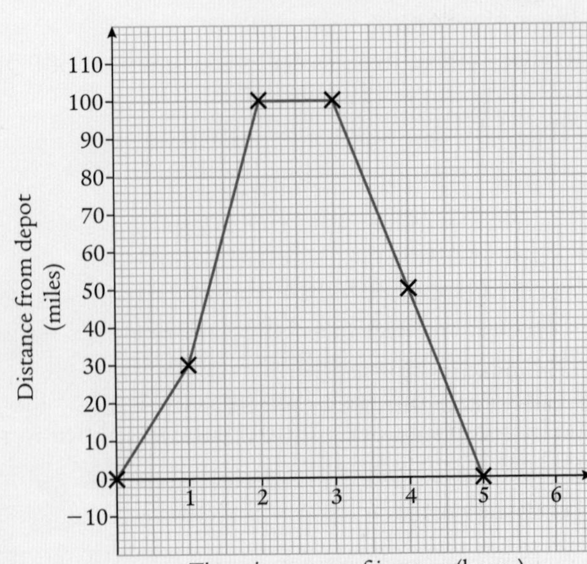

7.1 The derived function

Consider the graph $y = f(x)$.
P is the point on the curve with x-coordinate x. So its y-coordinate is $f(x)$.
Q is the point on the curve with x-coordinate $x + h$. So its y-coordinate is $f(x + h)$.

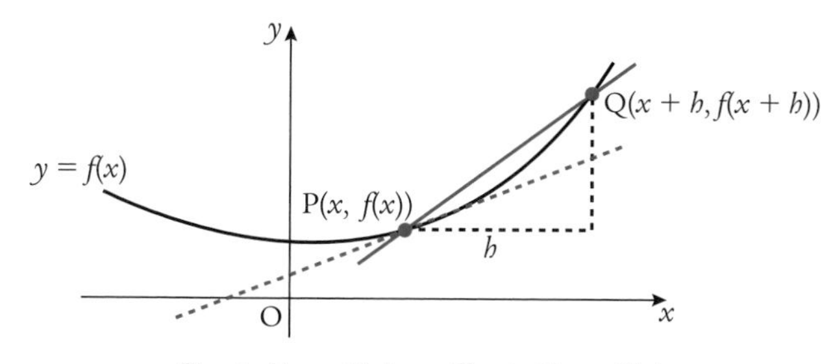

$$m_{PQ} = \frac{y_Q - y_P}{x_Q - x_P} = \frac{f(x+h) - f(x)}{(x+h) - x} = \frac{f(x+h) - f(x)}{h}$$

As h gets smaller, Q approaches P, and the secant[1] passing through P and Q will tend towards the tangent to the curve at P (the purple broken line).

Notations

Functional notation

The gradient of the curve at the point P is defined to be the gradient of the tangent to the curve at P.

It is denoted by $f'(x)$ ... and is called 'the derived function'.

Leibniz notation

In this notation, Δx rather than h represents the difference in x-values between P and Q.

Δy is the corresponding y-difference.

$$m_{PQ} = \frac{y_Q - y_P}{x_Q - x_P} = \frac{\Delta y}{\Delta x}$$

In this notation the derived function is denoted by $\frac{dy}{dx}$.[2]

If $y = f(x)$ then $\frac{dy}{dx} = f'(x)$.

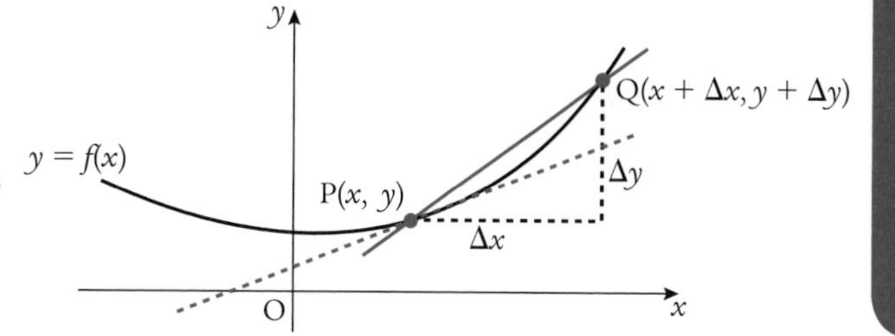

Definition

$$f'(x) = \lim_{h \to 0} \frac{f(x+h) - f(x)}{h}.$$

Notes:

- The process of finding the derived function is called differentiation.
- The derived function is also called the derivative.
- The derivative provides a formula for the **gradient** of the curve $y = f(x)$ at $P(x, f(x))$.
- $\lim_{h \to 0}$ (expression) means 'the limit to which the expression tends as h tends to zero'.
- h is infinitesimally small, but not zero. [In maths we have no definition for division by zero.]
- In Leibniz notation, the definition of the derived function reads:
  $$\frac{dy}{dx} = \lim_{\Delta x \to 0} \frac{\Delta y}{\Delta x}$$
- We will use whichever notation is the most appropriate at the time.

---

[1] A secant is a line cutting a curve at two places.

[2] Although $\frac{\Delta y}{\Delta x}$ is a division, $\frac{dy}{dx}$ is not. It is a single symbol representing the gradient at a point on a curve.

Example 1

Differentiate $3x^2$.

$f(x) = 3x^2$

$\Rightarrow f'(x) = \lim_{h\to 0} \dfrac{3(x+h)^2 - 3x^2}{h}$

$\Rightarrow f'(x) = \lim_{h\to 0} \dfrac{3x^2 + 6xh + 3h^2 - 3x^2}{h} = \lim_{h\to 0} \dfrac{6xh + 3h^2}{h} = \lim_{h\to 0}(6xh + 3h)$

$\Rightarrow f'(x) = 6x$

Example 2

A function is defined by $f(x) = x^5$.

a Find the derived function.

b Find the gradient of the curve $y = f(x)$ when $x = 2$.

c Find the equation of the tangent to the curve at $x = 2$.

Remember $a^5 - b^5 = (a - b)(a^4 + a^3b + a^2b^2 + ab^3 + b^4)$ ... ①

a $f(x) = x^5$

$\Rightarrow f'(x) = \lim_{h\to 0} \dfrac{(x+h)^5 - x^5}{h}$

Letting $x + h$ replace a and x replace b in the identity ①, we get

$\Rightarrow f'(x) = \lim_{h\to 0} \dfrac{((x+h) - x)((x+h)^4 + (x+h)^3x + (x+h)^2x^2 + (x+h)x^3 + x^4)}{h}$

$\Rightarrow f'(x) = \lim_{h\to 0} \dfrac{h((x+h)^4 + (x+h)^3x + (x+h)^2x^2 + (x+h)x^3 + x^4)}{h}$

$\Rightarrow f'(x) = \lim_{h\to 0}((x+h)^4 + (x+h)^3x + (x+h)^2x^2 + (x+h)x^3 + x^4)$

$\Rightarrow f'(x) = x^4 + x^3 . x + x^2 . x^2 + x . x^3 + x^4$

$\Rightarrow f'(x) = 5x^4$

b When $x = 2, f'(2) = 5 \times 2^4 = 80$, i.e. the gradient of the curve when $x = 2$ is 80.

c The gradient of the tangent at this point is also 80.
The y-coordinate when x is 2 is $y = 2^5 = 32$.
The point (2, 32) lies on $y = 80x + c$.
$\Rightarrow 32 = 80 . 2 + c$
$\Rightarrow c = 32 - 160 = -128$
The equation of the tangent at (2, 32) is $y = 80x - 128$.

Exercise 7.1

1 **a** Differentiate:

i $3x^2$ **ii** x^3 **iii** $2x^3$ **iv** x^4 **v** $3x^4$.

b Find the derived function of:

i $f(x) = 5x$ **ii** $f(x) = 3x$ **iii** $f(x) = -x$ **iv** $f(x) = 3$ **v** $f(x) = 5$.

2 Find the gradient of the curve when $x = 5$ in each case.

a $y = x^2$ **b** $y = 3x^2$ **c** $y = x^3$

3 Find the equation of the tangent to the curve $y = 2x^4$ when $x = 1$.

7.2 Simplifying the method

A general algorithm for terms of the form ax^n

If we consider the function $f(x) = ax^n$

$$\Rightarrow f'(x) = \lim_{h\to 0}\frac{f(x+h)-f(x)}{h} = \lim_{h\to 0}\frac{a(x+h)^n - ax^n}{h} = a\lim_{h\to 0}\frac{(x+h)^n - x^n}{h}$$

$$\Rightarrow f'(x) = a\lim_{h\to 0}\frac{((x+h)-x)((x+h)^{n-1}+(x+h)^{n-2}x+(x+h)^{n-3}x^2+\ldots+(x+h)x^{n-2}+x^{n-1})}{h}$$

$$\Rightarrow f'(x) = a\lim_{h\to 0}((x+h)^{n-1}+(x+h)^{n-2}x+(x+h)^{n-3}x^2+\ldots+(x+h)x^{n-2}+x^{n-1})$$

$$\Rightarrow f'(x) = anx^{n-1}$$

Although we have just proved $f(x) = ax^n \Rightarrow f'(x) = anx^{n-1}$ for all whole numbers n, it is in fact true for all real n. The proof is beyond the scope of this course.

Thus, if we wish to differentiate a term of the form ax^n, we need only:

- multiply by the power
- reduce the power by 1.

Notes:

1 If $k(x) = f(x) + g(x)$ then $k'(x) = f'(x) + g'(x)$.

Proof: $k'(x) = \lim_{h\to 0}\frac{k(x+h)-k(x)}{h} = \lim_{h\to 0}\frac{(f(x+h)+g(x+h))-(f(x)+g(x))}{h}$

$$\Rightarrow k'(x) = \lim_{h\to 0}\frac{f(x+h)-f(x)}{h} + \lim_{h\to 0}\frac{g(x+h)-g(x)}{h}$$

$$\Rightarrow k'(x) = f'(x) + g'(x).$$

2 Similarly we can show that if $k(x) = af(x)$ then $k'(x) = af'(x)$ ... where a is a constant.

3 If $f(x)$ is a constant function, then $f'(x) = 0$,

e.g. $f(x) = 3 \quad \Rightarrow f(x) = 3x^0 \quad \Rightarrow f'(x) = 0 \times 3x^{-1} = 0$.

Unit 2: Relationships and calculus

Example 1

Differentiate:

a x^5

b x^{-3}

c $\sqrt{x}$

d $x^{\frac{2}{3}}$

e $2x^3 + 5x^2 + 4x + 1$

f $x + \frac{1}{x}$.

a $f(x) = x^5 \Rightarrow f'(x) = 5x^4$ ... multiply by power; reduce power by 1

b $f(x) = x^{-3} \Rightarrow f'(x) = -3x^{-4}$ ... $-3 - 1 = -4$

c $f(x) = \sqrt{x} = x^{\frac{1}{2}} \Rightarrow f'(x) = \frac{1}{2}x^{-\frac{1}{2}}$ ... first express in the form ax^n

d $f(x) = x^{\frac{2}{3}} \Rightarrow f'(x) = \frac{2}{3}x^{-\frac{1}{3}}$ ... $\frac{2}{3} - 1 = -\frac{1}{3}$

e $f(x) = 2x^3 + 5x^2 + 4x + 1 \Rightarrow f'(x) = 6x^2 + 10x + 4$ ... differentiate each term separately

f $f(x) = x + \frac{1}{x} = x + x^{-1} \Rightarrow f'(x) = 1 - x^{-2} = 1 - \frac{1}{x^2}$ ... the last step is optional

Example 2

Using suitable axes, a droplet in a jet of water has been modelled by $h(x) = 2 + 20x - 5x^2$, where h feet is the height of the droplet, x seconds after leaving the nozzle of the hose.

a How high is the droplet after:
 i 1 second ii 2 seconds iii 3 seconds?

b In a time–distance graph, the speed of the droplet can be found by calculating the gradient.
 i Differentiate to find $h'(x)$.
 ii What is the droplet's speed after 3 seconds?
 iii What is its speed after 2 seconds? Comment.

a i 17 feet ii 22 feet iii 17 feet

b i $h'(x) = 20 - 10x$
 ii Speed after 3 seconds (gradient when $x = 3$): $h'(3) = 20 - 10.3 = -10$ feet per second.
 The negative sign indicates the droplet is coming down.
 iii Speed after 2 seconds (gradient when $x = 2$): $h'(2) = 20 - 10.2 = 0$ feet per second.
 The droplet is at the top of its flight and has zero velocity.

Exercise 7.2

1 Differentiate, having expressed each term in the form ax^n,

a $3x^2$ b $7x^5$ c $-3x^4$ d $-x + 5x^2$

e $8x^{-2}$ f x^{-1} g $-6x^{-5}$ h $4 - 2x + 9x^{-4}$

i $\frac{4}{x^2}$ j $\frac{7}{x^3}$ k $\frac{1}{x^5}$ l $\frac{1}{x} + \frac{3}{x^2} + \frac{5}{x^3}$

m $4x^{\frac{1}{2}}$ n $5x^{\frac{3}{2}}$ o $-4x^{\frac{2}{7}}$ p $1 + 2x^{\frac{1}{2}} + 3x^{\frac{1}{3}} + 4x^{\frac{1}{4}}$

q $3\sqrt{x}$ r $7\sqrt{x}$ s $-2\sqrt[3]{x^2}$ t $\sqrt{2} + \sqrt{x} + \sqrt[3]{x} + \sqrt[5]{x}$.

2 Find the gradient of the curve $y = 3x^2 - 12x - 3$ when:

a $x = -1$ b $x = 1$ c $x = 2$ d $x = 5$.

3 Which curve is steeper, $y_1 = 2x^3 + 4x^2 + 10$ or $y_2 = x^4 - 7x + 20$, when:

i $x = 2$ ii $x = 3$ iii $x = -1$? [Careful.]

4 A quadratic function has a graph $y = 3x^2 + 5x - 4$.

a Calculate $\frac{dy}{dx}$.

b Find when the gradient is zero.

c Find the equation of the tangent to the parabola when $x = 0$.

5 A cubic function has a graph $y = 2x^3 - 3x^2 - 12x + 1$.

a What is the gradient of the curve when $x = 3$?

b For what values of x is the gradient zero?

c For what values of x is the gradient negative?

d Find the equation of the tangent to the curve when $x = 1$.

6 A function has equation $f(x) = 5x + \frac{27}{x}$.

a Find $f'(x)$.

b Calculate the gradient of the curve $y = f(x)$ when: i $x = 1$ ii $x = 6$.

c For what values of x is the gradient 2?

d Find, correct to 2 decimal places, the values of x that make the gradient zero.

7 The Clyde Arc is a bridge in Glasgow.
The arc itself is a parabola with equation $y = -\frac{1}{4}x^2 + 8x - 55$, where, using suitable axes and units, y units is the height of the arc, x units from the end of the bridge.

a Calculate $\frac{dy}{dx}$.

b Find when the gradient of the arc is zero.

c Find the equation of the tangent to the parabola when $x = 2$.

7.3 Examples requiring manipulation

Before we differentiate, each term in the expression for $f(x)$ or y must be of the form ax^n so that we can use our rules.

We must get rid of brackets and split up fractions.

Example 1

Given that $y = (x + 1)(x - 2)(x + 3)$, find $\frac{dy}{dx}$

$$\begin{aligned} y &= (x + 1)(x - 2)(x + 3) \\ &= (x^2 - x - 2)(x + 3) \\ &= x^3 + 2x^2 - 5x - 6 \end{aligned}$$ ... it is now in a suitable form

$$\Rightarrow \frac{dy}{dx} = 3x^2 + 4x - 5$$

Example 2

A function is defined by $f(x) = \frac{x^2 + 1}{\sqrt{x}}$, find $f'(x)$.

$$f(x) = \frac{x^2 + 1}{\sqrt{x}} = \frac{x^2 + 1}{x^{\frac{1}{2}}} = \frac{x^2}{x^{\frac{1}{2}}} + \frac{1}{x^{\frac{1}{2}}}$$

$$= x^{2-\frac{1}{2}} + x^{-\frac{1}{2}} = x^{\frac{3}{2}} + x^{-\frac{1}{2}}$$ ... it is now in a suitable form

$$\Rightarrow f'(x) = \tfrac{3}{2}x^{\frac{1}{2}} - \tfrac{1}{2}x^{-\frac{3}{2}}$$

Exercise 7.3

1 Express each term in the form ax^n, then differentiate the function.

a $3x^2 - \frac{2}{x}$

b $\frac{3}{x} + \frac{1}{x^2}$

c $\frac{1}{3x^2} - \frac{3}{2x}$

d $\frac{1}{5x^4} + \frac{1}{7x^3} + \frac{1}{9x}$

2 Find the derivative of:

a $(x + 1)^2$

b $(4 - x)^2$

c $(2x + 1)(x + 1)$

d $x\left(x^{\frac{1}{2}} + x^{-\frac{1}{2}}\right)$.

3 Differentiate:

a $\left(x - \frac{1}{x}\right)^2$

b $\sqrt{x}(2x - 1)$

c $\sqrt[3]{x} + x(x^2 + 2x - 1)$.

4 Find $\frac{dy}{dx}$ when:

a $y = \frac{x + 1}{x}$

b $y = \frac{x^2 + x + 1}{x}$

c $y = \frac{(x + 1)^2}{x^3}$

d $y = \frac{x^{\frac{1}{2}} + x^{\frac{3}{2}}}{x^2}$.

5 Find $f'(x)$ when:

a $f(x) = \frac{\sqrt{x} + 3}{\sqrt{x}}$

b $f(x) = \frac{2x^2 + 3x}{2\sqrt{x}}$

c $f(x) = \frac{\sqrt{x} - x + x^2}{\sqrt{x}}$

d $f(x) = \frac{1}{\sqrt{x}} + \frac{1}{\sqrt[3]{x^2}}$.

6 Find the derived function when the function, in a suitable domain, is defined by:

a $\frac{(2x + 1)(x + 1)}{x^2}$

b $\frac{x(x + 4)}{\sqrt{x}}$

c $\frac{(x - 1)(x + 1)}{x^{-2}}$

d $\frac{x^{\frac{2}{3}}\left(x^2 + 2x^{\frac{1}{2}}\right)}{x^{\frac{1}{3}}}$.

7 a Show that $1 - x$ can be expressed as $(1 + \sqrt{x})(1 - \sqrt{x})$.

b Differentiate:

i $\frac{1 - x}{1 + \sqrt{x}}$

ii $\frac{(1 - x)(1 + 2x)}{1 + \sqrt{x}}$

iii $\frac{x^2 + 2x - 3}{\sqrt{x} - 1}$.

8 Initially a painting when bought might lose value. However, over the years it is hoped that it will appreciate in value. Its value in hundreds of pounds is represented by $V(x)$, where x is the number of years since the painting was bought.

$$V(x) = \frac{4(x^2 - 2x + 5)}{\sqrt{x}}$$

a Calculate $V(4)$.

b Find an expression for $V'(x)$.

c The gradient in this context will tell you how fast the painting is increasing in value in hundreds of pounds per year. How fast is the value increasing when: **i** $x = 1$ **ii** $x = 4$? Interpret your answers.

9 A drinks company wish a rectangular carton to hold 1000 ml.
The base of the carton is a square of side x cm. The height of the carton is h cm.

a Express h in terms of x.

b Express the surface area of the carton, k cm^2, in terms of:
  i x and h ii x only.

c A graph of the function $y = k(x)$ is drawn.

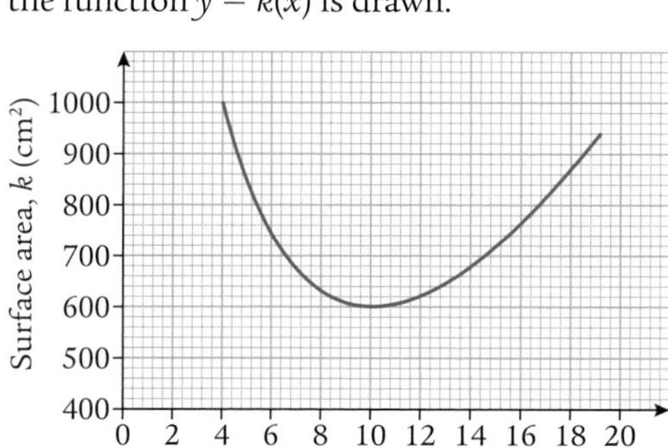

i Calculate $k'(x)$.

ii Evaluate $k'(5)$ and $k(5)$ and hence find the equation of the tangent to the curve at $x = 5$.

iii For what value of x is the gradient of the curve $y = k(x)$ zero?

iv State the equation of the tangent when the gradient is zero.

v By examining the graph, can you say what is the importance of the point with a gradient of zero?

7.4 Increasing and decreasing functions

Definitions

Examine the sketch of the cubic function $y = f(x)$.

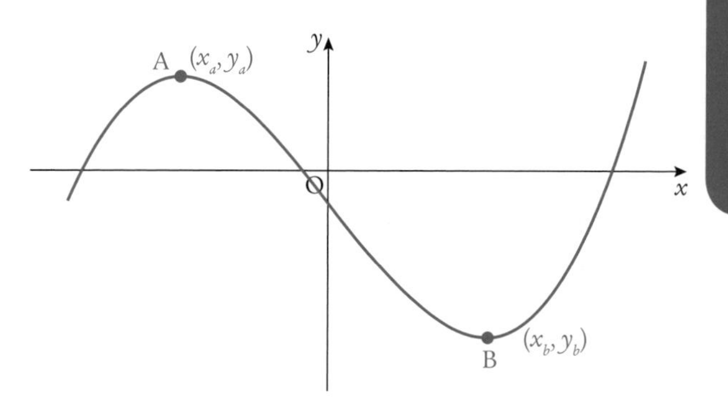

It has a maximum turning point at $A(x_a, y_a)$.

It has a minimum turning point at $B(x_b, y_b)$.

In the region where $x < x_a$ and where $x > x_b$, you can see that the gradient is **positive**.

We say that in these regions the function is strictly **increasing**.

In the region where $x_a < x < x_b$, you can see that the gradient is **negative**.

We say that in these regions the function is strictly **decreasing**.

At A and B the gradient is zero ... the function is neither increasing nor decreasing.

These are called **stationary** points.

We might describe the region where $x_a \leqslant x \leqslant x_b$ as where the function is **not** increasing.

In general, for any function $f(x)$,

- $f'(x) > 0 \Leftrightarrow$ function is strictly increasing
- $f'(x) < 0 \Leftrightarrow$ function is strictly decreasing
- $f'(x) = 0 \Leftrightarrow$ function is stationary.

Unit 2: Relationships and calculus

Example 1

Prove that $f(x) = x^3 + 9x^2 + 27x - 1$ is never decreasing.

$f(x) = x^3 + 9x^2 + 27x - 1$

$\Rightarrow f'(x) = 3x^2 + 18x + 27 = 3(x^2 + 6x + 9) = 3(x + 3)^2$

Now, being a perfect square, $(x + 3)^2 \geqslant 0$ for all values of x.

Thus $f'(x) \geqslant 0$ for all values of x.

So the function is increasing or stationary at all values of x.

Example 2

For what region(s) is the function $f(x) = 2x^3 + 3x^2 - 36x + 4$ a decreasing function?

$f(x) = 2x^3 + 3x^2 - 36x + 4$

$\Rightarrow f'(x) = 6x^2 + 6x - 36 = 6(x^2 + x - 6) = 6(x + 3)(x - 2)$

The function is decreasing when $f'(x) < 0$.

First consider when $f'(x) = 0$, i.e. $6(x + 3)(x - 2) = 0 \Rightarrow x = -3$ or $x = 2$.

Since the coefficient of x^2 is positive, the curve is concave down.

From the sketch we see that $f'(x) < 0$ when $-3 < x < 2$,
i.e. $f(x)$ is strictly decreasing in the region $-3 < x < 2$.

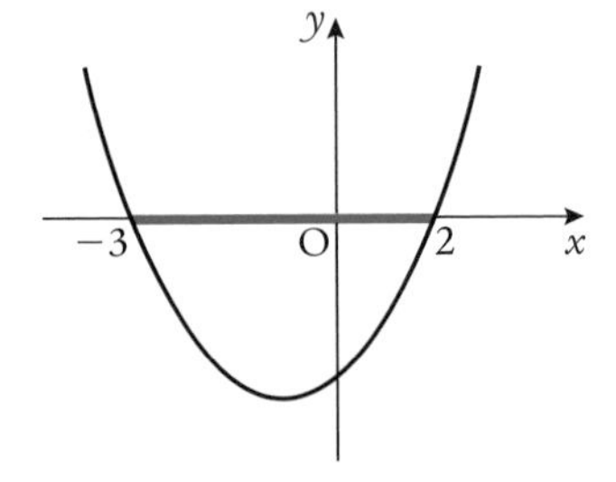

Stationary points

We can determine the **nature** of a stationary point (SP) by considering the gradient either side of it.

- A **maximum** turning point (TP): f will be increasing to the left of it, decreasing to the right.
- A **minimum** turning point: f will be decreasing to the left of it and increasing to the right.
- Where a function is increasing on both sides of a stationary point, it is called a **horizontal point of inflexion** (PI) on an increasing function.
- Where a function is decreasing on both sides of a stationary point, it is called a **horizontal point of inflexion** on a decreasing function.

Example 3

A quartic equation is defined by $f(x) = x^4 - 6x^2 - 8x + 1$.

Identify the stationary points and determine their nature.

$f(x) = x^4 - 6x^2 - 8x + 1 \Rightarrow f'(x) = 4x^3 - 12x - 8$

Stationary points occur when $f'(x) = 0$.

$\Rightarrow 4(x^3 - 3x - 2) = 0$

$\Rightarrow 4(x + 1)(x + 1)(x - 2) = 0$ ... using factor theorem

$\Rightarrow x = -1$(twice) or $x = 2$

Draw a 'table of signs' ...

x	→	−1	→	2	→
$x + 1$	−	0	+	+	+
$x + 1$	−	0	+	+	+
$x - 2$	−	−	−	0	+
$f'(x)$	−	0	−	0	+
Tangent	╲	—	╲	—	╱

- row 1 shows x below −1, at −1, between −1 and 2, at 2, and beyond 2
- rows 2, 3, 4 give the sign of each factor of $f'(x)$ as x varies
- row 5 shows the sign of $f'(x)$ based on the signs of its factors. ... $f'(x)$ is the product of its factors
- row 6 shows the 'tangent' corresponding to the sign of $f'(x)$.

From this we can 'see' the nature of the stationary points, viz.

at $x = -1$ the curve has a point of inflexion $(-1, 4)$ on a decreasing function ... $y = f(-1) = 4$

at $x = 2$, the curve has a minimum turning point, $(2, -23)$. ... $y = f(2) = -23$

Exercise 7.4

1 Say whether the function is increasing, decreasing or stationary at the given value of x.

a i $f(x) = 2x + 1; x = 4$ ii $f(x) = 2 - 3x; x = 1$

b i $f(x) = x^2 - 3x + 2; x = 0$ ii $f(x) = 1 - 3x - 2x^2; x = -1$

c i $f(x) = x^4 - 3x^3; x = 1$ ii $f(x) = x^5 - 5x + 1; x = 1$

d i $f(x) = x + \frac{1}{x}; x = 2$ ii $f(x) = \frac{x}{2} + \frac{1}{x^2} + \frac{2}{3}; x = 2$

e i $f(x) = \frac{x + 1}{\sqrt{x}}; x = 0{\cdot}25$ ii $f(x) = \frac{1 - x}{1 + \sqrt{x}}; x = 1$

2 By solving a suitable quadratic inequality, describe the regions of increase and decrease in each of these cubic functions.

a $f(x) = \frac{x^3}{3} - \frac{3x^2}{2} - 10x + 3$ b $f(x) = x^3 - 3x + 6$ c $f(x) = 1 + 2x + \frac{x^2}{2} - \frac{x^3}{3}$

3 a Prove that $f(x) = x^3 + 3x^2 + 3x + 2, x \geqslant 0$ is an ever increasing function.

b Prove that $g(x) = 3x^3 - 3x^2 + x - 2$ is never decreasing.

c i By considering the discriminant of the derived function, show that the function $f(x) = x^3 + 2x^2 + 2x - 5$, has no stationary points.

ii By considering the derivative at $x = 0$ decide whether the function is always increasing or always decreasing.

d $f(x) = 2 - x + x^2 - \frac{1}{3}x^3$.
Show that if $x_a < x_b$ then $f(x_a) > f(x_b)$.

4 Here are the derivatives of four functions in their factorised form.

i State the x-coordinates of the stationary points in each function.

ii Use a table of signs to establish the nature of each stationary point.

a $f'(x) = (x - 2)(x + 5)$ b $g'(x) = (x - 1)(x + 1)(x + 2)$

c $h'(x) = (x - 2)(x - 2)(x + 1)$ d $k'(x) = (1 - x)(2 + x)(x - 3)$

5 Find the stationary points of each function and determine their nature.

a $f(x) = 3x^3 + 3x^2 + x - 3$ b $f(x) = x^3 - x^2 - x - 2$

c $f(x) = \frac{x^4}{4} - \frac{x^3}{3} - \frac{9x^2}{2} + 9x - 2$ d $f(x) = 4x + \frac{1}{\sqrt{x}}; x > 0$

7.5 Sketching $y = f'(x)$ given $y = f(x)$

From a sketch of $y = f(x)$, we can produce a rough sketch of $y = f'(x)$.

Consider the cubic equation $f(x) = ax^3 + bx^2 + cx + d$ and its derivative $f'(x) = 3ax^2 + 2bx + c$.

- Both sketches should start and finish at the same x-values. [See purple broken guide lines.]
- Where there are stationary points on $y = f(x)$, there will be zeros on $y = f'(x)$. [These have been 'transferred' from the top sketch to the bottom ... black broken lines.]
- Where the gradient of $f(x) > 0$, the curve of $y = f'(x)$ will be above the axis. [See parts corresponding to + signs.]
- Where the gradient of $f(x) < 0$, the curve of $y = f'(x)$ will be below the axis. [See parts corresponding to − signs.]

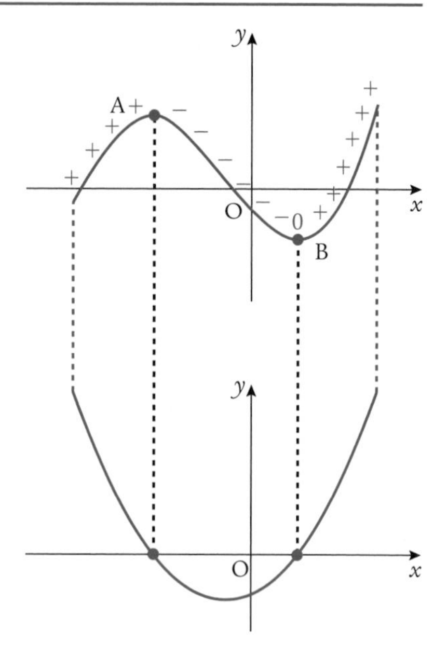

Example 1

The diagram shows $y = f(x)$ close to the origin.

The curve has stationary points A, B and C.

Make a sketch of $y = f'(x)$.

Draw a new set of axes below the original sketch.

- Transfer the domain of the sketch [purple broken lines].
- Transfer the stationary points to new x-axis.
- Identify each region as 'increasing' or 'decreasing'.
- Draw a curve passing through A, B and C, being above x-axis when function is increasing and below x-axis when function is decreasing.

[Note that no inference can be made as to the scale on the y-axis. Sometimes, to help position the curve, further information will be given (e.g. the y-intercept).]

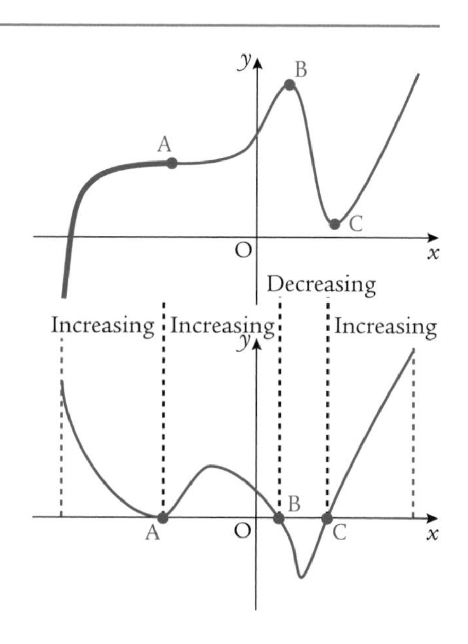

Exercise 7.5

1 There follows sketches of $y = f(x)$ for various functions.
One extra piece of information is also given to help scale the y-axis.
Make a sketch of $y = f'(x)$ in each case.

a Minimum TP is $(-1, -3)$; $f'(0) = 1$

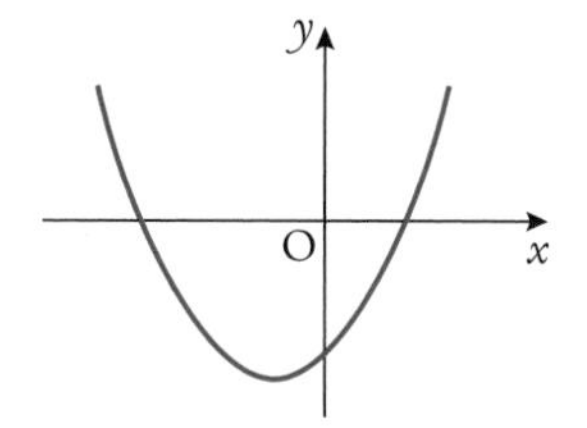

b Maximum TP is $(1, 5)$; $f'(0) = 1$

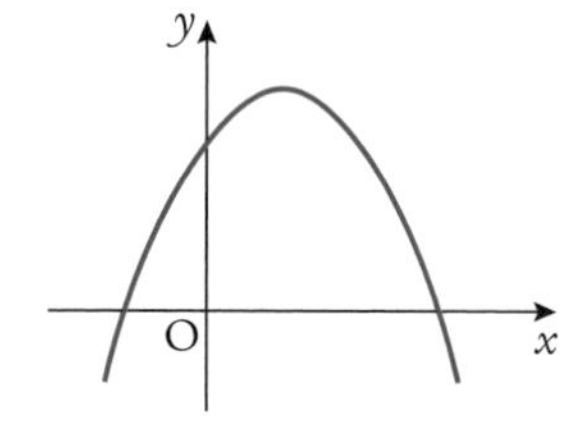

Unit 2: Relationships and calculus

c TPs at $(-4, 3)$, $(2, -5)$; $f'(0) = -1$

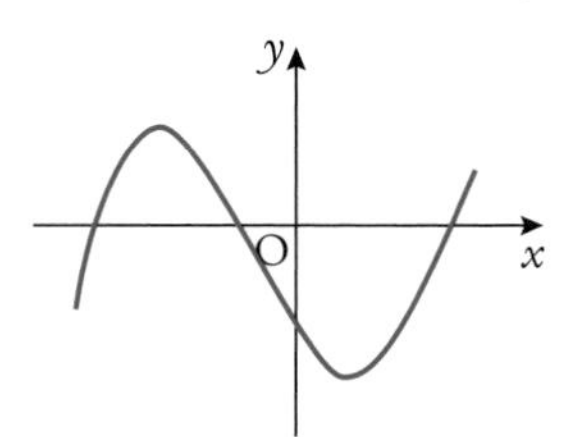

d TPs at $(2, -6)$, $(6, 1)$; $f'(0) = -2$

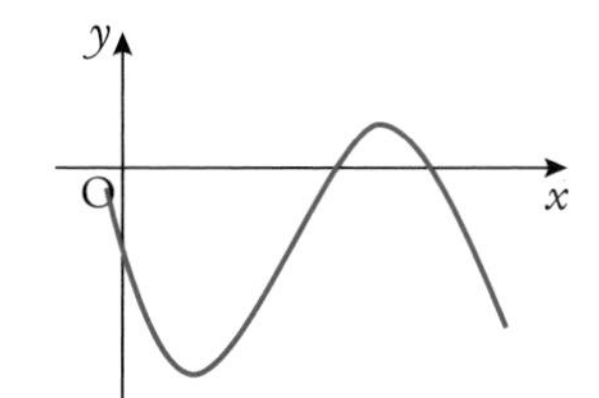

e PI at $(-2, 1)$; $f'(0) = 1$

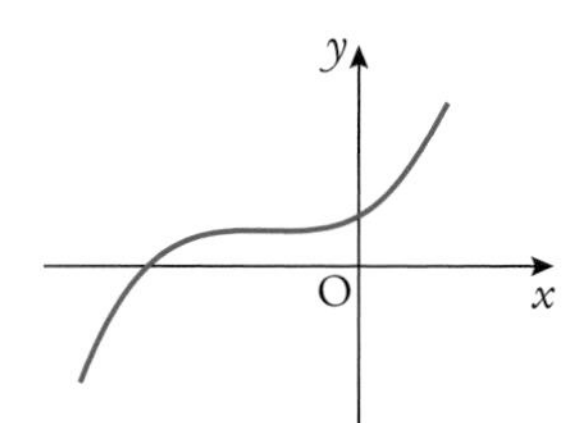

f PI at $(2, 1)$; $f'(0) = -0{\cdot}5$

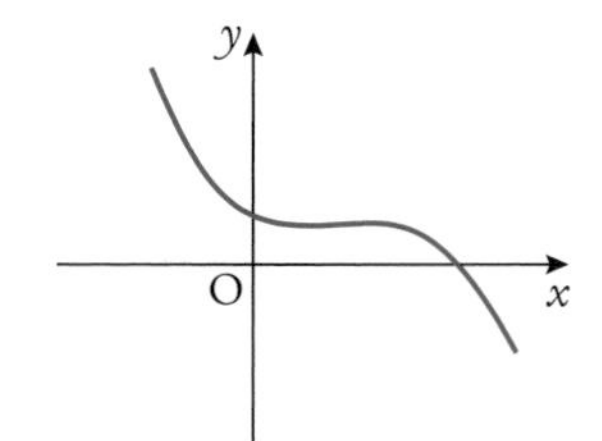

2 Produce sketches of the graph of the derived function from the following. The stationary points have been indicated.

a

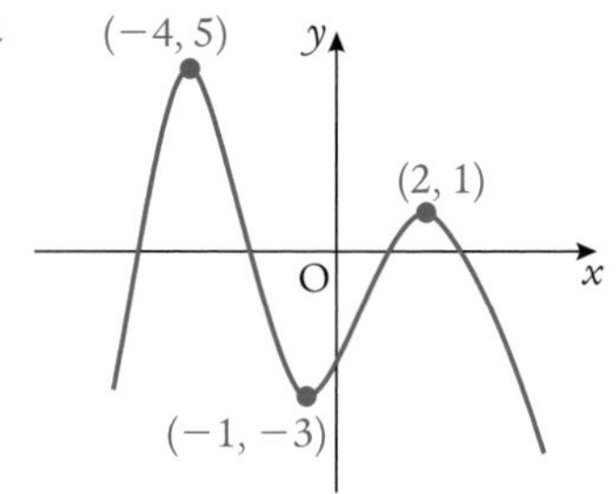

b

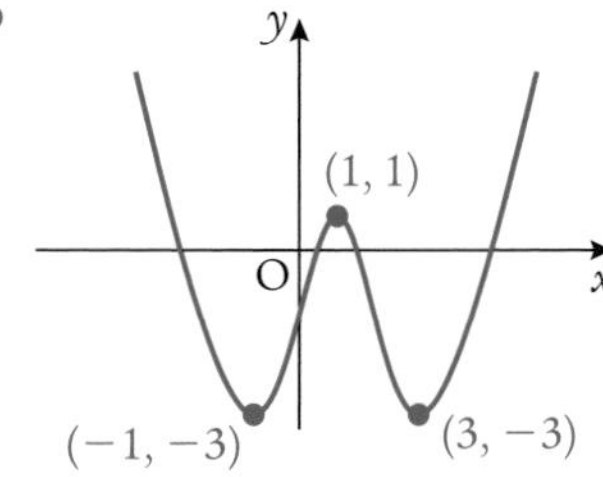

c

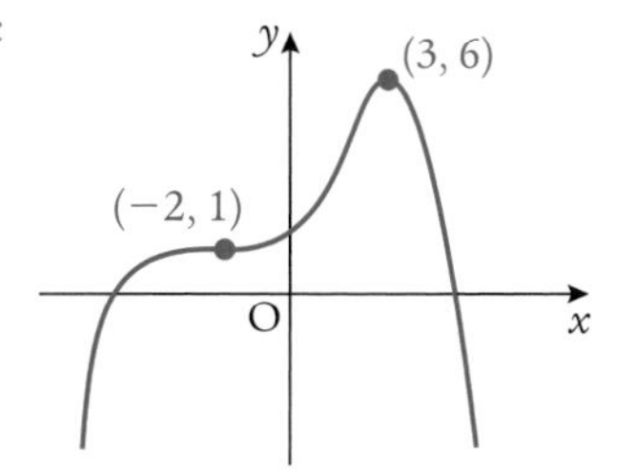

d

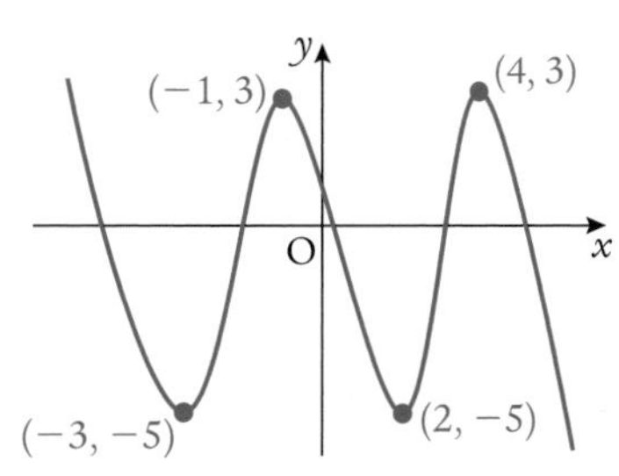

e

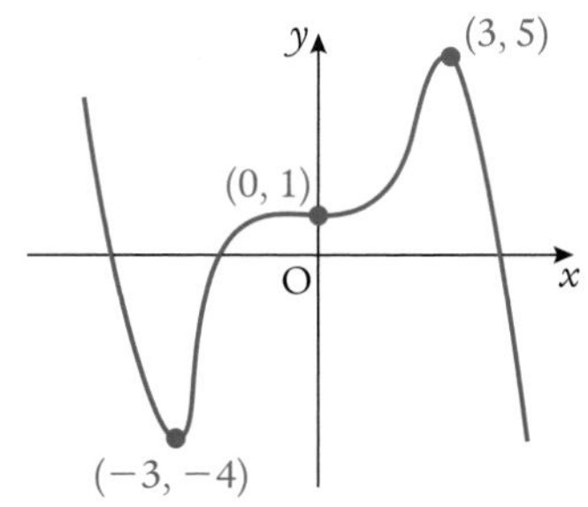

f

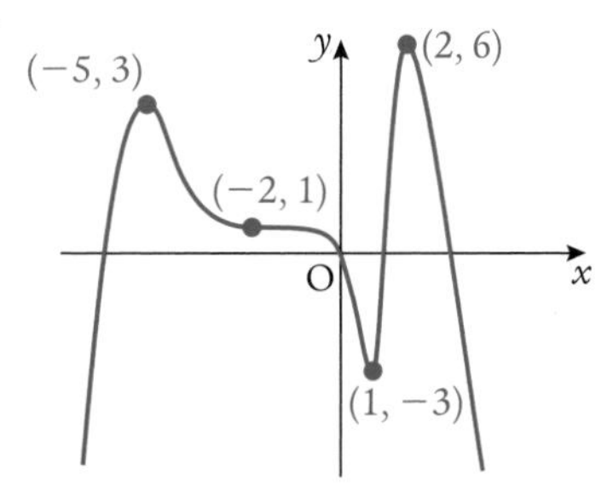

3 The function $f(x) = x + \frac{1}{x}, x > 0$ has a stationary point at $(1, 2)$.

As x gets larger, the function starts behaving like $y = x$ (since $\frac{1}{x}$ becomes insignificant compared to x).

Use this sketch of $y = f(x)$ to help you sketch $y = f'(x)$.

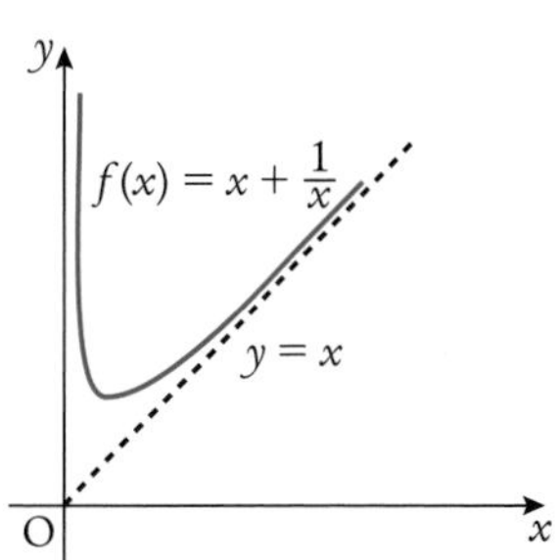

4 a The function $f(x) = \sin x$, $-\pi \leqslant x \leqslant \pi$ is drawn.

There is a minimum at $\left(-\frac{\pi}{2}, -1\right)$ and a maximum at $\left(\frac{\pi}{2}, 1\right)$.

When $x = -\pi$ and $x = \pi$ the gradient is -1.

When $x = 0$ the gradient is 1.

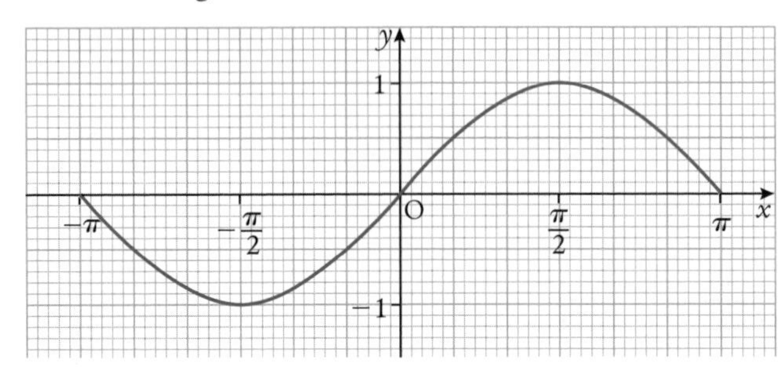

Use this information to help you make a sketch of the derivative of the sine function.

b Make a sketch of the derivative of the cosine function in the domain $-\pi \leqslant x \leqslant \pi$.

5 The relationship between a function and its derivative can be explored using the graphing facilities of a spreadsheet.

In row 1 enter the headings: A1 'x', B1 'f(x)', C1 'f'(x)', D1 'x+h', E1 'h'

In A2 enter a start to the domain e.g. -10

In A3 enter =A2+1 ... fill this down to row 26.

In B2 enter a function to study, say, $x^3 + x^2 + x - 50$ =A2^3+A2^2+A2-50 ... fill this down to row 26. Then copy B2 to B26 into E2 to E26.

In F2 enter a value for 'h'; you can change this and see how its size affects the approximation you get for the derivative. Let's say h is 0.01

In D2 enter =A2+$F$2 ... fill this down to row 26. This creates values for $x + h$ in column D and values for $f(x + h)$ in column E.

Finally, in C2 enter =(E2-B2)/$F$2 to calculate estimates for the derivative at x ... fill this down to row 26.

Select cells A1 to C26 and pick **Charts** from the menu; choose 'Scatter' then 'Scatter with Smooth Lines'. You'll get a graph similar to this. Using 'Chart Layout' the scales have been manipulated to make them more suited to the chosen function.

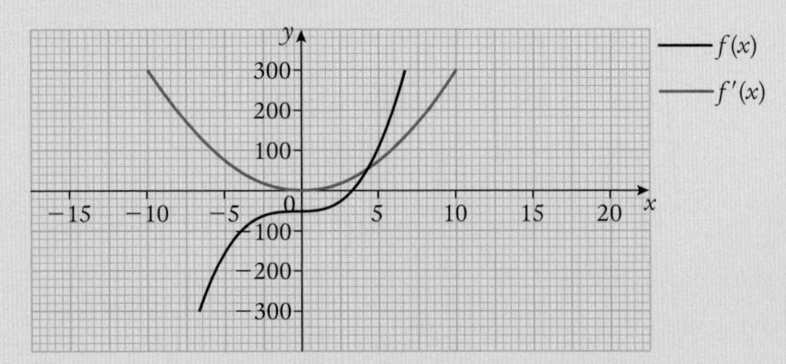

With slight alterations, other functions can be explored.

For example, in A2 enter a start to the domain, e.g. -PI()

In A3 enter =A2+PI()/12 ... fill this down to row 26. The domain is now $-\pi \leqslant x \leqslant \pi$.

In B2 enter the function =SIN(A2) ... fill down and copy B2 to B26 into E2 to E26 as before.

After adjusting the scales, the chart will look like:

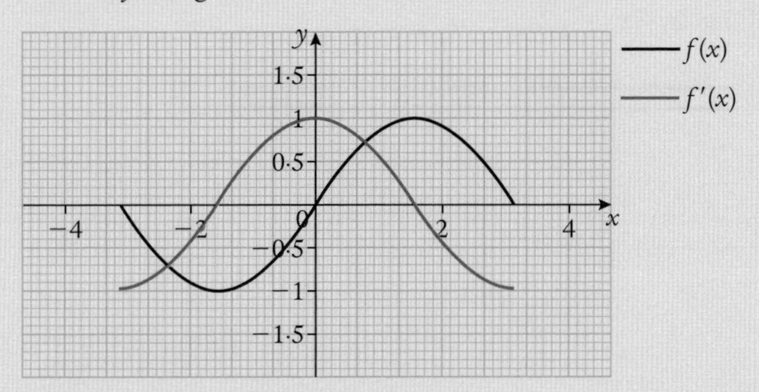

Explore other functions.

Try $y = \sin x°$: let the domain be $-180 \leqslant x \leqslant 180$ [in steps of 15].

The function in A2 must be =SIN(RADIANS(A2)) since the spreadsheet works in radians.

Do you still get a graph like that above?

7.6 The chain rule

Consider the composite function $y = g(f(x))$.

For the sake of reference we will call g the **outside** function and f the **inside** function.

Let $u = f(x)$; then $y = g(u)$.

Using Leibniz notation:

$$y = g(u) \Rightarrow \frac{dy}{du} = \lim_{\Delta u \to 0} \frac{\Delta y}{\Delta u}$$ ... the derivative of the *outside* function: $g'(u)$.

$$u = f(x) \Rightarrow \frac{du}{dx} = \lim_{\Delta u \to 0} \frac{\Delta u}{\Delta x}$$ ... the derivative of the *inside* function: $f'(x)$.

Multiplying these two derivatives produces a useful result:

$$\frac{dy}{du} \cdot \frac{du}{dx} = \lim_{\Delta u \to 0} \frac{\Delta y}{\Delta u} \times \lim_{\Delta u \to 0} \frac{\Delta u}{\Delta x} = \lim_{\Delta u \to 0, \Delta x \to 0} \frac{\Delta y}{\Delta u} \cdot \frac{\Delta u}{\Delta x} = \lim_{\Delta x \to 0} \frac{\Delta y}{\Delta x} = \frac{dy}{dx}$$

This result is known as the chain rule, viz.

$$\frac{dy}{dx} = \frac{dy}{du} \cdot \frac{du}{dx}$$

In functional notation this result reads:

If $h(x) = g(f(x))$ then $h'(x) = g'(u) \,.\, f'(x) = g'(f(x)) \,.\, f'(x)$

Example 1

Differentiate $y = (3x + 4)^6$.

It is impractical to expand this and then differentiate, but using the chain rule ...

Let $u = 3x + 4 \Rightarrow y = u^6$ ... give the inner function a symbol u

$$\Rightarrow \frac{du}{dx} = 3 \text{ and } \frac{dy}{du} = 6u^5$$

$$\Rightarrow \frac{dy}{dx} = \frac{dy}{du} \cdot \frac{du}{dx} = 6u^5 \,.\, 3 = 18u^5$$

$$\Rightarrow \frac{dy}{dx} = 18(3x + 4)^5$$

Example 2

Find the derived function when the function is defined by $f(x) = \sqrt{2x^3 + 5x + 1}$.

Let $u = 2x^3 + 5x + 1 \Rightarrow y = \sqrt{u} = u^{\frac{1}{2}}$ ... give the inner function a symbol u

$\Rightarrow \frac{du}{dx} = 6x^2 + 5$ and $\frac{dy}{du} = \frac{1}{2}u^{-\frac{1}{2}}$

$\Rightarrow \frac{dy}{dx} = \frac{dy}{du}.\frac{du}{dx} = \frac{1}{2}u^{-\frac{1}{2}}.(6x^2 + 5) = \frac{(6x^2 + 5)}{2u^{\frac{1}{2}}} = \frac{(6x^2 + 5)}{2\sqrt{u}}$

$\Rightarrow \frac{dy}{dx} = \frac{(6x^2 + 5)}{2\sqrt{2x^3 + 5x + 1}}$.

Note: This process can be shortened by remembering

$h'(x) = g'(f(x)).f'(x)$ ... i.e. derivative of *outside* × derivative of *inside*

Thus in the case of $f(x) = \sqrt{2x^3 + 5x + 1}$ ... we mentally identify the inside function as $2x^3 + 5x + 1$

- differentiate **outside** ... $\frac{1}{2}(2x^3 + 5x + 1)^{-\frac{1}{2}}$
- differentiate **inside** ... $(6x^2 + 5)$
- and multiply to get directly to $\frac{dy}{dx} = \frac{1}{2}(2x^3 + 5x + 1)^{-\frac{1}{2}}.(6x^2 + 5)$ in one line.

Exercise 7.6

1 Differentiate each of the following using the chain rule.

a $(2x + 1)^3$ b $(3x - 2)^5$ c $(x^2 + x)^4$ d $(x + 1)^{-1}$

e $(5 - x)^6$ f $(1 + 6x)^{-3}$ g $\left(\frac{x}{2} - 1\right)^2$ h $(2x^2 + 3x + 1)^5$

i $(x^3 + 2x^2 - 2)^{-2}$ j $(4 - 3x)^{-1}$

2 Find the derived function by first changing the function from fractional form, e.g. express $\frac{1}{2x - 1}$ as $(2x - 1)^{-1}$.

a $\frac{1}{2x + 1}$ b $\frac{4}{3x + 2}$ c $\frac{6}{1 - 2x}$ d $\frac{1}{(3x - 1)^2}$

e $\frac{1}{x^2 + 2x + 1}$ f $\frac{1}{(1 - x)^3}$ g $\frac{5}{(2x^2 - 3x + 1)}$ h $\frac{3}{x} + \frac{2}{x + 1}$

i $\frac{2}{(x + 1)^2} + \frac{1}{(x - 1)^2}$ j $\frac{4}{3x^3 + 2x^2 + x}$

3 Find the derivative of:

a $\sqrt{2x + 3}$ b $\sqrt{1 + 5x}$ c $\sqrt{3 + x^2}$ d $\sqrt[3]{3x + 7}$

e $\sqrt[4]{x^2 - 3x + 1}$ f $\frac{1}{\sqrt{3x + 1}}$ g $\frac{3}{\sqrt[3]{4x - 3}}$ h $\frac{1}{\sqrt{x}} + \frac{1}{\sqrt{x + 1}}$

i $\sqrt{2x + 1} + \frac{1}{\sqrt{2x + 1}}$ j $\sqrt[3]{(x - 1)(x + 1)}$.

4 Given that $f(x) = (3x + 2)^3$, find the value of: a $f'(0)$ b $f'(1)$ c $f'(2)$.

5 The function g is defined by $g(x) = \frac{36}{\sqrt{x + 1}}, x > -1$.

a Calculate the gradient of the curve $y = g(x)$ when: i $x = 0$ ii $x = 3$ iii $x = 8$.

b How can you tell that g is a decreasing function for all x in the domain?

6 The profile of a dome has been modelled by the function $h(x) = \sqrt{25 - x^2}$, $-5 \leqslant x \leqslant 5$. Its graph looks like:

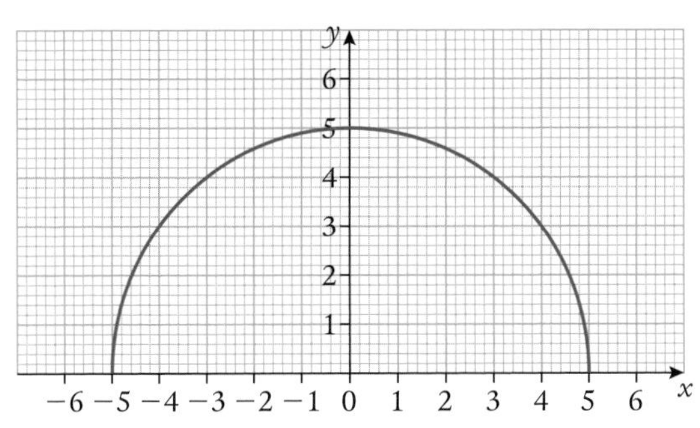

a Find an expression for $h'(x)$.
b Find the gradient of the tangent to the curve at: **i** $x = -3$ **ii** $x = 0$ **iii** $x = 4$.
c Find the equation of the tangent to the curve at $x = 3$.
d At $x = 5$ and $x = -5$ the profile is vertical. Sketch the graph $y = h'(x)$.

7 The time it takes to pump liquid from a pit is a function of the diameter of the pipe used, $t(x) = 20x + \dfrac{810}{(2x + 1)}$, where t is the time and x is the diameter, both measured in convenient units.
a Find $t'(x)$.
b Find the gradient of the curve $y = t(x)$ when $x = 7$.
c Find the equation of the tangent to the curve at $x = 1$.
d For what value of x is the gradient zero?
e Find the nature of this stationary point.

7.7 Trigonometric functions

Consider the function $f(x) = \sin x$, where x is measured in radians.

$$f'(x) = \lim_{h \to 0} \frac{f(x + h) - f(x)}{h}$$

$$\Rightarrow f'(x) = \lim_{h \to 0} \frac{\sin(x + h) - \sin x}{h}$$

$$\Rightarrow f'(x) = \lim_{h \to 0} \frac{\sin x \cos h + \cos x \sin h - \sin x}{h}$$

We learned in Chapter 2 that if h is measured in radians and h is small then $\cos h \approx 1$ and $\sin h \approx h$ and $\dfrac{\sin x \cos h + \cos x \sin h - \sin x}{h} \approx \dfrac{\sin x \,.\, 1 + \cos x \,.\, h - \sin x}{h} \approx \dfrac{\cos x \,.\, h}{h} \approx \cos x$.

This approximation becomes more and more accurate as h approaches zero.

Thus $\Rightarrow f'(x) = \lim_{h \to 0} \dfrac{\sin x \cos h + \cos x \sin h - \sin x}{h} = \cos x$.

So, $y = \sin x \Rightarrow \dfrac{dy}{dx} = \cos x$ ... when x is measured in radians

Similarly it can be shown that:

$y = \cos x \Rightarrow \dfrac{dy}{dx} = -\sin x$ ... when x is measured in radians

Using $\dfrac{d(\cos x)}{dx}$ to represent the derivative of $\cos x$, we have

$$\frac{d(\sin x)}{dx} = \cos x \text{ and } \frac{d(\cos x)}{dx} = -\sin x$$

Example 1

$f(x) = 2 \sin x - \sqrt{3} \cos x$

a Find $f'(x)$. b Evaluate $f'\left(\frac{\pi}{3}\right)$.

a $f(x) = 2 \sin x - \sqrt{3} \cos x; f'(x) = 2 \cos x + \sqrt{3} \sin x$

b $f'\left(\frac{\pi}{3}\right) = 2 \cos \frac{\pi}{3} + \sqrt{3} \sin \frac{\pi}{3} = 2 . \frac{1}{2} + \sqrt{3} . \frac{\sqrt{3}}{2} = 1 + \frac{3}{2} = \frac{5}{2}$

Example 2

Use the chain rule to differentiate: a $\sin^2 x$ b $\frac{\tan x + 1}{\sin x}$.

a $\sin^2 x = (\sin x)^2$ ... the inner function is $\sin x$

$\frac{d(\sin^2 x)}{dx} = 2 \sin x . \cos x = 2 \sin x \cos x$ ... derivative of **outside** × derivative of **inside**

b $\frac{\tan x + 1}{\sin x} = \frac{\tan x}{\sin x} + \frac{1}{\sin x} = \frac{1}{\cos x} + \frac{1}{\sin x} = (\cos x)^{-1} + (\sin x)^{-1}$

$\frac{d((\cos x)^{-1} + (\sin x)^{-1})}{dx} = [-(\cos x)^{-2} . -\sin x] + [-(\sin x)^{-2} . \cos x] = \frac{\sin x}{\cos^2 x} - \frac{\cos x}{\sin^2 x}$

Exercise 7.7

1 Starting with the definition $f'(x) = \lim_{h \to 0} \frac{\cos(x + h) - \cos x}{h}$, show that $\frac{d(\cos x)}{dx} = -\sin x$.

2 Differentiate:

a $3 \sin x$ b $2 \cos x$

c $\cos x + \sin x$ d $3 \cos x + 4 \sin x$.

3 A function has equation $y = 5 \sin x + 12 \cos x$.

a State $\frac{dy}{dx}$.

b Evaluate $\frac{dy}{dx}$ when: i $x = \frac{\pi}{3}$ ii $x = \frac{\pi}{2}$ iii $x = \pi$.

c Find the equation of the tangent to the curve at $x = \frac{\pi}{2}$.

4 a $f(x) = 2 \sin x + \sqrt{3} \cos x$; find $f'\left(\frac{\pi}{3}\right)$.

b $g(x) = 4 \sin x - 3 \cos x$; find $g'\left(\frac{\pi}{6}\right)$.

c $h(x) = 1 - 2 \cos x$; find $h'\left(\frac{\pi}{6}\right)$.

5 Differentiate the following using the chain rule.

a $\sin 3x$ b $\cos 5x$ c $2 \sin 4x$ d $5 \cos(-x)$

e $\sin(2x + 1)$ f $\cos(3x - 1)$ g $3 \sin(1 + 5x)$ h $2 \cos(3 - x)$

i $\cos^2 x$ j $x - \sin^2 x$ k $1 + \cos^2 x$ l $\cos^2 x - \sin^2 x$

m $\frac{3}{\cos x}$ n $\frac{2}{\sin x}$ o $\frac{1}{\cos^2 x}$ p $\frac{1 + \sin x}{\sin^3 x}$

q $\sqrt{\sin x}, 0 \leqslant x \leqslant \pi$ r $\sin(\cos x)$ s $\frac{1}{\sqrt{\cos x}}$ t $\frac{1}{\sqrt{\sin x + \cos x}}$

6 a Differentiate $\sin(3x - 1)$.

b Hence differentiate $\sin^2(3x - 1)$. [Note the composition of 3 functions $f(g(h(x)))$.]

c Similarly find the derivative of:

i $\cos^2(2x + 3)$ ii $\cos(2x + 3)^2$ iii $\sqrt{\sin(2x - 1)}, \frac{1}{2} \leqslant x \leqslant \frac{\pi + 1}{2}$.

7 It must be stressed that $\frac{\mathrm{d} \sin x}{\mathrm{d}x} = \cos x$, **only when the units used are radians**.

Remember that $x° = \frac{\pi}{180}x$ radians so $\sin x° = \sin\left(\frac{\pi}{180}x\right)$.

a Use the chain rule to help you differentiate $\sin x°$. Express your answer in degrees.

b Similarly differentiate $\cos x°$.

8 A function is defined by $f(x) = \frac{1}{\sin(2x)}, 0 \leqslant x \leqslant \frac{\pi}{2}$.

a Find $f'(x)$.

b For what value of x is $f'(x) = 0$?

c By considering values of $f'(x)$ either side of this, decide the nature of the turning point.

d What is the minimum value of the function in the given domain?

9 As a water wheel rotates the height of a point on its circumference can be modelled by $h(t) = 2\sin\left(\frac{\pi t}{3}\right)$ where h is the height in metres after t seconds has elapsed.

a In a time-distance relationship, the derivative gives a formula for the velocity of the point. Find a formula for $h'(t)$.

b What is the velocity of the point on the wheel after:

i 1 second ii 2 seconds?

c For what value(s) of t, in the first 4 seconds, is the velocity zero?

10 a Show that the function $x + \sin x$ is never decreasing.

b Determine whether $\cos x - x$ is never decreasing or is never increasing.

11 In Chapter 3 we learned the double angle formulae.

a Make use of these formulae to work out:

i $\frac{\mathrm{d}(2\sin x \cos x)}{\mathrm{d}x}$ ii $\frac{\mathrm{d}\left(\sin\frac{x}{2}\cos\frac{x}{2}\right)}{\mathrm{d}x}$.

b $\cos 2x = \cos^2 x - \sin^2 x$.

i Differentiate $\cos 2x$ **and** $\cos^2 x - \sin^2 x$.

ii Equate your two answers and comment.

iii Check that the derivative of $\cos^2 x - \sin^2 x$, $2\cos^2 x - 1$, and $1 - 2\sin^2 x$ are the same.

Preparation for assessment

1 Differentiate:

a $3x^3 + 2x^2 + x - 1$ b $\dfrac{1 + 3x}{2\sqrt{x}}$ c $(x + 1)(x - 2)$ d $x - x^{-\frac{2}{3}}$

e $2 \sin x$ f $5 \cos x$ g $3 - 2 \cos x$ h $1 + 2 \sin x - 7 \cos x$.

2 A function is defined by $f(x) = 2x^3 - 12x^2 - 30x + 5$.

a Calculate: i $f'(0)$ ii $f'(1)$.

b For what values of x is the gradient of the curve $y = f(x)$ zero?

c Describe the part of the domain where $f(x)$ is a decreasing function.

3 A parabola has equation $f(x) = 3x^2 + kx + 1$.

a For what value of k is $f'(1) = 8$?

b When k has this value, what is the equation of the tangent to the parabola at $x = 2$?

4 Differentiate:

a $\sqrt{x^2 + 2x - 1}$ b $\dfrac{1}{x - 3}$ c $(3x + 7)^{\frac{3}{4}}$

d $(2 \sin x - 1)^5$ e $\sin^3 x - \cos^3 x$ f $\dfrac{3}{\sin 2x}$.

5 The sketch shows $y = f(x)$ close to the origin.
$f'(0) = 1$
Make a sketch of $y = f'(x)$.

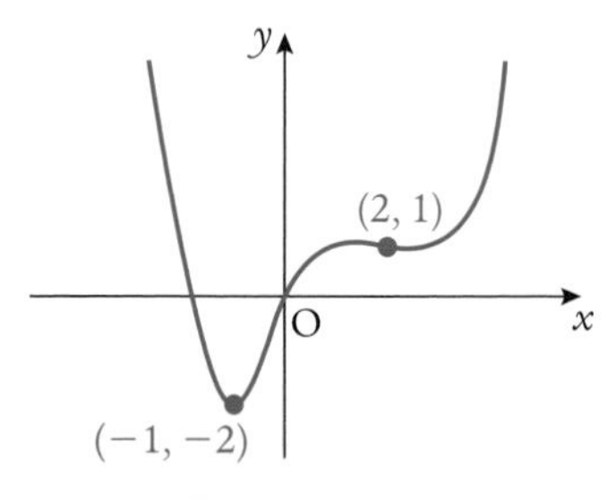

6 The depth of water in the harbour, d metres, can be modelled by
$d(x) = 4 \sin\left(\frac{\pi}{6}x + 5\right) + 6, 0 \leq x \leq 24$
where x is the number of hours that have elapsed since midnight.

a Find an expression for $d'(x)$.

b The derived function gives a formula for the rate at which the harbour is filling in metres per hour.
At what rate is it filling when: i $x = 1$ ii $x = 6$?
Give your answers correct to 2 decimal places.

c For what values of x in the domain is the derivative zero?
Give your answer to 2 decimal places.

Summary

1 $f'(x) = \lim\limits_{h \to 0} \dfrac{f(x+h) - f(x)}{h}$

- The process of finding the derived function is called differentiation.
- The derived function is also called the derivative.
- The derivative provides a formula for the **gradient** of the curve $y = f(x)$ at $P(x, f(x))$.
- The gradient at a point on a curve is defined to be the gradient of the tangent to the curve at that point.
- $\lim\limits_{h \to 0}$(expression) means 'the limit to which the expression tends as h tends to zero'.
- h is infinitesimally small, but not zero.
- In Leibniz notation, the definition of the derived function reads:
  $\dfrac{dy}{dx} = \lim\limits_{\Delta x \to 0} \dfrac{\Delta y}{\Delta x}$

2 $f(x) = ax^n \Rightarrow f'(x) = nax^{n-1}$

Thus, if we wish to differentiate a term of the form ax^n, we need only:

- multiply by the power
- reduce the power by 1.

3 i $k(x) = f(x) + g(x) \Rightarrow k'(x) = f'(x) + g'(x)$

ii $k(x) = af(x) \Rightarrow k'(x) = af'(x)$

iii $f(x) = a$ (where a is a constant) $\Rightarrow f'(x) = 0$

4 In general, for any function $f(x)$,

- $f'(x) > 0 \Leftrightarrow$ function is strictly increasing
- $f'(x) < 0 \Leftrightarrow$ function is strictly decreasing
- $f'(x) = 0 \Leftrightarrow$ function is stationary.

5 Given the graph $y = f(x)$ you should be able to sketch $y = f'(x)$.

- Both sketches should start and finish at the same x-values.
- Where there are stationary points on $y = f(x)$, there will be zeros on $y = f'(x)$.
- Where the gradient of $f(x) > 0$, the curve of $y = f'(x)$ will be above the axis.
- Where the gradient of $f(x) < 0$, the curve of $y = f'(x)$ will be below the axis.

6 The chain rule

Leibniz notation: $\dfrac{dy}{dx} = \dfrac{dy}{du} \cdot \dfrac{du}{dx}$.

Functional notation: $h(x) = g(f(x)) \Rightarrow h'(x) = g'(f(x)) \cdot f'(x)$.

In words: derivative of **outside** $\times$ derivative of **inside**.

7

- $\dfrac{d(\sin x)}{dx} = \cos x$
- $\dfrac{d(\cos x)}{dx} = -\sin x$

Unit 2: Relationships and calculus

8 Trigonometry 3 – equations

Before we start...

Trigonometry developed from the needs of astronomers as they tried to make sense of the movements of planets among the stars and their distances from each other and the Earth.

The ancient astronomers discovered that equations of the form $y = a\sin(bx + c) + d$ could be used to model periodic motion. They believed that when two heavenly bodies occupied the same part of the heavens at the same time (known as a **conjunction**), that these were auspicious occasions of good omen and could decide whether new ventures should or should not be undertaken. Solving equations of the form $a\sin(bx + c) + d = e\sin(fx + g) + h$ would allow the astronomer to predict such moments of conjunction.

This chapter looks at the solutions to various forms of trigonometric equation. Be aware, however, that not all trigonometric equations will yield to these basic algebraic methods.

Consider a classic problem often referred to as 'The tethered goat'.

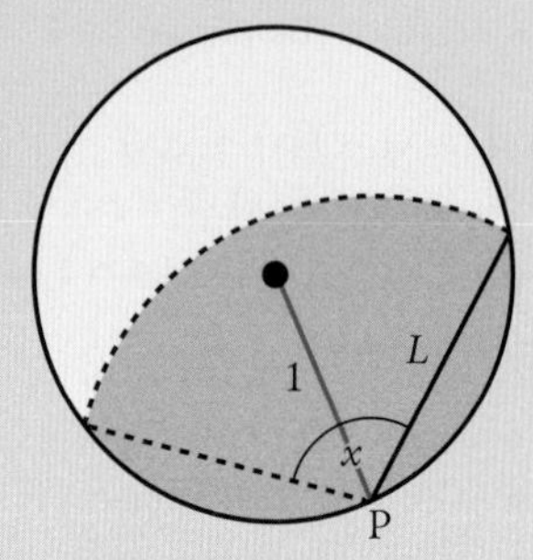

A goat is tethered inside a circular field of radius 1 unit. It is attached to a post, P, on the circumference of the field by a rope of length L units.

How long has the rope got to be so that the goat has access to half of the field?

If the angle between the extreme positions of the rope is x radians, then you can solve this problem if you can solve the equation:

$$\sin x - x\cos x - \frac{\pi}{2} = 0.$$

Here are a few challenges to tackle before the end of the chapter.

i Can you find an expression for the area accessible to the goat in terms of L and x?
ii Can you derive the above equation?
iii Can you solve the above equation ... knowing that L must lie between 1 and 2 units?

What you need to know

1 Solve the following equations in the domain $0 \leqslant x \leqslant 360$, correct to 1 decimal place where appropriate.

a $3\sin x° = 2$ **b** $2\cos x° + 1 = 0$

2 When you type $\sin^{-1}(0{\cdot}8)$ into a calculator, it returns the value 53° to the nearest degree. What other angles from 0° to 720° have a sine of 0·8?

3 Solve: **a** $3\sin 2x° + 1 = 0; 0 \leqslant x \leqslant 360$ **b** $1 - 2\cos 4x° = 0; 0 \leqslant x \leqslant 90$.

4 For what values of x between 0 and 360 is $(2\sin x° - 1)(\sqrt{3}\cos x° - 1) = 0$?

5 a Factorise: **i** $2a^2 - a - 1$ ii $2\sin^2 x° - \sin x° - 1$.

b Hence solve: $2\sin^2 x° - \sin x° - 1 = 0$.

8.1 Appropriate solutions

Domains and units

Propagation of solutions

When we solve an equation of the form $\sin x^\circ = a$, or $\cos x^\circ = a$, or $\tan x^\circ = a$, where $-1 \leqslant a \leqslant 1$, a is a constant, we must remember that there is an infinite set of solutions.

- The first solution we get from the calculator and we denote it by $\sin^{-1} a$, $\cos^{-1} a$, or $\tan^{-1} a$, as appropriate.
- A second solution comes from the symmetry of the function
  ... if $\sin^{-1} a$ is a solution then so is $180 - \sin^{-1} a$
  ... if $\cos^{-1} a$ is a solution then so is $360 - \cos^{-1} a$
  ... if $\tan^{-1} a$ is a solution then so is $180 + \tan^{-1} a$.
- All other solutions come from the fact that the trig. functions are periodic
  ... if x is a solution then so is $x \pm 360n$, where n is a whole number.

Care must be taken when solving equations to remember that the propagation of solutions occurs at the point where the inverse trig. function is being found, not later after further manipulation.

Domain

Normally the number of solutions will be restricted by the context, or by geometric consideration, or by the person constructing the question.

- It is common that the first revolution is used: $0 \leqslant x \leqslant 360$ or $0 \leqslant x \leqslant 2\pi$.
- If the angle is acute: $0 \leqslant x \leqslant 90$ or $0 \leqslant x \leqslant \frac{\pi}{2}$.
- Often a particular quadrant is specified. [See Examples 1, 2 and 3 where the compound angle formulae are exploited.]

Units

Care must be taken to work in appropriate units ... degrees or radians.

- Remember, $\sin x^\circ$ for degrees; $\sin x$ for radians.
- The given domain may also dictate the most appropriate unit.
- If you wish to use $\frac{d}{dx}\sin x = \cos x$, then radians is the only unit for which this is true.

Example 1

Solve $3\sin(2x + 30)^\circ = 1$; $0 \leqslant x \leqslant 360$, giving your answers to the nearest degree.

$3\sin(2x + 30)^\circ = 1$

$\Rightarrow \sin(2x + 30)^\circ = \frac{1}{3}$

$\Rightarrow \quad 2x + 30 = \sin^{-1}\frac{1}{3}$ or $180 - \sin^{-1}\frac{1}{3}$ or $\sin^{-1}\frac{1}{3} \pm 360n$ or $(180 - \sin^{-1}\frac{1}{3}) \pm 360n$

... propagation of solutions occurs at this step[1]

$\Rightarrow \quad 2x + 30 = \ldots -340{\cdot}5, -199{\cdot}5, 19{\cdot}5, 160{\cdot}5, 379{\cdot}5, 520{\cdot}5, 739{\cdot}5, 880{\cdot}5, \ldots$

... if you don't include enough items here, some solutions may go missing at the next step

$\Rightarrow \quad 2x = \ldots -370{\cdot}5, -229{\cdot}5, -10{\cdot}5, 130{\cdot}5, 349{\cdot}5, 490{\cdot}5, 709{\cdot}5, 850{\cdot}5, \ldots$

... when you divide by 2 you are going to pull more solutions into the domain

$\Rightarrow \quad x = -185{\cdot}25, -114{\cdot}75, -5{\cdot}25, 65{\cdot}25, 174{\cdot}75, 245{\cdot}25, 354{\cdot}75, 425{\cdot}25, \ldots$

... list the solutions in the desired domain, units and to the desired accuracy

$\Rightarrow \quad x^\circ = 65^\circ, 175^\circ, 245^\circ$, or 355°

[Note: In equations where the angle is 'compound' we have to guard against possible solutions being missed. Always include some possibilities either side of the desired domain in the initial 'propagation' of solutions.]

---

[1] The above is written to reflect the order in which you would find the solutions. Generally the solutions are then listed in numerical order with the two principal solutions in the middle of the list, i.e.
$\sin^{-1}\frac{1}{3} - 360, (180 - \sin^{-1}\frac{1}{3}) - 360, \sin^{-1}\frac{1}{3}, 180 - \sin^{-1}\frac{1}{3}, \sin^{-1}\frac{1}{3} + 360, (180 - \sin^{-1}\frac{1}{3}) + 360, \ldots$

Example 2

Solve $2\cos^2 4x + 3\cos 4x - 2 = 0; 0 \leq x \leq \pi$.

$2\cos^2 4x + 3\cos 4x - 2 = 0$

$\Rightarrow (2\cos 4x - 1)(\cos 4x + 2) = 0$ ... note the quadratic form $2a^2 + 3a - 2 = (2a - 1)(a + 2)$

So $2\cos 4x - 1 = 0$ or $\cos 4x + 2 = 0$

$\Rightarrow \cos 4x = \frac{1}{2}$ or $\cos 4x = -2$ ... there are no solutions to the second option: $-1 \leq \cos a \leq 1$

$\Rightarrow 4x = \frac{\pi}{3}$, or $2\pi - \frac{\pi}{3}$, or $\frac{\pi}{3} \pm 2\pi n$, or $\left(2\pi - \frac{\pi}{3}\right) \pm 2\pi n$

$\Rightarrow 4x = \ldots -\frac{5\pi}{3}, -\frac{\pi}{3}, \frac{\pi}{3}, \frac{5\pi}{3}, \frac{7\pi}{3}, \frac{11\pi}{3}, \frac{13\pi}{3}, \frac{17\pi}{3}$ ... π is the upper end of the domain so have to go to 4π

$\Rightarrow \quad x = -\frac{5\pi}{12}, -\frac{\pi}{12}, \frac{\pi}{12}, \frac{5\pi}{12}, \frac{7\pi}{12}, \frac{11\pi}{12}, \frac{13\pi}{12}, \frac{17\pi}{12}$ ... because we are dividing by 4

In the desired domain, $x = \frac{\pi}{12}, \frac{5\pi}{12}, \frac{7\pi}{12}, \frac{11\pi}{12}$.

Example 3

The graph shows the curves $y = \sin\left(2x + \frac{\pi}{3}\right)$ and $y = \frac{1}{2}\sin 2x$; $-\pi \leq x \leq \pi$.

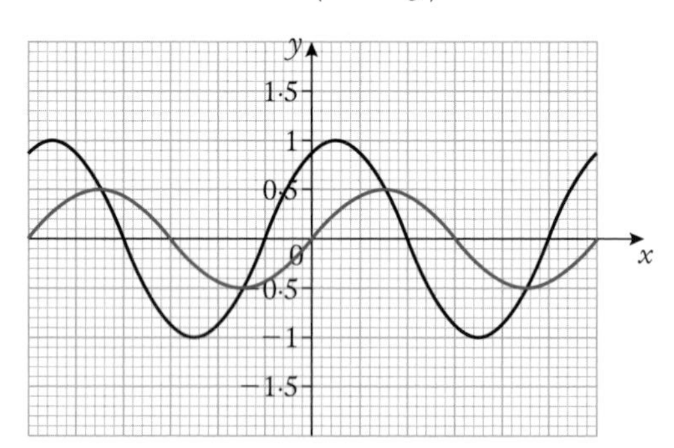

Find their points of intersection in this domain.

The curves intersect where $\sin\left(2x + \frac{\pi}{3}\right) = \frac{1}{2}\sin 2x$ ... in this form we don't have a technique for solving the equation ... so expand the compound angle

$\Rightarrow \sin 2x \cos\frac{\pi}{3} + \cos 2x \sin\frac{\pi}{3} = \frac{1}{2}\sin 2x$

$\Rightarrow \quad \sin 2x \,.\, \frac{1}{2} + \cos 2x \,.\, \frac{\sqrt{3}}{2} = \frac{1}{2}\sin 2x$

$\Rightarrow \cos 2x = 0$ ... we can solve equations of this form

$\Rightarrow 2x = \frac{\pi}{2}$ or $2\pi - \frac{\pi}{2}$ or $\frac{\pi}{2} \pm 2\pi n$ or $\left(2\pi - \frac{\pi}{2}\right) \pm 2\pi n$

$\Rightarrow 2x = \ldots -\frac{5\pi}{2}, -\frac{3\pi}{2}, -\frac{\pi}{2}, \frac{\pi}{2}, \frac{3\pi}{2}, \frac{5\pi}{2}, \ldots$

$\Rightarrow \quad x = -\frac{5\pi}{4}, -\frac{3\pi}{4}, -\frac{\pi}{4}, \frac{\pi}{4}, \frac{3\pi}{4}, \frac{5\pi}{4}, \ldots$

In the domain stated, $\Rightarrow x = -\frac{3\pi}{4}, -\frac{\pi}{4}, \frac{\pi}{4}, \frac{3\pi}{4}$.

Exercise 8.1

1 Solve each equation in the domain specified.

a $4 \sin 3x° = 1; 0 \leqslant x \leqslant 360$

b $5 \cos 2x° = 2; 0 \leqslant x \leqslant 360$

c $2 \sin (3x + 30)° = 1; 0 \leqslant x \leqslant 360$

d $3 \sin (4x + 90)° = 1; 0 \leqslant x \leqslant 180$

e $4 + 3 \cos (3x - 60)° = 6; -180 \leqslant x \leqslant 180$

f $3 - 2 \sin (5x + 100)° = 2; 0 \leqslant x \leqslant 180$

2 Solve each equation in the domain specified. Note that the angles are measured in radians.

a $5 \sin 2x = 4; 0 \leqslant x \leqslant 2\pi$

b $3 \cos 3x = 2; 0 \leqslant x \leqslant 2\pi$

c $2 \sin \left(3x + \frac{\pi}{3}\right) = 1; 0 \leqslant x \leqslant 2\pi$

d $2\sin \left(2x + \frac{\pi}{6}\right) = \sqrt{3}; 0 \leqslant x \leqslant 2\pi$

e $1 - 2 \cos \left(3x - \frac{\pi}{3}\right) = 2; -\pi \leqslant x \leqslant \pi$

f $5 + \sin (5x + \pi) = 6; 0 \leqslant x \leqslant \pi$

3 Solve, by first factorising a quadratic expression:

a $2 \sin^2 2x° - \sin 2x° - 1 = 0; 0 \leqslant x \leqslant 360$

b $6 \cos^2 3x° - \cos 3x° - 1 = 0; 0 \leqslant x \leqslant 360$

c $4 \sin^2 3x° + 11 \sin 3x° + 6 = 0; 0 \leqslant x \leqslant 180$

d $\cos^2 2x° + \cos 2x° - 2 = 0; 0 \leqslant x \leqslant 360$

e $\cos^2 3x = 3 \cos 3x + 4; -180 \leqslant x \leqslant 180$

f $\sin^2 2x° + 3 = -4 \sin 2x°; -180 \leqslant x \leqslant 180.$

4 Find the solution to each equation in the stated range.

a $\sin^2 3x - 1 = 0; 0 \leqslant x \leqslant 2\pi$

b $2 \cos^2 2x - 3 \cos 2x + 1 = 0; 0 \leqslant x \leqslant 2\pi$

c $\sin^2 4x - 3 \sin 4x + 2 = 0; 0 \leqslant x \leqslant \pi$

d $\cos^2 3x - 2 \cos 3x - 3 = 0; -\pi \leqslant x \leqslant \pi$

e $(2 \sin 2x - \sqrt{3})(\sin 2x + 2) = 0; -\pi \leqslant x \leqslant \pi$

f $(\sqrt{2} \cos 3x - 1)(\cos 3x + 2) = 0; 0 \leqslant x \leqslant 2\pi$

5 The sketch shows $y = 2 \sin^2 2x$ and $y = 3 \sin 2x - 1$ between 0 and π.

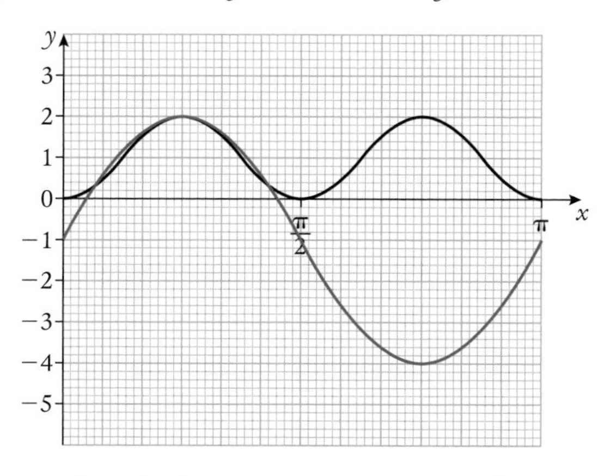

a Where do the two curves intersect in this interval?

b Where do they intersect between π and 2π?

6 The depth of water at the berth of the tall ship in Glasgow can be modelled by $d(t) = 0{\cdot}6 \sin(15t + 60)° + 1{\cdot}5$, where it is d metres deep t hours after midnight.

a How deep is the water at 4 a.m.?

b What is the difference between the water at low tide and at high tide?

c At what times in the first 24 hours was the depth 1·5 m?

8.2 Using double angle formulae

For certain forms of trigonometric equations the **double angle** formulae can be exploited.

These are especially useful when the equation contains a trigonometric term involving the double angle and a trigonometric term involving the single angle.

Remember: $\sin 2x = 2\sin x\cos x$

$$\begin{aligned}\cos 2x &= \cos^2 x - \sin^2 x\\ &= 2\cos^2 x - 1\\ &= 1 - 2\sin^2 x\end{aligned}$$

Example 1

Solve $\sin 2x - \cos x = 0; 0 \leqslant x \leqslant 2\pi$.

$$\sin 2x - \cos x = 0$$

$\Rightarrow \quad 2\sin x\cos x - \cos x = 0 \quad$ ... using $\sin 2x = 2\sin x\cos x$

$\Rightarrow \quad \cos x(2\sin x - 1) = 0$

$\Rightarrow \cos x = 0$ or $2\sin x - 1 = 0$

$\Rightarrow \quad x = \frac{\pi}{2}$ or $2\pi - \frac{\pi}{2}$ or $x = \frac{\pi}{6}$ or $\pi - \frac{\pi}{6}$

$\Rightarrow \quad x = \frac{\pi}{2}, \frac{3\pi}{2}, \frac{\pi}{6}$, or $\frac{5\pi}{6}$

Example 2

Solve: **a** $\cos 2x + \sin x = 0; 0 \leqslant x \leqslant 2\pi$ **b** $3\cos 2x° - 14\cos x° + 7 = 0; 0 \leqslant x \leqslant 360$.

a Note that the trig. term involving the single angle is a **sine** ... use $\cos 2x = 1 - 2\sin^2 x$

$$\cos 2x + \sin x = 0$$

$\Rightarrow \quad (1 - 2\sin^2 x) + \sin x = 0$

$\Rightarrow \quad 2\sin^2 x - \sin x - 1 = 0$

$\Rightarrow (2\sin x + 1)(\sin x - 1) = 0$

$\Rightarrow 2\sin x + 1 = 0$ or $\sin x - 1 = 0$

$\Rightarrow \sin x = -\frac{1}{2}$ or $\sin x = 1$

$\Rightarrow x = -\frac{\pi}{6}, \pi - \left(\frac{-\pi}{6}\right), \left(\frac{-\pi}{6}\right) \pm 2\pi n, \left(\pi - \left(-\frac{\pi}{6}\right)\right) \pm 2\pi n$ or $x = \frac{\pi}{2}, \pi - \frac{\pi}{2}$

$\Rightarrow x = \frac{7\pi}{6}, \frac{11\pi}{6}$, or $\frac{\pi}{2}$ in the region $0 \leqslant x \leqslant 2\pi$

b Note that the trig. term involving the single angle is a **cosine** ... use $\cos 2x = 2\cos^2 x - 1$

$$3\cos 2x° - 14\cos x° + 7 = 0$$

$\Rightarrow 3(2\cos^2 x° - 1) - 14\cos x° + 7 = 0$

$\Rightarrow \quad 6\cos^2 x° - 14\cos x° + 4 = 0 = 2(3\cos^2 x° - 7\cos x° + 2)$

$\Rightarrow \quad 2(\cos x - 2)(3\cos x - 1) = 0$

$\Rightarrow \cos x° = 2$ (no solutions) or $\cos x° = \frac{1}{3}$

$\Rightarrow x = 70{\cdot}5$ or $360 - 70{\cdot}5 = 289{\cdot}5$ (to 1 d.p.)

Example 3

Solve $\sin x - \sin\frac{x}{2} = 0; 0 \leqslant x \leqslant 2\pi$.

$$\sin x - \sin\frac{x}{2} = 0$$

$\Rightarrow 2\sin\frac{x}{2}\cos\frac{x}{2} - \sin\frac{x}{2} = 0$ ... using $\sin 2A = 2\sin A\cos A$, where $A = \frac{x}{2}$

$\Rightarrow \sin\frac{x}{2}\left(2\cos\frac{x}{2} - 1\right) = 0$

$\Rightarrow \sin\frac{x}{2} = 0$ or $2\cos\frac{x}{2} - 1 = 0 \Rightarrow \sin\frac{x}{2} = 0$ or $\cos\frac{x}{2} = \frac{1}{2}$

$\Rightarrow \frac{x}{2} = 0, \pi, 2\pi, \ldots$ or $\frac{\pi}{3}, \frac{5\pi}{3}, \ldots$

$\Rightarrow x = 0, 2\pi, 4\pi, \ldots$ or $\frac{2\pi}{3}, \frac{10\pi}{3}, \ldots$

$\Rightarrow x = 0, 2\pi$, or $\frac{2\pi}{3}$ in the desired domain.

Exercise 8.2

1 Solve in the stated domain.

a $\sin 2x° + \cos x° = 0; 0 \leqslant x \leqslant 360$

b $3\sin 2x° - 4\cos x° = 0; 0 \leqslant x \leqslant 360$

c $3\sin 2x + 15\sin x = 0; 0 \leqslant x \leqslant 360$

d $2\sin 2x + 3\sin x = 0; -180 \leqslant x \leqslant 180$

e $\sin 2x° - 8\cos x° = 0; -180 \leqslant x \leqslant 180$

f $15\sin x° - 5\sin 2x° = 0; 0 \leqslant x \leqslant 360$

g $3\sin x° - 2\sin\frac{x°}{2} = 0; 0 \leqslant x \leqslant 720$

h $\cos\frac{x°}{2} - \sin x° = 0; 0 \leqslant x \leqslant 360$

2 Solve in the stated domain, leaving π in your answers where possible.

a $\sqrt{2}\sin 2x - 2\cos x = 0; 0 \leqslant x \leqslant 2\pi$

b $\sin 2x - \sqrt{3}\sin x = 0; 0 \leqslant x \leqslant 2\pi$

c $\sin 2x - 2\sin x = 0; 0 \leqslant x \leqslant 2\pi$

d $2\cos x - \sin 2x = 0; -\pi \leqslant x \leqslant \pi$

e $\sqrt{6}\cos x + \sqrt{2}\sin 2x = 0; 0 \leqslant x \leqslant \pi$

f $2\sin\frac{x}{2} + \sqrt{2}\sin x = 0; 0 \leqslant x \leqslant 4\pi$

3 The linkages in a locomotive move in a periodic fashion.
Two points on the assembly are modelled by

$$h(x) = 2\sin 2x° \text{ and } g(x) = \cos x°$$

where h and g are the heights of the two points above the centre of the large wheel in convenient units and $x°$ is the angle turned by the wheel.
The graph charts the two points for $0 \leqslant x \leqslant 360$.

Unit 2: Relationships and calculus

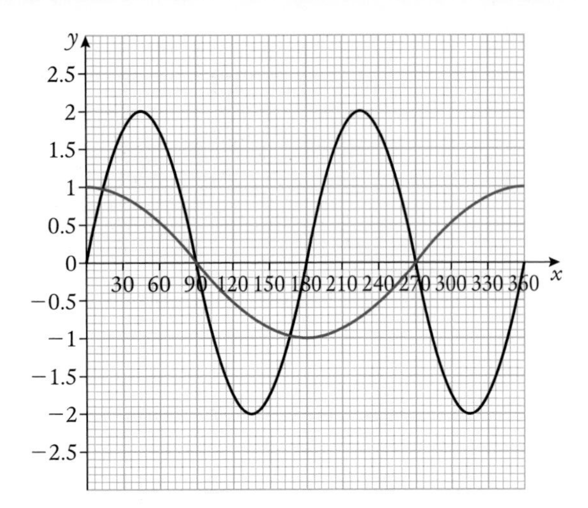

Find the four values of x in this interval that the two points are at the same height.

4 Solve each equation where $0 \leqslant x \leqslant 360$.

a $\cos 2x° - 3\cos x° + 2 = 0$

b $2 + \sin x° - 3\cos 2x° = 0$

c $3\cos 2x° + 8\cos x° - 5 = 0$

d $11 - 9\sin x° + 2\cos 2x° = 0$

e $3 - 4\cos x° + 2\cos 2x°$

f $3\cos x° + 10\cos \dfrac{x°}{2} + 7 = 0$

5 Solve each equation, giving your solutions correct to 3 significant figures.

a $3 + 8\sin x - 5\cos 2x = 0; 0 \leqslant x \leqslant 2\pi$

b $3\cos 2x - 10\cos x + 7 = 0; 0 \leqslant x \leqslant 2\pi$

c $7\sin x - \cos 2x - 3 = 0; -\pi \leqslant x \leqslant \pi$

d $4\cos 2x + 10\cos x + 1 = 0; 0 \leqslant x \leqslant \pi$

e $8 + 15\sin \dfrac{x}{2} - \cos x = 0; 0 \leqslant x \leqslant 4\pi$

f $9\cos x - 12\cos \dfrac{x}{2} - 7 = 0; 0 \leqslant x \leqslant 2\pi$

6 Two different studies of the population of a rabbit warren were conducted.

Both studies felt the numbers followed a cycle with a period of about 6 years.

Study A modelled the population by $P = 3\cos 2x + \cos x + 5$, where P represents the numbers in hundreds of rabbits and x represents the year in the 6-year cycle.

Study B models the situation by $P = 2\cos x + 3$.

The chart gives an indication of how the models differ.

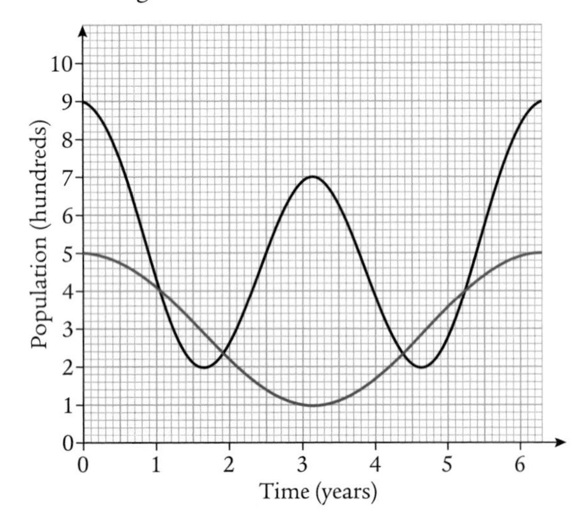

For what values of x and P do the models agree?

7 The postage stamp shows a ladies' bike. The back wheel and front wheel turn at different rates.

The height, h units, of a point on the front wheel can be modelled by $h = 2(\cos 2x° + 1)$ and the height of a point on the large wheel by $h = 4\cos x° + 4$, where $x°$ is the angle through which the large wheel has turned.

a Find the values of x, $0 \leqslant x \leqslant 360$, which make both heights the same.

b What is the value of h that corresponds to these values of x?

8 Using suitable units the light coming from an orbiting binary system of stars can be modelled by the equation $y = \cos 2x + \sin x + 3$, where y is a measure of the light at time x.

The graph shows the function in the region $0 \leqslant x \leqslant 2\pi$.

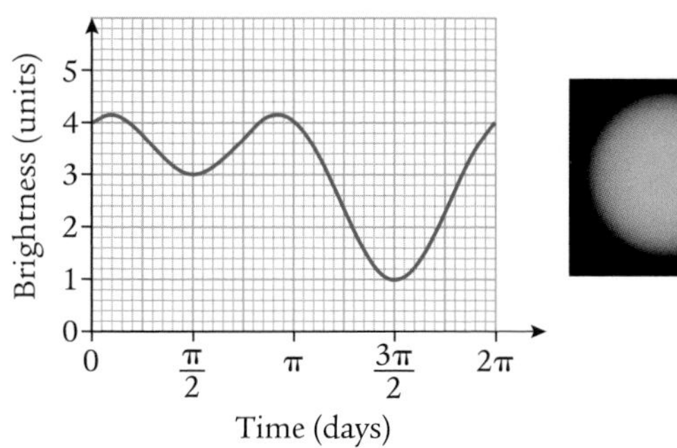

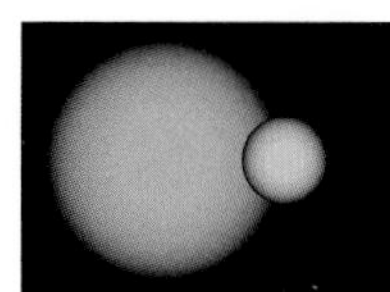

The smaller dip is when the smaller star goes in front of the larger one. The larger dip is when the larger star goes in front of the smaller. The maximums are when the stars are both distinct.

a Find $\frac{dy}{dx}$.

b Solve $\frac{dy}{dx} = 0$ to find when the gradient of the curve is zero in the domain $0 \leqslant x \leqslant 2\pi$.

c Hence find the maximum and minimum brightness.

8.3 Using compound angle formulae

In Chapter 3 we discovered that expressions of the form $a\sin x \pm b\cos x$ can be expressed in the form $p\sin(x \pm q)$ or $p\cos(x \pm q)$, where p and q are positive numbers.

This manipulation has the benefit that an expression with two trigonometric terms is reduced to an expression with only one such term.

This is very useful when solving equations.

There are four possible forms into which you can change $a\sin x \pm b\cos x$ ... if you have the freedom of choice, use $p\sin(x \pm q)$, picking the sign that agrees with the original expression.

Example 1

Solve the equation $4 \sin x° + 3 \cos x° = 1$, for $0 \leq x \leq 360$.

$4 \sin x° + 3 \cos x° = 1$

Expanding $p \sin (x + q)° = p \sin x° \cos q° + p \cos x° \sin q°$ ... '+' selected to agree with original

Equate the coefficients of $\sin x°$: $p \cos q° = 4$ ... ① ... $q°$ is in 1st or 4th quadrant

Equate the coefficients of $\cos x°$: $p \sin q° = 3$ ... ② ... $q°$ is in 1st or 2nd quadrant

... so $q°$ is in 1st quadrant

② ÷ ①: $\frac{p \sin q°}{p \cos q°} = \tan q° = \frac{3}{4}$ and $q°$ is in 1st quadrant

$\Rightarrow \quad q° = 36{\cdot}9°$

②² + ①²: $p^2 \cos^2 q° + p^2 \sin^2 q° = 4^2 + 3^2 = 25$

$\Rightarrow \quad p^2 (\cos^2 q° + \sin^2 q°) = 25$

$\Rightarrow \quad p = 5$

So, $4 \sin x° + 3 \cos x° = 1$

$\Rightarrow \quad 5 \sin (x + 36{\cdot}9)° = 1$

$\Rightarrow \quad \sin (x + 36{\cdot}9)° = 0{\cdot}2$

$\Rightarrow x + 36{\cdot}9 = \sin^{-1}(0{\cdot}2)$ or $180 - \sin^{-1}(0{\cdot}2)$ or $\sin^{-1}(0{\cdot}2) \pm 360n$ or $180 - \sin^{-1}(0{\cdot}2) \pm 360n$

$\Rightarrow x + 36{\cdot}9 = 11{\cdot}5$ or $168{\cdot}5$ or $371{\cdot}5$ or ...

$\Rightarrow x = 11{\cdot}5 - 36{\cdot}9$ or $168{\cdot}5 - 36{\cdot}9$ or $371{\cdot}5 - 36{\cdot}9$ or ...

$\Rightarrow x = -25{\cdot}4$ or $131{\cdot}6$ or $334{\cdot}6$ or ...

$\Rightarrow x = 131{\cdot}6$ or $334{\cdot}6$ ... domain is given as $0 \leq x \leq 360$

Example 2

a Express $5 \cos x + 12 \sin x$ in the form $k \cos (x + \alpha)$.

b Hence solve the equation $5 \cos x + 12 \sin x = 8$; $0 \leq x \leq 2\pi$.

a Expand: $k \cos (x + \alpha) = k \cos x \cos \alpha - k \sin x \sin \alpha$ ... the required form is 'forced' on you

Compare with $5 \cos x + 12 \sin x$.

Equate the coefficients of $\sin x$: $-k \sin \alpha = 12$ ... ① ... α is in 3rd or 4th quadrant

Equate the coefficients of $\cos x$: $k \cos \alpha = 5$ ... ② ... α is in 1st or 4th quadrant

... so α is in 4th quadrant

① ÷ ②: $\frac{k \sin \alpha}{k \cos \alpha} = \tan \alpha = -\frac{12}{5}$ and α is in 4th quadrant

$\Rightarrow \quad \alpha = -1{\cdot}176$ or $\pi + (-1{\cdot}176)$ or $2\pi + (-1{\cdot}176) = -1{\cdot}176$ or $1{\cdot}966$ or $5{\cdot}107$

Since angle is in 4th quadrant we pick $\alpha = 5{\cdot}107$

②² + ①²: $k^2 \cos^2 \alpha + k^2 \sin^2 \alpha = 5^2 + 12^2 = 169$

$\Rightarrow \quad k^2 (\cos^2 \alpha + \sin^2 \alpha) = 169$

$\Rightarrow \quad k = 13$

b $5 \cos x + 12 \sin x = 8$

$\Rightarrow 13 \cos (x + 5{\cdot}107) = 8$

$\Rightarrow \quad \cos (x + 5{\cdot}107) = 8 \div 13$

$\Rightarrow x + 5{\cdot}107 = \cos^{-1}\left(\frac{8}{13}\right)$ or $2\pi - \cos^{-1}\left(\frac{8}{13}\right)$ or $\cos^{-1}\left(\frac{8}{13}\right) \pm 2\pi n$ or $2\pi - \cos^{-1}\left(\frac{8}{13}\right) \pm 2\pi n$

$\Rightarrow x + 5{\cdot}107 = 0{\cdot}908, 5{\cdot}375, 7{\cdot}191, 11{\cdot}658, 13{\cdot}474, \ldots$

$\Rightarrow \quad x = -4{\cdot}199, 0{\cdot}268, 2{\cdot}084, 6{\cdot}551, 8{\cdot}367, \ldots$

$\Rightarrow \quad x = 0{\cdot}268, 2{\cdot}084$ ... the domain is $0 \leq x \leq 2\pi$

Exercise 8.3

1 a Solve for x in the domain $0 \leqslant x \leqslant 360$.

i $\sin x° + \cos x° = 1$ ii $4 \sin x° + 3 \cos x° = 2$

iii $8 \sin x° - 15 \cos x° = 10$ iv $7 \cos x° - 24 \sin x° = 5$

v $20 \sin x° + 21 \cos x° = 6$ vi $2 \sin x° - 3 \cos x° = -1$

b Solve for x in the domain $0 \leqslant x \leqslant 2\pi$.

i $\sin x - \cos x = 0{\cdot}5$ ii $3 \cos x - 4 \sin x = 3$

iii $\cos x + \sqrt{2} \sin x = 1$ iv $2 \sin x + \sqrt{3} \cos x = -2$

2 a Express $8 \cos x + 15 \sin x$ in the form $k \sin (x - \alpha)$.

b Hence solve the equation $8 \cos x + 15 \sin x = 12$; $0 \leqslant x \leqslant 2\pi$.

3 a Express $24 \sin x° + 7 \cos x°$ in the form $k \cos (x + \alpha)°$.

b Hence solve the equation $24 \sin x° + 7 \cos x° = 1$; $0 \leqslant x \leqslant 360$.

4 a Express $\cos x - 2 \sin x$ in the form $k \sin (x + \alpha)$.

b Hence solve the equation $\cos x - 2 \sin x = 1$; $0 \leqslant x \leqslant 2\pi$.

5 a Express $\sqrt{2} \sin 2x° - 4 \cos 2x°$ in the form $k \cos (2x - \alpha)°$.

b Hence solve the equation $\sqrt{2} \sin 2x° - 4 \cos 2x° = 3$; $0 \leqslant x \leqslant 360$.

6 Two communication satellites, *Beacon 3* and *Hermes*, orbit the Earth.
At a particular latitude their orbits have been modelled by a set of trig. equations.
Using convenient units and axes, these equations are:
Beacon 3: $b(x) = 24 \sin x° - 4$ *Hermes*: $h(x) = 16 - 7 \cos x°$
where b units and h units are the heights of the satellites above the horizon and x is the number of minutes since observations began.

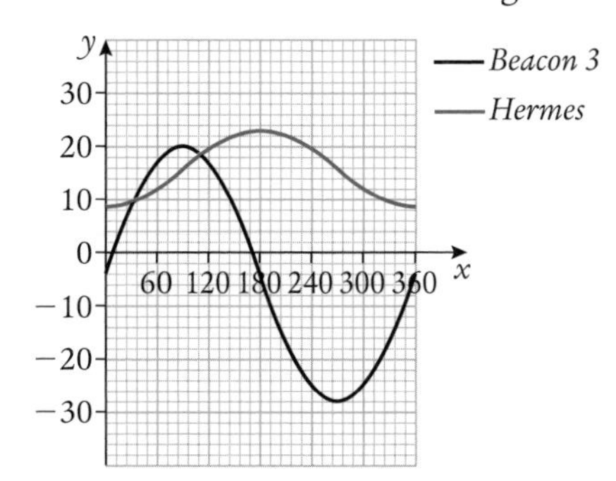

a Between what times is *Beacon 3* above the horizon?

b At what times are the two satellites at the same height?

7 A rectangular plot of land, ABCD, is to have a house, AEFD, built at one end.
The diagonal AC makes an angle of $x°$ with the side DC.
AG = 12 m and GC = 23 m.

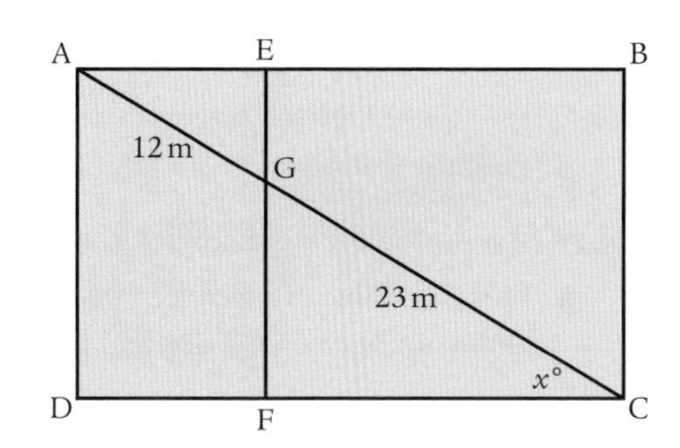

a Express:

i the length of the house, AD, in terms of x

ii the breadth of the house, AE, in terms of x.

b Find an expression for the perimeter of the house in terms of x.

c Calculate the value of x when the perimeter is 66 metres.

8.4 Approximate solutions

When an equation does not lend itself to any of the above strategies we can still make use of the approximation methods developed in Chapter 6 for the polynomials.

Example 1

In the introduction we talked of the tethered goat problem and the need to solve the equation $\sin x - x\cos x - \frac{\pi}{2} = 0$.

We know the rope must be a bit longer than 1 unit, the radius of the field.

A rough sketch lets us see that x must be a bit greater than $\frac{\pi}{2} = 1{\cdot}570...$ and less than $\frac{2\pi}{3} = 2{\cdot}09...$[2]

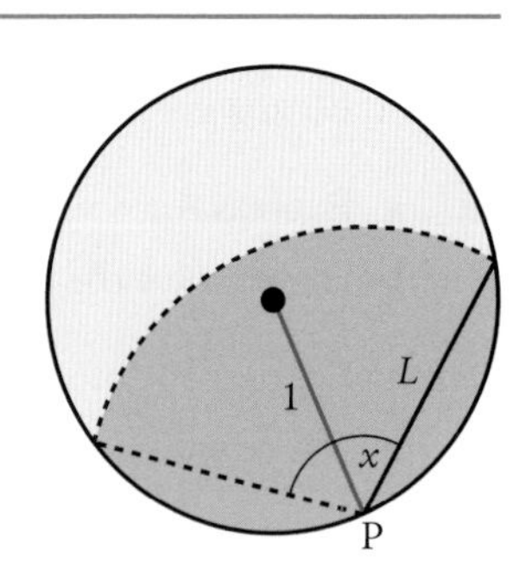

a Show that a solution lies in the region $1{\cdot}9 < x < 2$.

b Find the solution correct to 2 decimal places.

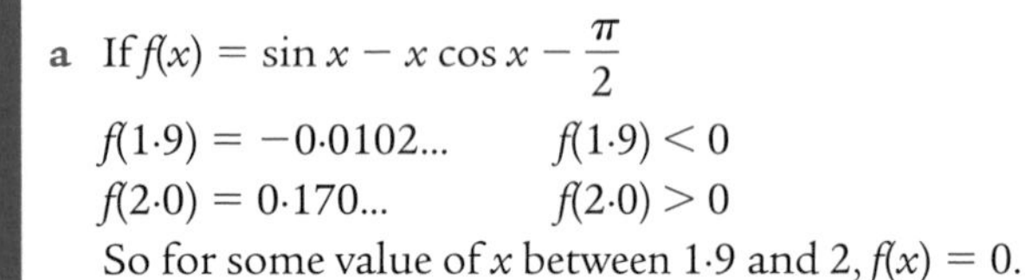

a If $f(x) = \sin x - x\cos x - \frac{\pi}{2}$

$f(1{\cdot}9) = -0{\cdot}0102...$ $\quad f(1{\cdot}9) < 0$

$f(2{\cdot}0) = 0{\cdot}170...$ $\quad f(2{\cdot}0) > 0$

So for some value of x between 1·9 and 2, $f(x) = 0$.

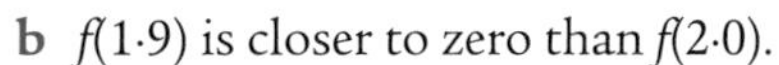

b $f(1{\cdot}9)$ is closer to zero than $f(2{\cdot}0)$.

- So try $f(1{\cdot}92)$ ... $f(1{\cdot}92) = 0{\cdot}02577...$ $\quad f(1{\cdot}92) > 0$
  $\Rightarrow 1{\cdot}9 < x < 1{\cdot}92$
- Try $f(1{\cdot}91)$ ... $f(1{\cdot}91) = 0{\cdot}00774...$ $\quad f(1{\cdot}91) > 0$
  $\Rightarrow 1{\cdot}9 < x < 1{\cdot}91$
- Try $f(1{\cdot}905)$ ... $f(1{\cdot}905) = -0{\cdot}00125...$ $\quad f(1{\cdot}905) < 0$
  $\Rightarrow 1{\cdot}905 < x < 1{\cdot}91$

So, $x = 1{\cdot}91$ (to 2 d.p.).

Exercise 8.4

1 The sketch shows $f(x) = x + \sin x$, with x measured in radians, close to the origin.

a Show that the function is never decreasing.

b By solving the equation $x + \sin x - 1 = 0$, find where $f(x) = 1$, correct to 2 decimal places.

c For what value of x is $f(x) = 3$ (to 2 d.p.)?

d Find the stationary point(s) in the given interval.

2 We know that for small angles, $\tan x \approx x$.

a For what value of x between 0 and 1 is $\tan x = \frac{1}{x}$? Give your answer to 2 decimal places.

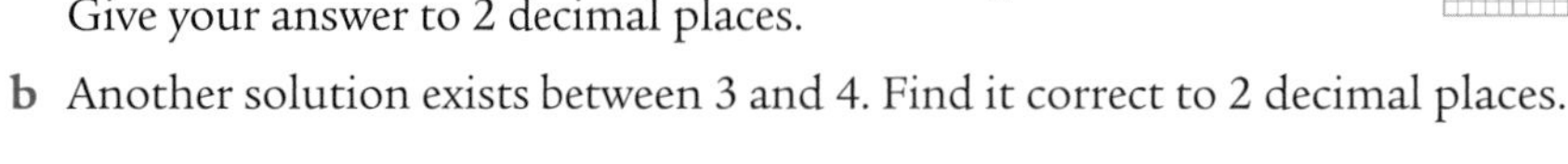

b Another solution exists between 3 and 4. Find it correct to 2 decimal places.

[2] $x = \frac{2\pi}{3}$ when the rope is the same length as the radius of the field.

3 a For angles between 0 radians and 1 radian, the function $c(x) = 1{\cdot}01 - 0{\cdot}5x^2$ is a good approximation for $\cos x$.
By considering the equation $\cos x - 1{\cdot}01 + 0{\cdot}5x^2 = 0$, find the value of x for which the approximation is best. (Calculate correct to 2 decimal places.)

b It has been suggested that $s(x) = \frac{1}{2}\left(1 - \frac{x^3}{3}\right)$ is an approximation for $\sin x$ for angles between 0 radians and 1 radian.
For what value of x is the approximation best?

4 The sketch shows the functions $y = \cos x$ and $y = x^2$ between $-\frac{\pi}{2}$ and $\frac{\pi}{2}$.

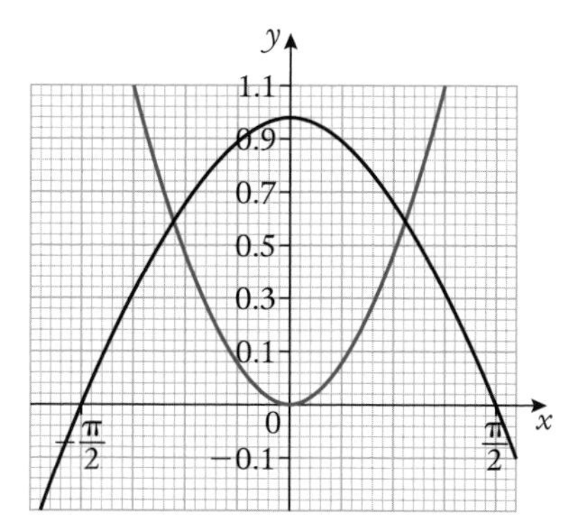

Find where the two curves intersect, giving your answers correct to 2 decimal places.

5 This sort of searching is greatly simplified by the use of spreadsheets.
Suppose we wish to solve $\sin x + x - 1 = 0$.
In row 1 type headings: x [in A1], $f(x)$ [in B1], precision [in C1].
In A2 type: 0
In A3 type: =A-1$C$2 ... fill this down to row 12.
In B2 type: =SIN(A2)+A2-1 ... fill this down to row 12.
Use the **developer** to get a **spinner**.
Format Control: min value 0; max value 10; increment 1; Cell link C3.
Click the spinner till 0 appears in C3.
In C2 type: =1/10^C3 ... this will initially be 1.
It is useful to ask for a graph of A2 to B12. It helps to find the search interval.
Your spreadsheet will look like this [with a bit of formatting]:

x	f(x)	precision
0	−1	1
1	0.84147098	
2	1.90929743	
3	2.14112001	
4	2.2431975	
5	3.04107573	
6	4.7205845	
7	6.6569866	
8	7.98935825	
9	8.41211849	
10	8.45597889	

Note that $f(x)$ changes sign in the interval $0 \leqslant x \leqslant 1$.
Click the precision button once.

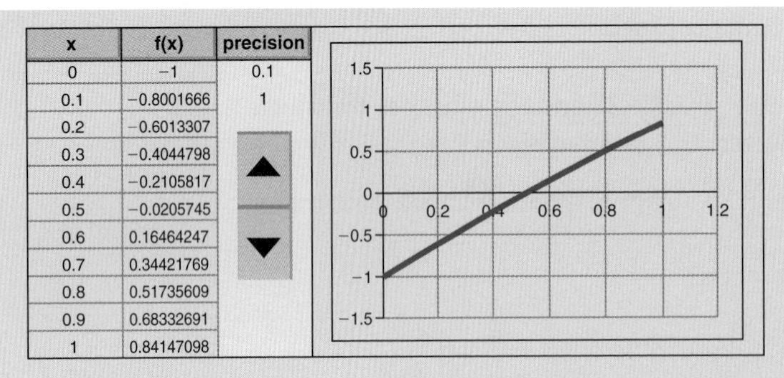

x	f(x)	precision
0	−1	0.1
0.1	−0.8001666	1
0.2	−0.6013307	
0.3	−0.4044798	
0.4	−0.2105817	
0.5	−0.0205745	
0.6	0.16464247	
0.7	0.34421769	
0.8	0.51735609	
0.9	0.68332691	
1	0.84147098	

We are now examining function values in the interval $0 \leqslant x \leqslant 1$ in steps of 0·1.

Note that $f(x)$ changes sign in the interval $0{\cdot}5 \leqslant x \leqslant 0{\cdot}6$.

In A2 overtype the contents with 0·5 ... and click the precision button once more.

x	f(x)	precision
0.5	−0.0205745	0.01
0.51	−0.0018228	
0.52	0.01688014	
0.53	0.03553334	
0.54	0.05413599	
0.55	0.07268723	
0.56	0.0911862	
0.57	0.10963205	
0.58	0.12802394	
0.59	0.14636102	
0.6	0.16464247	

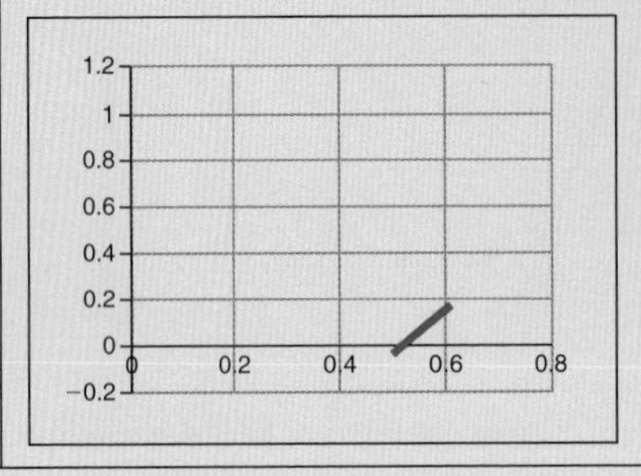

Note that $f(x)$ changes sign in the interval $0{\cdot}51 \leqslant x \leqslant 0{\cdot}52$.

In A2 overtype the contents with 0·51 ... and click the precision button once more.

Continue with this cycle of operations, always putting the x-value before the sign switch in A2 and increasing the precision, till you get the accuracy you require.

Use this method to solve the following to 4 decimal places:

a $x \sin x = 0;\ 6 < x < 7$ [In B2 type: =A2*SIN(A2) ... and fill down to row 12]

b $\sin 3x - \cos x = 0;\ 0 < x < 1$ [there are two answers to find]

c $x^2 \cos 3x = 0;\ 4 < x < 5$

d $4 \cos^3 x - 1 = 0;\ 0 < x < 1$

Preparation for assessment

1 Solve $3 \sin (2x + 60)° + 1 = 0; 0 \leqslant x \leqslant 360$, to the nearest degree.

2 The sketch shows the graphs of $y = x + \sin\left(2x - \frac{\pi}{6}\right)$ and $y = x - 0{\cdot}5$.

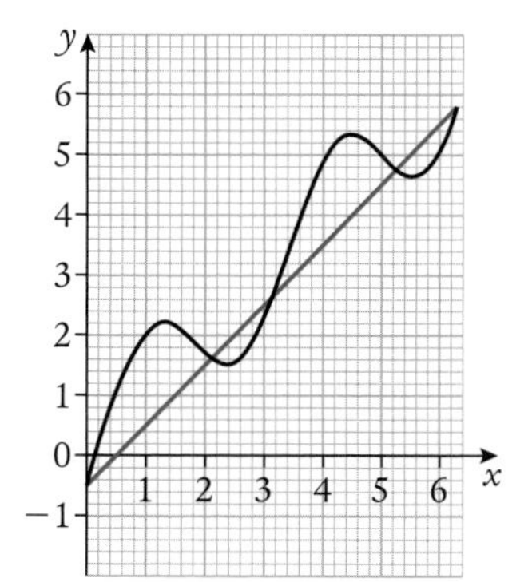

Where do they intersect in the interval $0 \leqslant x \leqslant 2\pi$?

3 Solve $2 \sin^2 x + 3 \sin x - 2 = 0; 0 \leqslant x \leqslant 2\pi$.

4 Solve:

a $\sin 2x - \sqrt{3} \sin x = 0; 0 \leqslant x \leqslant 2\pi$ **b** $\sin x° - \sin \frac{x°}{2} = 0; 0 \leqslant x \leqslant 360$.

5 A lighthouse and a beacon both flash a light throughout the night to warn of dangerous reefs. The brightness, L units, of the lighthouse can be modelled by $L(x) = 3 \cos 2x° + 3$, where x is the number of seconds since observations began.
The beacon's brightness can be modelled by $B(x) = \sin x° + 2$.

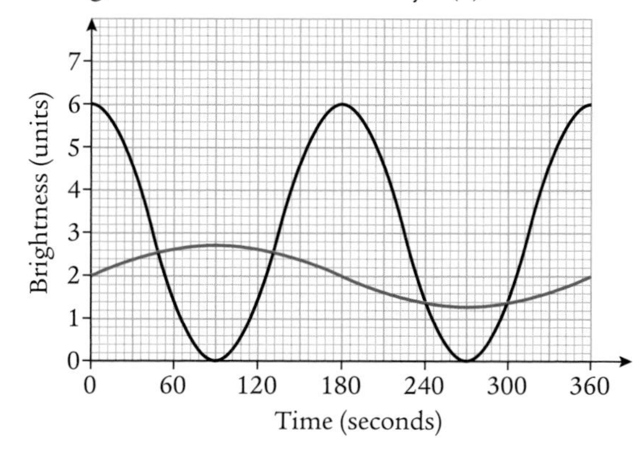

a Which curve represents which light?

b When is L equal to B?

6 The height, h metres, of a particular pod on the London Eye can be modelled by $h(x) = 56 \cos x° + 33 \sin x° + 70$, where $x°$ is the angle the wheel has turned through.

a How high is the pod when the wheel begins to turn ($x = 0$)?

b You get into a pod from a platform which is 5 metres high. For what value of x is $h = 5$?

c For what values of x in the first turn of the wheel is the pod 100 m high?

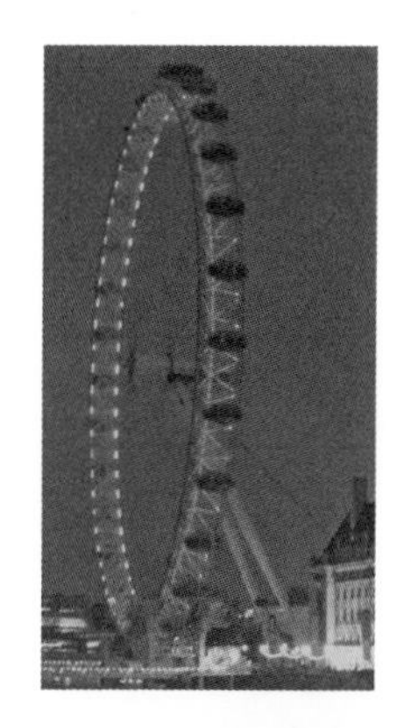

7 **a** Express $5 \sin x + 12 \cos x$ in the form $k \cos (x + a)$, where $k > 0$ and $0 \leqslant a \leqslant 2\pi$.

b Hence solve the equation $5 \sin x + 12 \cos x = 10; 0 \leqslant x \leqslant 2\pi$.

Summary

1 When we solve an equation of the form $\sin x° = a$, or $\cos x° = a$, or $\tan x° = a$, where $-1 \leqslant a \leqslant 1$, a is a constant, there is an infinite number of solutions.

- The first solution we get from the calculator. We denote it by $\sin^{-1} a$, $\cos^{-1} a$, or $\tan^{-1} a$.
- A second solution comes from the symmetry of the function:
  ... if $\sin^{-1} a$ is a solution then so is $180 - \sin^{-1} a$
  ... if $\cos^{-1} a$ is a solution then so is $360 - \cos^{-1} a$
  ... if $\tan^{-1} a$ is a solution then so is $180 + \tan^{-1} a$.
- All other solutions come from the fact that the trig. functions are periodic
  ... if x is a solution then so is $x \pm 360n$, where n is a whole number.

Care must be taken when solving equations to remember that the propagation of solutions occurs at the point where the inverse trig. function is being found, not later after further manipulation.

2 For certain forms of trigonometric equations the **double angle** formulae can be exploited.

$$\sin 2x = 2 \sin x \cos x$$

$$\begin{aligned} \cos 2x &= \cos^2 x - \sin^2 x \\ &= 2\cos^2 x - 1 \\ &= 1 - 2\sin^2 x \end{aligned}$$

3 To solve an equation of the form $a \sin x \pm b \cos x = 0$, express it in the form $p \sin(x \pm q)$ or $p \cos(x \pm q)$, where p and q are positive numbers ... if you have the freedom of choice, use $p \sin(x \pm q)$, picking the sign that agrees with the original expression.

4 When an equation does not lend itself to any of the above strategies make use of the approximation methods developed in Chapter 6 for the polynomials.

9 Logs and exponential functions

Before we start...

If you had to work without a calculator, finding products, quotients, powers and roots would be a laborious or even impossible task for you.

In 1614, a merchant of Edinburgh, John Napier, published a set of tables that made performing these calculations relatively simple.

He referred in his book to the 'the wonderful rule of logarithms'.

This is a very simplified version of the tables.

	0	1	2	3	4	5	6	7	8	9
1	0·000	0·041	0·079	0·114	0·146	0·176	0·204	0·230	0·255	0·279
2	0·301	0·322	0·342	0·362	0·380	0·398	0·415	0·431	0·447	0·462
3	0·477	0·491	0·505	0·519	0·531	0·544	0·556	0·568	0·580	0·591
4	0·602	0·613	0·623	0·633	0·643	0·653	0·663	0·672	0·681	0·690
5	0·699	0·708	0·716	0·724	0·732	0·740	0·748	0·756	0·763	0·771
6	0·778	0·785	0·792	0·799	0·806	0·813	0·820	0·826	0·833	0·839
7	0·845	0·851	0·857	0·863	0·869	0·875	0·881	0·886	0·892	0·898
8	0·903	0·908	0·914	0·919	0·924	0·929	0·934	0·940	0·944	0·949
9	0·954	0·959	0·964	0·968	0·973	0·978	0·982	0·987	0·991	0·996
10	1·000	1·004	1·009	1·013	1·017	1·021	1·025	1·029	1·033	1·037

The table lets us turn purple numbers into black numbers. ... which Napier called logarithms ... 'logs' for short.

Check from the table that $\log(1{\cdot}7) = 0{\cdot}230$ and that $\log(5{\cdot}3) = 0{\cdot}724$

What is 'wonderful' about the logs is that instead of **multiplying** the purple numbers we need only **add** the logs.

i Using the table, the calculation $1{\cdot}7 \times 5{\cdot}3$ turns into $0{\cdot}230 + 0{\cdot}724 = 0{\cdot}954$.
The purple number that matches 0·954 is 9·0, i.e. $\log(9{\cdot}0) = 0{\cdot}954$.
- Using a calculator ... $1{\cdot}7 \times 5{\cdot}3 = 9{\cdot}0$ (to 1 d.p.).

ii The calculation $5{\cdot}3 \div 1{\cdot}7$ becomes $0{\cdot}724 - 0{\cdot}230 = 0{\cdot}494$.
The purple number that matches the black number 0·494 is about 3·1.
- Using a calculator, $5{\cdot}3 \div 1{\cdot}7 = 3{\cdot}1$ (to 1 d.p.).

iii Even better, the calculation $1{\cdot}7^3$ becomes $0{\cdot}230 \times 3 = 0{\cdot}69$ [think about it!]
The purple number which has 0·69 for a log is 4·9 ... $1{\cdot}7^3 = 4{\cdot}9$ [without a calculator!]

iv Note that if we want $\sqrt[3]{4{\cdot}9}$, we use $0{\cdot}690 \div 3 = 0{\cdot}230$.
The purple number which has 0·23 for a log is 1·7 ... $\sqrt[3]{4{\cdot}9} = 1{\cdot}7$ [without a calculator!]

This chapter hopes to show how Napier's 'wonderful' log tables work.

What you need to know

1 Find the value of:

a 3^4 b 2^{-1} c $\sqrt[3]{8}$ d 5^0 e $27^{\frac{1}{3}}$ f $27^{\frac{2}{3}}$.

2 Simplify:

a $a^5 \times a^3$ b $a^5 \div a^3$ c $(a^3)^5$ d $\frac{a^7}{a^3}$ e $\frac{a^3b^{-1}}{a^{-2}b^0}$.

3 Use Napier's log tables (remember: no calculator) to estimate, to 1 decimal place, the value of:

a i $4{\cdot}2 \times 2{\cdot}5$ ii $1{\cdot}9 \times 2{\cdot}9$ iii $6{\cdot}4 \times 1{\cdot}3$ iv $5{\cdot}7 \times 1{\cdot}6$

b i $6{\cdot}7 \div 2{\cdot}9$ ii $8{\cdot}7 \div 3{\cdot}2$ iii $9{\cdot}3 \div 1{\cdot}6$ iv $\frac{4{\cdot}3}{1{\cdot}1}$

c i $2{\cdot}9^2$ ii $2{\cdot}1^3$ iii $1{\cdot}5^5$ iv $1{\cdot}3^9$

d i $\sqrt{9{\cdot}7}$ ii $\sqrt[3]{10{\cdot}6}$ iii $\sqrt[4]{5{\cdot}1}$ iv $\sqrt[10]{6{\cdot}2}$.

4 a Evaluate, correct to 3 significant figures:

i $10^{0{\cdot}301}$ ii $10^{0{\cdot}477}$ iii $10^{0{\cdot}602}$ iv $10^{0{\cdot}699}$ v $10^{0{\cdot}778}$.

b Compare your results for a with Napier's table then use the table to find, to 2 significant figures,

i $10^{0{\cdot}380}$ ii $10^{0{\cdot}748}$ iii $10^{0{\cdot}851}$ iv $10^{0{\cdot}949}$ v $10^{0{\cdot}968}$.

5 In each of the these, follow the argument then repeat it for the new data.

a $4{\cdot}2 \times 2{\cdot}5 = 10^{0{\cdot}623} \times 10^{0{\cdot}398} = 10^{0{\cdot}623 + 0{\cdot}398} = 10^{1{\cdot}021} = 10{\cdot}5$

i $1{\cdot}9 \times 2{\cdot}9$ ii $6{\cdot}4 \times 1{\cdot}3$

b $6{\cdot}7 \div 2{\cdot}9 = 10^{0{\cdot}826} \div 10^{0{\cdot}462} = 10^{0{\cdot}826 - 0{\cdot}462} = 10^{0{\cdot}364} = 2{\cdot}31$ (to 3 s.f.)

i $8{\cdot}7 \div 3{\cdot}2$ ii $9{\cdot}3 \div 1{\cdot}6$

c $2{\cdot}9^2 = (10^{0{\cdot}462})^2 = 10^{0{\cdot}462 \times 2} = 10^{0{\cdot}924} = 8{\cdot}39$ (to 3 s.f.)

i $2{\cdot}1^3$ ii $1{\cdot}5^5$

d $\sqrt{9{\cdot}7} = \sqrt{10^{0{\cdot}987}} = (10^{0{\cdot}987})^{\frac{1}{2}} = 10^{0{\cdot}987 \times \frac{1}{2}} = 10^{0{\cdot}494} = 3{\cdot}12$ (to 3 s.f.)

i $\sqrt[3]{10{\cdot}6}$ ii $\sqrt[4]{5{\cdot}1}$

9.1 The logarithm of a number (base 10)

If $y = 10^x$, then the logarithm of y is x to the base 10 ... and we write ... $\log(y) = x$

$$y = 10^x \Leftrightarrow \log(y) = x$$

e.g. since $2 = 10^{0{\cdot}301}$, we say that the log of 2 to the base 10 is 0·301 ... $\log(2) = 0{\cdot}301$

and, since $1000 = 10^3$, we say that the log of 1000 to the base 10 is 3 ... $\log(1000) = 3$.

Some rules

Let us say that $p = 10^a$ and $q = 10^b$ ... giving $\log p = a$ and $\log q = b$ ... ①

- $pq = 10^a \times 10^b = 10^{a+b}$ ... giving $\log(pq) = a + b$ ... ②

  Substituting the expressions for a and b in ① into ②

  Law 1: $\log pq = \log p + \log q$

- $\frac{p}{q} = p \div q = 10^a \div 10^b = 10^{a-b}$ ... giving $\log(p \div q) = a - b$ ... ③

  Substituting the expressions for a and b into ③

  Law 2: $\log(p \div q) = \log p - \log q$

- $p^k = (10^a)^k = 10^{ka}$ ... giving $\log p^k = ka$ ... ④
  Substituting the expressions for a into ④
  Law 3: $\log p^k = k \log p$

Note that since $10^0 = 1$ and $10^1 = 10$ we also have the facts:

$\log 1 = 0$ and $\log 10 = 1$

Example 1

Simplify: $3 \log 4 - \log 8$.

$3 \log 4 - \log 8$
$= \log 4^3 - \log 8 = \log 64 - \log 8$ ... using law 3
$= \log (64 \div 8)$ ... using law 2
$= \log 8$

Example 2

Simplify $\log\left(\frac{1}{2}\right) + \log 2$.

$\log\left(\frac{1}{2}\right) + \log 2 = \log 1 - \log 2 + \log 2$ ... using law 2
$= \log 1$
$= 0$

Example 3

Simplify: $\dfrac{\log 125}{\log 25}$.

$\dfrac{\log 125}{\log 25} = \dfrac{\log 5^3}{\log 5^2}$

$= \dfrac{3 \log 5}{2 \log 5} = \dfrac{3}{2}$

Example 4

a Express 45 000 in standard form.

b By using Napier's tables and the laws of indices, find log 45 000.

a $45\,000 = 4{\cdot}5 \times 10^4$

b Using Napier's tables, this becomes $10^{0{\cdot}653} \times 10^4 = 10^{4{\cdot}653}$.
Thus $45\,000 = 10^{4{\cdot}653}$.
So $\log 45\,000 = 4{\cdot}653$.

Exercise 9.1

1 State the log, to the base 10, of:

a 1 **b** 10 **c** 100 **d** 1000 **e** 10 000 **f** $\frac{1}{10}$

g $\frac{1}{100}$ **h** $\frac{1}{1000}$ **i** $\sqrt{10}$ **j** $\sqrt{1000}$ **k** $\sqrt[3]{100}$.

2 a How are log 3 and $\log\left(\frac{1}{3}\right)$ related?

b Knowing that $81 = 3^4$, state the relationship between log 81 and log 3.

c Simplify: i log 81 ÷ log 3 ii log 4 ÷ log 32. [Hint: think powers of 2.]

3 Simplify:

a $\log 3 + \log 2$ b $\log 4 + \log 6$ c $\log 2 + \log 3 + \log 7$

d $\log 12 - \log 6$ e $\log 24 - \log 8$ f $\log 5 + \log 6 - \log 15$

g $\log 7 + \log 1$ h $\log 7 - \log 1$ i $\log\left(\frac{1}{2}\right) + \log\left(\frac{2}{3}\right)$

j $\log\left(\frac{1}{3}\right) - \log\left(\frac{2}{3}\right)$ k $\log 5 - \log\left(\frac{5}{6}\right)$ l $\log(0{\cdot}2) + \log(5)$.

4 Use the laws of logs to help you simplify:

a $\log 3 + 2\log 2$ b $\log 16 - 3\log 2$ c $2\log 5 + 3\log 4$

d $\frac{1}{2}\log 4 + \log 2$ e $\frac{1}{2}\log 16 + \frac{1}{3}\log 8$ f $\frac{2}{3}\log 8 + \log 4$.

5 i Express each number in standard form.

ii Using Napier's tables, find the log of the number.

a 56 b 560 c 5600 d 56 000

e 87 f 350 g 1900 h 81 000

6 In chemistry, the pH of a substance is defined by the formula:

pH = −log (hydrogen ion concentration measured in moles per litre).

The hydrogen ion concentration of vinegar is $3{\cdot}9 \times 10^{-3}$ moles per litre.

Its pH is therefore $-\log(3{\cdot}9 \times 10^{-3}) = -(\log 3{\cdot}9 + \log 10^{-3}) = -[0{\cdot}591 + (-3)] = 2{\cdot}4$ (to 2 s.f.).

Calculate, to 2 significant figures, the pH of:

a lemon juice with hydrogen ion concentration of $6{\cdot}3 \times 10^{-3}$ moles per litre

b coffee with a concentration of $1{\cdot}6 \times 10^{-3}$ moles per litre

c rain water ... concentration of $3{\cdot}2 \times 10^{-6}$ moles per litre

d milk ... concentration of $2{\cdot}5 \times 10^{-7}$ moles per litre.

9.2 Other bases

For historical reasons we introduced this topic using base 10.

However, the definition of a logarithm can be extended for any positive base:

If $y = a^x$, then the logarithm of y to the base a is x ...

$$y = a^x \Leftrightarrow \log_a y = x$$

When using log notation, we indicate the base by using a subscript. In general, if no base is mentioned, base 10 is assumed.

e.g. since $8 = 2^3$ then $\log_2 8 = 3$ ... $8 = 2^3 \Leftrightarrow \log_2 8 = 3$

also $49 = 7^2$ then $\log_7 49 = 2$ ... $49 = 7^2 \Leftrightarrow \log_7 49 = 2$

The laws already encountered work whatever the base. In addition:

Law 4: $\log_a a = 1$ ... $a = a^1 \Leftrightarrow \log_a a = 1$

Law 5: $\log_a 1 = 0$ ... $1 = a^0 \Leftrightarrow \log_a 1 = 0$

Example 1

Simplify and evaluate $\log_6 2 + \log_6 3$.

$\log_6 2 + \log_6 3 = \log_6(2 \times 3)$ ... law 1

$= \log_6 6 = 1$ ... law 4

Example 2

Evaluate $\log_2 8 + \log_2 32$.

$$\begin{aligned}\log_2 8 + \log_2 32 &= \log_2 2^3 + \log_2 2^5 \\ &= 3\log_2 2 + 5\log_2 2 && \text{... law 3} \\ &= 3 + 5 && \text{... law 4} \\ &= 8\end{aligned}$$

Exercise 9.2

1 Evaluate:

a $\log_2 16$ **b** $\log_3 9$ **c** $\log_5\left(\frac{1}{5}\right)$ **d** $\log_4 2$

e $\log_8 2$ **f** $\log_6\left(\frac{1}{36}\right)$ **g** $\log_5\left(\frac{1}{125}\right)$.

2 Simplify and evaluate:

a $\log_{10}5 + \log_{10}2$ **b** $\log_4 2 + \log_4 8$ **c** $2\log_9 3 + \log_9 81$

d $2\log_6 3 + 2\log_6 2$ **e** $2\log_{16}10 - 2\log_{16}5$ **f** $\log_2 6 - \log_2 12$

g $\frac{1}{2}\log_4 81 - 2\log_4 12$ **h** $2\log_7 14 - \frac{1}{2}\log_7 16$.

3 $y = \log_3 4 + \log_3 5 = \log_3 20$ ... and so $20 = 3^y$

Repeat this argument to produce a simple equation with the variable in the *exponent* (power) in each of the following cases.

a $x = 3\log_4 6$ **b** $m = \log_4 3 + \log_4 6$

c $n = 2\log_5 2 - \frac{1}{2}\log_5 4$ **d** $p = 4\log_7 2 - \log_7 32$

4 Solve each equation by using the definition of a log, viz $\log_a x = b \Leftrightarrow x = a^b$.

a $\log_3 x = 4$ **b** $2\log_4 x = 4$ **c** $\log_2 2x = 2$ **d** $3\log_5 x + 1 = 4$

5 During an earthquake, scientists use seismographs to measure the energy released.

This is worked out in TNT equivalent, T grams.

It is then converted to a Richter number, R, and reported in the news.

R can be computed, to 1 decimal place, using the formula:

$R = 0{\cdot}67\log_{10} T - 0{\cdot}8$.

a Work out the Richter number for a quake that was equivalent to TNT of weight:

i 30 grams **ii** 2700 grams **iii** 85 kg

iv 2·7 tonnes **v** 500 tonnes.

b Work out the weight of TNT equivalent to a Richter number of 2. Give your answer to 2 significant figures.

6 A, B and x are related by the formula $A = B^x$.

a **i** Take the log, base c, of both sides.

ii Use the laws of logs to make x the subject of the formula.

b If $c = B$, express x in terms of A and B in its simplest form.

c Equating part **a ii** and **b** you should find that $\log_B A = \frac{\log_c A}{\log_c B}$, where c is any base you wish to use. Use this and your calculator to find the value of:

i $\log_7 12$ **ii** $\log_2 27$ **iii** $\log_{12} 6$ **iv** $\log_\pi 20$ **v** $\log_\pi \pi$.

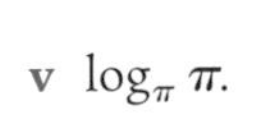

9.3 The logarithmic function

The log function, $f(x) = \log_a x, x > 0$, has properties that make it useful for modelling many situations.

The graph

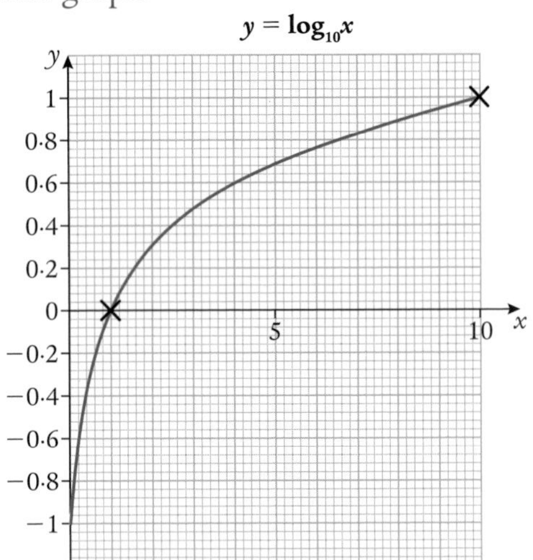

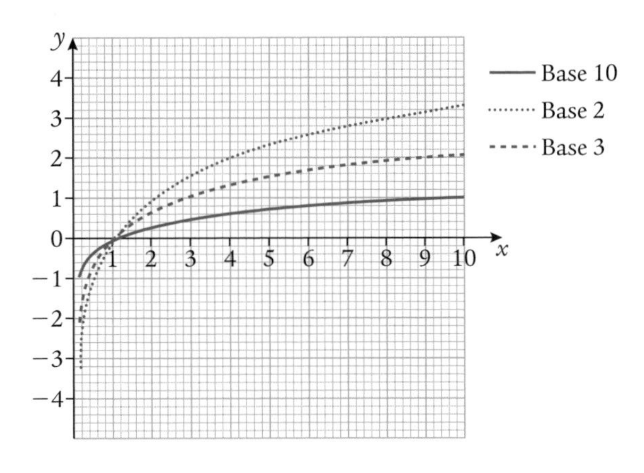

Note that $y = \log_a x$

- passes through the point (1, 0) ... $\log_a 1 = 0$ [law 5]
- passes through the point $(a, 1)$ ... $\log_a a = 1$ [law 4]

Note also from the graph that if $\log A = \log B$, then $A = B$

Example 1

The graphs have equations of the form $y = \log_q x$, $y = p\log_q x$ and $y = p\log_q x + r$, respectively.

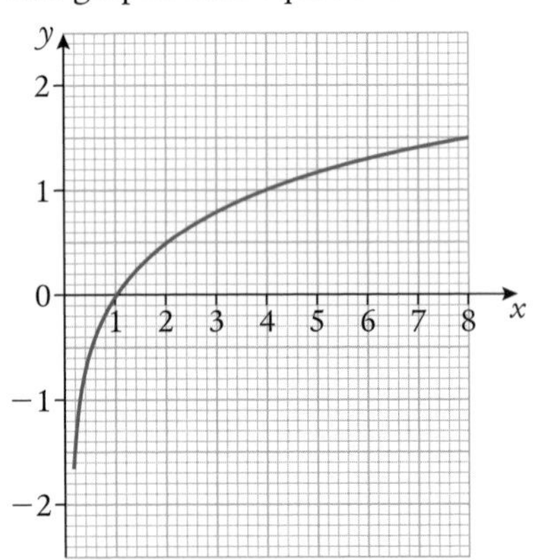

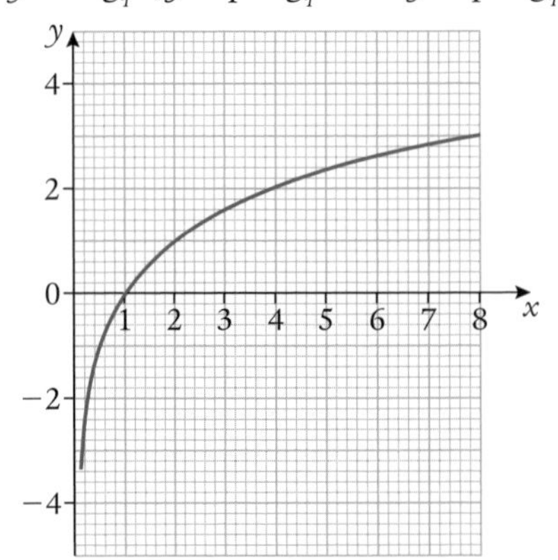

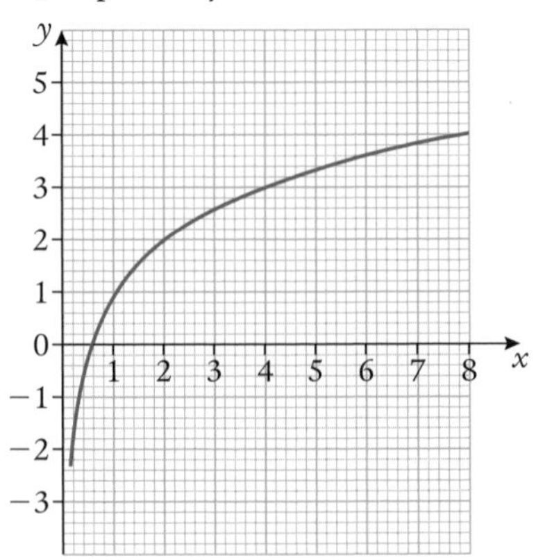

State the value of p, q and r, assuming they are the same in each case.

The first graph passes through (1, 0) and (4, 1) and is of the form $y = \log_q x$.

Using (4, 1) we get $1 = \log_q 4 \quad \Rightarrow q^1 = 4$

thus $q = 4$ ... we have the graph $y = \log_4 x$

The second graph passes through (1, 0) and (4, 2) ... and is of form $y = p\log_4 x$.

Using (4, 2) we get $2 = p\log_4 4$

$\Rightarrow \quad 2 = p$ ... $\log_4 4 = 1$

$\Rightarrow \quad p = 2$ ... we have the graph $y = 2\log_4 x$

The third graph passes through (1, 1) and (4, 3) ... and is of form $y = 2\log_4 x + r$.

Using (4, 3) we get $3 = 2\log_4 4 + r$

$\Rightarrow \quad r = 3 - 2\log_4 4 = 3 - 2 = 1$ ... we have the graph $y = 2\log_4 x + 1$

Example 2

Where does $y = \log_{10} x$ intersect $y = 2\log_{10}(x - 1), x > 1$.

The curves intersect when:

$\log_a x = 2\log_a(x - 1)$

$\Rightarrow \log_a x = \log_a(x - 1)^2$ ... using law 3

$\Rightarrow x = (x - 1)^2$ if $\log A = \log B$, then $A = B$

$\Rightarrow x^2 - 3x + 1 = 0$

$\Rightarrow x = 2{\cdot}618$ or ~~0·381~~ (to 3 d.p.) ... $0{\cdot}381 < 1$ and we know $x > 1$

$\Rightarrow y = \log_{10} 2{\cdot}618$

$\Rightarrow y = 0{\cdot}418$

The curves intersect at (2·618, 0·418).

Exercise 9.3

1 Make a sketch of the following graphs, identifying two points in each.

a $y = \log_3 x$ **b** $y = \log_5 x$ **c** $y = 2\log_6 x$ **d** $y = 3\log_{10} x$

e $y = \log_3 x^2$ **f** $y = \log_5 x^3$ **g** $y = \log_6\left(\frac{1}{x}\right)$ **h** $y = \log_{10}\sqrt{x}$

2 The diagram shows a sketch of $y = \log_a x, x > 0$.
Draw sketches of:

a $y = 2\log_a x$ **b** $y = \log_a x^3$

c $y = \log_a\left(\frac{1}{x}\right)$ **d** $y = \log_a(x - 4), x > 4$

e $y = \log_a x + 3$ **f** $y = 2\log_a x + 1$.

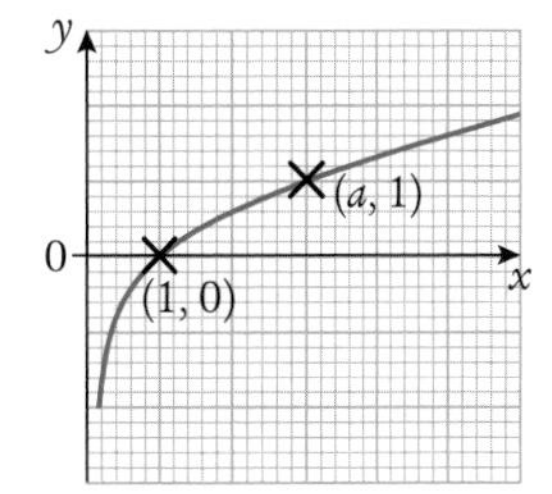

3 In acoustics (the science of sound), the loudness, L, of a noise measured in decibels (dB) is given by $L = 20\log_{10}(0{\cdot}05P)$, where P is the sound pressure in the air measured in micropascals (mPa).

a The quietest audible sound has a pressure of 20 mPa.
How many decibels is this?

b Calculate the decibels for each of the following:

i leaf rustling ... 63 mPa

ii TV in room ... 20 000 mPa

iii permanent damage ... $3{\cdot}56 \times 10^5$ mPa.

c Traffic has a loudness of 90 dB. How many micropascals does this relate to?

4 A psychologist had a theory that a candidate's score in an exam was a function of the time between studying and sitting the exam. He reckoned $P(t) = 70 - k\log_{10}(t + 1)$, where P% was the score, t was the time lapse measured in weeks and k was a constant.

a When $t = 1$, the score in the test was 55%.
What is the value of the constant, k, to the nearest whole number?

b What does the model predict the score will be after 5 weeks?

c Write an expression for the difference between the score in the third week ($t = 3$) and the seventh week.

d Can you deduce in which week the score will be 20%?

9.4 The inverse log function ... exponential functions

When Napier produced his table he called the process of turning a purple number into a black number ... getting the **log** of the number.

He called the reverse process, the inverse, ... getting the anti-log.

What **is** the inverse of the function $f(x) = \log_a x$?

If $y = \log_a x, x > 0$, then the inverse function must be defined by:

$x = \log_a y, y > 0$

$\Leftrightarrow y = a^x$

Note that the domain is all real numbers and the range is all positive numbers.

The function $f(x) = a^x$ is referred to as an **exponential function**.

It is the inverse of the log function, so

$a^{\log_a x} = x$ and $\log_a (a^x) = x$

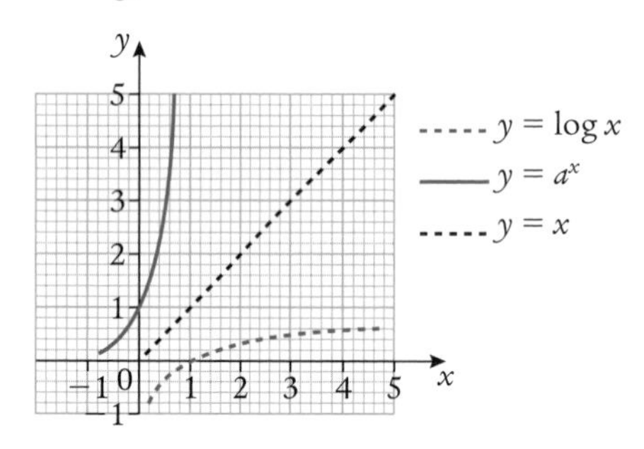

The laws of the exponential functions are the laws of indices.

A very useful base

The result is not required by this course, but consider finding the derivative of the function $f(x) = a^x$.

$$y = f(x) \Rightarrow \frac{dy}{dx} = \lim_{h \to 0} \frac{f(x+h) - f(x)}{(x+h) - x}$$

$$\frac{d(a^x)}{dx} = \lim_{h \to 0} \frac{a^{x+h} - a^x}{h} = \lim_{h \to 0} \frac{a^x a^h - a^x}{h}$$

$$= \lim_{h \to 0} a^x \left(\frac{a^h - 1}{h}\right) = a^x \lim_{h \to 0} \left(\frac{a^h - 1}{h}\right).$$

So the derivative of $a^x = ka^x$, where $k = \lim_{h \to 0} \left(\frac{a^h - 1}{h}\right)$.

If you investigate the value of this limit, you will see that it depends on the value of a. Using a suitably small value of h will give you an impression of the value of the limit:

$$a = 2, \quad \left(\frac{2^{0\cdot0001} - 1}{0\cdot0001}\right) = 0\cdot693... \quad \Rightarrow \frac{d(2^x)}{dx} \approx 0\cdot693 \times 2^x$$

$$a = 3, \quad \left(\frac{3^{0\cdot0001} - 1}{0\cdot0001}\right) = 1\cdot098... \quad \Rightarrow \frac{d(3^x)}{dx} \approx 1\cdot099 \times 3^x$$

Somewhere between $a = 2$ and $a = 3$, $\lim_{h \to 0} \left(\frac{a^h - 1}{h}\right)$ takes the value 1

... and the derivative of a^x will itself be a^x.

The mathematician Euler chose the letter e to stand for this particular base.

So e is defined by $\lim_{h \to 0}\left(\frac{e^h - 1}{h}\right) = 1$.

Using trial and improvement on a calculator, you will find that e = 2·71828...

When we use e as the base of an exponential function, it is referred to as **the** exponential function.

So working with **the** exponential function, a lot of maths becomes much easier.

$$f(x) = e^x \Rightarrow f'(x) = e^x$$

Note: $\log_e x$ is often written as $\ln x$, and referred to as the **natural** log of x.

In a spreadsheet: $\log_e x$ is given by LN(x) and e^x is given by EXP(x).

Example 1

£P is deposited in the bank. The amount in the bank, £A, after x years of getting 4% interest, can be calculated from the formula $A = P \times 1{\cdot}04^x$.

a Express 1·04 as a power of e.

b Express A in terms of P and x using e as a base.

a $1{\cdot}04 = e^{\ln 1{\cdot}04} = e^{0{\cdot}0392}$ ... using the function on your calculator

b $A = P \times (e^{0{\cdot}0392})^x = P \times e^{0{\cdot}0392x}$ ... A is now in terms of the exponential function

Exercise 9.4

1 State the inverse of the following functions.

a $f(x) = \log_5 x$ **b** $f(x) = \log_7 x$ **c** $f(x) = \ln x$ **d** $f(x) = 2\log_5 x$

e $f(x) = 3\log_4 x$ **f** $f(x) = 4\ln x$ **g** $f(x) = \log_3 x + 2$ **h** $f(x) = \log_7 x - 1$

2 **a** When $f(x) = 4^x$ find: **i** $f(2)$ **ii** $f(0{\cdot}5)$ **iii** $f(-1)$.

b When $f(x) = e^x$ find: **i** $f(4)$ **ii** $f(0{\cdot}693)$ **iii** $f(1{\cdot}099)$.

3 For each exponential function, make e the base.

a $f(x) = 2{\cdot}21^x$ **b** $f(x) = 32^x$ **c** $f(x) = 240^x$ **d** $f(x) = 5 \times 4^x$

4 When a chain hangs between two supports we get a curve known as a **catenary** ... it is **not** a parabola.

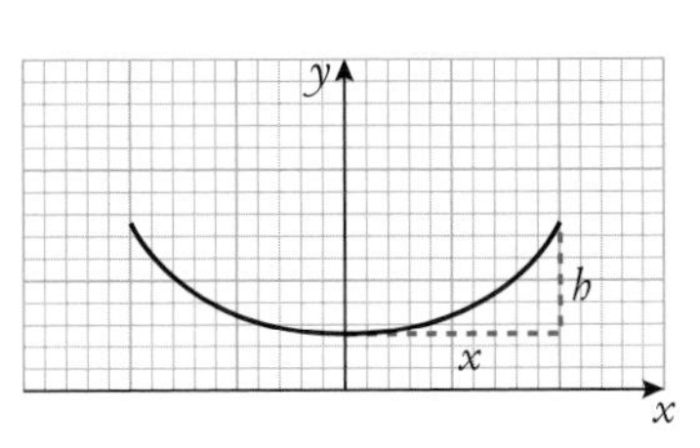

Using suitable units and axes, the chains of the suspension bridge in Glasgow can be modelled by

$$h = 4\left(e^{\frac{x}{4}} + e^{-\frac{x}{4}}\right)$$

where h units is the height of the chain measured x units horizontally from the lowest point.

Calculate h when:

a $x = 0$ **b** $x = 2$ **c** $x = 4$.

5 At the whisky distillery 2% of the content of a cask evaporates each year. This is known as the angels' share. The cask is initially filled with V litres of whisky. The amount in the cask, B litres, x years after it was filled, can be calculated using the formula $B = V \times 0{\cdot}98^x$.

a Initially a cask is filled with 200 litres. How much is in it after

i 5 years **ii** 7 years?

b The whisky is ready for bottling after 12 years.

i How much is in the cask after 12 years?

ii What is the angel's share?

c Find a formula for B of the form $B = Ve^{Ax}$.

6 A radioactive isotope decays over time.

This decay can be modelled by $A_t = A_0(1 - r)^t$

where A_0 is the amount of radioactive material we start with, A_t is the amount remaining after t years when the fraction it loses per year is r.

For a particular material, $r = 0{\cdot}007$.

a If we start with 10 g of radioactive material, how much is left after:

i 5 years **ii** 50 years?

b Scientists talk of the half-life of such a material. They mean the time it takes for the radioactive material to reduce to half what it was, i.e. $A_t = 0{\cdot}5A_0$.

i A substance has a half-life of 10 years, calculate the value of r.

ii Model this decay using an equation of the form $A_t = A_0e^{at}$.

c A living entity will absorb carbon-14 as it lives. When it dies no more carbon-14 can be absorbed. The existing carbon-14 decays. The half-life of carbon-14 is 5720 years.

i Construct a model of the form $A_t = A_0(1 - r)^t$ and of the form $A_t = A_0e^{at}$.

ii A piece of wooden post is found at an archaeological dig. It is found to contain just 20% of the carbon-14 it would have at the time it was chopped down. Estimate the age of the post.

7 The curving edge of the Eiffel Tower can be roughly modelled by the log function, $h = -90\ln\left(\frac{x}{60}\right)$, where h metres is the vertical height of the edge and x metres is its distance from the centre line.

a Calculate the height of a point on the curved edge which is the following distance from the centre line:

i 20 m **ii** 30 m **iii** 60 m.

b How far from the centre line is a point on the curve 50 m high?

c Make x the subject of the formula.

8 Two functions f and g are defined by $f(x) = 4x + 3$ and $g(x) = e^x$.

a Write an expression for $f(g(x))$.

b **i** Write an expression for $g(f(x))$.

ii Express it in the form Ae^{Bx} giving the value of A to the nearest whole number.

9 Two functions f and g are defined by $f(x) = x^2$ and $g(x) = e^x$.

a Give simplified expressions for: **i** $f(g(x))$ and **ii** $g(f(x))$.

b Express both in the form $(e^x)^A$.

c Hence find for which values of x, $f(g(x)) = g(f(x))$.

9.5 Solving equations

Example 1

Solve to 3 significant figures:

a $3e^x = 2$ b $4e^{5x} + 1 = 2$ c $3 \times 4^x = 18$.

a $3e^x = 2$

$\Rightarrow e^x = \frac{2}{3}$ $\Rightarrow \ln(e^x) = \ln\left(\frac{2}{3}\right)$

$\Rightarrow x = \ln\left(\frac{2}{3}\right) = -0{\cdot}405$ (to 3 s.f.)

b $4e^{5x} + 1 = 2$

$\Rightarrow 4e^{5x} = 1$

$\Rightarrow e^{5x} = 1 \div 4 = 0{\cdot}25$

$\Rightarrow 5x = \ln 0{\cdot}25 = -1{\cdot}38629...$

$\Rightarrow x = -0{\cdot}277$

c $3 \times 4^x = 18$

$\Rightarrow 4^x = 6$

$\Rightarrow \ln 4^x = \ln 6$

$\Rightarrow x \ln 4 = \ln 6$

$\Rightarrow x = \ln 6 \div \ln 4 = 1{\cdot}2924...$

$\Rightarrow x = 1{\cdot}29$ (to 3 s.f.)

Example 2

The curve $y = ax^b$, where a and b are constants, passes through the points (3, 2) and (4, 5).

Find a, b and state the equation of the curve.

Passes through (3, 2) $\Rightarrow 2 = a \times 3^b$ ... ①

Passes through (4, 5) $\Rightarrow 5 = a \times 4^b$ ... ②

② ÷ ①: $\frac{5}{2} = \frac{a \times 4^b}{a \times 3^b} = \left(\frac{4}{3}\right)^b$

$\Rightarrow \ln\left(\frac{5}{2}\right) = \ln\left(\frac{4}{3}\right)^b$

$\Rightarrow \ln\left(\frac{5}{2}\right) = b \ln\left(\frac{4}{3}\right)$

$\Rightarrow b = \frac{\ln\left(\frac{5}{2}\right)}{\ln\left(\frac{4}{3}\right)} = \frac{0{\cdot}9162...}{0{\cdot}2876...} = 3{\cdot}185... = 3{\cdot}19$ (to 3 s.f.)

Substitute in ①: $2 = a \times 3^{3{\cdot}185...}$

$\Rightarrow 2 = a \times 33{\cdot}087...$

$\Rightarrow a = 0{\cdot}060$ (to 3 s.f.)

The equation of the curve is $y = 0{\cdot}060\, x^{3{\cdot}19}$.

Example 3

The variables x and y are related by the equation $\log y = b \log x + \log a$, where a and b are constants.

When x is 2, y is 5 and when $x = 7, y = 9$.

Find the values of a and b.

When x is 2, y is 5: $\log 5 = b \log 2 + \log a$ ... ①

when $x = 7, y = 9$: $\log 9 = b \log 7 + \log a$ ... ②

② − ①: $\log 9 - \log 5 = b(\log 7 - \log 2)$

$\Rightarrow \quad \log 1{\cdot}8 = b \log 3{\cdot}5$

$\Rightarrow \quad b = \dfrac{\log 1{\cdot}8}{\log 3{\cdot}5} = 0{\cdot}469$ (to 3 s.f.)

Substitute in ①: $\log 5 = 0{\cdot}469 \times \log 2 + \log a$

$\Rightarrow \log a = \log 5 - 0{\cdot}469 \times \log 2 = 0{\cdot}558$ (to 3 s.f.)

$\Rightarrow \quad a = 10^{0{\cdot}558} = 3{\cdot}61$ (to 3 s.f.)

Exercise 9.5

1 Solve for x giving your answer correct to 3 significant figures.

a $3e^x = 7$ b $4e^x = 9$ c $7e^x + 4 = 27$ d $0{\cdot}5e^x + 1{\cdot}5 = 10$

e $4^x = 12$ f $5 \times 3^x = 50$ g $12 \times 2^x = 45$ h $7 \times 3^x + 1 = 50$

2 Solve: a $\log_4 x = 7$ b $\log_x 4 = 7$ c $\log_4 7 = x$.

3 Simplify using the laws of logs then solve for x.

a $\log_3 (x + 8) - \log_3 (x - 8) = 2$

b $2 \log_2 x - \log_2 (x + 6) = 3$

4 Solve: a $5^{x+1} = 2^{6-x}$ b $3^x = 2^{x+1}$.

5 The curve $y = ax^b$, where a and b are constants, passes through the points (2, 7) and (5, 12). Find a, b and state the equation of the curve.

6 A function is defined by $f(x) = ax^b$. Given that $f(5) = 13$ and $f(8) = 35$:

a form two equations in a and b

b find the function.

7 The astronomer Kepler believed the relationship between the time it took a planet to go round the Sun, y years, was related to the distance from the Sun, x astronomical units. He believed the relationship was of the form $y = ax^b$.
Mercury took 0·24 year to orbit at a distance of 0·39 AU from the Sun.
Mars took 1·88 years at a distance of 1·52 AU.

a Form two equations in a and b.

b Solve the equations simultaneously to find the values of a and b to 1 decimal place.

c What is the law relating the time to orbit and the distance from the Sun?

8 The variables x and y are related by the equation $\log y = b \log x + \log a$, where a and b are constants.
When x is 3, y is 8 and when $x = 10$, $y = 14$.
Find the values of a and b.

9 The distance to the horizon, d km, is related to the height your eyes are above sea-level, h m. The relation can be expressed in the form $\log d = b \log h + \log a$, where a and b are constants.
Standing on a rock at the beach, $h = 2$ m, the distance to the horizon is 5 km.
Standing on Goat Fell, $h = 874$ m, the distance to the horizon is 106 km.

a Form two equations in a and b.

b Solve them to find a and b correct to 1 decimal place.

c What is the distance to the horizon from the top of the Eiffel Tower where $h = 280$ m?

10 The curve $y = ab^x$, where a and b are constants, passes through the points (2, 3) and (6, 8). Find a, b and state the equation of the curve.

11 Peter put some money into a savings account. The relation between what is in the account at any time £y is related to how long it has been left in the bank, x years, by the formula $y = ab^x$, where a and b are constants.
After 2 years there was £1061 in the account and after 5 years there was £1159.

a Calculate the value of a and b in the formula.

b If Peter leaves the money in for 10 years, how much will there be in the account?

12 The variables x and y are related by the equation $\log y = x \log b + \log a$, where a and b are constants.
When x is 2, y is 1000 and when $x = 4$, $y = 10\,000$.
Find the values of a and b.

13 As a ball continually bounces, the height to which it goes, h cm, is related to which bounce it is, k, by the formula $\log h = k \log b + \log a$, where a and b are constants.
On the third bounce, $k = 3$, it reached a height of 108 cm and on the fifth bounce, it reached 39 cm.

a Find the formula that relates k and h.

b After how many bounces will it reach approximately 5 cm?

9.6 Making the model linear

In science, data are collected in an experiment to examine the relation between two variables.

Graphs can be drawn to help point the way to a conjectured model.

For example:
The purple profile hints at a relationship $y = ab^x$; note that it doesn't pass through the origin.

The black profile suggests $y = ax^b$; note that it does pass through the origin.

The purple broken line suggests a fractional constant power; the black broken line, a negative constant power.

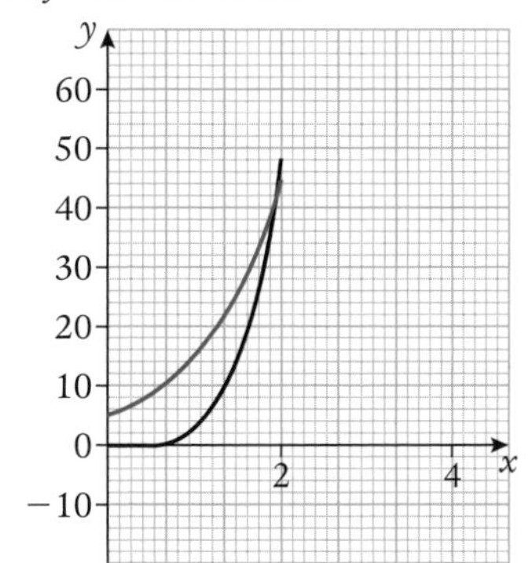

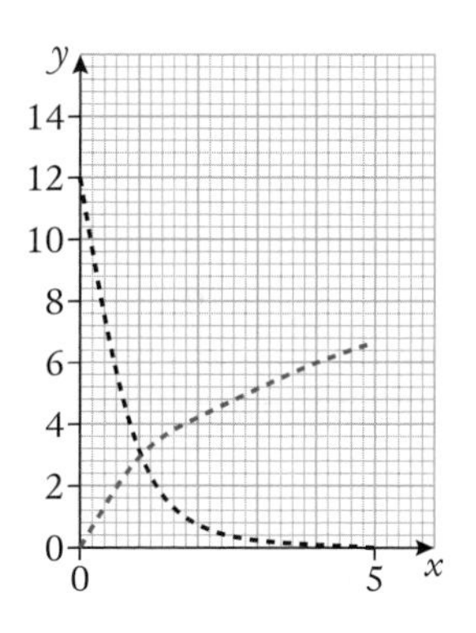

We can use logs to help confirm the conjectured model.

Consider $y = ax^b$

$y = ax^b \Rightarrow \log y = \log (ax^b)$

$\Rightarrow \log y = \log a + \log (x^b) = \log a + b \log x$

$\Rightarrow \log y = b \log x + \log a$ ... note that b and $\log a$ are constants

If we drew a graph of $\log y$ against $\log x$, you would get a graph with equation of the form $Y = mX + c$ ... a straight line: $\log a$ would be the y-intercept and b would be the gradient.

Consider $y = ab^x$

$y = ab^x \Rightarrow \log y = \log (ab^x)$

$\Rightarrow \log y = \log a + \log (b^x) = \log a + x \log b$

$\Rightarrow \log y = x \log b + \log a$ ... note that $\log b$ and $\log a$ are constants

If we drew a graph of $\log y$ against x, you would get a graph with equation of the form $Y = mx + c$ ... a straight line: $\log a$ would be the y-intercept and $\log b$ would be the gradient.

Example 1

A zoologist weighs and tags different animals to study the size of their territorial area.

Body weight, x (kg)	10	20	30	40	50	60	70
Territorial area, y (km²)	15	35	80	115	155	175	230

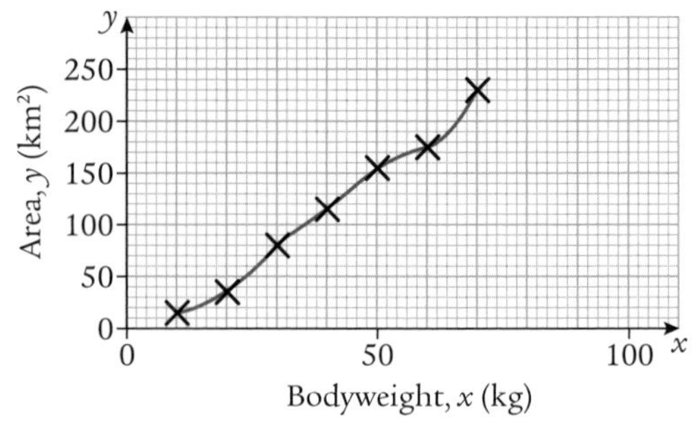

Find a formula that models the data.

From a quick sketch, we conjecture that the situation can be modelled by $y = ax^b$, where y km² is the area of the territory, and x kg is the body weight of the animal.

Take the log of both sets of data:

Log (weight)	1·00	1·30	1·48	1·60	1·70	1·78	1·85
Log (territorial area)	1·18	1·54	1·90	2·06	2·19	2·24	2·36

Plot the points and draw a line of best fit.

The fit is good enough to make us feel the choice of model, $y = ax^b$, is appropriate.

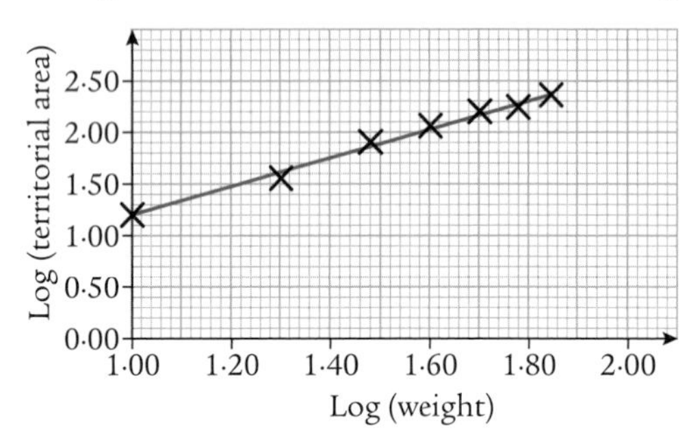

The line passes through (1·00, 1·18) and (1·70, 2·19).

The equation is of the form $\log y = b \log x + \log a$.

It passes through (1·00, 1·18) giving us: $1{\cdot}18 = 1{\cdot}00b + \log a$ ... ①

It passes through (1·70, 2·19) giving us: $2{\cdot}19 = 1{\cdot}70b + \log a$ ... ②

② − ①: $1{\cdot}01 = 0{\cdot}7b$

$\Rightarrow b = 1{\cdot}01 \div 0{\cdot}7 = 1{\cdot}44$ (to 3 s.f.)

Substitute in ①: $1{\cdot}18 = 1{\cdot}44 + \log a$

$\Rightarrow \log a = -0{\cdot}26$

$\Rightarrow \quad a = 10^{-0{\cdot}26} = 0{\cdot}550$ (to 3 s.f.)

So the relationship between body weight, x kg, and territorial area, y km², is $y = 0{\cdot}55x^{1{\cdot}44}$.

Example 2

When $\log y$ is plotted against x, a straight line graph is produced passing through (1, 2) and (3, 8). Find the values of the constants, a and b, in the relationship $y = ab^x$.

The gradient of the line $= \dfrac{8-2}{3-1} = 3$.

The line has equation $\log y = mx + c$.

Using (1, 2): $\log y = 2$ when $x = 1$, so $2 = 3 \times 1 + c \Rightarrow c = -1$

The line has equation $\log y = 3x - 1$ ... ①

A relationship of the form $y = ab^x$ leads to the linear relation $\log y = x \log b + \log a$ ...②

Comparing ① and ② we get $\log b = 3 \Rightarrow b = 10^3$; $\log a = -1 \Rightarrow a = 10^{-1}$.

The relationship is $y = \frac{1}{10} \times 1000^x$.

Exercise 9.6

1 Use the laws of logs to express each of the following in a form that discloses the linear relation between $\log y$ and $\log x$ or between $\log y$ and x.

a $y = ax^b$ **b** $y = ab^x$ **c** $y = 3x^4$ **d** $y = 3 \times 4^x$

2 The graph of $\log y$ against $\log x$ produces a straight line passing through the points (3, 9) and (1, 5).
Express the relationship between x and y in the form $y = ax^b$.

log y
(3, 9)
(1, 5)
O
log x

3 The graph of $\log_e y$ against $\log_e x$ produces a straight line passing through the points (1, 2) and (5, 7).
Express y in terms of x.

4 When $\log y$ is graphed against x we get a straight line passing through (2, 4) and (3, 8).
Express the relationship between x and y in the form $y = ab^x$, where a and b are constants.

5 When $\log_3 y$ is plotted against x, a straight line graph is produced passing through (1, 3) and (3, 5). Find the values of the constants a and b in the relationship $y = ab^x$.

6 In an experiment on gravity, pebbles are timed as they drop from fixed heights.

Time (seconds)	1	2	3	4	5	6
Drop (metres)	5	20	45	80	125	180

a Copy and complete this table of logs.

log (Time)						
log (Drop)						

b Plot the points and draw a best fitting straight line using the x-axis for 'Time'.

c Pick two points on the line to help you work out the relationship between Time and the distance dropped in the form $y = ax^b$.

7 It was feared that the population of pheasants in the local woods was falling.
Over the past five years a count has been made of breeding pairs.

Year	1	2	3	4	5
Number	43	36	32	27	24

The gamekeeper thought it was dropping exponentially and wanted to model it as $N = ae^{bY}$, where N is the number of breeding pairs Y years after records were kept and a and b are constants. He found the natural log of each number, producing this table:

Year	1	2	3	4	5
ln (Number)	3·76	3·58	3·47	3·30	3·18

a Check that graphing $\log_e$ (number) against year is a straight line which passes through (1, 3·76) and (5, 3·18).

b Calculate the values of the constants in the model.

c When is the population expected to drop below 10 breeding pairs?

Preparation for assessment

1 The relation between the variables x and y is $y = 3 \times 2^x$. Make x the subject of the formula.

2 The length of time £5000 has to stay in the bank to grow to £A is Y years.

Y can be estimated from the formula $Y = 33 \ln\left(\frac{A}{5000}\right)$. Make A the subject of the formula.

3 This is a sketch of the function $f(x) = \log_3 (x - 2), x > 2$ near the origin.

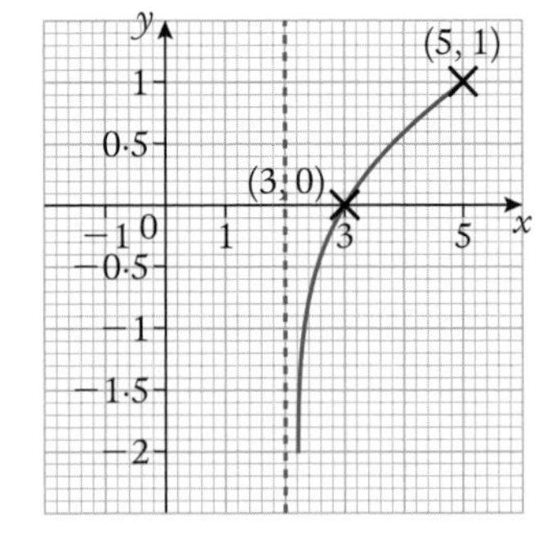

a Make a sketch of the inverse function.

b State the equation of the inverse function.

4 Simplify and evaluate:

$\log_5 2 + 2 \log_5 10 - \frac{1}{2} \log_5 64$.

5 Solve:

a $4 \times 2^{x+1} = 8$

b $\log_4 (x + 1) + \log_4 (2x - 5) = 1, x > 2{\cdot}5$.

6 Two functions f and g are defined by $f(x) = 4x - 1$ and $g(x) = e^x$.

a Write an expression for $f(g(x))$.

b i Write an expression for $g(f(x))$.

ii Express it in the form Ae^{Bx} giving the value of A to 2 significant figures.

7 Variables x and y have a relationship of the form $y = ax^b$, where a and b are constants. When x is 4, y is 6 and when x is 16, y is 12.
Find the values of a and b.

8 The equation $\log y = x \log b + \log a$ describes the relationship between variables x and y, where a and b are constants. When $x = 2, y = 1$ and when $x = 3, y = 2$.
Find the value of the constants a and b **exactly**.

9 The variables x and y form a relationship $y = ax^b$, where a and b are constants.

a Show that there is a linear relationship between $\log x$ and $\log y$.

b Some data of related values of x and y have been collected.

x	2·40	2·60	2·80	3·00
y	22·65	27·01	31·79	37·00
log x	0·38	0·41	0·45	0·48
log y	1·35	1·43	1·50	1·57

Use data from the last two rows of the table to help you evaluate a and b.

c What is the value of y corresponding to an x-value of 10?

10 The population of a village is being studied. They think it can be modelled by $P = ab^t$, where P is the number of residents in the village t years after records began.

a Show that, if this is the case, there is a linear relationship between t and $\log P$.

b In year 1 the population was 4200, in year 4 it was 4862 persons, in year 8 it was 5910. Show that this supports the linear conjecture.

c Calculate the values of a and b in the model.

d In what year will the population be double that of year zero ($t = 0$)?

Summary

1 If $a = 10^x$, then we say the logarithm of a is x and we write $\log a = x$.

2 If $a = b^x$. then the logarithm of a to the base b is x. We write $\log_b a = x$.
In general, $y = a^x \Leftrightarrow \log_a y = x$.

3 $f(x) = a^x$ is called an exponential function.

4 $f(x) = e^x$ is called **the** exponential function, where e = 2·718 correct to 4 significant figures.
[Note: by choosing e as a base, future work in calculus will be eased.]

5 $\log_e x$ is often denoted by $\ln x$.

6 The log function and the exponential function are the inverse of each other.
$\log_a (a^x) = x$ and $a^{\log_a x} = x$

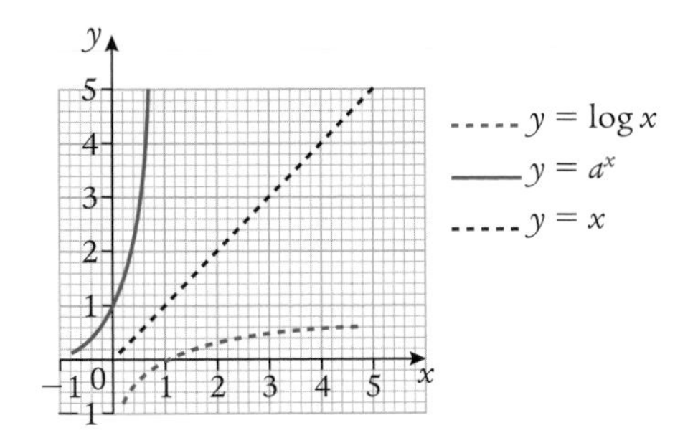

7 The rules of logarithms:
i $\log_a xy = \log_a x + \log_a y$
ii $\log_a \frac{x}{y} = \log_a x - \log_a y$
iii $\log_a x^y = y \log_a x$.

8 Useful facts arriving from the definition:
i $\log_a 1 = 0$ ii $\log_a a = 1$.

9 Solve exponential equations by taking the log to a suitable base of both sides.
i $3^x = 5$ ... take log base 10 to get ... $x \log 3 = \log 5$.
ii $10 = 2e^{1·5x}$ ... take log base e to get ... $\ln 10 = \ln 2 + 1·5x$.

10 i If we graph $\log y$ against $\log x$ and get a straight line, then $y = ax^b$ (a, b constant).
ii If we graph $\log y$ against x and get a straight line, then $y = ab^x$ (a, b constant).
iii In either case, finding two points on the line will allow us to form two equations and solve for a and b.

10 Differential calculus 2 – applications

Before we start...

Necessity is the mother of Invention.

Differential calculus was invented by Newton and Leibniz because they both needed a method for finding the gradient at any point on a curve.

They knew that the gradient gave a measure of how one variable was changing with respect to the other, for example, the gradient in a distance-time graph gives the rate at which distance changes with time ... the speed.

There are many applications of differential calculus.

This chapter explores some of them.

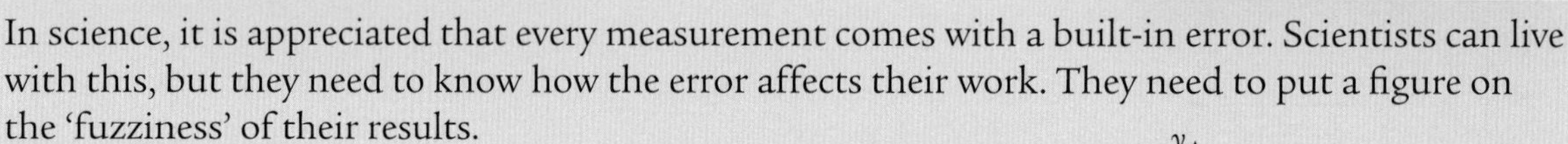

In science, it is appreciated that every measurement comes with a built-in error. Scientists can live with this, but they need to know how the error affects their work. They need to put a figure on the 'fuzziness' of their results.

Calculus can help here.

Using Leibniz notation, we have the definition: $\dfrac{dy}{dx} = \lim\limits_{\Delta x \to 0} \dfrac{\Delta y}{\Delta x}$

When Δx is very small then we can say $\dfrac{dy}{dx} \approx \dfrac{\Delta y}{\Delta x}$

... and so $\Delta y \approx \dfrac{dy}{dx} \Delta x$

If we are given that $\sin 1 = 0{\cdot}84147$ and that $\cos 1 = 0{\cdot}54030$, can we calculate $\sin 1{\cdot}01$ without the use of a calculator?

In this question we have $\Delta x = 1{\cdot}01 - 1 = 0{\cdot}01$.

We know that $\dfrac{d(\sin x)}{dx} = \cos x$.

So $\Delta y \approx \dfrac{dy}{dx} \Delta x = \cos 1 \times 0{\cdot}01 = 0{\cdot}54030 \times 0{\cdot}01 = 0{\cdot}0054030$.

So $\sin(1{\cdot}01) \approx \sin 1 + \Delta y = 0{\cdot}84147 + 0{\cdot}00540 = 0{\cdot}84687$.

The actual answer is 0·846831845... They agree to 4 decimal places.

This technique can be very handy in analysing errors.

An error of e radians in measuring an angle x will lead to an error of $e \cos x$ when calculating $\sin x$.

This particular application is not required for the course but it is only one of many that you may find useful as you pursue a career in maths or in most of the sciences.

The Exam syllabus is the mother of Necessity.
Invention is the mother of not studying for the exam.

What you need to know

1 A function is defined by $f(x) = 2x^4 + x^2 + 1$.

a Differentiate the function.

b Evaluate: **i** $f'(-1)$ **ii** $f'(0)$ **iii** $f'(1)$.

c Show that if $x < 0$ then $f'(x) < 0$ and that if $x > 0$ then $f'(x) > 0$.

d Identify the parts of the domain where $f(x)$ is: **i** increasing **ii** decreasing **iii** stationary.

e Make a sketch of $y = f'(x)$.

2 The sketch is of $y = \sin\left(2x + \frac{\pi}{3}\right); 0 \leqslant x \leqslant 2\pi$.

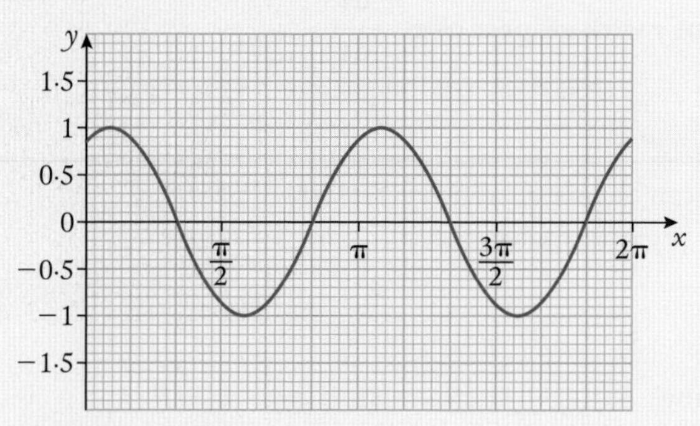

a $\frac{dy}{dx} = f'(x)$. Find $f'(x)$.

b Calculate: **i** $f'\left(\frac{\pi}{2}\right)$ **ii** $f'(0)$.

c Find the x-coordinates of the stationary points.

d Use the graph to help you state the nature of these points.

3 A cubic function is defined by $f(x) = x^3 - 3x^2 - 9x + 1$.

a Differentiate the function.

b Find the stationary points and determine their nature.

c By solving a quadratic inequation, determine the region of the domain where the function is decreasing.

4 In this course you are expected to remember various facts you learned in National 5, e.g.

Volume of a cylinder: $V = \pi r^2 h$. Curved surface area of a cylinder: $A = 2\pi rh$.

Area of a circle: $A = \pi r^2$. Circumference of a circle: $C = 2\pi r$.
[r represents radius and h height.]

A manufacturer wishes to make a cylindrical can to hold 250 ml of juice.

Let x cm be its radius and h cm be its height.

a Express h in terms of x.

b Express the curved surface area in terms of x.

c Hence find an expression for $f(x)$, the total surface area of the can.

d Find $f'(x)$.

e Find the stationary point and show that it is a minimum turning point.

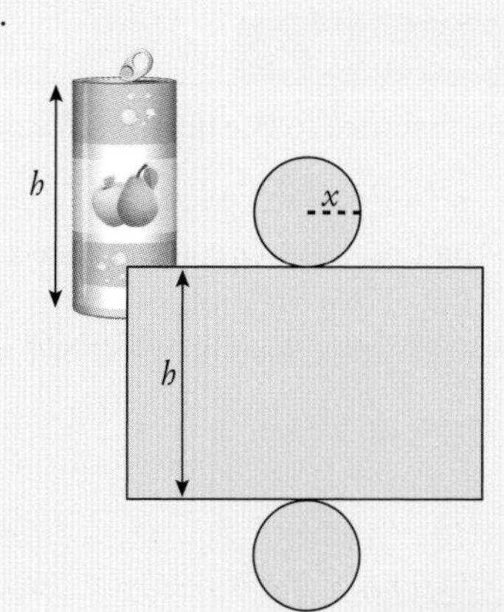

10.1 Rates of change

In National 5 you learned that the gradient was a measure of how the y-variable changed as the x-variable changed. This measure is known as the **rate of change**.

Using calculus, we differentiate to find a formula for the gradient of a curve at a point.

In any context, this can be interpreted as a rate of change.

For example, if you graph the amount of fuel used, F litres, against the distance travelled, x miles, the gradient will give you the **fuel consumption** in litres per mile ... $\frac{dF}{dx}$.

If you graphed the market price of a commodity, £P, against time, t days, the gradient would be interpreted as the **rate of change of the price** in pounds per day. ... $\frac{dP}{dx}$.

In general the derived function[1] gives **the rate of change of y with respect to x ...** $\frac{dy}{dx}$.

Example 1

Historical records suggest that the population in a village can be modelled by $P(x) = 500 + 35x^2$, where P is the number of residents x decades after records began in the year 1800.

a What was the population in 1800? $[x = 0]$

b What is the rate of growth of the population in 1850? $[x = 5]$

c What is the population in 1900 and how fast is it growing?

a $P(0) = 500 + 35 . 0^2 = 500.$
In 1800 the population was 500 people.

b $\frac{dP}{dx} = P'(x) = 70x \Rightarrow P'(5) = 70 \times 5 = 350.$
In 1850 the population is growing at a rate of 350 people per decade.

c In 1900, $x = 10$ and $P = 500 + 35 \times 10^2 = 4000$; $P'(10) = 70 \times 10 = 700.$
In 1900 there were 4000 people in the village and it was growing at the rate of 700 people per decade.

The **chain rule** can be used to combine rates of change.

For example,

rate of change of y with time = rate of change of y with x × rate of change of x with time

This is easier seen in Leibniz notation:

$$\frac{dy}{dt} = \frac{dy}{dx} \cdot \frac{dx}{dt}.$$

Example 2

The distance to the horizon, K kilometres, is a function of the height, x metres, of the observer

$$K(x) = 3{\cdot}57\sqrt{x}.$$

In the Eiffel Tower you can take a lift from the 2nd platform (115 m above the ground) to the 3rd platform (279 m above the ground) at a constant speed of 2 m/s.

a How do we know $\frac{dx}{dt} = 2$ where t is the time in seconds?

b Find $K'(x)$.

[1] The derived function is the result of differentiating the function.

c At what rate, in kilometres per metre, is the distance to the horizon changing when the lift is 144 m up?

d At what rate, in **kilometres per second**, is the distance to the horizon changing when the lift is 144 m up?

a We are given the **rate of change of height with time** is 2 metres per second. This is $\frac{dx}{dt}$.

b $K(x) = 3{\cdot}57(x)^{\frac{1}{2}} \Rightarrow K'(x) = \frac{1}{2} \times 3{\cdot}57\,(x)^{-\frac{1}{2}} = \frac{1{\cdot}785}{\sqrt{x}}$.

c $K'(144) = \frac{1{\cdot}785}{\sqrt{144}} = \frac{1{\cdot}785}{12} = 0{\cdot}14875$. When the lift is 144 m up, the distance to the horizon is increasing at the rate of 0·149 kilometres for each metre of rise.

d Part **c** gives us that when $x = 144$, $\frac{dK}{dx} = 0{\cdot}14875$; we are given that $\frac{dx}{dt} = 2$.

By the chain rule: $\frac{dK}{dt} = \frac{dK}{dx} \times \frac{dx}{dt} = 0{\cdot}14875 \times 2 = 0{\cdot}2975$.

When the lift is 144 m up, the distance to the horizon is growing at the rate of 0·2975 km per second.

Exercise 10.1

1 We can convert pounds to dollars using the formula $\$(x) = 1{\cdot}6x$, where \$ is the number of dollars you get for x pounds.

a Find $\frac{d\$}{dx}$.

b What is the rate of change in dollars per pound when x is: **i** 1 **ii** 10?

c **i** Verify that a linear function has a constant rate of change.

ii What is the rate of change of a constant function?

2 Archie puts £100 in a saving scheme that pays simple interest. The amount of money, £A, in the scheme after t years is given by $A(t) = 100 + 3t$.

a Find $\frac{dA}{dt}$.

b What is the rate of change in pounds per year when: **i** $t = 1$ **ii** $t = 5$?

c **i** What is $A(1)$?

ii What interest is paid at the end of year 1?

iii What is the rate of interest?

3 The number of pheasants in a wood has been modelled by $P(x) = 0{\cdot}4x^2 + 4x + 20$, where P is the number of pheasants in the xth year since the start of records.

a How many pheasants were in the wood when records began?

b What is the rate of growth of the population at this time (in pheasants per year)?

c **i** What is the population after 5 years?

ii What is the rate of growth after 5 years?

d After 10 years the actual population was 120 pheasants growing at a rate of 18 pheasants per year. How does this compare with the model?

4 A scientist, after several experiments of bringing a liquid up to its boiling point, modelled the **heating curve** of the liquid by $C(t) = 75 - \frac{55}{t+1}$, where C is the temperature in degrees Celsius and t is the time in seconds since heat was first applied. The model is only useful for $0 \leqslant t \leqslant 120$.

a The experiment was started at room temperature. What was room temperature?

b Calculate, when $t = 4$,

  i the temperature of the liquid

  ii the rate at which the temperature is increasing.

c What is the rate of change of temperature at the 9th second?

5 When a stone is dropped in a still pond, ripples move out from the point of entry in ever increasing circles.
The radius of one particular ripple can be calculated using $r(t) = 3t$, where the radius is r cm at time t seconds, measured from the moment the stone touched the water.

a At what rate in cm/s is the radius increasing? $\left[\frac{dr}{dt}\right]$

b The area of a circle is given by $A(r) = \pi r^2$.
Find the rate of change of the area in cm² per cm at the moment the radius is 4 cm.
$\left[\frac{dA}{dr} \text{ when } r = 4\right]$

c **i** Explain why A can be expressed in terms of t by $A(t) = 9\pi t^2$.

  ii Find an expression for the rate of change of area in cm² per second at time t seconds.

  iii How fast is the area changing at the 5th second?

d Verify that the answer to **cii** can be obtained using the chain rule, i.e. that $\frac{dA}{dt} = \frac{dA}{dr} \cdot \frac{dr}{dt}$.

6 The percentage of the Moon that is lit up, P%, can be calculated using the formula $P(t) = 50 - 50\cos\left(\frac{\pi}{15}t\right)$, where t days have elapsed since the last new Moon.

a Find a formula for the rate of change of P with time.

b Calculate how fast the illuminated area is growing on the 5th day after the new Moon in units of **% per day**.

c What is the rate of change of P with respect to t when

  i $t = 15$ **ii** $t = 20$?

d In the context, how do we spot from our answers when the Moon is:

  i waxing **ii** waning?

7 A rumour is started by four students in a school.
The number of people that have heard the rumour t days after it was started is modelled by:
$N(t) = 4(1 + t + 0{\cdot}9t^2 + 0{\cdot}4t^3)$.

a At what rate is the rumour spreading on: **i** day 2 **ii** day 5? (Give your answers to the nearest whole number.)

b The model is only valid when $0 \leq t \leq 8$, after which the whole school has heard it.
How fast is the rumour spreading when $t = 8$?

8 In the early 1600s, Johannes Kepler discovered the relation between the period of a planet and its distance from the Sun.
He realised that $P^2 = R^3$, where P is the period measure in years and R is the distance to the Sun measured in astronomical units. (Earth is 1 AU from the Sun.)

a Express P as a function of R.

b **i** Find an expression for $\frac{dP}{dR}$.

  ii Find the rate at which the period changes with respect to its distance from the Sun when this distance is 25 AU.

c Express R as a function of P.

d At what rate is the distance from the Sun changing when the period is 8 years?

10.2 Motion

A very familiar rate of change is **velocity**[2] ... the rate at which distance is changing with time.

If the **displacement**[3] of a moving object, s metres, measured from some origin is a function of time, t seconds, i.e. displacement $= s(t)$ metres, then the velocity is also a function of t and is the rate of change of displacement with time:

$v(t) = s'(t)$ metres per second (m s^{-1}).

In the same context, the rate of change of velocity with time is known as **acceleration**[4]

$a(t) = v'(t)$ metres per second per second (m s^{-2}).

Example 1

The vertical displacement of a golf ball can be modelled by $s(t) = 20t - 5t^2$, where s metres is the height of the ball t seconds after being struck.

a Find an expression for the velocity of the ball at time t.

b Calculate the velocity when **i** $t = 1$ **ii** $t = 3$, and interpret your answers in the context.

c Explain why $s(t)$ is zero when $t = 0$ and $t = 4$.

d Calculate the acceleration of the ball.

a $s(t) = 20t - 5t^2 \Rightarrow v(t) = s'(t) = 20 - 10t$.

b **i** $v(1) = 20 - 10 . 1 = 10$ metres per second.
[Velocity positive ... ball moving away from its origin.]

ii $v(3) = 20 - 10 . 3 = -10$ metres per second.
[Velocity negative ... ball moving back to its origin.]

c Ball has zero displacement at the beginning (0 seconds) and again when it lands (after 4 seconds).

d $a(t) = v'(t) = -10$ metres per second per second.
The negative sign tells us that the velocity of the ball **away from the ground** is getting less, i.e. the ball is slowing down **(decelerating)** as it rises.

Example 2

As the waves come into the bay, a boat tied up in the harbour bobs up and down.

The height of a point on the deck is s metres vertically from the platform of the pier.

This can be modelled by $s(t) = 0{\cdot}2 \sin(\pi t) + 0{\cdot}1$, where t seconds have elapsed since measurements began.

a What is the displacement of the point when $t = 1{\cdot}5$?

b What is the velocity of the point at this time?

c Calculate the acceleration when $t = 1{\cdot}5$.

a $s(t) = 0{\cdot}2 \sin(\pi t) + 0{\cdot}1$. So $s(1{\cdot}5) = 0{\cdot}2 \sin(\pi \times 1{\cdot}5) + 0{\cdot}1 = -0{\cdot}1$.
The point is 0·1 metre *below* the pier.

b $v(t) = s'(t) = 0{\cdot}2\pi \cos(\pi t)$.
$\Rightarrow v(1{\cdot}5) = 0{\cdot}2\pi \cos(\pi \times 1{\cdot}5) = 0$ ... the velocity is 0 m s^{-1}.

c $a(t) = v'(t) = -0{\cdot}2\pi^2 \sin(\pi t)$.
$\Rightarrow a(1{\cdot}5) = -0{\cdot}2\pi^2 \sin(\pi \times 1{\cdot}5) = 1{\cdot}97$ ...
The acceleration is 1·97 m s^{-2} (to 2 d.p.).

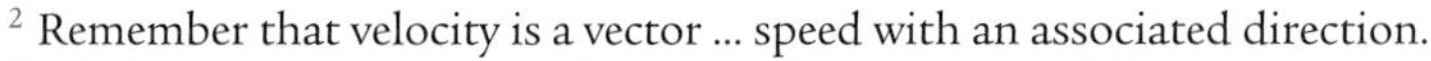

[2] Remember that velocity is a vector ... speed with an associated direction.
[3] Displacement is a vector ... distance with an associated direction.
[4] Acceleration is also a vector.

Exercise 10.2

1 A particle moves in a straight line so that its displacement from the origin after t seconds can be calculated from $s(t) = 1 - 3t + 2t^2$, where s is measured in metres.

a Find a formula for the velocity after t seconds.

b What is the velocity when: **i** $t = 1$ **ii** $t = 3$?

c Find a formula for the acceleration after t seconds. Comment.

2 A company experimenting with fireworks, models the height of one type of rocket before it runs out of fuel by $s(t) = 2t + 10t^2 + 5t^3$, where the rocket covers s metres in t seconds.

a How far does the rocket climb between the 2nd and 3rd second?

b Express its velocity in terms of t.

c Calculate: **i** $v(1)$ **ii** $v(2)$.

d Express its acceleration in terms of t.

e What is the acceleration of the rocket when $t = 2$?

3 A golf ball flies upwards from the tee. Its vertical height is modelled by $s(t) = 5 + 24t - 5t^2$, where s is measured in metres above the level fairway after time, t seconds.

a Find expressions for $v(t)$ and $a(t)$.

b The tee is on a raised embankment. How high off the fairway is the tee?

c When is the ball at its highest? [Hint: its velocity will be zero.]

d What is its acceleration at this point?

e What is the velocity and acceleration of the ball when it strikes the ground?

4 A car is watched from the top of a tower which is 100 m high.
It is moving away from the foot of the tower at speed of $1\ \text{m s}^{-1}$, i.e. at time t seconds it is t metres from the foot of the tower.

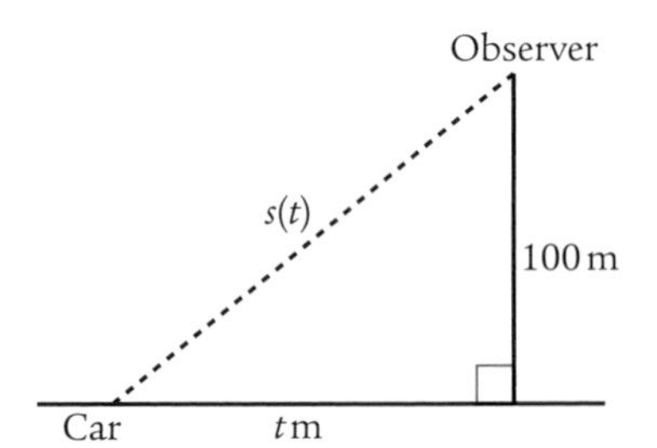

a Show that the displacement of the car from an observer on the top of the tower is modelled by:

$$s(t) = \sqrt{t^2 + 10\,000}.$$

b Find an expression for the corresponding velocity of the car. (How fast is the car receding from the observer?)

c What is this velocity after a minute, correct to 2 significant figures?

d An approximation for the acceleration at time t can be found using:

$$a(t) \approx \frac{v(t + 0{\cdot}01) - v(t)}{0{\cdot}01}.$$

Estimate the acceleration, correct to 2 significant figures, when $t = 1$.

5 A pump is driven by a piston. The movement of the piston head can be modelled by:

$$s(t) = 3\sin\left(\frac{\pi t}{6}\right) + 5$$

where s cm is the displacement of the piston head from the crankshaft at time t seconds.

a Find an expression for the velocity of the piston head.

b Calculate the velocity when $t = 10$.

c Find a formula for the acceleration of the piston head.

d What is its value when $t = 1$?

6 When learning the laws of motion we discover $s(t) = ut + \frac{1}{2}at^2$, where u and a are constants and s is the displacement in metres after t seconds.

a Find an expression for the velocity at time t, $v(t)$.

b What is the velocity when $t = 0$, i.e. the initial velocity?

c i Find an expression for the acceleration at time t, $a(t)$.

ii What can be said for the acceleration in this context?

10.3 Curve sketching

Calculus can be used to glean enough information from the equation of a function to enable you to make a quick sketch of the curve.

You should identify:

- where the curve cuts the y-axis ... where $x = 0$
- where the curve cuts the x-axis ... where $y = 0$
- i where the curve is stationary ... where $\frac{dy}{dx} = 0$
  ii the nature of the stationary points
- how the 'tails' of the curve behave ...
  i as x gets very large, is y positive or negative
  ii as x gets very small, is y positive or negative?

Example 1

Draw a sketch of $y = x^3 - 3x^2 - 144x - 140$ given that $(x + 1)$ is a A_f.

Using synthetic division we find the other factors:

$$\begin{array}{r|rrrr} -1 & 1 & -3 & -144 & -140 \\ & & -1 & 4 & 140 \\ \hline & 1 & -4 & -140 & 0 \end{array}$$

$y = (x + 1)(x^2 - 4x - 140) = (x + 1)(x + 10)(x - 14)$.

- The curve crosses the x-axis when $y = 0$ ... i.e. at $x = -1, -10$ or 14.
- The curve crosses the y-axis when $x = 0$ ... i.e. when $y = -140$.

Differentiating we get:

$\frac{dy}{dx} = 3x^2 - 6x - 144 = 3(x^2 - 2x - 48) = 3(x - 8)(x + 6)$.

- Stationary points occur when $\frac{dy}{dx} = 0$ i.e. when $x = 8$ or $x = -6$.
- Their nature can be found from a table of signs:
  Max. TP at $(-6, 400)$.
  Min. TP at $(8, -972)$.

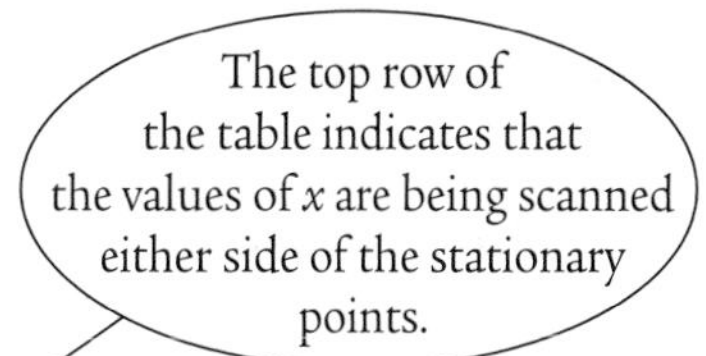

x	→	−6	→	8	→
−8	–	–	–	0	+
$x + 6$	–	0	+	+	+
$\frac{dy}{dx}$	+	0	–	0	+
Gradient	/	–	\	–	/
Nature		max.		min.	
y		400		−972	

- Finally, consider the behaviour of the function as x gets very large and i positive ii negative.
  In factorised form $y = (x + 1)(x + 10)(x - 14)$.
  When x is very large and positive:
  y = (large +ve factor) × (large +ve factor) × (large +ve factor) = (large +ve value).
  So the right-hand 'tail' of the curve goes off into the top right of the 1st quadrant.
  When x is very large and negative:
  y = (large −ve factor) × (large −ve factor) × (large −ve factor) = (large −ve value).
  So the left-hand 'tail' of the curve goes off into the bottom left of 3rd quadrant.

Pulling all this evidence together allows us to sketch the graph of the function.

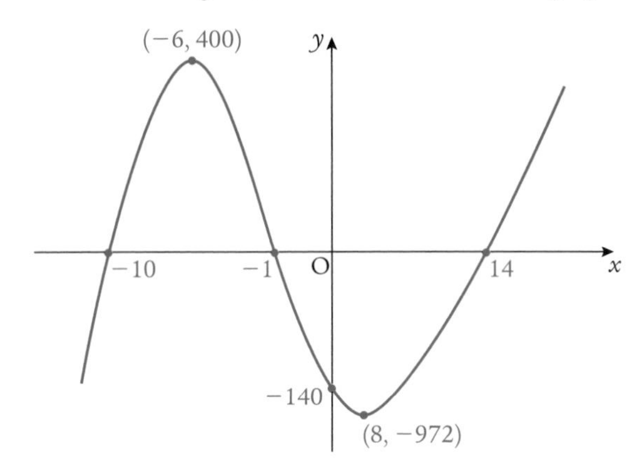

Exercise 10.3

1 Sketch each of these cubic functions.

a $y = (x + 5)(x - 10)(x + 14)$

b $y = (x + 5)(x - 7)^2$

c $y = (x - 8)(x + 13)(x + 37)$

d $y = x^3 - 3x + 2$

e $y = 98 - 21x - 12x^2 - x^3$ [Hint: $x - 2$ is a factor.]

f $y = x^3 - 12x^2 - 99x + 238$ [Hint: $x + 7$ is a factor.]

2 Make a sketch of each of the quartic functions.

a $y = (x^2 - 4)(x^2 - 1)$

b $y = x^4 - 50x^2 + 49$

c $y = 12x^3 - 3x^4$

3 One factor of the expression $x^4 + 4x^3 + 4x^2 - 9$ is $(x + 3)$.

a Completely factorise the expression.

b Find the stationary points on the curve $y = x^4 + 4x^3 + 4x^2 - 9$ and find their nature.

c Draw a sketch of $y = x^4 + 4x^3 + 4x^2 - 9$, labelling where it cuts the coordinate axes and the stationary points.

4 Often, in real context, a function may have a closed domain: then you might not get intercepts on the axes.
On these occasions you should look for some useful points on the curve to guide you.
The cost of a job, C, in thousands of pound, is modelled by:

$C(x) = 10x + \frac{250}{x}$

where x days is the time taken to complete the job.
The formula is only sound in the domain $1 \leqslant x \leqslant 10$.

a Find the stationary point and determine its nature.

b Calculate: **i** $C(1)$ **ii** $C(10)$.

c Is the function increasing or decreasing at: **i** $C(1{\cdot}01)$ **ii** $C(9{\cdot}99)$?

d Sketch the curve $y = 10x + \frac{250}{x}$; $1 \leqslant x \leqslant 10$.

Unit 3: Applications

10.4 Optimisation

Closed intervals

As mentioned in Exercise 10.3, Question 4, in real contexts, the function, $f(x)$,will often be defined in a **closed interval**. By that we mean there is a smallest and a largest value for x. We call these the 'endpoints' of the interval.

If we wish to find the maximum and/or minimum **value** of such a function then they must occur at either an endpoint or at a stationary point.[5]

Example 1

A function is defined by $f(x) = 2x^3 + 3x^2 - 12x + 1; -3 \leqslant x \leqslant 3$.

a Find $f'(x)$ and hence the x-coordinates of the stationary points in the interval $-3 \leqslant x \leqslant 3$.

b Calculate: **i** the minimum value and **ii** the maximum value of the function.

a $f'(x) = 6x^2 + 6x - 12$.

At a SP, $f'(x) = 0 \Rightarrow 6x^2 + 6x - 12 = 0$

$\Rightarrow \quad x^2 + x - 2 = 0$

$\Rightarrow \quad (x - 1)(x + 2) = 0$

$\Rightarrow \quad x = 1 \text{ or } -2.$

Stationary points occur when $x = 1$ and when $x = -2$.

b Evaluate the function at the endpoints, $x = -3$ and $x = 3$, and at stationary points $x = -2$ and $x = 1$: $f(-3) = 10, f(-2) = 21, f(1) = -6, f(3) = 46$.

i The largest of these values is the maximum value of the function in the interval, viz. 46.

ii The smallest of them is -6 ... so it is the minimum value of the function in the interval.[6]

In other contexts you may still be required to establish the nature of the SP by considering the behaviour of the gradient either side of it. For example, $(-2, 21)$ is a stationary point on $f(x)$ and $f(-2) = 21$ is neither the largest nor smallest value of the function. What is its nature?

Optimisation

Finding the maximum or minimum value of a function can be very useful in many real contexts. Business and industry are always looking for ways of being more efficient.

Example 2

A manufacturer takes a piece of card (20 cm by 15 cm). He folds its edges and glues it to make a small tray as shown. Each main fold is made x cm from the edge of the card.

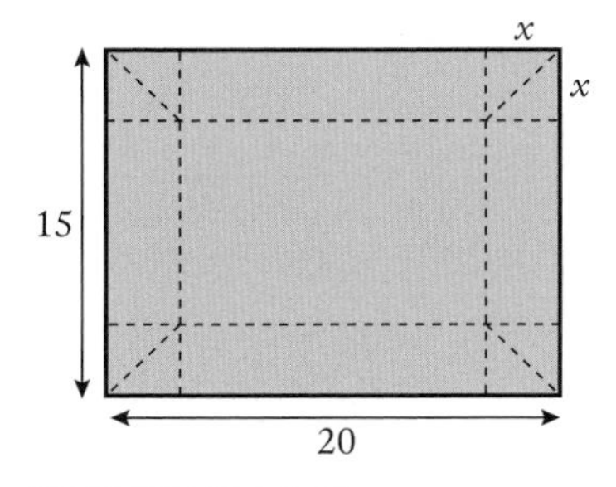

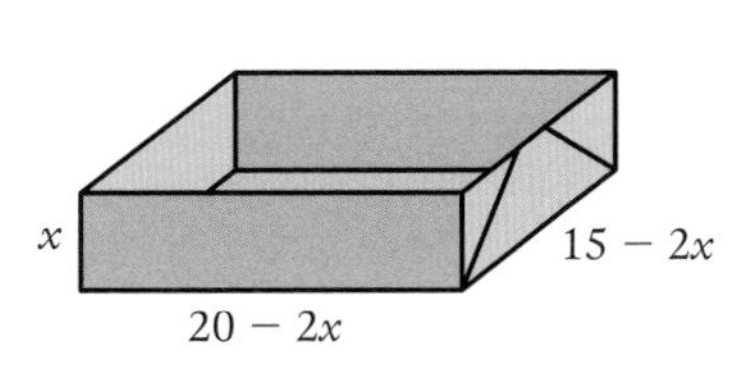

[5] In this course we will not be examining functions that contain 'breaks' within the interval, e.g. in the interval $0 \leqslant x \leqslant \pi$, $\tan x$ is undefined at $\frac{\pi}{2}$ ... there is a 'break' in the continuity of the curve.

[6] We did not have to determine the nature of the SP $(1, -6)$ using a sign table to know that it is the minimum value of the function in the closed interval.

Unit 3: Applications

What size should x be in order to maximise the volume of the tray?

We wish to optimise the **volume** ... so express the **volume** in terms of x:

$V(x) = x(20 - 2x)(15 - 2x)$

$\Rightarrow V(x) = 4x^3 - 70x^2 + 300x$ ... multiplying out the brackets

$\Rightarrow V'(x) = 12x^2 - 140x + 300$ ... the rate of change of V with respect to x

For any stationary point, $V'(x) = 0$. ... true only at stationary points

$12x^2 - 140x + 300 = 0 \Rightarrow x = 2{\cdot}829$ or $x = 8{\cdot}839$ (to 4 s.f.) ... using the quadratic formula

From practical considerations the endpoints of this function are $x = 0$ and $x = \frac{15}{2} = 7{\cdot}5$. ... think about it

Thus $x = 8{\cdot}839$ is outside the practical interval.

$V(0) = 0$, $V(2{\cdot}829) = 379$ (to 3 s.f.), $V(7{\cdot}5) = 0$. ... calculated from formula for $V(x)$

Thus the maximum value of the volume is 379 cm^3, occurring when the folds are 2·83 cm from the edge (to 3 s.f.). ... answer in the context

Note: in other circumstances you may have to justify the nature of the SP.

Exercise 10.4

1 Find the maximum and minimum value of each function in the given interval.

a $f(x) = 2x^2 - 12x + 5; 0 \leqslant x \leqslant 7$

b $f(x) = 1 - 3x - x^2; -3 \leqslant x \leqslant 2$

c $f(x) = 2x^3 - 3x^2 - 12x - 6; -3 \leqslant x \leqslant 3$

d $f(x) = 5 - 9x - 6x^2 - x^3; -4 \leqslant x \leqslant 0$

e $f(x) = 20x^3 - 3x^5; 0 \geqslant x \geqslant 3$

f $f(x) = 4x^3 - 3x^4; -1 \leqslant x \leqslant 2$

g $f(x) = x + \dfrac{36}{x}; -6 \leqslant x \leqslant 18$

h $f(x) = x^2 + \dfrac{256}{x^2}; 1 \leqslant x \leqslant 8$

i $f(x) = \sin\left(x + \dfrac{\pi}{3}\right); -\dfrac{\pi}{3} \leqslant x \leqslant \dfrac{\pi}{2}$

j $f(x) = 1 - 3\cos(2x); \dfrac{\pi}{3} \leqslant x \leqslant \dfrac{5\pi}{3}$

2 The sum of two numbers is 60.

a Let one of the numbers be x. Find an expression in x for the other number.

b Hence state an expression for the product of the two numbers.

c For what value of x is the product of the two numbers maximised? [The nature of the SP must be examined ... the interval is not closed.]

3 Two numbers multiply to make 49.

a Let one of the numbers be x. Find an expression in x for the other number.

b Hence state an expression for the sum of the two numbers.

c For what value of x is the sum of the two numbers smallest? [The nature of the SP must be examined.]

4 A piece of rope 50 metres long is pegged out to form a rectangle.

a Let x m be the length of the rectangle.
Find an expression in x for the breadth of the rectangle.

b Hence find an expression in x for the area of the rectangle.

c Find the value of x which maximises the area enclosed. (Show all your working.)

5 A farmer has 200 m of flexible fencing.
He uses it to form three sides of a rectangle against an existing stretch of wall as the fourth side.
Let the distance out from the wall be x m.

a Form an expression for the length of the rectangle in terms of x.

b Find the length and breadth of the space that encloses the most area.

c Suppose that the farmer had used the fencing to form two sides of a rectangular enclosure, the other two sides being formed by the corner of an existing wall.
What then would be the dimensions of the largest possible enclosed area?

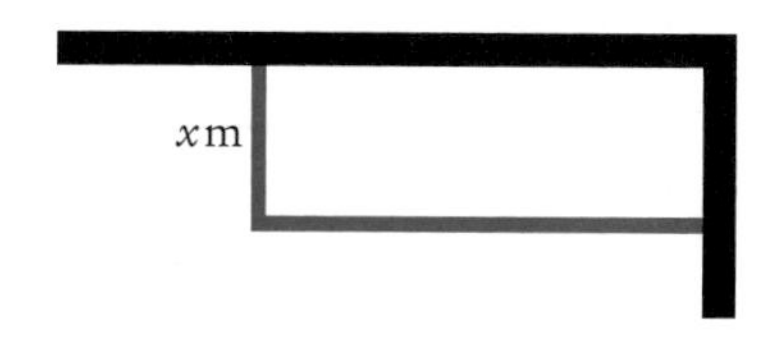

6 A rectangular card which is 9 cm by 24 cm is folded as in Example 2 above to make a box.

a Find an expression, $V(x)$, for the volume of the box in terms of x.

b Find the stationary points on the function $V(x)$ and describe their nature.

c Find the value of x which will maximise the volume.

d Solve the problem when the card is 8 cm by 15 cm.

7 A jogger is trying to reach a hotel.
He is 1 mile south of a crossroads and the hotel is 10 miles east of it.
The jogger can average 5 mph on the road or 4 mph off-road.
He decides to cut across country.
He aims for a point x miles east of the crossroads.

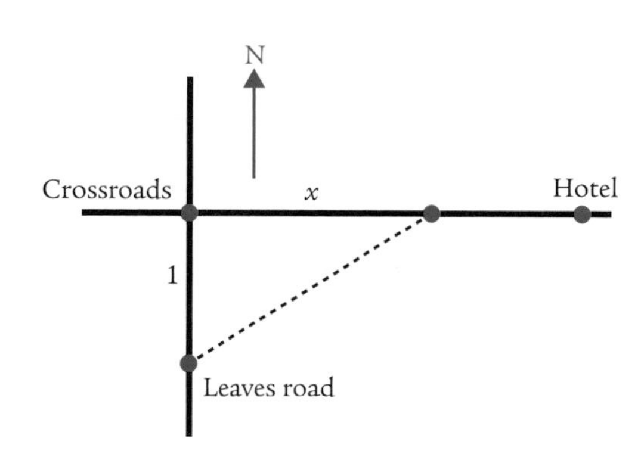

a Find an expression in x for:

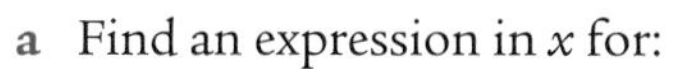

i the distance he does off-road
ii the time it takes him off-road
iii the distance he does on-road
iv the time it takes him on-road
v the total time of his run from the moment he left the road to his arrival at the hotel.

b Calculate the value of x that minimises this time.

c What is this minimum time?

8 Usually a manufacturer will know what he wants a box to contain before he designs it.
A box is to have a capacity of 1000 ml.
The box is to be a cuboid with a square base. Let the side of the square be x cm.

a Find an expression in terms of x for the height, h cm, of the box.

b **i** Show that the surface area of the box is $s(x) = 2x^2 + \dfrac{4000}{x}$.
ii Hence find the dimensions of the box which will minimise the surface area.

9 A length of stick is split in two and used to form the diagonals of a kite.
The area of the kite is 5000 cm^2.

a Let x be the length of one diagonal. Find the length of the other in terms of x.

b Write an expression in x for the total length of the original stick.

c What is the shortest the stick can be?

10 A cylindrical can is to hold 500 ml of juice.
The volume of a cylinder is $V(x) = \pi x^2 h$, where x cm is the radius and h cm is the height of the cylinder.

a Express h in terms of the radius.

b The surface area of a cylinder is $2\pi r^2 + 2\pi rh$. Express the surface area in terms of r.

c For what value of r will the surface area be a minimum (... and hence cost the least to make)?

11 a A 9-metre rope is used to form the two shorter sides of a right-angled triangle.
Let one side have the length x cm.
What length do you need to make x for the hypotenuse to be as short as possible?

b The same rope can be used to form the hypotenuse and one shorter side of a right-angled triangle.
Let the shorter side have a length x cm.

i Show that the area of the triangle is:

$$A(x) = \tfrac{1}{2}\sqrt{81x^2 - 18x^3}.$$

ii Find the value of x that maximises this area.

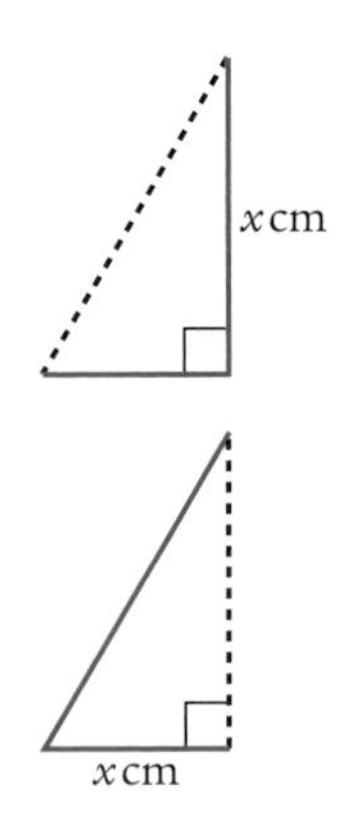

12 The arch of a bridge is modelled by $y = 36 - \frac{1}{12}x^2$, where y metres is the height of the arch x metres from its centre. A rectangular netting, ABCD, is hung from the central axis, AD to a point, C, on the arch.

a Taking the x-coordinate of C as x, state the y-coordinate of C in terms of x.

b Express the area of the netting in terms of x.

c For what value of x is this area a maximum?

d State the dimensions of the netting at this point.

13 The equipment to extract ore costs £180 000 to hire for a month.

The labour cost for extracting a tonne of ore is £50.

If x tonnes are extracted in a month, the hire cost can be divided by x to find the part of the cost that can be attributed to the x tonnes.

So the total cost of shifting x tonnes in a month is £$C(x) = 50x + \dfrac{180\,000}{x}$.

a Calculate $C'(x)$.

b How many tonnes should the mining company aim to extract to minimise the total cost?

c What is this minimum cost?

Preparation for assessment

1 A sphere is growing at a rate of 2 litres per minute.

a If initially the sphere had zero volume, express the volume, V litres, in terms of t, the number of minutes it has been growing.

b Express V in terms of r, the radius of the sphere.

c Find the rate of change of V with respect to r in litres per centimetre when the radius is:
i 1 cm ii 5 cm.

d Express r in terms of V.

e Find the rate of change of r with respect to V in centimetres per litre when the volume is:
i 1 litre ii 5 litres.

f By the chain rule we know $\frac{dr}{dt} = \frac{dr}{dV} \times \frac{dV}{dt}$.
Use this to find the rate of change of r with respect to time, in centimetres per minute.

g How fast is the radius growing when it is 6 cm in size?

2 A particle is moving away from an observer in a straight line.
Its displacement at time t seconds after a timer is set is given by $s(t) = t^3 + 5t^2$.

a Express v, the velocity of the particle, in terms of t.

b When is the velocity equal to: i $13\ \text{m s}^{-1}$ ii $32\ \text{m s}^{-1}$?

c Express a, the acceleration of the particle, in terms of t.

d What was the acceleration at the 4th second?

e When was the acceleration equal to $70\ \text{m s}^{-2}$

f What were the initial velocity and acceleration?

3 Relative to a suitable set of axes, the movement of a particular point in an engine has been modelled by $s(t) = \sin 2t + 2\cos t$, where s is the displacement in centimetres from the origin t seconds after the engine was started.

a What is the displacement after π seconds?

b When is the displacement zero in the interval $0 \leqslant t \leqslant 2\pi$?

c Express the velocity in terms of t.

d What is the velocity after π seconds?

e Show that the acceleration, velocity and displacement all have a value of zero when $t = \frac{3\pi}{2}$.

4 Sketch the function $f(x) = (x + 4)(x - 20)(x - 5)$ given that when expanded it takes the form $f(x) = x^3 - 21x^2 + 400$.

5 A 16-metre length of rope is to be used to mark off the sector of a circle.
Working in radians,

- the length of an arc, $L = x\theta$, where x is the radius of the circle
- the area of the segment is $A = \frac{x^2\theta}{2}$.

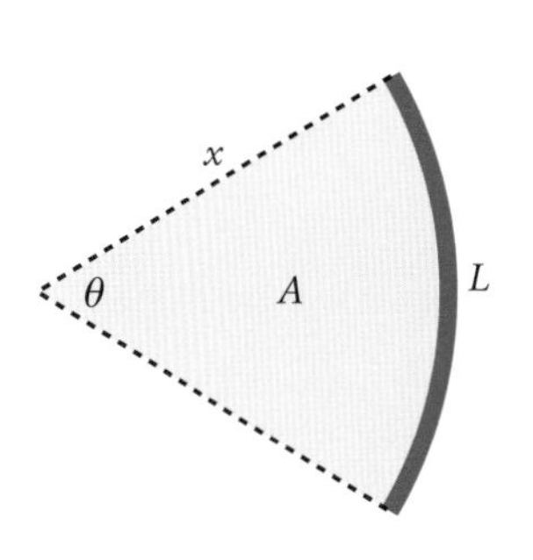

a Write down an expression for the perimeter of the sector in terms of x and θ.

b Knowing that the perimeter is 16 m, express θ in terms of x.

c i Express the area of the sector in terms of x alone.
ii Hence find the value of x which will maximise the area of the sector.

Summary

1 $\frac{dy}{dx}$ gives us the **rate of change of y with respect to x.**

For example, if F represents the number of litres used in x miles and if F is a function of x, $F(x)$, then $\frac{dF}{dx} = F'(x)$ will give you the **fuel consumption** in litres per mile.

2 The **displacement** of a point, s metres, is the distance the object is from the origin in a specified direction. It is usually expressed in this context as a function of time, $s(t)$ metres.

Velocity is the rate of change of displacement with time $v(t) = s'(t)$ metres per second [m s^{-1}].

Acceleration is the rate of change of velocity with time $a(t) = v'(t)$ metres per second per second [m s^{-2}].

3 To make a quick sketch of a function, given its equation, identify:

- where the curve cuts the y-axis ... where $x = 0$
- where the curve cuts the x-axis ... where $y = 0$
- **i** where the curve is stationary ... where $\frac{dy}{dx} = 0$
  ii the nature of the stationary points
- how the 'tails' of the curve behave ...
  i as x gets very large, is y positive or negative
  ii as x gets very small, is y positive or negative?

If the curve doesn't cut an axis, then some specific points might help, e.g. in a closed interval, the endpoints should be calculated.

4 In a **closed interval** the maximum and/or minimum **value** of a function must occur at either an endpoint or at a stationary point. This is useful when trying to optimise a solution to a problem.

11 Integral calculus

Before we start...

1 In the 3rd century BC, Archimedes explored methods of finding the area of a circle and of a parabola by slicing them up into thin strips. Each of the thin strips can then be approximated by a rectangle and its area found. Summing all the strips will produce an approximation for the area of the shape that has been dissected. The thinner the strips, the more accurate is the answer.

In the 16th century AD, Cavalieri and Fermat extended this research, developing the forerunner to what has become known as **integral calculus**.

However, it wasn't until the 17th century that Newton and Leibniz discovered the link between this problem and their differential calculus.

It is this link that the chapter will investigate and exploit.

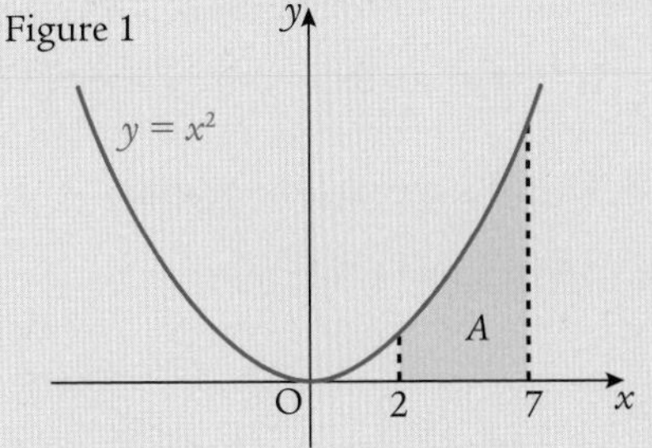

Figure 1

2 We know the area of a trapezium $= \frac{h}{2}(p_1 + p_2)$, where p_1 and p_2 are the lengths of the parallel sides and h is the distance between them.

Consider the curve $y = x^2$.

Suppose we wish to find the area under the curve between $x = 2$ and $x = 7$ (A in Figure 1).

We could cut it into vertical strips of equal thickness, h units, and treat each strip as a trapezium.

Figure 2 shows a typical strip.

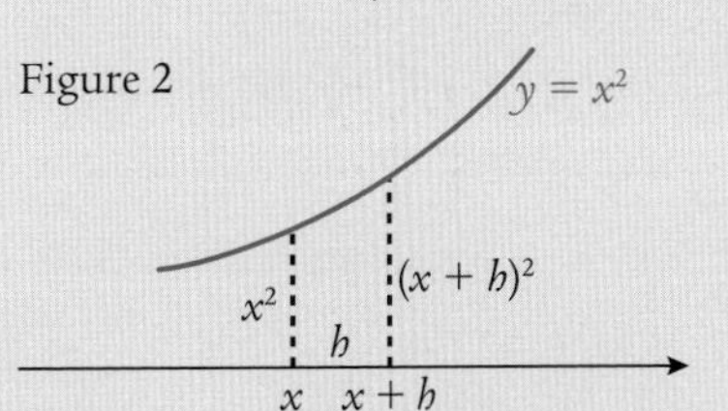

Figure 2

The area of the strip would be $\frac{h}{2}(x^2 + (x + h)^2)$.

If we used 5 strips then $h = (7 - 2) \div 5 = 1$.

We would calculate the area of the strip when $x = 2, 3, 4, 5$ and 6 and sum our answers.

$A \approx \frac{1}{2}(2^2 + (2 + 1)^2) + \frac{1}{2}(3^2 + (3 + 1)^2) + \frac{1}{2}(4^2 + (4 + 1)^2) + \frac{1}{2}(5^2 + (5 + 1)^2) + \frac{1}{2}(6^2 + (6 + 1)^2)$

$A \approx \frac{1}{2}(4 + 9) + \frac{1}{2}(9 + 16) + \frac{1}{2}(16 + 25) + \frac{1}{2}(25 + 36) + \frac{1}{2}(36 + 49)$

$A \approx 112{\cdot}5$

Do the same with 10 strips, $h = 0{\cdot}5$, and $A \approx 111{\cdot}875$.

With 20 strips, $h = 0{\cdot}25$, $A \approx 111{\cdot}71875$.

With 50 strips, $h = 0{\cdot}1$, $A \approx 111{\cdot}675$.

For the sake of comparison, the actual answer is $A \approx 111\frac{2}{3}$.

The above results were obtained using a spreadsheet ... example when $h = 0.1$ and A1 holds 2

The first column lists x-values from 2 to 7 in steps of h ... A2 holds: =A1+0.1

The second column tabulates the squares of the x-values ... B1 holds: =A1^2

The third column evaluates the area of the trapezium between x and $x + h$

... C1 holds: =(B1+B2)*0.1/2

The fourth column holds the sum of the third column ... and the required area

... =SUM(C1:C50)

Note that the 51st row doesn't provide the area of a strip to be included in the sum.

Perhaps you could explore the technique on a spreadsheet, experimenting with different functions [column B].

What you need to know

1 Differentiate:

a $3x^3 + 4x^2 - 5x + 4$

b i $\sqrt{x}$ ii $\frac{1}{x}$ iii $\frac{1}{x^3}$

c $\frac{1 + x^2}{\sqrt{x}}$

d i $\sin x$ ii $\cos x$.

2 Find $\frac{dy}{dx}$ in each case:

a i $y = (3x + 1)^2$ ii $y = \frac{1}{(3x + 1)^2}$ iii $y = \sqrt{(3x + 1)}$

b i $y = \sin 3x$ ii $y = \sin(3x - 4)$ iii $y = \cos(5x - 1)$ iv $y = \cos(1 - 5x)$.

3 This is a speed–time graph:

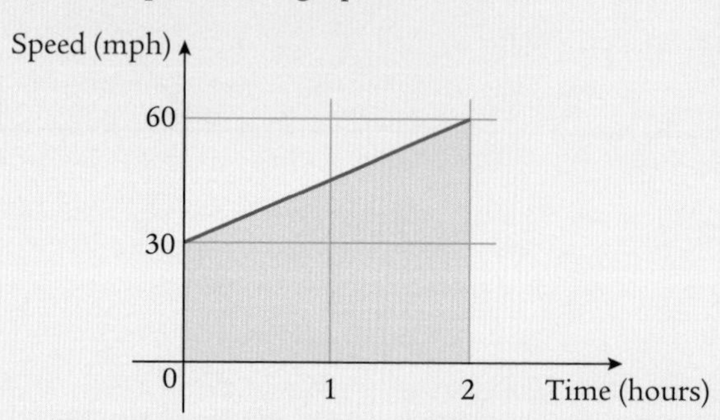

It shows a particle moving so that its speed steadily increases from 30 mph to 60 mph.

When you find the area under the line, you are multiplying speed with time ... so you are calculating the total distance covered. ($D = S \times T$)

How far do you travel when you accelerate from 30 mph to 60 mph over 2 hours?

4 The diagram shows a quarter circle of radius 5 units.

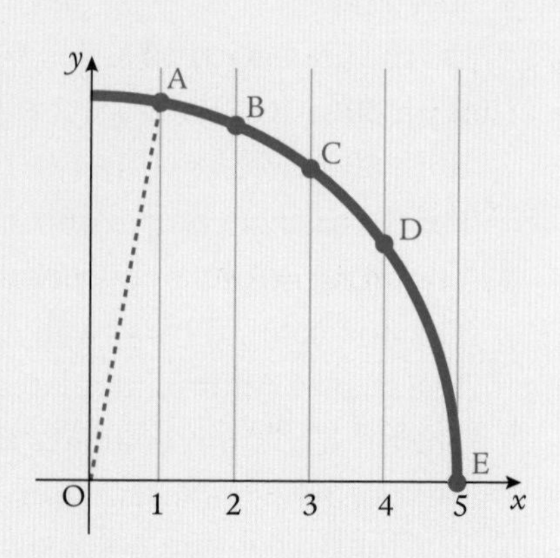

a The distance OA = 5 units. [It is a radius.] Use Pythagoras' theorem to calculate the perpendicular distance from A to the x-axis, i.e. its y-coordinate.

b Hence estimate the area of the 1st strip by taking it as a trapezium.

c Repeat this for the other 4 strips.

d Sum your answers to obtain an approximation for the area of the quarter circle.

e Compare your answer with that obtained using $A = \frac{\pi r^2}{4}$.

11.1 Reversing a process

If a function is defined by $y = 3x^2 + 2x + 1$, we can find the derived function $\frac{dy}{dx} = 6x + 2$.

If we are given $\frac{dy}{dx} = 6x + 2$, is this enough information to find y?

- To differentiate, for each term we **i** multiplied by the power **ii** reduced the power by 1.
- To reverse this process we **i** add 1 to the power **ii** divide by the new power.

This gives $y = \frac{6x^2}{2} + \frac{2x^1}{1} = 3x^2 + 2x$ ... but unfortunately, the constant term in the original function has not reappeared. We have to acknowledge that some constant term will be there (even if it is zero) by including '$+ c$' in the answer, i.e.

$$\frac{dy}{dx} = 6x + 2 \Rightarrow y = 3x^2 + 2x + c.$$

This solution[1] defines a family of curves, infinite in number. The sketch shows only four.

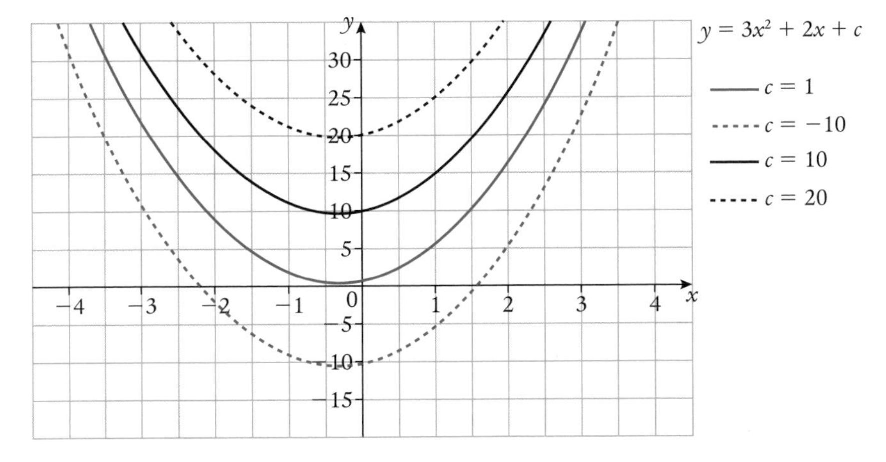

If we wish to pinpoint a particular curve then we must have more information, say a point that lies on the curve ... usually referred to as an **initial condition**.

So if we know $\frac{dy}{dx} = 6x + 2$ and $f(1) = 6$ then ...

- $\frac{dy}{dx} = 6x + 2 \Rightarrow y = 3x^2 + 2x + c$
- $f(1) = 6 \Rightarrow 3 \cdot 1^2 + 2 \cdot 1 + c = 6 \Rightarrow 5 + c = 6 \Rightarrow c = 1$.
- Thus $y = 3x^2 + 2x + 1$.

Definitions

1 If $F'(x) = f(x)$, then $F(x) + c$ is called an **anti-derivative** of $f(x)$.

2 $F(x)$ is called the **integral** of $f(x)$ with respect to x.

3 Using notation devised by Leibniz, 'integrate $f(x)$ with respect to x' is denoted by ... $\int f(x)\,dx$.

4 The process of finding an anti-derivative is called **integration**.

5 $\int f(x)\,dx = F(x) + c \Leftrightarrow F'(x) = f(x)$.

6 The constant generated, usually denoted by c, is called the **constant of integration**.

[1] An equation of the form $\frac{dy}{dx} = 6x + 2$ is a differential equation and $y = 3x^2 + 2x + c$ is called a **general solution**.
When we evaluate c from particular initial conditions, it is called a **particular solution**.

Basic rules

From definition 5 above we can generate various rules from known derivatives.

1 $\frac{d}{dx}\left(\frac{ax^{n+1}}{n+1}\right) = ax^n \Rightarrow \int ax^n \, dx \quad = \frac{ax^{n+1}}{n+1} + c, \quad n \neq -1$

... add 1 to power and divide by the new power ...

2 $\frac{d}{dx}(\sin x) = \cos x \Rightarrow \int \cos x \, dx \quad = \sin x + c$

3 $\frac{d}{dx}(-\cos x) = \sin x \Rightarrow \int \sin x \, dx \quad = -\cos x + c$

remember that this is only true if you work in radians

4 $\int af(x) \, dx = a\int f(x) \, dx$ where a is a constant.

5 $\int (f(x) + g(x)) \, dx = \int f(x) \, dx + \int g(x) \, dx$

6 In the context of displacement, velocity, and acceleration

i $s'(t) = v(t) \Rightarrow s(t) = \int v(t) \, dt + c_1$

ii $v'(t) = a(t) \Rightarrow v(t) = \int a(t) \, dt + c_2$

Initial conditions will help you determine c in the context.

Example 1

Integrate: a $5x^3 + 6x^2 + 2x + 4$ b $\frac{1}{x^2} + \sqrt{x}$ c $3 - 2\cos x$

a $\int 5x^3 + 6x^2 + 2x + 4 \, dx \quad = \frac{5x^4}{4} + \frac{6x^3}{3} + \frac{2x^2}{2} + \frac{4x^1}{1} + c = \frac{5}{4}x^4 + 2x^3 + x^2 + 4x + c$

b $\int \frac{1}{x^2} + \sqrt{x} \, dx \quad = \int x^{-2} + x^{\frac{1}{2}} \, dx = \frac{x^{-1}}{-1} + \frac{x^{\frac{3}{2}}}{\frac{3}{2}} + c = -x^{-1} + \frac{2}{3}x^{\frac{3}{2}} + c = -\frac{1}{x} + \frac{2}{3}\sqrt{x^3} + c$

c $\int 3 - 2\cos x \, dx \quad = \frac{3x^1}{1} - 2\sin x + c = 3x - 2\sin x + c$

Example 2

a Find the general solution to the differential equation $\frac{dy}{dx} = 3x^2 + 4x - 5$.

b Find the particular solution if $f(1) = 7$.

a $\frac{dy}{dx} = 3x^2 + 4x - 5 \Rightarrow y = \frac{3x^3}{3} + \frac{4x^2}{2} - \frac{5x^1}{1} + c$

$\Rightarrow y = x^3 + 2x^2 - 5x + c$

b $f(1) = 7 \Rightarrow 7 = 1^3 + 2.1^2 - 5.1 + c = -2 + c \Rightarrow c = 9$

So the particular solution is $y = x^3 + 2x^2 - 5x + 9$.

Example 3

A particle moves so that the acceleration, a m s^{-2}, can be modelled by $a(t) = 6t + 2$.

a i Find an expression for the velocity given that initially the particle was at rest.

ii How fast is the particle moving in the 3rd second?

b i Find an expression for the displacement given that initially the particle was 2 m from the origin.

ii How far from the origin is the particle at the 3rd second?

a i $v(t) = \int a(t)\,dt = \int 6t + 2\,dt = 3t^2 + 2t + c$
Initially the particle is at rest, which means $v(0) = 0$.
So $3.0^2 + 2.0 + c = 0 \Rightarrow c = 0$, so $v(t) = 3t^2 + 2t$.
ii $v(3) = 3.3^2 + 2.3 = 27 + 6 = 33$,
so velocity in the 3rd second = $33\ \text{m s}^{-1}$.

b i $s(t) = \int v(t)\,dt = \int 3t^2 + 2t\,dt = t^3 + t^2 + c$.
Initially the particle is 2 m from origin, which means $s(0) = 2$.
So $0^3 + 0^2 + c = 2 \Rightarrow c = 2$,
so $s(t) = t^3 + t^2 + 2$.
ii $s(3) = 3^3 + 3^2 + 2 = 27 + 9 + 2 = 38$
At the 3rd second the particle is 38 m from the origin.

Exercise 11.1

1 Find the integral in each case ... remember to include the constant of integration.

a i 4 ii $2x$ iii $6x^2$ iv $2x^4$ v $3x^5$

b i $2x + 1$ ii $7x - 2$ iii $3x^2 + 2x - 1$ iv $x^3 + 4x^2 + x - 5$ v $1 - 4x$ vi $1 - 4x - 3x^2$ vii $9x^2 - 4x^3 - 3$ viii $-(1 - x)$

2 Integrate:

a i $2\sin x$ ii $3\cos x$ iii $1 + 3\sin x$ iv $2 - 5\sin x$ v $x^2 - \cos x$ vi $\cos\left(\frac{\pi}{3}\right)$ vii $x + \sin\left(\frac{\pi}{2}\right)$ viii $\cos x + \sin x$ ix $2\cos x - 5\sin x$

b i $x^{\frac{1}{2}}$ ii $3x^{-2}$ iii $5x^{\frac{2}{3}}$ iv $x^{-\frac{1}{3}}$ v $3x^{\frac{3}{2}} - 2x^{\frac{1}{2}}$ vi $4x^{-2} - 14x^{\frac{3}{4}}$ vii $\sqrt{x}$ viii $8\sqrt[3]{x}$ ix $1 + \sqrt{x^3}$ x $\frac{1}{\sqrt{x}}$ xi $\sqrt{x} + \frac{3}{\sqrt{x}}$ xii $\frac{1}{x\sqrt{x}}$.

3 Find:

a i $\int x\,dx$ ii $\int 1\,dx$ iii $\int (x + 1)(x + 2)\,dx$ iv $\int (\sqrt{x} + \sqrt[3]{x})\,dx$ v $\int \frac{1}{x^2} - \frac{1}{x^3}\,dx$ vi $\int x(3x + 2)\,dx$ vii $\int (2x + 1)(3x - 1)\,dx$

b i $\int \frac{1 + x}{x^3}\,dx$ ii $\int \frac{x^2 - 1}{\sqrt{x}}\,dx$ iii $\int \frac{x^2 + 3\sqrt{x} + 2}{x^2}\,dx$ iv $\int 1 + \frac{3}{\sqrt{x}}\,dx$

c i $\int \frac{x - 1}{x - \sqrt{x}}\,dx$ ii $\int \frac{x - 1}{x + \sqrt{x}}\,dx$.

[Hint: For **ci**: rationalise the denominator by multiplying top and bottom by $x + \sqrt{x}$.]

4 Find the general solution of each differential equation.

a $\frac{dy}{dx} = 6x^2 + 1$ b $\frac{dy}{dx} = 12x^3 + 9x^2 + 1$ c $\frac{dy}{dx} = 7$ d $\frac{dy}{dx} = 0$

e $\frac{dy}{dx} = 4\sqrt{x}$ f $\frac{dy}{dx} = 1 - 2x^2 - 3x^3$ g $\frac{dy}{dx} = 1 + x + \frac{x^2}{2} + \frac{x^3}{6} + \frac{x^4}{24}$

h $\frac{dy}{dx} = 2 + 3\sin x$ i $\frac{dy}{dx} = 1 + \sin x - 5\cos x$ j $\frac{dy}{dx} = 3\cos x + 4\sin x + \pi$

5 Find the particular solution of each differential equation where an initial condition has been set.

a $\frac{dy}{dx} = 2x^2 - 1$ and when $x = 0, y = 4$. b $\frac{dy}{dx} = 3\sqrt{x}$ and when $x = 1, y = 5$.

c $\frac{dy}{dx} = 6x(x + 1)$ and when $x = 2, y = 29$. d $\frac{dy}{dx} = 3$ and when $x = -1, y = 1$.

e $\dfrac{dy}{dx} = 0$ and when $x = 5, y = 10$.

f $\dfrac{dy}{dx} = 1 + x + \dfrac{x^2}{2} + \dfrac{x^3}{6} + \dfrac{x^4}{24}$ and when $x = 0, y = 1$.

g $\dfrac{dy}{dx} = \dfrac{1 + x}{\sqrt{x}}$ and when $x = 4, y = 10$.

h $\dfrac{dy}{dx} = 3 \sin x$ and when $x = \dfrac{\pi}{2}, y = 4$.

6 **a** Find the equation of the curve which has a gradient at x given by $2x + 1$ passing through the point (3, 1).

b To find the gradient at the point (x, y) on a curve you need to calculate $3 \sin x + 1$.
The curve passes through the point (0, 2).
Find the equation of the curve.

7 A particle of water leaves the lip of Glenashdale Falls, accelerating downwards at 10 m s^{-2}.

a Initially the particle was moving at 5 m s^{-1}. [$v(0) = 5$.]
Express the velocity, $v \text{ m s}^{-1}$, in terms of t, the time in seconds.

b How fast was the particle moving after 1 second?

c The distance, s metres, the particle has fallen after t seconds is initially zero.
Express s in terms of t.

d The drop is 30 m. [The particle falls 30 m in total.]
i How long does the particle take to fall?
ii At what speed is it travelling when it lands?

8 At a firework display, a rocket accelerates at $3t \text{ m s}^{-2}$ vertically upwards.
Initially the rocket is at rest being launched from a platform 2 m above level ground.

a Find an expression for the velocity t seconds after launch.

b What is the velocity of the rocket 2 seconds into its flight?

c Express the height of the rocket above the ground in terms of t.

d The gunpowder burned out after 5 seconds and produced a starburst.
i How high did the rocket reach?
ii At what speed was it travelling at this moment?

9 A point on the rim of a waterwheel is accelerating at a rate of $(5 \cos t) \text{ m s}^{-2}$.
Initially the velocity, $v \text{ m s}^{-1}$, and displacement, s m, were both zero.

a Express v in terms of t and give its value after $\dfrac{\pi}{2}$ seconds.

b Find an expression for the displacement.

c **i** What is its displacement after π seconds?
ii What is the greatest value that the displacement takes?

10 The population of a small Ayrshire town at the beginning of the 18th century has been modelled by $\dfrac{dP}{dx} = 0{\cdot}7x$, where P is the number of inhabitants x years after 1700.
In the year 1700, there were 400 inhabitants.

a Express P in terms of x.

b What is the population in 1750 according to the model?

c By how much did the population increase in 1710? [Hint: $P(10) - P(9)$.]

11 A research programme modelled the growth in the number of fish in a loch by

$$\frac{dF}{dx} = 0{\cdot}9x - 2{\cdot}5$$

where F is the population of fish in thousands and x is the number of years since the programme began.
Initially there was a stock of 5000 fish in the loch [$F = 5$].

a Express F in terms of x.

b Why is there concern by year 3?

c When does the model predict there will be 25 000 fish?

12 On the day that a health centre became aware of a measles outbreak they treated 20 cases.
From previous outbreaks they felt they could model the spread of the outbreak over the first 10 weeks by $\frac{dC}{dt} = 3(1 + 2t + 1{\cdot}2t^2)$; $0 \leq t \leq 10$
where C is the number of cases t weeks after the outbreak. [Note: $C(0) = 20$.]

a Find $C(t)$.

b How many cases were there in week 3?

c How many new cases occurred in week 4?

13 A cygnet weighs 0·5 lb when it first hatches.
Using available data an ornithologist modelled its rate of growth by

$$\frac{dw}{dx} = \frac{1}{\sqrt{x}}$$

where w pounds is the weight x days after it hatched.

a Express w as a function of x.

b What was the weight of the cygnet in day 9?

c What weight did it put on between day 16 and day 25?

d The model breaks down in day 100 when the cygnet has reached its adult weight. What is this weight?

11.2 Composite functions

$(ax + b)^n$

$$\frac{d}{dx}(ax+b)^{n+1} = a(n+1)(ax+b)^n \Rightarrow \frac{1}{a(n+1)}\frac{d}{dx}(ax+b)^{n+1} = (ax+b)^n$$

$$\Rightarrow \frac{d}{dx}\frac{(ax+b)^{n+1}}{a(n+1)} = (ax+b)^n$$

$$\Rightarrow \int (ax+b)^n\,dx = \frac{(ax+b)^{n+1}}{a(n+1)} + c$$

sin $(ax + b)$ and cos $(ax + b)$

$$\frac{d}{dx}\cos(ax+b) = -a\sin(ax+b) \Rightarrow -\frac{1}{a}\frac{d}{dx}\cos(ax+b) = \sin(ax+b)$$

$$\Rightarrow \frac{d}{dx}\left(-\frac{1}{a}\cos(ax+b)\right) = \sin(ax+b)$$

$$\Rightarrow \int \sin(ax+b)\,dx = -\frac{1}{a}\cos(ax+b) + c$$

Similarly

$$\int \cos(ax+b)\,dx = \frac{1}{a}\sin(ax+b) + c$$

Example 1

Integrate: **a** $(3x + 1)^5$ **b** $\dfrac{2}{(x + 4)^2}$ **c** $\sqrt{1 - 2x}$.

a $\displaystyle\int (3x + 1)^5\,dx = \frac{(3x + 1)^6}{3.6} + c = \frac{(3x + 1)^6}{18} + c$

b $\displaystyle\int \frac{2}{(x + 4)^2}\,dx = \int 2(x + 4)^{-2}\,dx = \frac{2(x + 4)^{-1}}{-1.1} + c = -\frac{2}{x + 4} + c$

c $\displaystyle\int \sqrt{1 - 2x}\,dx = \int (1 - 2x)^{\frac{1}{2}}\,dx = \frac{(1 - 2x)^{\frac{3}{2}}}{\frac{3}{2}.-2} + c = -\frac{(1 - 2x)^{\frac{3}{2}}}{3} + c$

Answers can be checked by differentiation. Does it take you back to the original expression?

Example 2

Find: **a** $\displaystyle\int \cos(x - 2)\,dx$ **b** $\displaystyle\int \sin(3x + 2)\,dx$ **c** $\displaystyle\int \sin(1 - 4x)\,dx$.

a $\displaystyle\int \cos(x - 2)\,dx = \frac{\sin(x - 2)}{1} + c = \sin(x - 2) + c$

b $\displaystyle\int \sin(3x + 2)\,dx = \frac{-\cos(3x + 2)}{3} + c$

c $\displaystyle\int \sin(1 - 4x)\,dx = \frac{-\cos(1 - 4x)}{-4} + c = \tfrac{1}{4}\cos(1 - 4x) + c$

Exercise 11.2

1 Integrate:

a $(3x + 1)^5$ **b** $(2x - 1)^3$ **c** $(3 - 2x)^4$ **d** $(1 - x)^3$ **e** $(1 - \frac{1}{2}x)^2$

f $(2x + 3)^{-2}$ **g** $(1 - x)^{-2}$ **h** $(\frac{1}{2}x - 2)^{-3}$ **i** $(\frac{1}{3}x + 7)^{-2}$ **j** $(4 - 2x)^{-4}$

k $(5x + 3)^{\frac{1}{2}}$ **l** $(6x - 1)^{-\frac{1}{2}}$ **m** $(1 - x)^{-\frac{1}{3}}$ **n** $(2x + 1)^{\frac{3}{4}}$ **o** $(2 - 3x)^{-\frac{3}{2}}$

p $\sqrt{7x}$ **q** $\sqrt{1 - x}$ **r** $\sqrt{2x + 5}$ **s** $\sqrt[3]{4x - 1}$ **t** $\sqrt[3]{(x + 1)^2}$.

2 Find:

a $\displaystyle\int \sqrt[3]{7x}\,dx$ **b** $\displaystyle\int \frac{1}{\sqrt{x + 1}}\,dx$ **c** $\displaystyle\int \frac{3}{\sqrt{2x + 1}}\,dx$ **d** $\displaystyle\int 1 - \frac{2}{\sqrt{1 - 2x}}\,dx$.

3 Find the integral of:

a $\cos(3x + 4)$ **b** $\sin(2x - 1)$ **c** $2\cos(5 - 2x)$

d $6\sin(2 - 3x)$ **e** $1 - \cos(1 - x)$ **f** $3 - \sin(5 - x)$

g $3\cos 2x$ **h** $\frac{1}{3}\sin(3x - 1)$ **i** $5\sin 7x$

j $\cos 2x - \sin 5x$ **k** $1 - 21\sin 7x + 8\cos 4x$ **l** $x^2 + \sin 3x$.

4 We know that $\cos 2x = 2\cos^2 x - 1$.

a Make $\cos^2 x$ the subject of the formula.

b Hence find $\displaystyle\int \cos^2 x\,dx$.

c Similarly find $\displaystyle\int \sin^2 x\,dx$.

d Find: **i** $\displaystyle\int (\cos^2 x - \sin^2 x)\,dx$ **ii** $\displaystyle\int (\cos^2 x + \sin^2 x)\,dx$.

e Find $\displaystyle\int \cos^2 \frac{x}{2}\,dx$.

f Find $\displaystyle\int \sin x \cos x\,dx$. [Hint: $\sin 2A = 2\sin A\cos A$.]

5 As the tide goes in and out of a harbour, the surface of water rises and falls vertically twice in a 24-hour period.
The movement can be examined by observing a tethered buoy.
The acceleration in m s^{-2} is modelled by

$a(t) = \frac{1}{2}\cos\left(\frac{\pi t}{6}\right)$

where t is the time in hours since low tide.

a At the moment of low tide the vertical velocity of the surface is zero.
Find an expression for $v(t)$, the velocity of the surface of the water.

b How fast is the water rising 3 hours after low tide?

c The water at low tide is 1 m deep.
Find an expression for $s(t)$, the depth of water in the harbour.

d **i** How deep is the water 2 hours after low tide?
ii What is the acceleration of the surface at this time?

6 **a** To turn degrees into radians we multiply by a factor of $\frac{\pi}{180}$, i.e. $x° = \frac{\pi}{180}x$ radians.
Use this to find $\int \sin x° \,\mathrm{d}x$, expressing your final answer in degrees.

b Find $\int \cos x° \,\mathrm{d}x$, expressing your final answer in degrees.

7 Find the particular solution to the differential equations with the given initial conditions.

a $\frac{\mathrm{d}y}{\mathrm{d}x} = (1 - 2x)^3$, when $x = 0, y = \frac{1}{8}$.

b $\frac{\mathrm{d}y}{\mathrm{d}x} = 6 \sin 2x$, when $x = \frac{\pi}{6}, y = \frac{1}{2}$.

8 In manufacturing, the **cost per item** is a variable dependent on how many items are made.
This is called the marginal cost.
If $C(x)$ represents the cost in £s of producing x items, then $\frac{\mathrm{d}C}{\mathrm{d}x}$ represents the marginal cost per item after x items have been made.
A particular cottage industry makes luxury booklets.
They model their marginal costs by $\frac{\mathrm{d}C}{\mathrm{d}x} = \frac{196}{(2x + 8)^{\frac{2}{3}}}$.

a Find an expression for C(x), given that the initial outlay was £600. [C(0) = 600.]

b What is the cost of making 1000 booklets to the nearest pound?

c What is the marginal cost for this number of booklets?

9 The number of minutes of daylight in Glasgow is a periodic function.
Its rate of change in minutes per day can be modelled by

$\frac{\mathrm{d}L}{\mathrm{d}t} = 1{\cdot}73\pi \sin\left(\frac{2\pi t}{365}\right)$

where L is the number of minutes of daylight on day t after 1st January.

a At what rate is the number of minutes of daylight changing 80 days after the start of the year?

b Express L as a function of t if on 1st January there were 425 minutes of daylight in Glasgow.

c How many minutes of daylight did Glasgow get on the 100th day of the year?

d How many minutes of daylight were there on the longest day in Glasgow, according to the model?

11.3 Area under a curve

Consider the function $y = f(x)$, which is defined at least in the interval $p \leqslant x \leqslant q$.

Let the area trapped between the curve, the x-axis, the ordinate[2] at p and the ordinate at the variable point x be denoted by $A(x)$ ... i.e. it is a function of x.

Figure 1

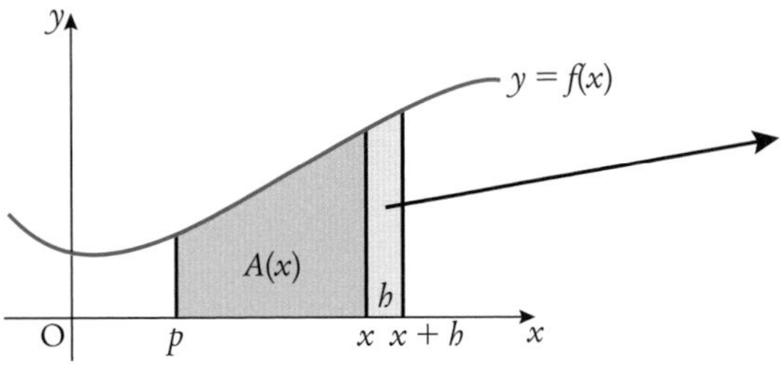

y, y = f(x), A(x), h, O, p, x, x + h, x

Figure 2

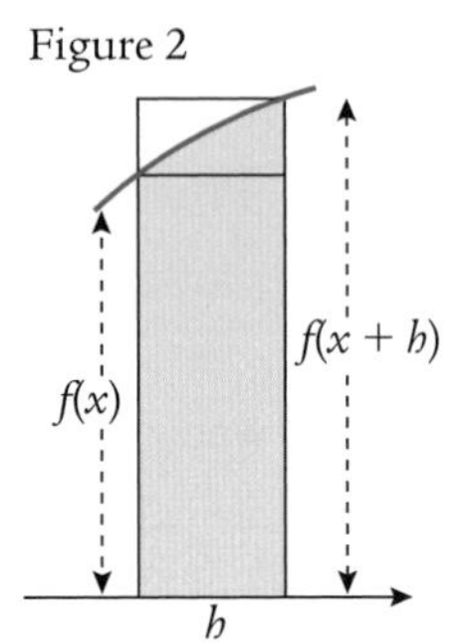

The area from p to x is denoted by $A(x)$; the area from p to $x + h$ is denoted by $A(x + h)$.

The area of the thin strip $= A(x + h) - A(x)$.

Considering Figure 2 we can see the area of the strip lies between the area of the rectangle of length $f(x)$ and the rectangle of length $f(x + h)$.

$$\Rightarrow h \times f(x) \leqslant A(x+h) - A(x) \leqslant h \times f(x+h)$$

$$\Rightarrow ff(x) \leqslant \frac{A(x+h) - A(x)}{h} \leqslant f(x+h),\ h \neq 0$$

As $h \to 0$, $\quad f(x) \leqslant \lim_{h \to 0} \frac{A(x+h) - A(x)}{h} \leqslant ff(x)$

$$\Rightarrow f(x) = A'(x) \qquad \text{... by the definition of the derivative}$$

$$\Rightarrow A(x) = \int f(x)\,dx \qquad \text{... by the definition of an integral}$$

If we wish to find the area bounded by $y = f(x)$, the x-axis, $x = a$ and $x = b$, we need only integrate $f(x)$ to get $A(x)$ and evaluate $A(b) - A(a)$ (as long as $a < b$ and both are in the interval $p \leqslant x \leqslant q$).

This area is denoted by: $\int_a^b f(x)\,dx$.

In general[3] if $\int f(x)\,dx = F(x) + c$, then

$$\int_a^b f(x)\,dx = F(b) - F(a).$$

Figure 3

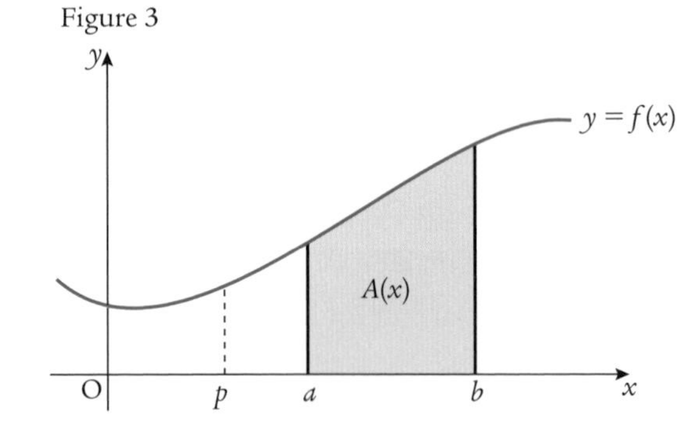

$\int_a^b f(x)\,dx$ is known as a **definite integral**, where a is the **lower limit** and b is the **upper limit**.

Example 1

Evaluate the following definite integrals.

a $\int_1^3 (3x^2 + 1)\,dx$ **b** $\int_0^9 (x + \sqrt{x})\,dx$ **c** $\int_0^{\pi} (1 + \sin x)\,dx$

[2] An ordinate is a line parallel to the y-axis.

[3] Note how the constants of integration sum to zero ... $F(b) + c - (F(a) + c) = F(b) - F(a)$.

a $\int_1^3 (3x^2 + 1)\,dx = [x^3 + x]_1^3 = (3^3 + 3) - (1^3 + 1) = 30 - 2 = 28.$

b $\int_0^9 \left(x + x^{\frac{1}{2}}\right) dx = \left[\frac{x^2}{2} + \frac{x^{\frac{3}{2}}}{\frac{3}{2}}\right]_0^9 = \left(\frac{9^2}{2} + \frac{2.9^{\frac{3}{2}}}{3}\right) - \left(\frac{0^2}{2} + \frac{2.0^{\frac{3}{2}}}{3}\right) = 40{\cdot}5 + 18 - 0 = 58{\cdot}5.$

c $\int_0^{\pi} (1 + \sin x)\,dx = \left[x - \cos x\right]_0^{\pi} = (\pi - \cos\pi) - (0 - \cos 0) = (\pi - (-1)) - (0 - 1) = \pi + 2.$

Example 2

Find each shaded area.

a

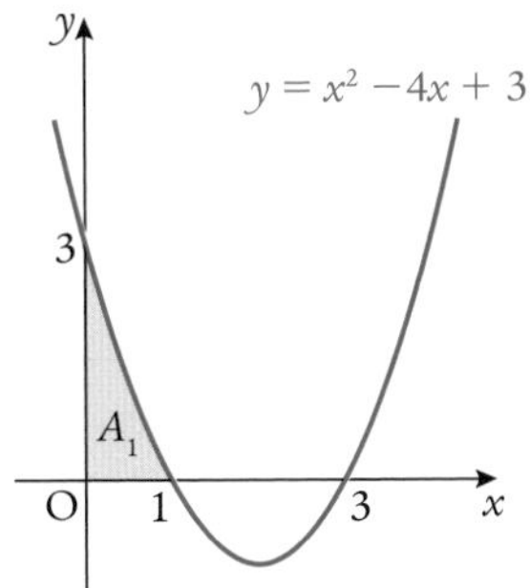

b

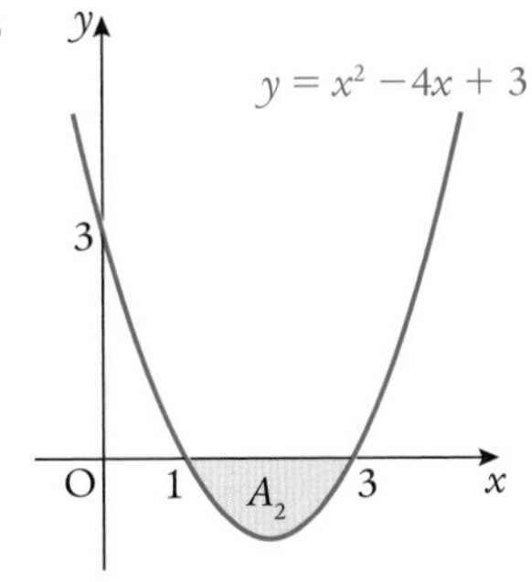

c

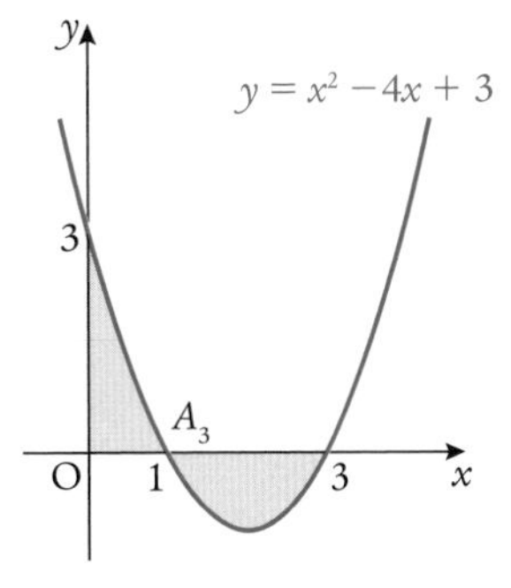

a $A_1 = \int_0^1 x^2 - 4x + 3\,dx \quad = \left[\frac{x^3}{3} - 2x^2 + 3x\right]_0^1 = \left(\frac{1}{3} - 2 + 3\right) - \left(\frac{0}{3} - 0 + 0\right) = 1\frac{1}{3}\text{ unit}^2.$

b $A_2 = \int_1^3 x^2 - 4x + 3\,dx \quad = \left[\frac{x^3}{3} - 2x^2 + 3x\right]_1^3 = (9 - 18 + 9) - \left(\frac{1}{3} - 2 + 3\right) = -1\frac{1}{3}\text{ unit}^2.$

The definite integral has a negative value ... indicating that it lies below the x-axis.
The area is just the magnitude of this, viz. $1\frac{1}{3}$ unit^2.

c Note that $\int_0^3 x^2 - 4x + 3\,dx \quad = \left[\frac{x^3}{3} - 2x^2 + 3x\right]_0^3 = (9 - 18 + 9) - \left(\frac{0}{3} - 0 + 0\right) = 0.$

We can see that the area is not zero. When part of the desired area lies above the x-axis and part below, we must work out each part separately and add magnitudes of the results.
$A_3 = A_1 + A_2 = 1\frac{1}{3} + 1\frac{1}{3} = 2\frac{2}{3}\text{ unit}^2.$

Exercise 11.3

1 Evaluate the following definite integrals.

a $\int_0^5 2x + 4\,dx$ b $\int_1^3 1 - 4x\,dx$ c $\int_{-1}^2 4x\,dx$

d $\int_{-3}^5 6x\,dx$ e $\int_1^8 3x^2\,dx$ f $\int_0^3 x^2 + 2x + 1\,dx$

g $\int_{-1}^1 1 - 6x^2\,dx$ h $\int_1^9 \sqrt{x}\,dx$ i $\int_{\sqrt{2}}^{\sqrt{3}} 4x^3\,dx$

j $\int_0^1 1 - 3x - x^2\,dx$ k $\int_{\frac{1}{2}}^1 \frac{1}{x^2}\,dx$ l $\int_{-1}^1 (2x + 1)^3\,dx$

m $\int_0^{\frac{\pi}{2}} \cos x\,dx$ n $\int_0^{\frac{\pi}{4}} \sin x\,dx$ o $\int_{-\pi}^{\pi} \cos x - \sin x\,dx$

p $\int_{\frac{\pi}{6}}^{\frac{\pi}{3}} 1 + \cos x\,dx$ q $\int_0^{\frac{\pi}{4}} \cos 2x\,dx$ r $\int_{\frac{\pi}{6}}^{\frac{\pi}{3}} 1 - \sin 3x\,dx$

s $\int_0^{\frac{\pi}{6}} \cos\left(2x + \frac{\pi}{3}\right) dx$ t $\int_{-\frac{\pi}{3}}^{\frac{\pi}{3}} \sin\left(3x - \frac{\pi}{2}\right) dx$

2 Calculate each of the shaded areas using integration.

a

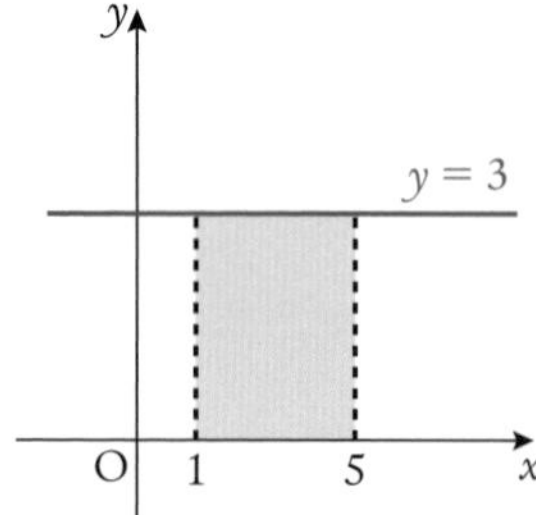

b

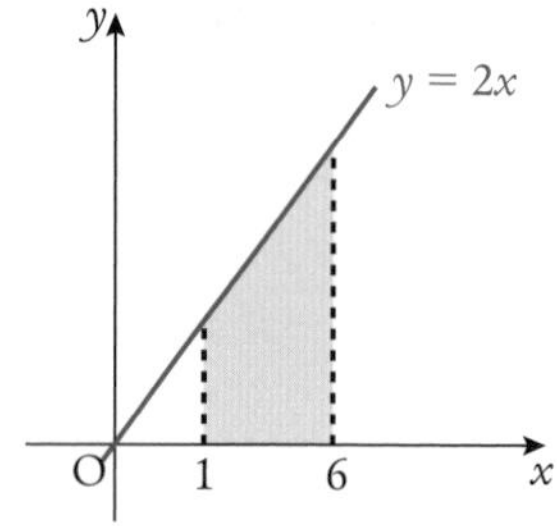

c

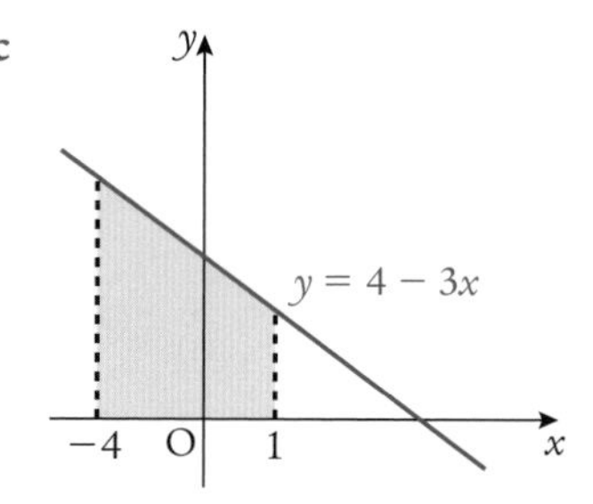

d

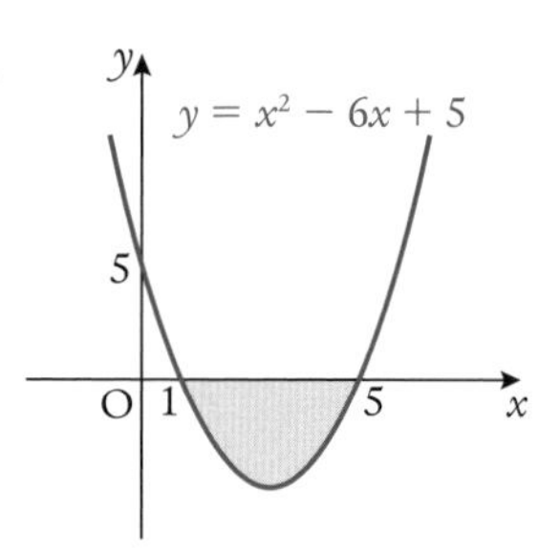

e

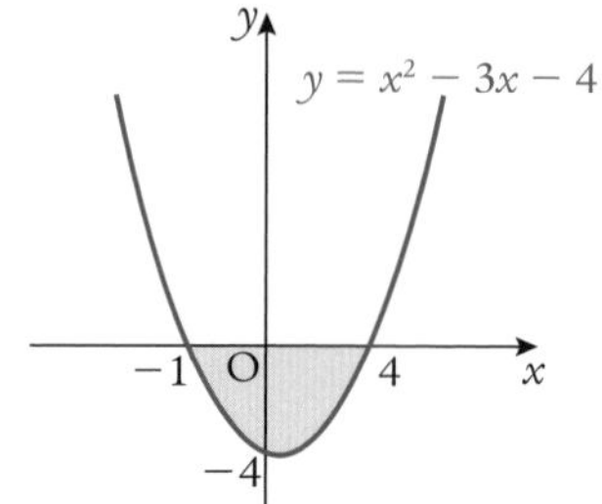

f

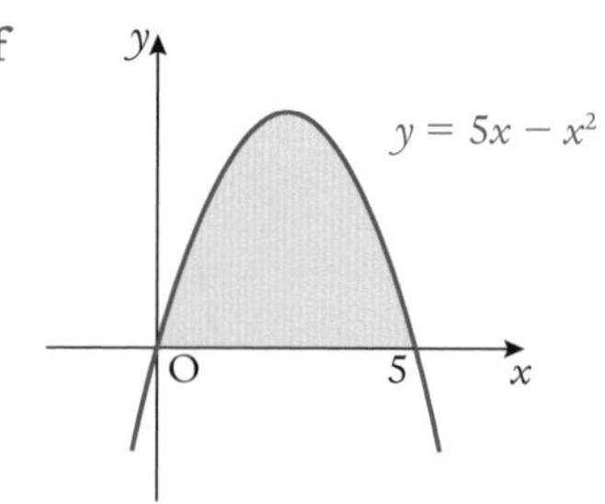

g

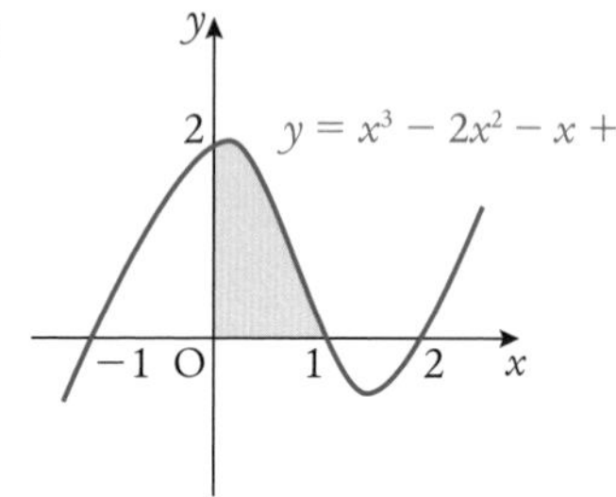

h

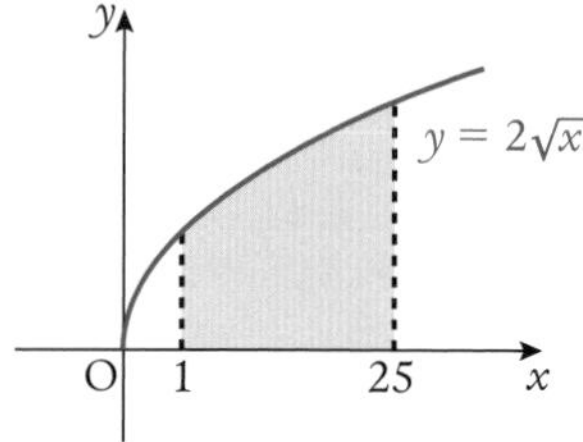

i

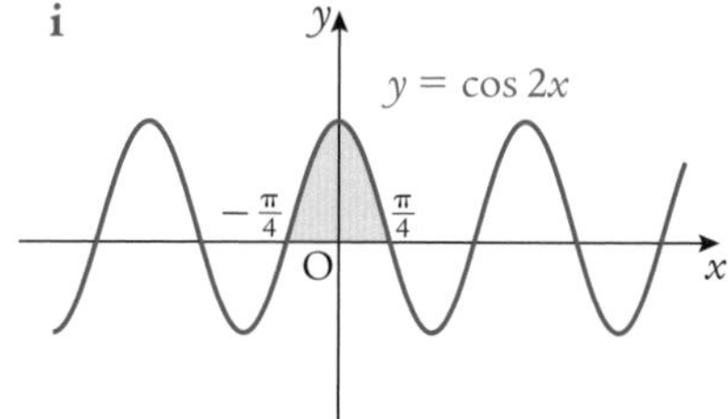

3 Find the total shaded area in each case. (Note that each part must be calculated separately.)

a

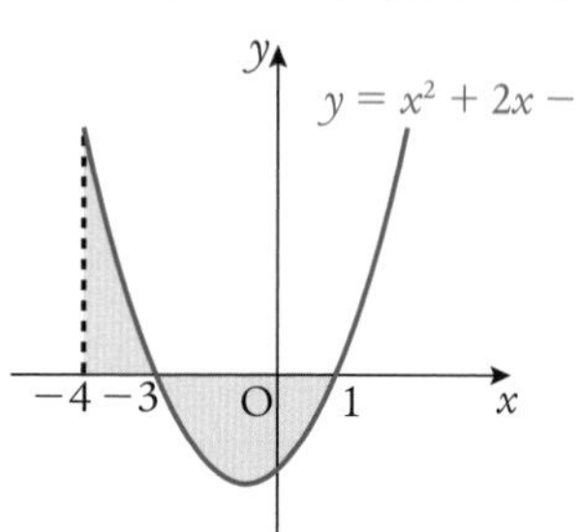

b

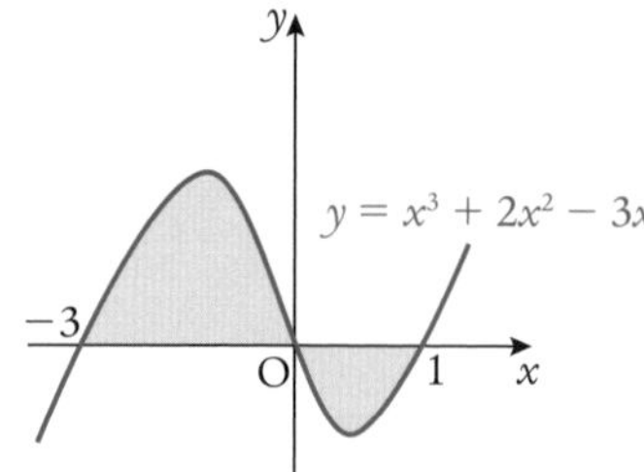

c

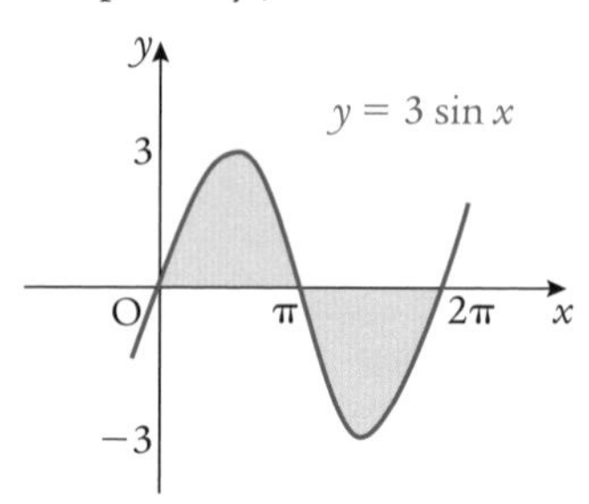

4 A function is defined by $y = x^2 + 2x - 15$.

a Where does the curve cut the x-axis?

b Calculate the area of the 'cap' trapped between the curve and the x-axis.

c Repeat this process for the function:

i $y = 2x^2 - 7x - 4$

ii $y = 21 - 4x - x^2$.

5 An area is described as being trapped between the curve $y = 3x^2$, the x-axis and the ordinates $x = 1$ and $x = a$, where $a > 1$.

a Sketch the area.

b Find the value of a if the area is to be 124 unit^2.

6 Each of the following functions is defined on the interval $-1 \leqslant x \leqslant 3$.

i Check if the function crosses the x-axis in the interval.

ii Make a sketch of the function in the interval.

iii Find the area trapped between the function and the x-axis in the interval.

a $y = x^2 - 4$ **b** $y = x^2 - x - 6$ **c** $y = x^2 - 2x$

7 The 'average value' of a function in the interval $a \leqslant x \leqslant b$ is defined as:

$$\text{average value} = \frac{\text{area under curve between } a \text{ and } b}{b - a}$$

Find the average value of each function in the given interval.

a $y = 2x + 3$ in the interval $1 \leqslant x \leqslant 5$.

b $y = (x - 4)(x + 5)$ in the interval $-3 \leqslant x \leqslant 2$.

c $y = x(x - 1)(x + 6)$ in the interval $0 \leqslant x \leqslant 2$.

d $y = \sin x$ in the interval $0 \leqslant x \leqslant \pi$.

8 Earlier we discovered that the number of minutes of daylight in Glasgow was a function of t, the number of days since 1st January. This function is described by:

$L(t) = 740 - 315 \cos\left(\frac{2\pi t}{365}\right)$.

What is the average number of minutes of daylight per day in Glasgow?

9 If we graph speed against time, the area under the curve gives the distance travelled.

a A particle moves so that its acceleration is modelled by $a(t) = 2\,\text{m s}^{-2}$.

i Find an expression for $v(t)$ given that initially the particle was at rest.

ii By considering the area under $y = v(t)$, find the distance the particle travelled between the 4th and 10th second.

b Repeat part **a** given that $a(t) = 3t\,\text{m s}^{-2}$ and that the particle had an initial velocity of $4\,\text{m s}^{-1}$.

11.4 Area between two curves

1 The area between two curves, $y = f(x)$ and $y = g(x)$, $a \leqslant x \leqslant b$, where $f(x) \geqslant g(x)$ in the interval, can be obtained by subtracting the area below $g(x)$ from the area below $f(x)$.

Thus in Figure 1,

$$A = \int_a^b f(x)\,dx - \int_a^b g(x)\,dx = \int_a^b (f(x) - g(x))\,dx.$$

Figure 1

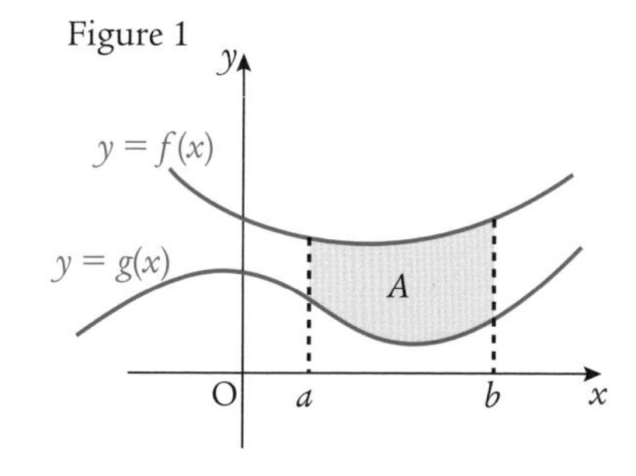

2 If two curves $y = f(x)$ and $y = g(x)$, intersect at $x = a$ and $x = b$ and $f(x) \geqslant g(x)$ in the interval $a \leqslant x \leqslant b$, then the area trapped between the curves (Figure 2) is also

$$A = \int_a^b f(x)\,dx - \int_a^b g(x)\,dx = \int_a^b (f(x) - g(x))\,dx.$$

Figure 2

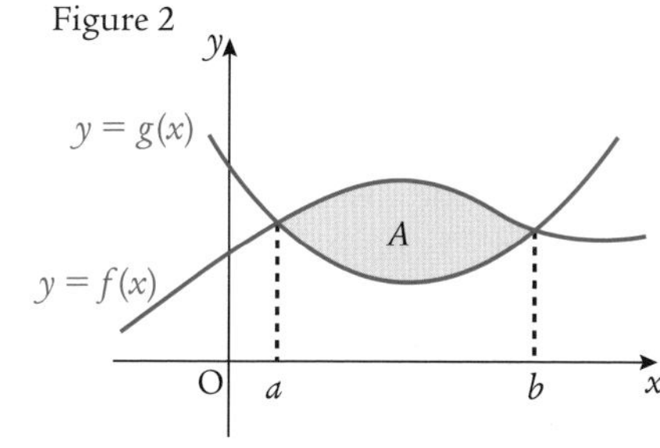

3 If we wish to find the area of more than one section between the curves, each should be evaluated separately and their magnitudes added since when $g(x) > f(x)$ we will get a negative answer (Figure 3).

Figure 3

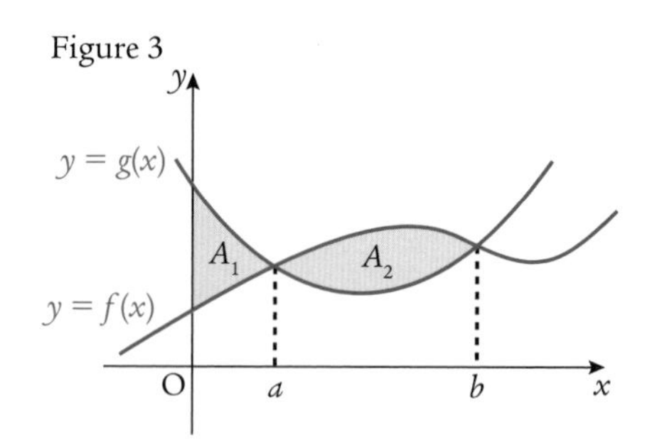

Example 1

Find the area trapped between the curves
$y = 2x^2 + 3x + 1$ and $y = x^2 + 2x + 3$.

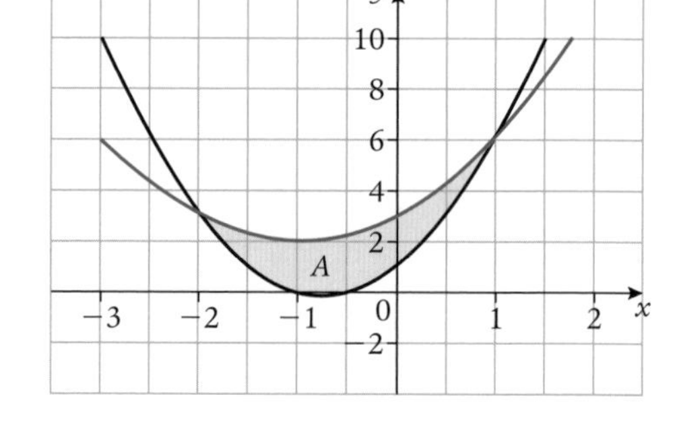

The curves intersect when:
$2x^2 + 3x + 1 = x^2 + 2x + 3$
$\Rightarrow x^2 + x - 2 = 0$
$\Rightarrow (x - 1)(x + 2) = 0$
$\Rightarrow x = 1$ or -2.

In this interval $y = x^2 + 2x + 3$ is the upper function.

$$\text{Area } A = \int_{-2}^{1} (x^2 + 2x + 3) - (2x^2 + 3x + 1)\,dx$$

$$\Rightarrow A = \int_{-2}^{1} (-x^2 - x + 2)\,dx = \left[-\frac{x^3}{3} - \frac{x^2}{2} + \frac{2x}{1}\right]_{-2}^{1}$$

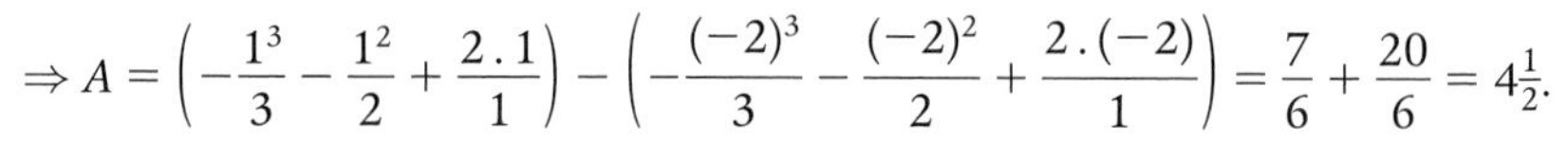

$$\Rightarrow A = \left(-\frac{1^3}{3} - \frac{1^2}{2} + \frac{2.1}{1}\right) - \left(-\frac{(-2)^3}{3} - \frac{(-2)^2}{2} + \frac{2.(-2)}{1}\right) = \frac{7}{6} + \frac{20}{6} = 4\tfrac{1}{2}.$$

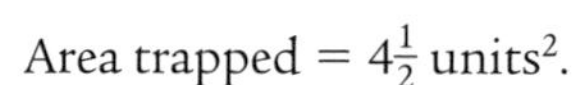

Area trapped $= 4\frac{1}{2}$ units^2.

Exercise 11.4

1 The sketch shows $y = x^2 + 2x$ and $y = 8$.

a By solving $x^2 + 2x = 8$, find where the two curves intersect.

b Hence find the area enclosed between the two curves.

c Find the area enclosed between $y = x^2 - 6x$ and $y = -5$.

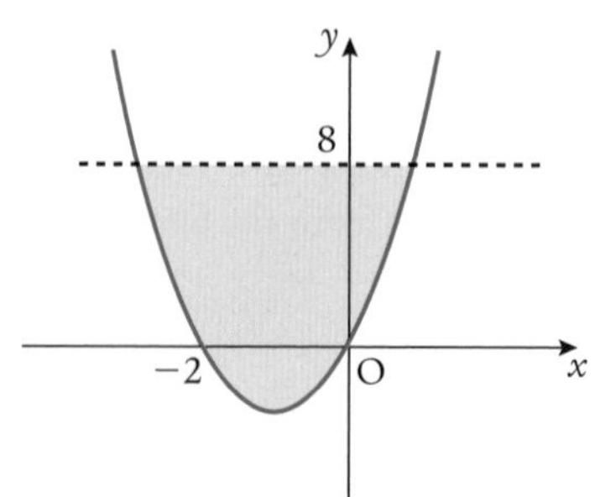

2 The sketch shows $y = x^2$ and $y = x + 6$

a Find where they intersect.

b Find the area enclosed between the two curves.

c Find the area enclosed between $y = x^2 - 3x - 5$ and $y = 2x + 1$.

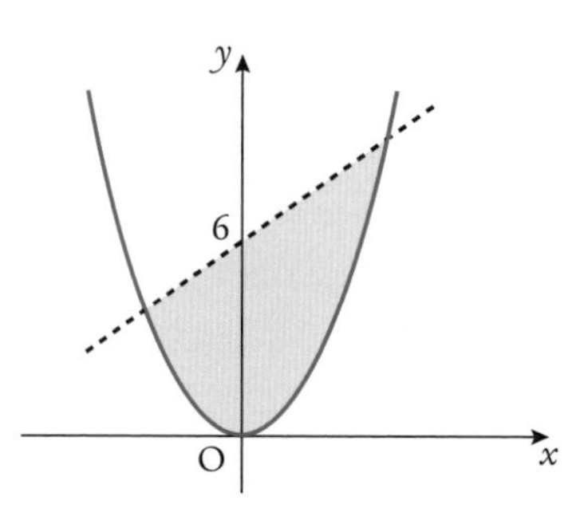

3 The sketch shows $y = x^2 + 3x + 1$ and $y = 4 - 2x - x^2$.

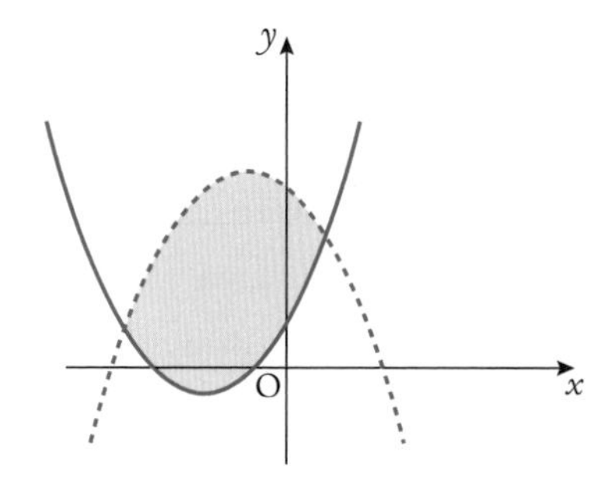

a Find where they intersect.

b Find the area enclosed between the two curves.

c Find the area enclosed between $y = x^2 - 5x - 12$ and $y = 8 - 2x - x^2$.

4 Two functions are defined by $y = x^3 + 3x^2 + 4x + 1$ and $y = 1 + x - x^2$.

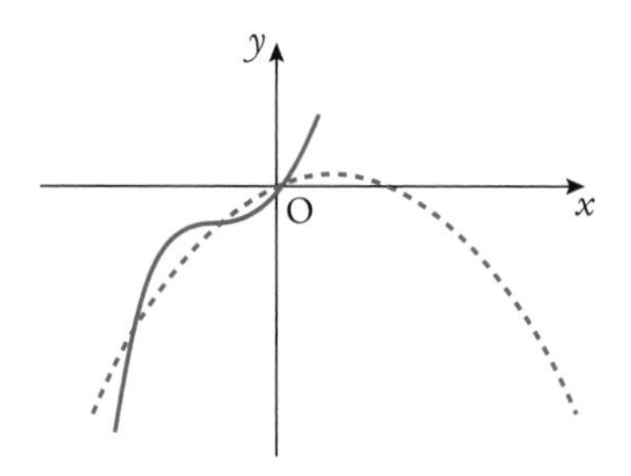

a Where do these two functions intersect?

b Two distinct areas are trapped between the functions. Find both, and hence the total area enclosed between the curves.

c Find the area enclosed between $y = x^3$ and $y = x^2 + 20x$.

5 a Where does $y = 2x^3 + x^2 - 2x - 1$ intersect $y = x^3 - x^2 - x + 1$?

b What area is trapped between them?

c Find the area between the curve $y = x^3 - 2x^2$ and the line $y = 11x - 12$.

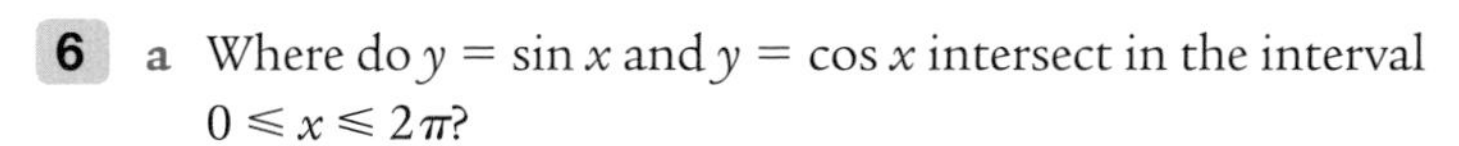

6 a Where do $y = \sin x$ and $y = \cos x$ intersect in the interval $0 \leqslant x \leqslant 2\pi$?

b Find the area trapped between $y = \sin x$ and $y = \cos x$ in the interval between the points of intersection.

7 The Clyde Arc is a bridge in Glasgow. It is a roadway suspended by a parabolic arch.

The roadway itself is a parabola.

Using convenient units and axes the arch is modelled by $h(x) = -\frac{1}{9}(x^2 - 36)$.

The road, using the same units and axes, is modelled by $h(x) = -\frac{1}{72}(x^2 - 36)$.

In both cases h units is the height of the parabola x units along the x-axis.

a Where do the curves intersect?

b Find the area enclosed between the arch and the roadway.

8 a Make a sketch of the area that lies between the curves $y = \sin x$ and $y = \sin x - 0{\cdot}5$.

b Write down a definite integral that can be used to represent the area trapped between these two curves and the lines $x = 0$ and $x = 2\pi$.

c Calculate this area.

d The Tradeston Bridge in Glasgow makes a sinusoidal[4] shape as it snakes across the river. It does this to make the incline more gentle for pedestrians.

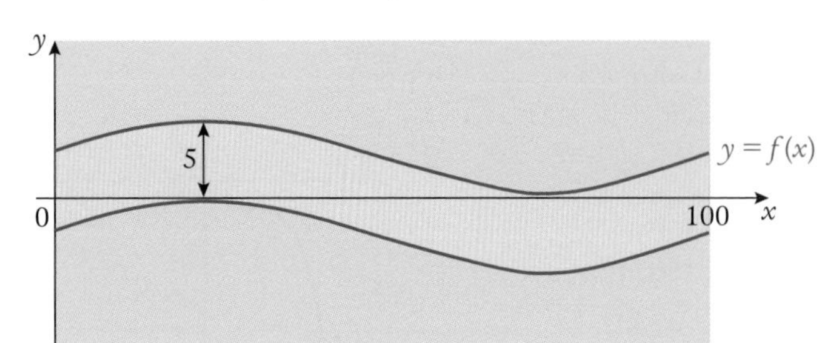

The diagram shows the structure of the walkway of the bridge.
In a straight line the bridge measures 100 m and is 5 m wide.
One edge is modelled by an equation $y = f(x)$ as shown.

i What is the equation of the other edge?

ii Write down a definite integral that gives you the area of the walkway.

iii What is the area of the walkway?

Preparation for assessment

1 Integrate $8x^3 + \frac{1}{x^3} + x^{\frac{2}{3}} + 1$ with respect to x.

2 a Solve the differential equation $\frac{dy}{dx} = x + \frac{1}{\sqrt{x}}$, if $y = 8$ when $x = 4$.

b Find a function, $y = f(x)$, that passes through the point $\left(\frac{\pi}{3}, 1\right)$ and where $f'(x) = \sin\left(2x + \frac{\pi}{3}\right)$.

3 Find:

a $\int \frac{x^3 + 1}{\sqrt{x}}\,dx$ b $\int (3x + 4)^{\frac{2}{3}}\,dx$ c $\int 3\cos(5 - 4x)\,dx$.

4 A particle moves in a straight line so that $v(t) = 2t^2 + 1$, where v m s^{-1} is the velocity at time t seconds.

a Express the acceleration a m s^{-2} in terms of t.

b Express the displacement, s metres, in terms of t given that initially the particle was 1 metre from the origin.

c What was: **i** the displacement **ii** the velocity **iii** the acceleration at the 3rd second?

[4] Shaped like a sine curve.

5 Using suitable axes and units, the upper edge of the window on the side of the Science Museum in Glasgow can be modelled by $y = -\frac{1}{200}x(x - 120)$. The lower edge is the x-axis.

a Find where the curve cuts the x-axis.

b Find the area trapped between the curve and the x-axis.

6 The cubic curve $y = x^3 - x^2 - 2x + 7$ intersects the parabola $y = 1 + 5x - x^2$ at three points.

a Show that they intersect at the point (1, 5) and find the x-coordinates of the other two points.

b Calculate the area enclosed between the two curves. [A sketch might help.]

Summary

1 If $F'(x) = f(x)$ then $F(x) + c$ is called an **anti-derivative** of $f(x)$.

2 $F(x)$ is called the **integral** of $f(x)$ with respect to x; c is called the **constant of integration**.

3 $\int f(x)\,dx = F(x) + c \Leftrightarrow F'(x) = f(x)$

4 $\int ax^n\,dx = \frac{ax^{n+1}}{n+1} + c, \; n \neq -1$

5 $\int \cos x\,dx = \sin x + c$

6 $\int \sin x\,dx = -\cos x + c$

7 $\int af(x)\,dx = a\int f(x)\,dx$ where a is a constant.

8 $\int (f(x) + g(x))\,dx = \int f(x)\,dx + \int g(x)\,dx$

9 In the context of displacement, velocity and acceleration

i $s'(t) = v(t) \Rightarrow s(t) = \int v(t)dt + c_1$

ii $v'(t) = a(t) \Rightarrow v(t) = \int a(t)dt + c_2$.

Initial conditions will help you to determine the constant c.

10 $\int (ax + b)^n\,dx = \frac{(ax+b)^{n+1}}{a(n+1)} + c$

11 $\int \sin(ax + b)\,dx = -\frac{1}{a}\cos(ax + b) + c$

12 $\int \cos(ax + b)\,dx \qquad = \frac{1}{a}\sin(ax + b) + c$

13 $\int_a^b f(x)\,dx$ is known as a **definite integral**, where a is the **lower limit** and b is the **upper limit**.

14 The area $A(x) = \int_a^b f(x)\,dx$.

If the area is below the x-axis, the definite integral will have a negative value. The area is the magnitude of this integral.

If the desired area contains parts above and below the x-axis, the parts must be considered separately and their magnitudes added.

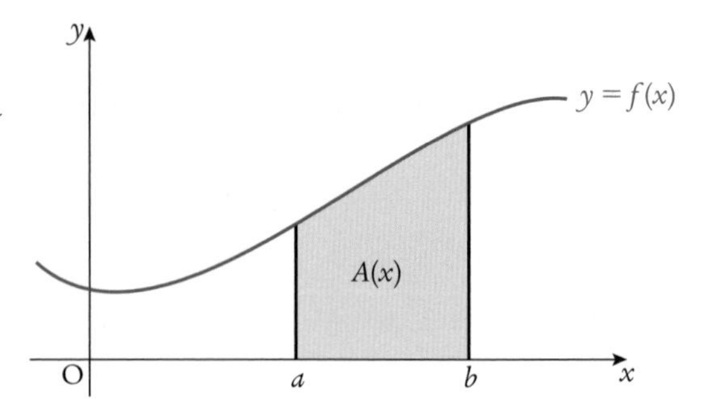

15 The area between two curves, $y = f(x)$ and $y = g(x)$, $a \leqslant x \leqslant b$, where $f(x) \geqslant g(x)$ in the interval, is $A = \int_a^b (f(x) - g(x))\,dx$.

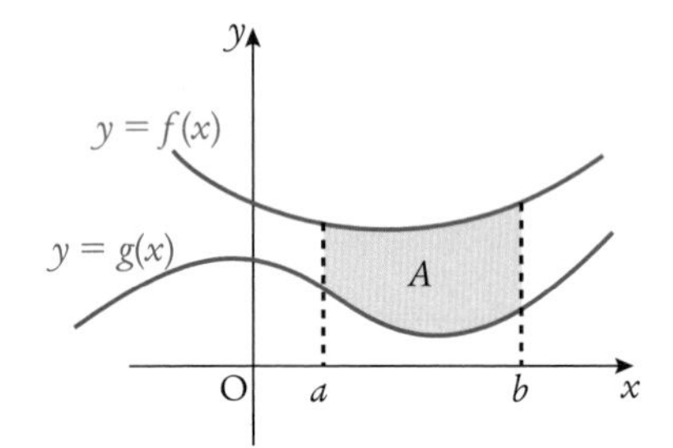

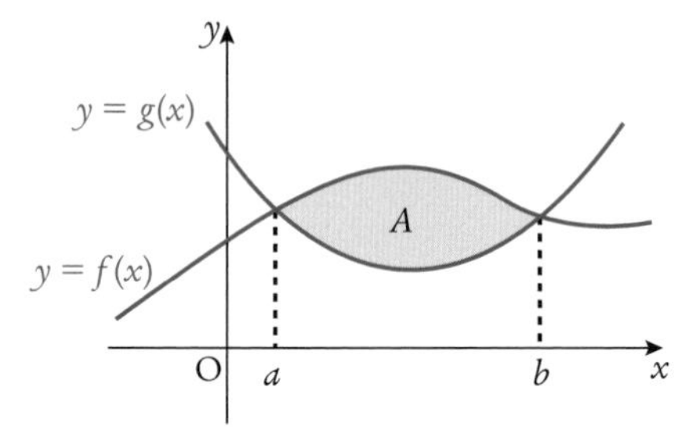

16 If the desired area contains parts where $f(x) > g(x)$ and where $f(x) < g(x)$, the parts must be considered separately and their magnitudes added.

12 Lines and circles

Before we start...

Euclid explored geometry and arithmetic around 300 BC, cataloguing the findings of others and adding his own proofs in 13 books. These books formed the work known as '*Elements*'. They formed the basis of the geometry taught in schools until the late 1960s.

Within these works were theorems about lines and circles associated with the triangle.

For this course some of the facts and vocabulary from this field of work are required.

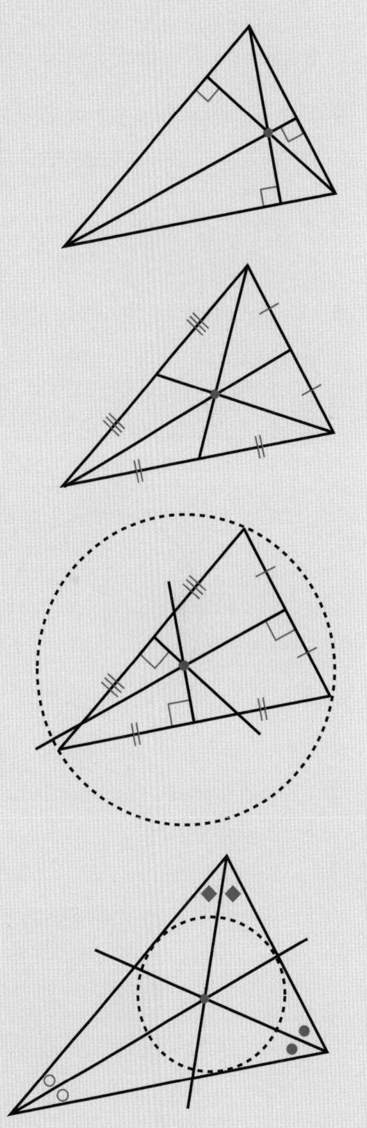

1 An **altitude** of a triangle runs from a vertex, perpendicular to the opposite side.
A triangle has three altitudes. These all run through a single point.
This property is called **concurrency** ... altitudes are concurrent.
They pass through a single point called the **orthocentre**.

2 A **median** of a triangle runs from a vertex to the midpoint of the opposite side.
A triangle has three medians ... medians are concurrent.
They pass through a single point called the **centroid**.
As a matter of interest, the centroid is the centre of gravity of the shape.

3 A **perpendicular bisector** of a side of a triangle runs from the midpoint of the side and is perpendicular to it.
A triangle has three perpendicular bisectors ... perpendicular bisectors are concurrent.
They pass through a single point called the **circumcentre**.
The circumcentre is the centre of the circle that passes through all three vertices.
This circle is called the **circumcircle**.
Using this property, given three points on a circle, we can always determine the centre and radius of the circle.

4 An **angle bisector** of a triangle runs from a vertex, bisecting the angle.
A triangle has three angle bisectors ... angle bisectors are concurrent.
They pass through a single point called the **incentre**.
The incentre is the centre of the circle to which all three sides of the triangle are tangents. This circle is called the **incircle**.

There's much to be found on this topic on the internet.

See what you can find out about the *Euler Line* (pronounced Oiler Line).

When René Descartes devised Cartesian[1] coordinates, he invented analytical geometry, which unified the branches of algebra and geometry. This chapter explores an area of this work.

[1] He didn't give them their name ... the name was chosen in his honour.

What you need to know

1 Use the formula $m = \dfrac{y_2 - y_1}{x_2 - x_1}$ to either:

find the gradient of the line that passes through both points, or help you make a comment.

a $(-2, -1)$ and $(2, 2)$ **b** $(-2, 2)$ and $(2, 1)$ **c** $(-2, 3)$ and $(4, 3)$ **d** $(-2, 1)$ and $(-2, 4)$

2 State the gradient and the y-intercept of each line or make a comment.

a $y = 3x - 1$ **b** $y = 3$ **c** $y - 1 = 3(x + 2)$ **d** $x = 3$ **e** $3x + 4y = 8$

3 Find the equation of the line:

a that passes through the point $(1, 5)$ and has a gradient of 2

b that passes through the points $(3, 3)$ and $(9, 1)$.

4 The sketch shows $y = x^2$ and $y = x^3$ close to the origin.

What are the coordinates of the points of intersection?

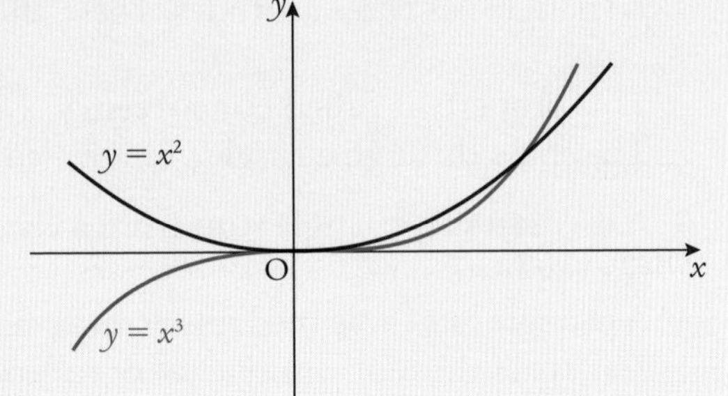

5 Where do the lines $x - 3y + 6 = 0$ and $x + 2y - 14 = 0$ intersect?

6 Use Pythagoras' theorem to find:

a the distance between

i A(2, 1) and B(7, 13) **ii** $A(x_1, y_1)$ and $B(x_2, y_2)$

b the value of r^2.

i

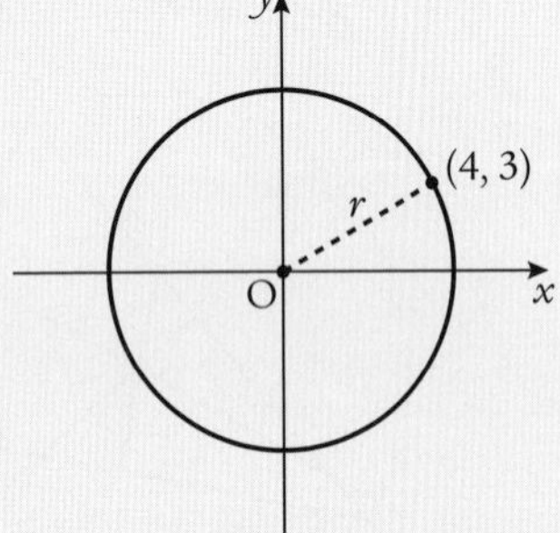

ii

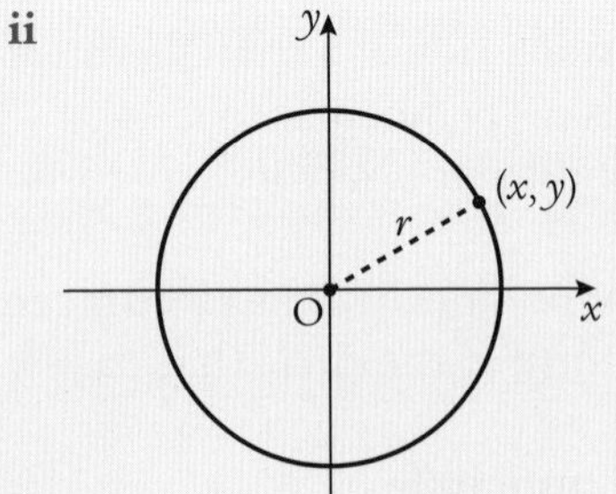

7 Consider that a tracing of the first quadrant is rotated about the origin until it lies in the second quadrant.

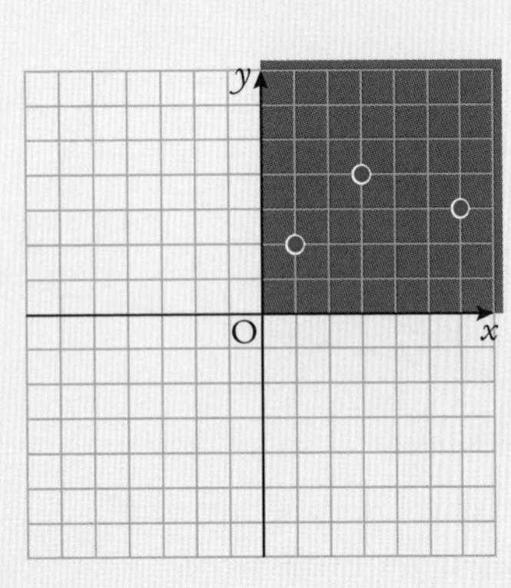

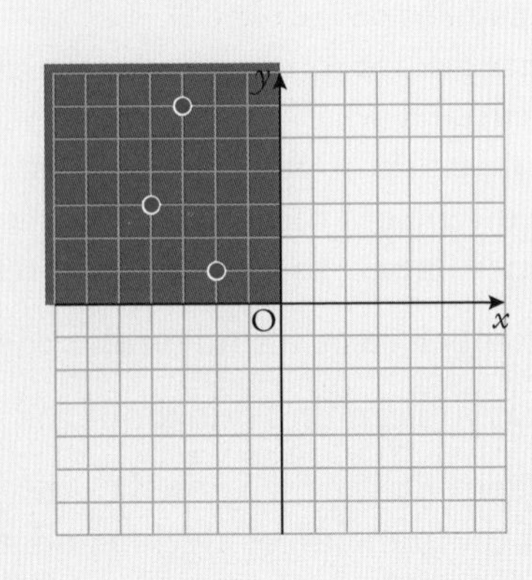

Where will the following points 'land'?

a $(1, 2)$ **b** $(3, 4)$ **c** (x, y)

12.1 The gradient revisited

We know that the gradient of a line can be calculated using $m = \dfrac{y_2 - y_1}{x_2 - x_1}$.
Figure 1 shows that this is the tangent of the angle the line makes with the positive x-direction, i.e. $m = \tan \theta°$.

Figure 1

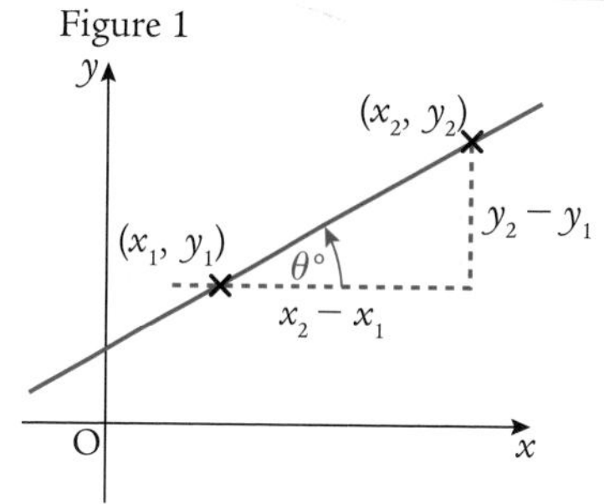

From this we can see that:

a parallel lines have the same gradient and ...
lines with the same gradient are parallel

b a line parallel to the x-axis has a gradient $m = \tan 0°$... $m = 0$

c a line perpendicular to the x-axis has a gradient $m = \tan 90°$ ... which is undefined

d when $\theta°$ is an obtuse angle, $m = \tan \theta°$ is negative (Figure 2) [the calculator gives a negative answer for $\tan^{-1} m$ ... so use $\theta = 180 + \tan^{-1} m$]

Figure 2

e if we know the gradients of two lines, we can deduce the angle, $a°$, at which the lines intersect (Figure 3), and by considering the angles in a triangle we can see easily that

$a = \theta_1 - \theta_2$ [check it out]

which, in terms of the gradients, becomes, e.g. $a = \tan^{-1}m_1 - \tan^{-1}m_2$ ... when both angles are acute (adjusted as in **d** when obtuse)

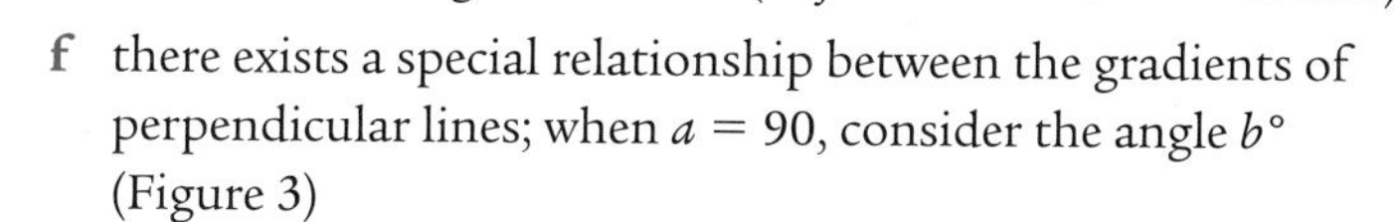

f there exists a special relationship between the gradients of perpendicular lines; when $a = 90$, consider the angle $b°$ (Figure 3)

Figure 3

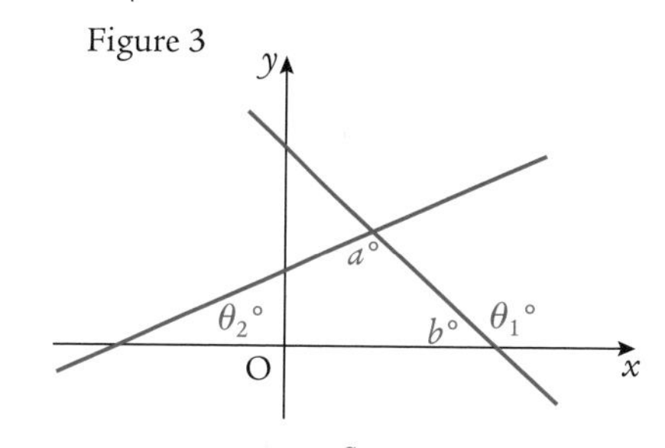

- $b = 180 - \theta_1 \Rightarrow \tan b° = \tan (180 - \theta_1)° = -\tan \theta_1° = -m_1$
- $b = 90 - \theta_2 \Rightarrow \tan b° = \tan (90 - \theta_2)° = \dfrac{1}{\tan \theta_2°} = \dfrac{1}{m_2}$ ...

Thus $-m_1 = \dfrac{1}{m_2} \Rightarrow m_1 m_2 = -1$.

When lines are perpendicular then $m_1 m_2 = -1$ ... the converse is also true[2].

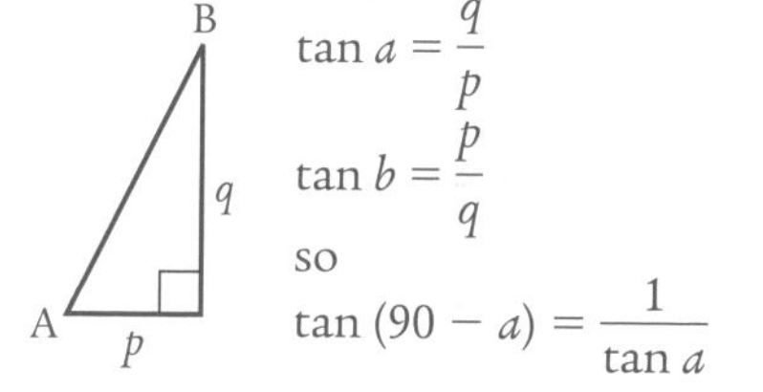

Example 1

A quadrilateral ABCD has vertices A$(-1, -1)$, B$(-2, 7)$, C$(5, 3)$ and D$(6, -5)$. The point E$(2, 1)$ lies inside the quadrilateral.

a Prove that AB is parallel to CD.

b Prove that A, E and C are collinear.

c Prove that AC is perpendicular to BD.

a $m_{AB} = \dfrac{7 - (-1)}{-2 - (-1)} = -8$; $m_{CD} = \dfrac{-5 - 3}{6 - 5} = -8$.

$\Rightarrow m_{AB} = m_{CD} \Rightarrow AB \| CD$ ... QED

b $m_{AE} = \dfrac{1 - (-1)}{2 - (-1)} = \dfrac{2}{3}$; $m_{EC} = \dfrac{3 - 1}{5 - 2} = \dfrac{2}{3}$; $m_{AE} = m_{EC} \Rightarrow AE \| EC$.

However E is a point common to both AE and EC. So A, E and C are collinear.

c $m_{AC} = \dfrac{3 - (-1)}{5 - (-1)} = \dfrac{2}{3}$; $m_{BD} = \dfrac{-5 - 7}{6 - (-2)} = -\dfrac{3}{2}$

$m_{AC} \times m_{BD} = \frac{2}{3} \times \left(-\frac{3}{2}\right) = -1 \Rightarrow AC \perp BD$ ... QED

---

[2] When $m_1 m_2 = -1$, then the lines are perpendicular.

Example 2

Calculate the angle between the lines with equations $y = 3x + 1$ and $y = 1 - x$.

The angle that $y = 3x + 1$ makes with the x-axis $= \tan^{-1} 3 = 71{\cdot}6°$ (to 1 d.p.).
The angle that $y = 1 - x$ makes with the x-axis $= \tan^{-1}(-1) = -45°$ or $180° + \tan^{-1}(-1) = 135°$ ... we want the positive answer, so the line makes an angle of $135°$ with the x-axis.

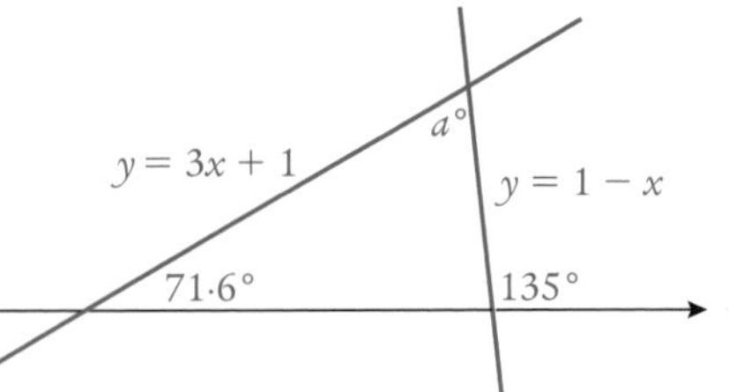

A quick sketch lets us see that $a = 135 - 71{\cdot}6 = 63{\cdot}4$.

The lines intersect at an angle of $63{\cdot}4°$.

Example 3

Find the equation of the line passing through (2, 3)

a parallel to $y = \frac{1}{2}x - 1$

b perpendicular to $y = \frac{1}{2}x - 1$.

a Being parallel to $y = \frac{1}{2}x - 1$, the required line has a gradient of $m = \frac{1}{2}$ and passes through (2, 3).
Substituting this into $y - b = m(x - a)$ gives $y - 3 = \frac{1}{2}(x - 2)$
... which can be 'tidied' to $x - 2y + 4 = 0$.

b Being perpendicular to $y = \frac{1}{2}x - 1$, the required line has a gradient m, where $m \times \frac{1}{2} = -1 \Rightarrow m = -2$ ... it also passes through (2, 3).
Again, substituting into $y - b = m(x - a)$ gives $y - 3 = -2(x - 2)$
... which can be 'tidied' to $2x + y - 7 = 0$.

Exercise 12.1

1 **a** Prove that the line joining A(−2, −1) and B(6, 3) and the line joining C(3, −2) and D(7, 0) are parallel.

b PQRS is a quadrilateral with vertices P(−4, 1), Q(−1, 3), R(4, 1) and S(1, −1).
Prove that PQRS is a parallelogram.

c Prove that each pair of lines are parallel.
  i $y = 3x + 4$ and $2y - 6x = 1$.
  ii $y = 3 - 4x$ and the line joining A(−1, 5) to B(1, −3).

d A, B and C are the points (−5, 7), (1, −3) and (−1, 6) respectively. D is the point (2, d).
  i Prove that AB is not parallel to CD if $d = 3$.
  ii For what value of d is AB parallel to CD?

2 **a** Prove that the line joining (−5, 6) to (3, 2) is perpendicular to $y = 2x + 1$.

b If a line is perpendicular to $y = 3x - 4$, what is its gradient?

c What is the gradient of a line perpendicular to $y = 1 - 5x$?

d EFGH is a quadrilateral with vertices E(−2, 1), F(−1, 8), G(4, 3) and H(3, −4).
  i Show that its diagonals intersect at right angles.
  ii Find the midpoints of both EG and FH.
  iii Hence say what kind of quadrilateral it is.

3 The four-figure grid references on a map are just coordinates using a different notation, e.g. a point on a map with reference 3417 is the point (34, 17).
Three hill tops on a map are Ben Dhu, 3417, Ben Glas, 5847 and Tor Fell, 3822.
Show that these three hill tops will be collinear on the map.

4 At what angle do each of the following pairs of lines intersect?

a $y = 2x + 5$ and $y = \frac{1}{2}x + 1$ b $y = 3x - 1$ and $y = 1 - 3x$

c $y = 4x + 3$ and $y = 1 - \frac{1}{4}x$ d $y = 1 - \frac{2}{3}x$ and $3x + 2y = 1$

5 a A$(-2, 3)$ and B$(6, -1)$ are joined to form a line segment.
Using $\left(\frac{x_2 + x_1}{2}, \frac{y_2 + y_1}{2}\right)$, find the midpoint of AB.

b Find the equation of AB.

c Hence find the equation of the perpendicular bisector of AB.

6 a What is the gradient of the line $2x - 3y + 9 = 0$?

b Find the equation of the line that passes through $(5, 2)$ and is perpendicular to $2x - 3y + 9 = 0$.

7 On a plan, a wall runs on a line modelled by $y = 5 - 2x$ where north is in the y-direction.
The wall ends at the point where $x = -3$.
Another wall is to be built from this point at right angles to the original wall.

a What is the equation of the line of this second wall?

b Are the following points north of, south of, or on the line of the second wall?
i A(2, 13) ii B(3, 14) iii C(4, 15)

8 A line passes through $(0, b)$ and $(a, 0)$.

a Show that its equation can be expressed as $\frac{x}{a} + \frac{y}{b} = 1$.

b A perpendicular to the line passes through $(0, a)$.
Express its equation in a similar form to that in part a.

12.2 Problem solving

The following facts are most useful for solving problems involving straight lines.

Let A be the point (x_a, y_a) and B the point (x_b, y_b).

- $m_{AB} = \frac{y_b - y_a}{x_b - x_a}$ The gradient of AB.
- $m_{\perp} = -\frac{1}{m_{AB}}$ The gradient perpendicular to AB.
- $\left(\frac{x_b + x_a}{2}, \frac{y_b + y_a}{2}\right)$ The midpoint of AB.
- $y - b = m(x - a)$ Equation of line, gradient m, passing through (a, b).
- $mx + c = ax + b$ Lines $y = mx + c$ and $y = ax + b$ intersect. [Solve for x.]
- $AB = \sqrt{(x_b - x_a)^2 + (y_b - y_a)^2}$ The distance AB.

You may also be called upon to remember any geometric fact or theorem from past experience.

Example 1

A triangle has vertices A(1, 0), B(9, 1) and C(2, 8).

a Find the equation of the median from C.

b Find the equation of the altitude from A.

c Find where the altitude and median intersect.

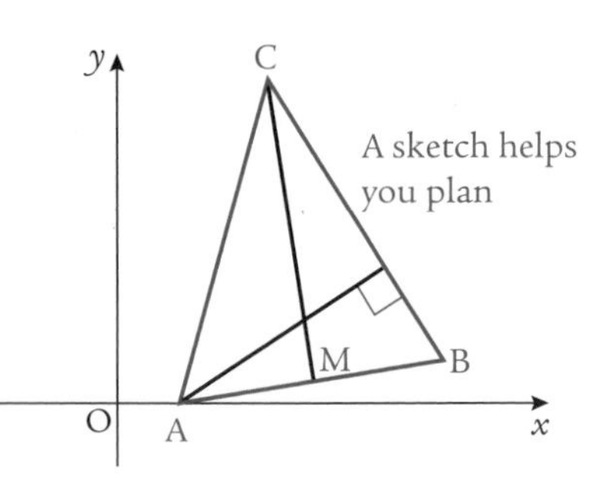

a The midpoint of AB is $\left(\frac{1+9}{2}, \frac{0+1}{2}\right) = \left(5, \frac{1}{2}\right)$ ... call it M.

The median has a gradient, $m_{CM} = \frac{8 - \frac{1}{2}}{2 - 5} = \frac{15}{-6} = -\frac{5}{2}$.

The equation of the median, using C: $y - 8 = -\frac{5}{2}(x - 2)$. ... ①

b $m_{BC} = \frac{8-1}{2-9} = \frac{7}{-7} = -1$; $m_{\perp} = \frac{-1}{m_{BC}} = 1$.

The equation of the altitude, using A: $y - 0 = 1(x - 1)$, i.e. $y = x - 1$ ... ②

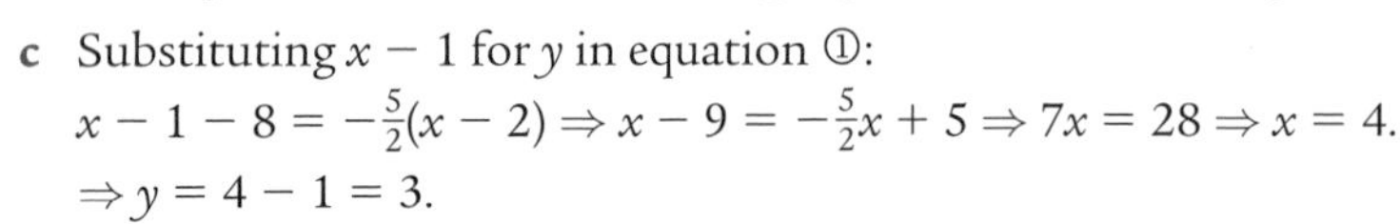

c Substituting $x - 1$ for y in equation ①:

$x - 1 - 8 = -\frac{5}{2}(x - 2) \Rightarrow x - 9 = -\frac{5}{2}x + 5 \Rightarrow 7x = 28 \Rightarrow x = 4$.

$\Rightarrow y = 4 - 1 = 3$.

The altitude and median intersect at the point (4, 3).

Example 2

Find the centre and radius of the circle that passes through the points A(8, 0), B(−1, 3) and C(1, 7).

We will use the geometric fact that **the perpendicular bisector of a chord passes through the centre of the circle**.

Consider the chord AB: $m_{AB} = \frac{3-0}{-1-8} = -\frac{1}{3} \Rightarrow m_{\perp} = 3$.

Midpoint of AB: $\left(\frac{-1+8}{2}, \frac{3+0}{2}\right) = \left(\frac{7}{2}, \frac{3}{2}\right)$.

Equation of perpendicular bisector of AB: $y - \frac{3}{2} = 3\left(x - \frac{7}{2}\right)$ i.e. $y = 3x - 9$ ... ①

Consider the chord AC: $m_{AC} = \frac{7-0}{1-8} = -\frac{7}{7} \Rightarrow m_{\perp} = 1$.

Midpoint of AC: $\left(\frac{1+8}{2}, \frac{7+0}{2}\right) = \left(\frac{9}{2}, \frac{7}{2}\right)$.

Equation of perpendicular bisector of AC: $y - \frac{7}{2} = 1\left(x - \frac{9}{2}\right)$ i.e. $y = x - 1$ ... ②

Substituting $y = x - 1$ into equation ① to find where the bisectors intersect:

$x - 1 = 3x - 9 \Rightarrow x = 4$

$\Rightarrow y = 4 - 1 = 3$.

The perpendicular bisectors of the chords intersect at (4, 3). This must be the centre of the circle.

Call it D. Then the radius can be calculated by finding the length of AD, BD or CD.

$AD = \sqrt{(4-8)^2 + (3-0)^2} = 5$.

The circle has a centre (4, 3) and a radius of 5 units.

Example 3

Find the distance from the point C(5, 1) to the line $y = 2x + 1$ correct to 2 decimal places.

The shortest distance between a point and a line can be measured by dropping a perpendicular from the point to the line. Any other line drawn from the point to the given line must be longer. [Think about it.]

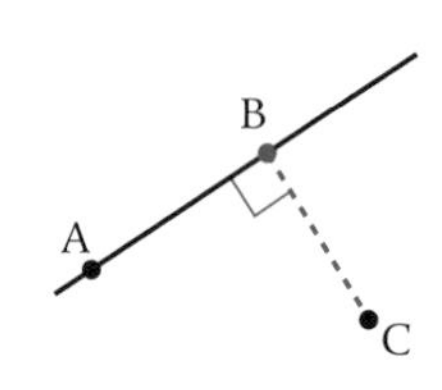

Let A be a point on the line. Let B be the foot of the perpendicular dropped from C.[3]

From the equation we have $m_{AB} = 2 \Rightarrow m_{BC} = -\frac{1}{2}$.

So the equation of BC is $y - 1 = -\frac{1}{2}(x - 5)$, i.e. $y = -\frac{1}{2}x + \frac{7}{2}$.

B is where BC intersects AB, i.e. where $2x + 1 = -\frac{1}{2}x + \frac{7}{2} \Rightarrow 4x + 2 = -x + 7 \Rightarrow x = 1$.

$\Rightarrow y = 3$.

B is the point (1, 3).

The distance between B and C, $BC = \sqrt{(5-1)^2 + (1-3)^2} = \sqrt{20}$.

The distance between the point C and the line AB is 4·47 units (to 2 d.p.).

Exercise 12.2

1 Triangle ABC has vertices A(−4, −3), B(0, 7) and C(10, −1).
P, Q and R are the midpoints of AB, BC and CA respectively.

- **a** State the coordinates of P, Q and R.
- **b** Find the equation of:
  - **i** the median PC **ii** the median through B.
- **c** Find the point S, where the two medians intersect.
- **d** **i** Find the equation of the median AQ.
  - **ii** Show that S also lies on AQ, thus showing the medians are concurrent.

2 The triangle EFG has vertices E(4, 2), F(13, 5) and G(8, 10).
The altitudes cut EF, FG and GE at L, M and N respectively.

- **a** Find the equation of: **i** EF **ii** GL.
- **b** Hence find the coordinates of L.
- **c** Find the equation of the altitude FN.
- **d** Hence find P, the point where the altitudes GL and FN intersect.
- **e** Find the equation of the altitude from E and prove that P lies on it too.

3 In prehistoric times, men built a circle of stones.
At the centre of the circle they buried an urn.
After thousands of years all that remains are three stones.
Using suitable axes, the coordinates of the stones are
A(−1, −7), B(1, 7) and C(7, 9).

- **a** Find the equation of the perpendicular bisector of:
  - **i** AC **ii** BC.
- **b** Hence find the coordinates of the centre of the original circle, i.e. where the perpendicular bisectors intersect.

4 An angle AOB has its vertex at the origin.
The line OA has equation $8x - 5y = 0$.
The line OB has equation $x - 4y = 0$.

- **a** Calculate, correct to 1 decimal place, the size of:
  - **i** angle xOB **ii** angle xOA.
- **b** Hence calculate the size of angle AOB.
- **c** If OC bisects AOB, find the equation of OC, giving the gradient, correct to 2 decimal places.

[3] This is often referred to as the **projection** of C on the line.

5 A line AB has equation $y = 4 - 4x$. The point P has coordinates $(-6, -6)$.

a Find the equation of the line perpendicular to AB passing through P.

b The perpendicular cuts AB at the point Q. What are the coordinates of Q?

c What is the shortest distance between the point P and the line AB?

6 Using convenient axes and units, a ship is steering a course that can be modelled by the line $y = 3x + 2$.

A reef is recorded as being at the point (7, 3).

How close to the reef will the ship get?

7 Triangle ABC has vertices A(4, 13), B(1, 2) and C(13, 6).

a Find the equation of the median through B.

b Find the equation of the altitude from A.

c Where do these two lines intersect?

d What is the distance of this point from the line BC?

e Find the equation of the median through A. Comment.

12.3 Circle centred on the origin

Consider the circle of radius r units, whose centre is the origin.

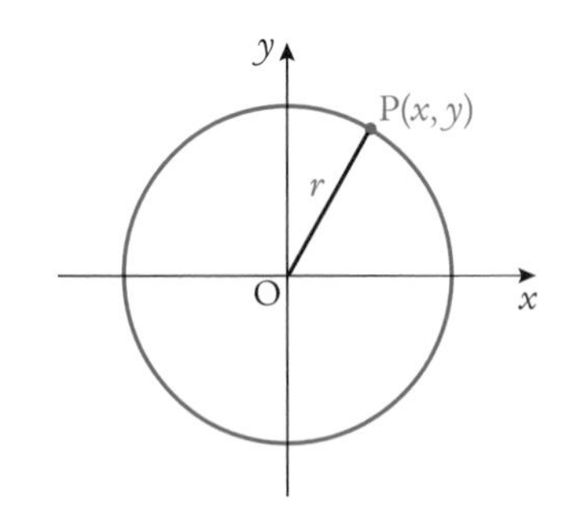

If P(x, y) is a point on the circle then using the distance formula $r^2 = OP^2 = (x - 0)^2 + (y - 0)^2 = x^2 + y^2$.

Since this is only true of points on the circle, we refer to $x^2 + y^2 = r^2$ as the equation of a circle of radius r and centre (0, 0).

Note that for a point $Q(x_q, y_q)$,

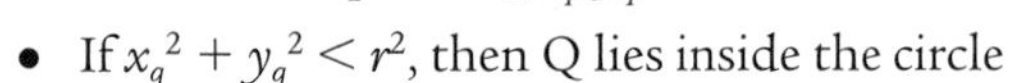

- If $x_q^2 + y_q^2 < r^2$, then Q lies inside the circle
- If $x_q^2 + y_q^2 > r^2$, then Q lies outside the circle
- If $x_q^2 + y_q^2 = r^2$, then Q lies on the circle.

Example 1

a Find the equation of the circle, centre the origin, passing through the point (5, 12).

b Decide whether each of the following points is on, inside or outside the circle.

i (4, 14) ii (6, 11) iii (−12, 5)

a $r^2 = (5 - 0)^2 + (12 - 0)^2 = 5^2 + 12^2 = 169$.
So the equation of the circle is $x^2 + y^2 = 169$.

b i $4^2 + 14^2 = 16 + 196 = 212$. Thus $4^2 + 14^2 > r^2$. So (4, 14) is outside the circle.

ii $6^2 + 11^2 = 36 + 121 = 157$. Thus $6^2 + 11^2 < r^2$. So (6, 11) is inside the circle.

iii $(-12)^2 + 5^2 = 144 + 25 = 169$. Thus $(-12)^2 + 5^2 = r^2$. So (−12, 5) is on the circle.

Example 2

a Describe the locus[4] of all points that fit the description $x^2 + y^2 = 49$.

b i Make y the subject of the formula.

ii What happens if you substitute a value for x greater than or equal to 7?

a The locus is a circle centre (0, 0) and radius $\sqrt{49} = 7$.

b i $y = \pm\sqrt{49 - x^2}$.

ii When $x > 7$, the expression under the square root is negative and there are no real values for y.

Example 3

Where does the line $y = 2x - 5$ intersect the circle $x^2 + y^2 = 10$?

Using the equation of the line, we know we can substitute $2x - 5$ for y in the equation of the circle. Where the two graphs intersect, the y-coordinates are the same.

$x^2 + (2x - 5)^2 = 10$

$\Rightarrow x^2 + 4x^2 - 20x + 25 = 10$

$\Rightarrow \quad 5x^2 - 20x + 15 = 0$

$\Rightarrow \quad x^2 - 4x + 3 = 0$

$\Rightarrow \quad (x - 3)(x - 1) = 0$

$\Rightarrow \quad x = 3 \text{ or } x = 1$

$\Rightarrow y = 1 \text{ or } y = -3$ ... using $y = 2x - 5$

The line cuts the circle at (3, 1) and (1, −3).

Exercise 12.3

1 State the centre and radius of the circle with equation:

a $x^2 + y^2 = 1$ b $x^2 + y^2 = 144$ c $x^2 + y^2 = 196$ d $2x^2 + 2y^2 = 50$

e $3x^2 + 3y^2 = \frac{1}{3}$ f $y^2 = 100 - x^2$ g $\frac{1}{3}x^2 + \frac{1}{3}y^2 = 12$ h $y^2 - 72 = 72 - x^2$.

2 An archery target has 10 rings, coloured in five pairs.
The pairs are coloured gold, red, blue, black, and white, starting with the innermost pair.
The innermost ring has a radius of 4 cm.
Thereafter, each radius is 4 cm bigger than the last.
So the innermost area, the gold area, has a radius of 8 cm.
It can be described by $0 \leqslant x^2 + y^2 < 64$.

a In a similar way describe:

i the red area ii the blue area iii the black area.

b The 'hits' are recorded as coordinates.
In which colour has each of these arrows fallen?

i (6, −10) ii (4, 6) iii (−12, 14) iv (20, 20) v (4, 14)

c What is the equation of the circle separating:

i red from black ii black and white?

y
x
gold
red
blue
black
white

3 Find the equation of the circle and state its radius given that its centre is the origin and that it passes through the point:

a (24, −7) b (−28, 45) c (60, 11) d (−10, −24).

[4] The figure formed by all points that meet specified conditions.

4 A circle whose centre is the origin has a diameter BC. A point A lies on the circumference so that the chord AB is 70 units long and the chord AC is 24 units long.
What is the equation of the circle?

5 A picture of the Moon has white marks on it. Some are stars and some are just flaws on the photograph.
The image of the Moon has a diameter of 20 cm.

a Using the centre of the Moon's disk as the origin, what is the equation of the outline of the image of the Moon?

b The coordinates of some of the marks have been noted.
Which ones can't possibly be stars?

i (8, 5) **ii** (−6, −9) **iii** (−4, 10)
iv (9, −4) **v** (−3, −11) **vi** (7, −7)

6 Find where each line cuts the given circle.

a $x^2 + y^2 = 45$ cut by $y = -3x - 15$. **b** $x^2 + y^2 = 40$ cut by $y = -2x - 2$.

c $x^2 + y^2 = 34$ cut by $y = 4x - 17$. **d** $x^2 + y^2 = 10$ cut by $y = \frac{1}{2}x + \frac{5}{2}$.

e $x^2 + y^2 = 40$ cut by $2y = x + 10$. [Hint: it may be easier to substitute for x instead.]

f $x^2 + y^2 = 20$ cut by $3y = x + 10$.

7 We can draw circles, centre the origin, on a spreadsheet.

Method 1 requires that you express the equation, say $x^2 + y^2 = 49$ as $y = \pm\sqrt{49 - x^2}$ and draw both $y = +\sqrt{49 - x^2}$ and $y = -\sqrt{49 - x^2}$ on the same chart. Together they make a reasonable approximation to the circle, centre the origin, of radius 7.

In A1 type: -7
In A2 type: =A1+0.4 ... and fill down to row 36.
In B1 type: =SQRT(49-A1^2) ... this will produce the upper semicircle.
In C1 type: =-SQRT(49-A1^2) ... this will produce the lower semicircle.

Fill both B and C down to row 36.

Highlight A1 to C36 and draw a chart > Scatter > Scatter with Smooth Lines.

Make both scales the same ... and reshape the frame so that the grid is squares (Figure 1).

Figure 1

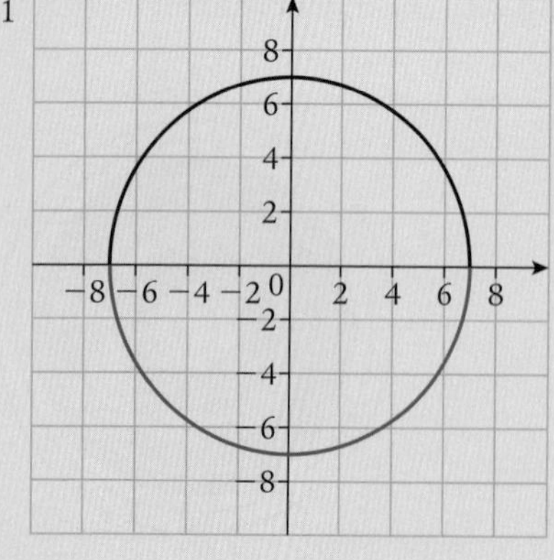

Figure 2

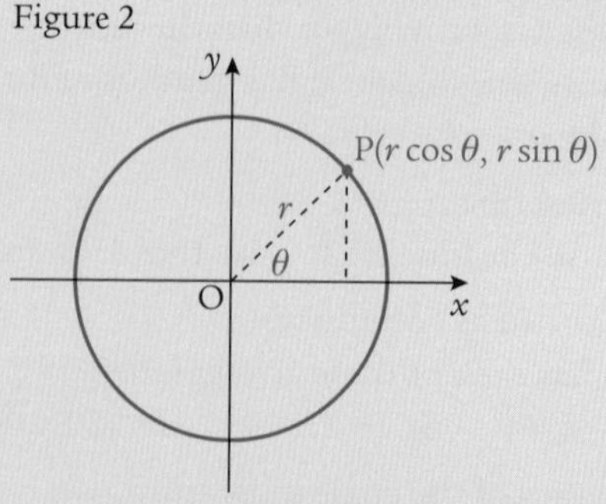

Method 2 produces better results. We use the connection illustrated in Figure 2.

The x-coordinate is $r\cos\theta$, the y-coordinate is $r\sin\theta$, so
$x^2 + y^2 = r^2\cos^2\theta + r^2\sin^2\theta = r^2(\cos^2\theta + \sin^2\theta) = r^2$.

In A1 type: 0 ... $\theta = 0$ radians.
In A2 type: =A1+PI()/12 ... increasing the angle by $\frac{\pi}{12}$.

Fill down to row 25 ... taking θ to 2π radians.
In B1 type: =7*(COS(A1)) ... the x-coordinate when $\theta = 0$ radians.
In C1 type: =7*(SIN(A1)) ... the y-coordinate when $\theta = 0$ radians.
Fill B and C down to row 25.

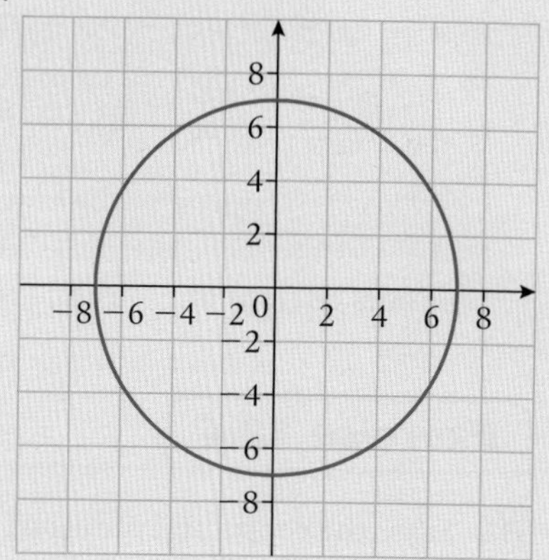

Select B1 to C25 and draw a chart > Scatter > Scatter with Smooth Lines.
[Don't include column A or you'll just get a sine and cosine curve.]
Again you need to resize the diagram so that the grid is squares.

12.4 When the centre of the circle is (a, b)

Consider the circle of radius r units, whose centre is C(a, b).

If P(x, y) is a point on the circle, then using the distance formula $r^2 = \text{CP}^2 = (x - a)^2 + (y - b)^2$.

Since this is only true of points on the circle, we refer to $(x - a)^2 + (y - b)^2 = r^2$ as the equation of a circle of radius r and centre (a, b).

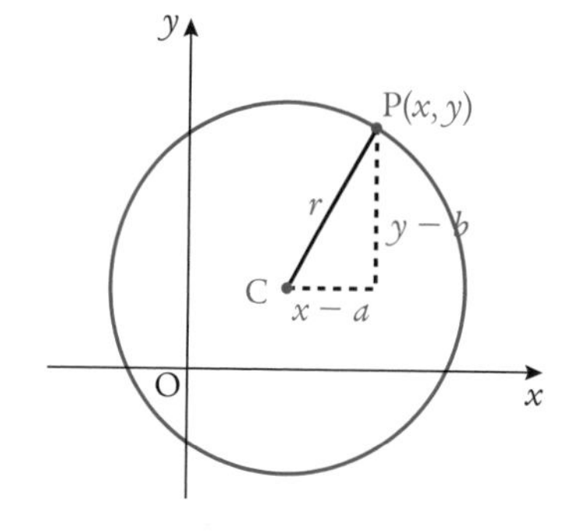

Note that, for a point Q(x_q, y_q),

- if $(x_q - a)^2 + (y_q - b)^2 < r^2$, then Q lies inside the circle
- if $(x_q - a)^2 + (y_q - b)^2 > r^2$, then Q lies outside the circle
- if $(x_q - a)^2 + (y_q - b)^2 = r^2$, then Q lies on the circle.

Example 1

a Find the equation of the circle with its centre at (1, 5) and which passes through (8, 3).

b Is the point (−6, 4) inside, outside or on the circle?

a Using the distance formula: $r^2 = (8 - 1)^2 + (3 - 5)^2 = 49 + 4 = 53$.
So the equation of the circle is: $(x - 1)^2 + (y - 5)^2 = 53$.

b Substituting (−6, 4) into the expression $(x - 1)^2 + (y - 5)^2$, we get
$(-6 - 1)^2 + (4 - 5)^2 = 49 + 1 = 50$.
Since $50 < 53$, the point (−6, 4) is inside the circle.

Example 2

a State the centre and radius of the circle with equation $(x - 1)^2 + (y - 2)^2 = 90$.

b Find where the line $y = 2x - 15$ cuts the circle.

a Comparing the given equation with the standard equation viz. $(x - a)^2 + (y - b)^2 = r^2$, we get the centre (1, 2) and the radius as $\sqrt{90}$ units $= 3\sqrt{10}$ units.[5]

[5] We tend to leave answers as simplified surds unless the question specifies a degree of accuracy.

Unit 3: Applications

b Substitute $2x - 15$ for y in the equation of the circle.

$(x - 1)^2 + [(2x - 15) - 2]^2 = 90$

$\Rightarrow \quad (x - 1)^2 + (2x - 17)^2 = 90$

$\Rightarrow x^2 - 2x + 1 + 4x^2 - 68x + 289 = 90$

$\Rightarrow \quad 5x^2 - 70x + 200 = 0$

$\Rightarrow \quad 5(x - 4)(x - 10) = 0$

$\Rightarrow \quad x = 4 \text{ or } 10$

$\Rightarrow y = -7 \text{ or } 5$

Line cuts circle at $(4, -7)$ and $(10, 5)$.

Exercise 12.4

1 State the centre and radius of each of the following circles.

a $(x - 5)^2 + (y - 3)^2 = 100$ **b** $(x - 3)^2 + (y + 2)^2 = 144$ **c** $(x + 1)^2 + (y - 1)^2 = 16$

d $(x + 2)^2 + (y + 5)^2 = 8$ **e** $(x + 6)^2 + (y - 7)^2 = 32$ **f** $x^2 + (y + 1)^2 = 27$

g $(x + 7)^2 + y^2 = 50$ **h** $(x + 1{\cdot}5)^2 + (y - 3{\cdot}5)^2 = 0{\cdot}49$ **i** $x^2 + (y - 8)^2 = 0{\cdot}5$

2 Give the equation of the circle with the given centre and radius.

a Centre (5, 4), radius 3. **b** Centre (1, 1), radius 10. **c** Centre $(-2, 1)$, radius 5.

d Centre $(-1, -3)$, radius $\sqrt{10}$. **e** Centre (0, 4), radius $\sqrt{2}$. **f** Centre (5, 0), radius 1.

3 Find the equation of the circle with the given centre passing through the given point.

a Centre (1, 4), passing through (4, 8). **b** Centre (5, 3), passing through (17, 8).

c Centre $(-3, 6)$, passing through (12, 14). **d** Centre $(-1, -1)$, passing through (19, 20).

e Centre (2, 0), passing through (5, 7). **f** Centre $(0, -3)$, passing through $(9, -1)$.

g Centre $(1, -2)$, passing through $(-4, 2)$. **h** Centre $(6, -9)$, passing through $(-4, -3)$.

4 An app for a tablet simulates a compass.
The needle, as it rotates, traces out a circle.
The needle is the diameter of the circle.
Its endpoints at the moment are (5, 11) and (9, 1).

a Calculate: **i** the centre **ii** the radius of the circle traced out.

b Hence find the equation of the circle traced out by the points of the needle.

c The dial in the compass has a diameter of 12 units.
What is the equation of the circumference of the dial?

5 A right-angled triangle has vertices $A(-14, -6)$, $B(10, 12)$ and $C(34, -20)$.

a Which two sides are perpendicular to each other?

b A circle is drawn which passes through the three vertices of the triangle.

i Where is the centre of this circle? [Hint: think of the angle in a semicircle.]

ii What is the radius of the circle?

iii What is the equation of the circle?

6 A square ABCD has vertices $A(-3, 4)$, $B(4, 5)$ and $C(5, -2)$.

a What are the coordinates of the fourth vertex, D?

b A circle is drawn through all four vertices. Find the equation of the circle.

c **i** Find the equation of the line AB.

ii Hence show that the point (1, 5) is inside the circle but outside the square.

7 A circle with centre (1, 2) passes through the point (6, −1).

a i Calculate the radius of the circle, leaving your answer in surd form.

ii Hence state the equation of the circle.

b The line $y = x + 3$ cuts the circle at two places. By substituting $x + 3$ for y in the equation of the circle, find the two points of intersection.

c In a similar fashion, find where the line with equation $y = x + 4$ cuts the circle with centre (5, 2) and radius 13 units.

8 A circle, A, has equation $(x - 4)^2 + (y + 1)^2 = 64$.

a Find the equation of the circle, B, which is concentric to this but has half the radius.

b Find the equations of the circles C and D, which are the images of A and B, respectively, reflected in the x-axis.

9 The movement of the arms of a cherry picker can be modelled by circles using suitable units and convenient axes.
As the arm AB turns, B sweeps through a circle $x^2 + y^2 = 100$.
As the arm BC moves, C sweeps through a circle with equation $(x - x_b)^2 + (y - y_b)^2 = 25$.

a When $x_b = 8$, calculate:

i y_b

ii the equation of the circle traced out by C

iii y_c, when $x_c = 12$.

b Repeat part **a** when $x_b = 6$ and $x_c = 6$.

c Repeat part **a** when $x_b = 2$ and $x_c = 4$.

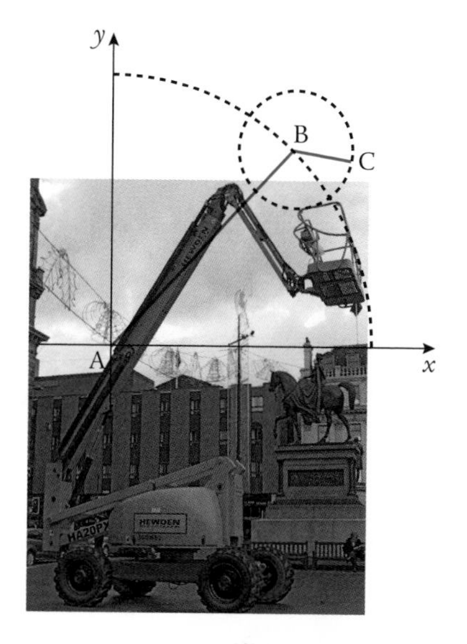

10 An aerial has two dishes. The face of each dish is circular.
The rim of the larger dish is modelled by the equation $(x - 3)^2 + (y + 8)^2 = 169$.

a Check that the point P(8, 4) lies on the rim.

b Find the point on the rim diametrically opposite P.

c The smaller dish is vertically above the larger and touching. Its radius is only a third of the larger dish.

i What is its equation?

ii What are the coordinates of the point of contact?

11 This form of the equation can be investigated on a spreadsheet, making use of the spinner controls.

In row 1 enter headings A1, Theta; B1, x; C1, y; D1, x-centre; E1, y-centre; F1, radius.

Using the developer, create a spinner to control D3: Minimum 0; maximum 20; increment 1.

Create a spinner to control E3: Minimum 0; maximum 20; increment 1.

Create a spinner to control F2: Minimum 0; maximum 10; increment 1 ... the radius runs from 0 to 10.

In D2 type: =D3-10 ... the x-centre now runs from −10 to 10 as you click the spinner.

In E2 type: =E3-10 ... the y-centre now runs from −10 to 10 as you click the spinner.

In A2 type: 0

In A3 type: =A2+PI()/12 ... and fill down to row 26 to get values from 0 to 2π.

In B2 type: =$F$2*COS(A2)+$D$2 ... and fill down to row 26.

In C2 type: =$F$2*SIN(A2)+$E$2 ... and fill down to row 26.

Select B2 to C26 and ask for a chart > Scatter > Scatter with Smooth Lines.

Format the axes so that they both run from −10 to 10.

Make sure the grid has both horizontal and vertical gridlines.

Reshape the frame to make the grid square.

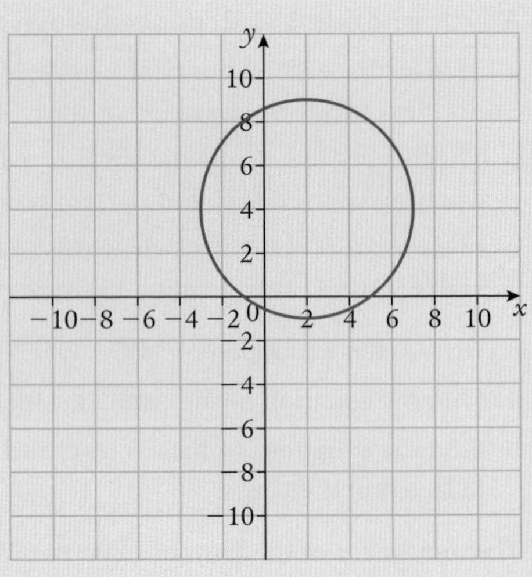

Clicking on the spinners gives you control of the centre, and the radius.

Changes to the chart are instantaneous.

12.5 An expanded form of the equation

Consider the equation $x^2 + y^2 + 2gx + 2fy + c = 0$.

$\Rightarrow (x^2 + 2gx) + (y^2 + 2fy) + c = 0$ ... rearranging the terms

$\Rightarrow [(x + g)^2 - g^2] + [(y + f)^2 - f^2] + c = 0$ ... completing the square in each bracket

$\Rightarrow (x + g)^2 + (y + f)^2 = g^2 + f^2 - c$

$\Rightarrow (x - (-g))^2 + (y - (-f))^2 = \left(\sqrt{g^2 + f^2 - c}\right)^2$

Comparing this with the equation of a circle, we see that, as long as $g^2 + f^2 - c > 0$,

$x^2 + y^2 + 2gx + 2fy + c = 0$ is a circle whose centre is $(-g, -f)$ and whose radius is $\sqrt{g^2 + f^2 - c}$.

Example 1

Show that $x^2 + y^2 + 4x - 6y + 12 = 0$ represents a circle. State its centre and radius.

Comparing this with $x^2 + y^2 + 2gx + 2fy + c = 0$, we see that $g = 2, f = -3$ and $c = 12$.

$g^2 + f^2 - c = 2^2 + (-3)^2 - 12 = 4 + 9 - 12 = 1$

Since $g^2 + f^2 - c > 0$, the equation represents a circle.

Centre is $(-g, -f) = (-2, 3)$. Radius is $\sqrt{g^2 + f^2 - c} = \sqrt{1} = 1$ unit.

Example 2

a Which of these equations represent a circle?

i $2x^2 + 2y^2 + 6x + 4y - 6 = 0$ **ii** $x^2 + y^2 + 2x + y + 6 = 0$

b Give the centre and the radius of the one that is a circle.

a To compare **i** with the general form we must make the coefficients of x^2 and y^2 equal to 1.

$2x^2 + 2y^2 + 6x + 4y - 6 = 0 \Rightarrow x^2 + y^2 + 3x + 2y - 3 = 0$.

Comparing this with the general form we find $g = \frac{3}{2}, f = 1$ and $c = -3$.

$g^2 + f^2 - c = \frac{9}{4} + 1 + 3 = \frac{25}{4}$

Since $g^2 + f^2 - c > 0$, then equation **i** is that of a circle.

Comparing **ii** with the general form gives $g = 1, f = \frac{1}{2}$ and $c = 6$.

$g^2 + f^2 - c = 1 + \frac{1}{4} - 6 = -\frac{19}{4}$

Since $g^2 + f^2 - c < 0$, then equation **ii** is NOT that of a circle.

b Equation **i** represents a circle centre $\left(-\frac{3}{2}, -1\right)$ and radius $\frac{5}{2}$ units.

Exercise 12.5

1 Find the centre and radius of each circle.

a $x^2 + y^2 + 10x + 2y + 1 = 0$

b $x^2 + y^2 + 4x - 18y - 15 = 0$

c $x^2 + y^2 - 40x - 12y + 267 = 0$

d $x^2 + y^2 - 14x - 36y + 84 = 0$

e $x^2 + y^2 - 14x - 24y - 207 = 0$

f $x^2 + y^2 - 58x - 18y + 297 = 0$

g $x^2 + y^2 - 2x - 6y - 4 = 0$

h $x^2 + y^2 + 8x - y + 5 = 0$

i $x^2 + y^2 - 5x + 3y - 9 = 0$

j $x^2 + y^2 + x + y - 1 = 0$

k $x^2 + y^2 - 6x + 6 = 0$

l $x^2 + y^2 + 2y - 3 = 0$

2 Find the equation of the circle with the given centre and radius, expressing your answer in the form $x^2 + y^2 + 2gx + 2fy + c = 0$.

a Centre (2, 3), radius = 6.

b Centre (−1, 5), radius = 10.

c Centre (7, −2), radius = 9.

d Centre (0, 1), radius = 1.

e Centre (4, 0), radius = 8.

f Centre (−5, 6), radius = $\sqrt{2}$.

g Centre (1, 1), radius = $\sqrt{3}$.

h Centre (−8, −1), radius = $\sqrt{41}$.

3 For what value(s) of k will each equation represent a circle?

a $x^2 + y^2 + 4x + y + k = 0$

b $x^2 + y^2 + kx + 2y + 5 = 0$

c $x^2 + y^2 + x + 3ky + 7 = 0$

d $x^2 + y^2 + 6kx + 8ky + 50k = 0$

4 The hands of a clock sweep out two concentric circles.

Using convenient units and axes, the circles have equations $x^2 + y^2 + 2ax - 8y + 4 = 0$ and $x^2 + y^2 - 4x + by - 16 = 0$.

a Calculate the values of: **i** a **ii** b.

b How much longer than the minute hand is the hour hand?

5 The diagonals of a square intersect at the point (2, 1) and the square has a side of length 6 units.

a What is the equation of the largest circle that can be drawn within the square? Express your answer in the form $x^2 + y^2 + 2gx + 2fy + c = 0$.

b The square undergoes an enlargement, scale factor 4, with the origin as the centre of enlargement, i.e. the coordinates of each point on it are multiplied by 4. What is the equation of the largest circle that can be drawn inside the enlarged square?

6 A circle with equation $x^2 + y^2 - 8x - 10y + 16 = 0$ is drawn.

a Where does it cut the x-axis?

b The circle is reflected in the y-axis.

i What is the equation of the image of the circle under reflection in the y-axis?

ii Where does the circle and its image intersect?

c For each of the following lines, find where they cut the original circle: **i** $y = x + 2$ **ii** $y = 8 - x$.

d These two lines intersect at C. Show that C and the two centres are collinear.

7 An isosceles triangle, ABC, has its base AB parallel to the x-axis.
A is the point $(-2, 1)$ and C is the point $(2, 9)$.

a What are the coordinates of B?

b Find the equation of the perpendicular bisector of: **i** AC **ii** AB.

c Hence find the centre of the circle that passes through A, B and C.

d Find the equation of this circle in the form $x^2 + y^2 + 2gx + 2fy + c = 0$.

e A second circle is drawn concentric with the first passing through the midpoints of both AC and BC.

i Find its equation.

ii By considering the distance from its centre to the line AB, show that this circle is NOT the incircle[6] of the triangle.

8 Initially the inventors of the bicycle made the front wheel very much larger than the back.
This gave the cyclist the mechanical advantage that modern gears now give.
On a computer model, the front wheel is given by $x^2 + y^2 - 8x - 16y + 44 = 0$.

a State the centre and radius of the front wheel.

b The road surface is modelled by the line $y = 2$.
Where does the front wheel touch the road?

c The back wheel is only a third the size of the front.
The leading edge of the back wheel is vertically below the trailing edge of the front.
Find the equation of the back wheel.

9 The cell phone system works because a series of transmitters has been set up all over the country, each with a particular range. It is hoped that when you move out of the range of one transmitter you move into that of another. For smooth transition between one cell and the next, there is some overlap.

Using convenient scales and units, the range of a particular transmitter, cell 1, is modelled by $x^2 + y^2 - 6x - 10y - 15 = 0$.

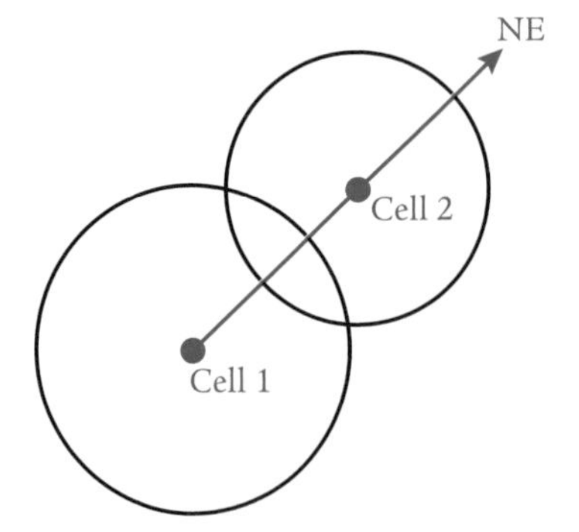

a State its centre and radius.

b Another transmitter is set up $8\sqrt{2}$ units north-east of the first. Its range however is one unit less.
Find its equation.

c Say whether the following points can be picked up by cell 1, cell 2, both or neither.
i $(5, 10)$ **ii** $(6, 11)$ **iii** $(7, 11)$ **iv** $(11, 6)$

[6] Circle to which the three sides of the triangle are tangents (see introduction to this chapter).

12.6 Intersections and tangents

Intersecting circles

Here is a series of 10 pictures of one circle, moving to the right across another.

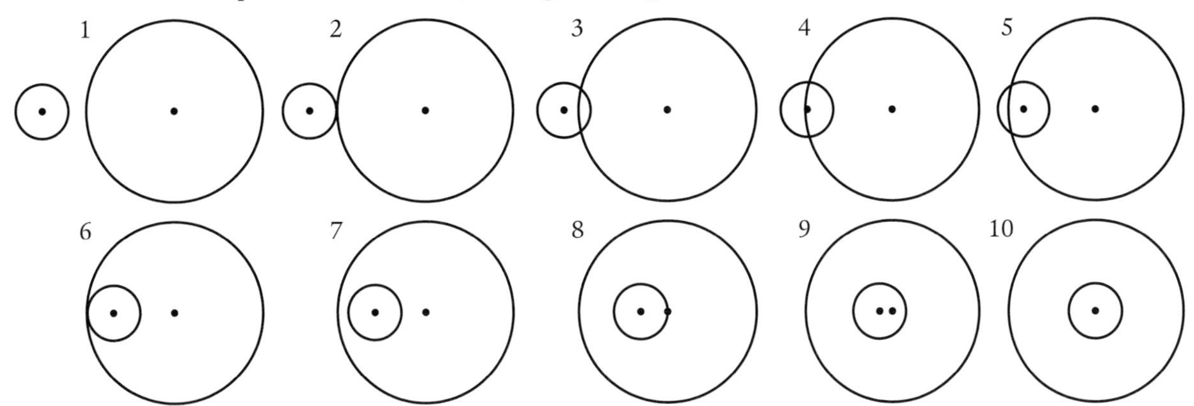

By considering the distance between the centres, d units, and the lengths of the radii, r_1 and r_2, we can determine whether the circles intersect or not. [Let r_1 be the larger radius.]

- $d > r_1 + r_2$ ... Figure 1 ... no intersections ... circles apart.
- $d = r_1 + r_2$ ... Figure 2 ... kissing circles ... smaller circle outside.
- $r_1 - r_2 < d < r_1 + r_2$ ... Figures 3, 4, 5 ... two intersections.
- $d = r_1 - r_2$ ... Figure 6 ... kissing circles ... smaller circle inside.
- $d < r_1 - r_2$ ... Figures 7, 8, 9, 10 ... no intersections ... smaller circle inside.

When the intersections exist, we can find them using a theorem:

Subtracting the equations of two intersecting circles we obtain the equation of the common chord.

[The proof is beyond the scope of the course ... but we can use the result.]

Example 1

Find where the circles $(x + 2)^2 + (y - 2)^2 = 50$ and $(x - 7)^2 + (y - 5)^2 = 20$ intersect.

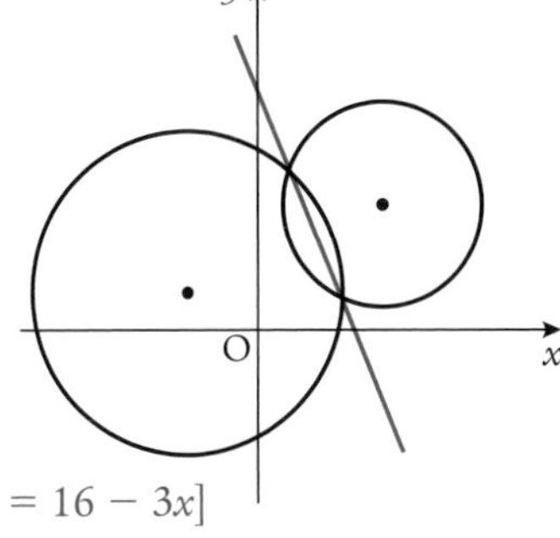

Expressing both equations in the form $x^2 + y^2 + 2gx + 2fy + c = 0$, we get

$x^2 + y^2 + 4x - 4y - 42 = 0$ ... ①

$x^2 + y^2 - 14x - 10y + 54 = 0$ ... ②

Subtracting ① − ②: $3x + y - 16 = 0$ ... the equation of the common chord [$y = 16 - 3x$]

Substitute $y = 16 - 3x$ into equation ① : $x^2 + (16 - 3x)^2 + 4x - 4(16 - 3x) - 42 = 0$.

This will give where the chord cuts the 1st circle ... and, since it is common to both circles, it will give where the two circles intersect.

$\Rightarrow x^2 + 256 - 96x + 9x^2 + 4x - 64 + 12x - 42 = 0$

$\Rightarrow \quad 10x^2 - 80x + 150 = 0$

$\Rightarrow \quad 10(x - 5)(x - 3) = 0$

$\Rightarrow \quad x = 5 \text{ or } 3$

$\Rightarrow y = 1 \text{ or } 7$ ... using the equation of the common chord.

Thus the two circles intersect at (5, 1) and (3, 7).

Tangents

- To find whether a line, $y = ax + b$, is a tangent to the circle $x^2 + y^2 + 2gx + 2fy + c = 0$, we substitute $ax + b$ for y in the equation of the circle ... obtaining a quadratic equation in x.
  Using the discriminant, we can find the conditions for the equation to have coincident roots ... i.e. the line touches the circle at one point.
- To find the tangent to a circle at a particular point, we make use of the fact that the tangent is perpendicular to the radius at the point of contact.

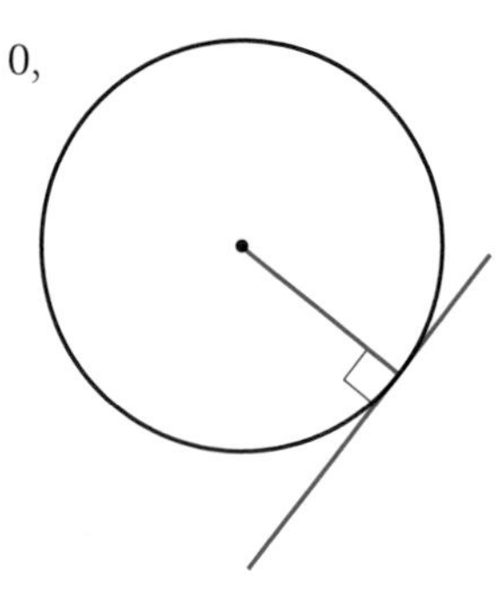

Example 2

Prove $x + 2y = 11$ is a tangent to the circle $x^2 + y^2 - 8x - 2y + 12 = 0$.

From the linear equation, $x = 11 - 2y$. Substitute into the circle:

$(11 - 2y)^2 + y^2 - 8(11 - 2y) - 2y + 12 = 0$

$\Rightarrow 121 - 44y + 4y^2 + y^2 - 88 + 16y - 2y + 12 = 0$

$\Rightarrow \quad 5y^2 - 30y + 45 = 0$

$\Rightarrow \quad y^2 - 6y + 9 = 0.$

The discriminant $= (-6)^2 - 4.1.9 = 0$.

Since the discriminant is zero, there are coincident roots, i.e. the line cuts the circle at only one point. So the line is a tangent to the circle.

Example 3

Find the equation of a tangent to the circle $x^2 + y^2 - 4y - 9 = 0$ at the point P(3, 0).

The centre of the circle C is the point (0, 2) ... by inspection.

Thus the gradient of the radius to P is $m_{rad} = \frac{0 - 2}{3 - 0} = -\frac{2}{3}$.

So the gradient of the tangent at (3, 0) is $m_{tan} = \frac{3}{2}$ ... perpendicular to radius

So the equation of the tangent at (3, 0) is $y - 0 = \frac{3}{2}(x - 3)$, which tidies to $3x - 2y = 9$.

Exercise 12.6

1 Prove each line is a tangent to the given circle and find the point of tangency.

- **a** Line: $y = 2x + 3$ and circle: $x^2 + y^2 - 12x - 10y + 41 = 0$
- **b** Line: $y = 2 - 3x$ and circle: $x^2 + y^2 + 8x + 12y + 12 = 0$
- **c** Line: $y = 4 - x$ and circle: $x^2 + y^2 + 18x + 6y - 38 = 0$
- **d** Line: $y = 3x + 7$ and circle: $x^2 + y^2 + 10x - 4y + 19 = 0$
- **e** Line: $x - 2y + 6 = 0$ and circle: $x^2 + y^2 - 12x - 2y + 17 = 0$
- **f** Line: $x + 3y = 15$ and circle: $x^2 + y^2 + 18x + 4y - 5 = 0$
- **g** Line: $3x + 4y + 4 = 0$ and circle: $x^2 + y^2 + 2x - 12y + 12 = 0$

2 Find the tangent to the given circle at the given point.

- **a** $x^2 + y^2 - 6x + 2y - 10 = 0$ at $(-1, 1)$
- **b** $x^2 + y^2 - 14x + 49 = 0$ at $(1, -2)$
- **c** $x^2 + y^2 + 12x - 14y + 17 = 0$ at $(2, 5)$
- **d** $x^2 + y^2 + 12x - 12y + 46 = 0$ at $(-1, 7)$

e $x^2 + y^2 - 20x + 4y + 84 = 0$ at $(8, 2)$
f $x^2 + y^2 + 18y - 9 = 0$ at $(-3, 0)$
g $x^2 + y^2 - 8x - 8y + 22 = 0$ at $(3, 1)$
h $x^2 + y^2 + 20x - 2y + 49 = 0$ at $(-6, -5)$

3 Two circles, A and B, have equations $x^2 + y^2 - 12x - 14y + 59 = 0$ and $x^2 + y^2 + 4x + 2y - 53 = 0$.

a Find the centres and radii of both circles.

b Calculate the distance between the centres.

c Calculate:

i the sum of the radii, $r_A + r_B$

ii the difference between the radii, $r_B - r_A$ [ignoring signs].

d Using your answers to **b** and **c**, prove that the two circles intersect at two places.

e Find the points of intersection.

4 Find the points of intersection between each pair of circles. [You may assume they do intersect.]

a $x^2 + y^2 - 4x - 12y + 30 = 0$ and $x^2 + y^2 + 6x - 2y - 30 = 0$
b $x^2 + y^2 - 6x + 8y - 25 = 0$ and $x^2 + y^2 + 4x - 12y + 15 = 0$
c $x^2 + y^2 - 12x + 2y + 17 = 0$ and $x^2 + y^2 - 10x + 15 = 0$
d $x^2 + y^2 - 16x + 4y + 43 = 0$ and $x^2 + y^2 - 20x + 6y + 59 = 0$

5 Two lines passing through the origin are tangents to the circle $x^2 + y^2 - 4x - 8y + 4 = 0$. Their equations are therefore of the form $y = mx$.

a i Substitute mx for y in the equation of the circle.
ii Rearrange the resultant equation into the general form of a quadratic.

b i Find the discriminant of the quadratic equation.
ii Equate it to zero to find the values of m, which will mean the lines are tangents to the circle.
iii State the equations of the two lines that pass through the origin and are tangents to the given circle.

c Repeat this for the circle $x^2 + y^2 - 4x - 8y + 10 = 0$.

6 The line $y = x + c$ is a tangent to the circle $x^2 + y^2 + 2x + 6y + 8 = 0$.

a i Substitute $x + c$ for y in the equation of the circle.
ii Rearrange the resultant equation into the general form of a quadratic.

b i Find the discriminant of the quadratic equation.
ii Equate it to zero to find the values of c, which will mean the lines are tangents to the circle.
iii Calculate the points of tangency.

c Repeat this for the circle $x^2 + y^2 + 2x + 6y + 2 = 0$.

7 A stone is dropped in a pond and ripples form concentric circles moving out from where it entered the water.

The bank of the pond is modelled by $y = \frac{3}{4}x + c$.

When the ripple touches the bank, the ripple has equation $x^2 + y^2 = 64$.

What are the possible values for c?

8 Two circles have equations $x^2 + y^2 = 100$ and $x^2 + y^2 - 6x - 8y = 0$.

a Calculate the centre and radius of each.

b How far apart are their centres?

c Show that the two circles touch at one point.

d Find this point of contact.

9 The line $y = 3 - 2x$ intersects the circle $x^2 + y^2 + 9x + 6y - 47 = 0$.

a Find the points of intersection.

b What is the length of the chord between these points?

10 At the control tower the air traffic control keeps track of planes moving in and out of their airspace using radar.

Let the range of the radar be modelled by the equation $x^2 + y^2 - 4x - 10y + 4 = 0$ using suitable axes and units.

The tower is modelled by a point at the centre of the circle.

a Give the centre and radius of the 'range' of the radar.

b A plane flies across the airspace on a path modelled by $y = 3x + 4$. At what points does the plane enter and leave the airspace?

c What was the closest the plane got to the tower?

Preparation for assessment

1 Find the equation of the line that passes through (0, 5) and makes an angle of 60° with the positive x-direction.

2 Find the acute angle between the lines $y = 2x + 3$ and $y = 1 - 3x$.

3 A line passes through A(2, −2) and B(−6, 10). Show that the point C(50, −74) is collinear with A and B.

4 By considering gradients, show that the triangle with vertices A(2, −5), B(−4, −1) and C(10, 7) is right-angled.

5 A triangle has vertices E(13, −2), F(−1, 6) and G(−4, −3).

a Find the equation of the altitude through E.

b What is the equation of the median that passes through G?

c Find the point where the median and altitude intersect.

6 Three villages need a communal aerial for terrestrial TV reception. With reference to a suitable set of axes, they are at positions Arddhu, A(10, 10), Bencru, B(10, −4) and Cairns, C(−2, 4).

a Considering triangle ABC,

i find the perpendicular bisectors of AB and AC

ii find the circumcentre (the point where the perpendicular bisectors intersect)

iii hence find the equation of the circle which passes through all three villages.

b Why is the circumcentre a suitable place to site the aerial?

c An inspector walks from Bencru to Cairns in a straight line.

i What is the equation of the straight line?

ii How close does he get to the site of the aerial?

7 A circle has an equation $x^2 + y^2 = 36$.

a Show that the line $4x + 3y = 30$ is a tangent to the circle.

b What is the equation of the radius to the point of tangency?

8 The circle $(x-4)^2+(y-2)^2=40$ cuts the line $y=-\frac{1}{2}x+9$ at A and B.

a Find the coordinates of A and B.

b Find the equations of the tangents to the circles at A and B.

c Find the points of intersection of the two tangents.

9 A circle passes through the origin cutting the axes at (10, 0) and (0, 24).

a Find the equation of the circle, expressing your answer in the form $x^2+y^2+2gx+2fy+c=0$.

b i Find where the line $y=x$ cuts the circle.

ii Find the equation of the tangents to the circle at these points.

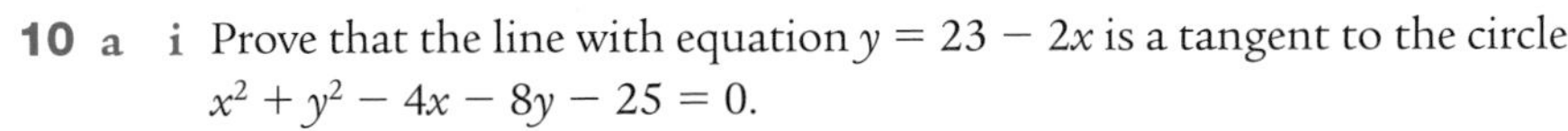

10 a i Prove that the line with equation $y=23-2x$ is a tangent to the circle $x^2+y^2-4x-8y-25=0$.

ii Calculate the coordinates of the point of tangency.

b A second circle is drawn inside the first, touching it so that the line $y=23-2x$ is a tangent to it also.
The second circle has a radius that is a third of the first circle.
Find the equation of the second circle.

11 Find where the circle $(x-3)^2+(y-1)^2=20$ cuts the circle $(x-12)^2+(y-4)^2=50$.

Summary

1 The gradient of a line, m, is equal to the tangent of the angle the line makes with the positive x-direction, i.e. $m=\tan\theta°$.

2 Parallel lines have the same gradient and ... lines with the same gradient are parallel.

3 Given the gradients of two lines, we can deduce the angle, $a°$ at which the lines intersect.
In terms of the gradients: $a°=\tan^{-1}m_1-\tan^{-1}m_2$. [Adjusted if an angle is obtuse.]

4 If lines are perpendicular, then $m_1m_2=-1$.

5 The following facts are most useful for solving problems.

- $m_{AB}=\dfrac{y_b-y_a}{x_b-x_a}$ — The gradient of AB.
- $m_{\perp}=-\dfrac{1}{m_{AB}}$ — The gradient perpendicular to AB.
- $\left(\dfrac{x_b+x_a}{2},\dfrac{y_b+y_a}{2}\right)$ — The midpoint of AB.
- $y-b=m(x-a)$ — Equation of line, gradient m passing through (a, b).
- $mx+c=ax+b$ — Lines $y=mx+c$ and $y=ax+b$ intersect. [Solve for x.]
- $AB=\sqrt{(x_b-x_a)^2+(y_b-y_a)^2}$ — The distance AB.

6 A circle centre the origin and radius r has equation $x^2+y^2=r^2$.
For a point $Q(x_q, y_q)$,

- if $x_q^{\ 2}+y_q^{\ 2}<r^2$, then Q lies inside the circle
- if $x_q+y_q^{\ 2}>r^2$, then Q lies outside the circle
- if $x_q^{\ 2}+y_q^{\ 2}=r^2$, then Q lies on the circle.

7 The circle, radius r units, whose centre is C(a, b) has equation $(x - a)^2 + (y - b)^2 = r^2$.
For a point Q(x_q, y_q),

- if $(x_q - a)^2 + (y_q - b)^2 < r^2$, then Q lies inside the circle
- if $(x_q - a)^2 + (y_q - b)^2 > r^2$, then Q lies outside the circle
- if $(x_q - a)^2 + (y_q - b)^2 = r^2$, then Q lies on the circle.

8 As long as $g^2 + f^2 - c > 0$, $x^2 + y^2 + 2gx + 2fy + c = 0$ is a circle with centre $(-g, -f)$ and radius $\sqrt{g^2 + f^2 - c}$.

9 By considering the distance between centres, d units, and the lengths of the radii, r_1 and r_2, we can determine whether the circles intersect or not.

- $d > r_1 + r_2$ ... no intersections ... circles apart.
- $d = r_1 + r_2$ ... kissing circles ... smaller circle outside.
- $r_1 - r_2 < d < r_1 + r_2$ ... two intersections.
- $d = r_1 - r_2$ ... kissing circles ... smaller circle inside.
- $d < r_1 - r_2$ ... no intersections ... smaller circle inside.

13 Recurrence relations

Before we start...

A **sequence** is an ordered list of numbers where a pattern can be ascribed ... e.g. 1, 3, 5, 7, 9, ...

A **series** or **progression** is a sum, whose terms form a sequence ... e.g. $1 + 3 + 5 + 7 + 9 + \ldots$

Problems involving sequences and series have been found in ancient texts such as the Rhind Papyrus, 1650 BC.

They were also included by Fibonacci in his book *Liber Abaci* in AD 1202 ... a problem on the breeding of rabbits led to the now famous Fibonacci sequence ... 1, 1, 2, 3, 5, 8, 13, ...

When studying sequences we always run the risk of ascribing patterns that just aren't there.

A student worked in a pizza shop. Using a blank pizza base, he put an olive on the circumference.

He then decided to add more olives, placing them one at a time on the circumference.

As he added an olive, he made a cut from that olive to every other already there.

He always made sure to detour to avoid going through any existing intersection.

He then counted the pieces of pizza he had made.

Here we see the state of the pizza after 1, 2, 3, 4, and 5 olives have been placed.

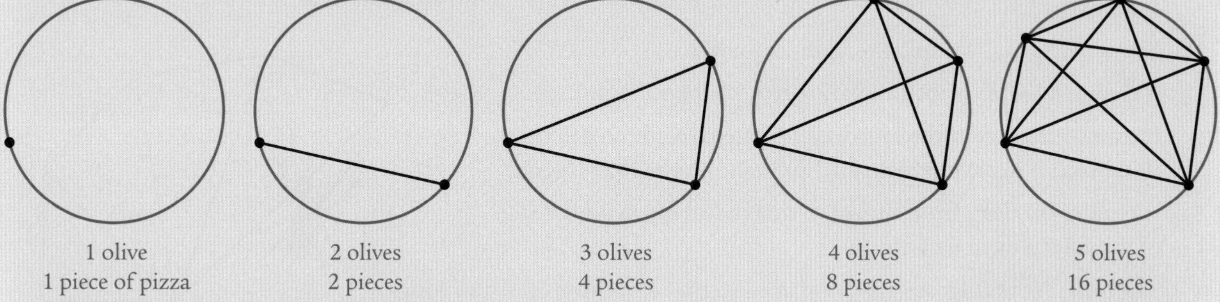

1 olive	2 olives	3 olives	4 olives	5 olives
1 piece of pizza	2 pieces	4 pieces	8 pieces	16 pieces

A sequence 1, 2, 4, 8, 16, ... is generated.

Perhaps you would like to hazard a guess as to the number of pieces obtained when there are 6 olives?

Draw it out and try it.

A warning: no matter how many cases seem to support a theory, the next case might just knock it down.

This chapter deals with sequences created when each term is generated from the term before it by a simple rule. Such a rule is called a **recurrence relation**.

If you have tackled the spreadsheet exercises in this book then you will have encountered recurrence relations already.

For example,

- In A1 type: 1
- In A2 type: =A1+3 ... add 3 to the previous term.
  Fill down as far as you like to generate the sequence 1, 4, 7, 10, ..., ... Does the start value matter? [Change the contents in A1 to 5 ... and without changing the rule, you generate 5, 8, 11, 14, ... *Same rule + different starting value = different sequence ... in this case!*]
- In A2 type: =3*A1+1 ... and fill down ... see how fast the terms grow ... does the start value matter?
- In A2 type: =0.5*A1+1 ... and fill down ... and something strange happens ... does the start value matter?

Unit 3: Applications

What you need to know

1 a A function is defined by $f(x) = 3x - 1$.
Find: i $f(4)$ ii $f(0)$ iii $f(-1)$ iv $f(f(0))$.

b A function is defined by $f(x) = 10 - x$.
Find: i $f(2)$ ii $f(0)$ iii $f(-2)$ iv $f(f(1))$.

2 A function is defined by $f(x) = x + 1$.

a Evaluate: i $f(1)$ ii $f(f(1))$ iii $f(f(f(1)))$ iv $f(f(f(f(1))))$.

b Comment on the set of answers.

3 a The nth term of a sequence can be calculated from the formula $u_n = 2n + 1$. List the first 5 terms.

b In each case below, the nth term is given.
List the first four terms. Say by what name the list is normally known.
i $u_n = 2n - 1$ ii $u_n = 2n$ iii $u_n = \frac{1}{2}n(n + 1)$

4 a $f(x) = x^3$. To what does $\dfrac{f(x + h) - f(x)}{h}$ tend as h tends to zero?

b $g(x) = \dfrac{1}{x}$. To what does $g(x)$ tend as x gets very large, i.e. as x tends to infinity?

c $h(x) = a^x$. If a is a proper fraction, i.e. $-1 < a < 1$, to what does $h(x)$ tend as x gets very large?

5 A very mathematical spider is spinning a web.
It creates a lot of threads radiating from a central point.
It starts 1 unit from the centre and jumps at right angles from the thread it is leaving to an anchor point on the next thread.
It repeats this leap from thread to thread leaving behind a trail of threads to form its web.
The initial radiating threads are spaced so that each jump is 1 unit long.
Explore the sequence created when considering the spider's distance from the centre of the web at each anchor point.

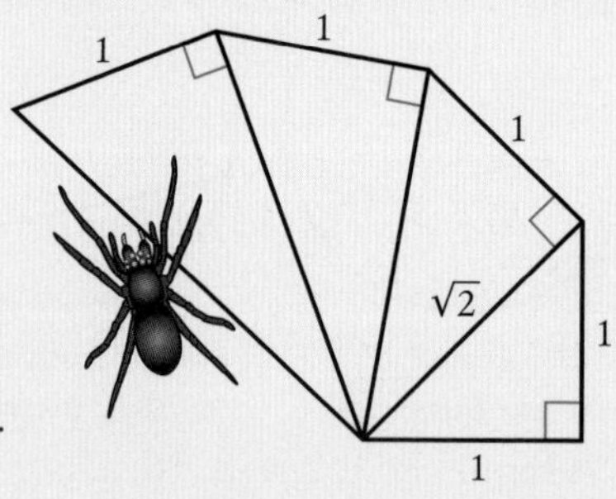

13.1 Recurrence relations 1: $u_{n+1} = u_n + b$

A sequence, say the odd numbers, can be described in various ways.

1 By referring to a well-known set of numbers, e.g. 'the odd numbers'.

2 By giving the type of the relationship and the first few terms, e.g. the linear[1] sequence 1, 3, 5, ...

3 By giving the nth term as a function of n, e.g. $u_n = 2n - 1$ ... note that by definition u_1 is the first term.

4 By giving the first term, u_1, and expressing the $(n + 1)$th term as a function of the nth term, e.g. $u_{n+1} = u_n + 2,\ u_1 = 1$.

[1] A *linear* sequence has an nth term of the form $an + b$. Knowing the 1st and 2nd terms allows us to use simultaneous equations to find a and b.

So the instructions to generate this particular sequence might be read, 'Starting with 1, keep adding 2 ..." giving 1, $1 + 2 = 3$, $3 + 2 = 5$, $5 + 2 = 7$, $7 + 2 = ...$

Warning

You can't just state, 'the sequence 1, 3, 5, ...' as this could be generated by an infinite number of formulae, although they would disagree as to what was the 4th term. (Remember in Chapter 6 we found that $u_n = x^3 - 6x^2 + 13x - 7$ generated 1, 3, 5 as the first three terms of a sequence.)

The argument applies no matter how many initial terms are given ... the pizza problem confirms this.

Our focus in this chapter is on the fourth type of description given above ... referred to as a **recurrence relation**.

Example 1

A sequence is described by the recurrence relation $u_{n+1} = u_n + 5,\ u_1 = 3$.

a List the first four terms in the sequence.

b Calculate: **i** $u_2 - u_1$ **ii** $u_4 - u_3$ **iii** $u_{n+1} = u_n$.

c A related sequence is described by $v_{n+1} = v_n + 5,\ v_1 = 1$.
By how much does: **i** v_3 differ from u_3 **ii** v_n differ from u_n?

a $u_1 = 3$ ... given
$u_2 = u_1 + 5 = 3 + 5 = 8$ ... from recurrence relation
$u_3 = u_2 + 5 = 8 + 5 = 13$; $u_4 = u_3 + 5 = 13 + 5 = 18$.
The sequence starts: 3, 8, 13, 18.

b **i** $u_2 - u_1 = 8 - 3 = 5$ **ii** $u_4 - u_3 = 18 - 13 = 5$ **iii** $u_{n+1} - u_n = 5$
For this reason, in the recurrence relation $u_{n+1} = u_n + b$, b is often referred to as the **common difference**.

c The second sequence starts 1, 6, 11, 16.
i $v_3 - u_3 = 11 - 13 = -2$ **ii** $v_n - u_n = v_1 - u_1 = 1 - 3 = -2$

Example 2

A sequence is described by the recurrence relation $u_{n+1} = u_n + b,\ u_1 = a$.
$u_2 = 5$ and $u_4 = 11$. What are the values of a and b?

Given that $u_{n+1} = u_n + b$, and that $u_1 = a$, the sequence begins: a, $a + b$, $a + b + b$, $a + b + b + b$, i.e. a, $a + b$, $a + 2b$, $a + 3b$, ...

$u_2 = 5$ gives us $\quad a + b = 5 \quad$ ... ①

$u_4 = 11$ gives us $\quad a + 3b = 11 \quad$ ... ②

② − ①: $\quad 2b = 6$

$\Rightarrow b = 3$

Substitute in ① $\quad \Rightarrow a = 2$

Thus the recurrence relation is $u_{n+1} = u_n + 3,\ u_1 = 2$.

Example 3

A man bought a computer on hire purchase. The amount he owes the company at the **start** of month n, £u_n, is modelled by the recurrence relation $u_{n+1} = u_n - 22,\ u_1 = 528$.

a What is: **i** $u_2 - u_1$ **ii** $u_3 - u_2$?

b What is the rate of repayment?

c What does he owe at the start of month 20?

d When will the repayment be complete?

a i $u_2 - u_1 = -22$ **ii** $u_3 - u_2 = -22$

b The man is repaying the debt at a rate of £22 per month.

c Look for a formula for the nth term ...
When $n = 1$ ... he has made no payments and owes £528.
When $n = 2$ ... he has made 1 payment and owes £$528 - 1 \times 22$.
When $n = 3$ ... he has made 2 payments and owes £$528 - 2 \times 22$.
When $n = 4$ ... he has made 3 payments and owes £$528 - 3 \times 22$.
When $n = n$ ... he has made $(n - 1)$ payments and owes £$528 - (n - 1) \times 22$.
Thus the nth term can be expressed as $u_n = 528 - 22(n - 1)$.
When $n = 20$, $u_n = 528 - 22(20 - 1) = 528 - 418 = 110$,
i.e. at the start of month 20, he still owes £110.

For this type of problem, a **formula** for the nth term, rather than the **recurrence relation** is more useful. We should note that this type of recurrence relation leads to a linear relationship between u_n and n, i.e. $u_n = pn + q$, where p and q are constants.

d The repayment will be complete when the amount owed is zero.
Using the nth term formula, $528 - 22(n - 1) = 0 \Rightarrow n - 1 = 528 \div 22 = 24 \Rightarrow n = 25$.
Thus he has nothing to pay at the start of the 25th month ... or the repayment is complete at the end of the 24th month.

Exercise 13.1

1 **i** List the first four terms of each sequence described by the recurrence relation.
ii Find a formula for the nth term in terms of n.
iii Use the formula to find the 100th term in the sequence.

a $u_{n+1} = u_n + 4,\ u_1 = 6$ **b** $u_{n+1} = u_n - 3,\ u_1 = 1$ **c** $u_{n+1} = u_n + 2,\ u_1 = -4$

d $u_{n+1} = u_n - 5,\ u_1 = -2$ **e** $u_{n+1} = u_n + 1{\cdot}5,\ u_1 = 0{\cdot}5$ **f** $u_{n+1} = u_n + b,\ u_1 = a$

2 Each of the following sequences has been generated from a recurrence relation of the form $u_{n+1} = u_n + b,\ u_1 = a$. For each sequence, establish the values of a and b.

a 2, 7, 12, ... **b** −4, 4, 12, ... **c** 7, 5, 3, ... **d** −1, −5, −9, ...

e $\frac{1}{2}, 1, \frac{3}{2}, 2, \ldots$ **f** $\frac{1}{2}, \frac{5}{6}, \frac{7}{6}, \frac{3}{2}, \ldots$ **g** $x, 2x, 3x, \ldots$

h $x, 2x - 1, 3x - 2, 4x - 3, \ldots$ **i** $1 - x, 3, x + 5, \ldots$

3 The following are the nth terms of sequences.
Express each sequence in terms of a recurrence relation.

a $u_n = 2n + 4$ **b** $u_n = 5n - 1$ **c** $u_n = 3 - n$ **d** $u_n = \frac{1}{2}n + 6$

e $u_n = 5 - 3n$ **f** $u_n = 6 - \frac{1}{3}n$ **g** $u_n = 3n + b$ **h** $u_n = an + b$

4 Sequences have been generated by a recurrence relation of the form $u_{n+1} = u_n + b,\ u_1 = a$.
For each sequence a couple of terms have been given.
Use this information to find the values of a and b in each and state the recurrence relation.

a $u_2 = 4$ and $u_4 = 10$ **b** $u_3 = 6$ and $u_5 = 14$ **c** $u_2 = 1$ and $u_3 = -1$

d $u_2 = 1$ and $u_4 = \frac{5}{2}$ **e** $u_2 = 2x + 1$ and $u_4 = 4x + 3$ **f** $u_2 = 2x + 1$ and $u_5 = 5x + 1$

5 A lorry carries a fixed volume of aggregate from the depot to the construction site.
The volume of aggregate, u m^3, at the depot **after** the nth trip can be modelled by the recurrence relation $u_{n+1} = u_n - 35,\ u_1 = 840$.

a What volume of aggregate was at the depot initially? [Hint: $u_1 = u_0 - 35$ and you want u_0.]

b What is the capacity of the lorry? [How much will it carry in one trip?]

c **i** How much will it have transported after 15 trips?
ii How much is left at the depot?

d **i** Write a recurrence relation to model the amount of aggregate at the **site** after the nth trip.
ii How many trips are needed before there is more than 500 m^3 at the site?

e How many trips are required to complete the job?

6 At the retail outlet, the queue for the latest model of phone was very long. It was so long that they had to close the queue when it reached a certain size.
The manager plans his day on the assumption that each assistant works at the same rate.
If he puts six assistants at the sales counter he can model the size of the queue (in people) after the nth hour of serving by $Q_{n+1} = Q_n - 90,\ Q_1 = q$. On the other hand, if he uses 10 assistants for sales he can model the queue length by $Q_{n+1} = Q_n - 150,\ Q_1 = q$.

a In each case, by how much does the queue size drop after one hour?

b At what rate is the manager assuming his assistants are working?

c Check that when six assistants are working the number of people in the queue after n hours is given by $Q_n = q - 90n$. Give a similar formula for when there are 10 assistants working.

d If the outlet is only open for 10 hours, what size can the queue be when the following number of assistants are on duty:
i 6 **ii** 10?

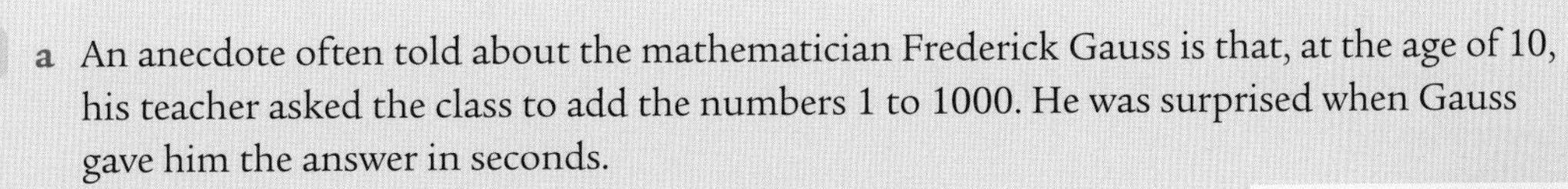

7 **a** An anecdote often told about the mathematician Frederick Gauss is that, at the age of 10, his teacher asked the class to add the numbers 1 to 1000. He was surprised when Gauss gave him the answer in seconds.

Gauss had come up with a shortcut to finding the sum to n terms of a recurrence relation of the form:

$u_{n+1} = u_n + b,\ u_1 = a$.

He did it when $a = 1$ and $b = 1$.

His method was to consider the sum,

$S = 1 + 2 + 3 + 4 + \ldots + 997 + 998 + 999 + 1000$

... and then consider it backwards

$S = 1000 + 999 + 998 + 997 + \ldots + 4 + 3 + 2 + 1$.

Adding the two sums he got:

$2S = 1001 + 1001 + 1001 + \ldots + 1001$
$\quad = 1000 \times 1001 = 1\,001\,000$

So $S = 1\,001\,000 \div 2 = 500\,500$.

Can you adapt Gauss' method to find the sum of the following:

i $1 + 3 + 5 + 7 + ... + 95 + 97 + 99$ [Hint: find the value of n first.]

ii $5 + 10 + 15 + ... + 95 + 100$

iii $a + (a + b) + (a + 2b) + ... + [a + (n - 2)b] + [a + (n - 1)b]$?

b Such sums can also be explored using a spreadsheet.

Say you wish explore the sum of the sequence derived from the recurrence relation $u_{n+1} = u_n + 4,\ u_1 = 7,$

In A1 type: 7 ... this is the first term of the sequence, a.

In A2 type: =A1+4 ... this is the recurrence formula.

Fill down to generate the sequence.

In B1 type: =A1 ... the sum to one term is the first term.

In B2 type: =B1+A2 ... adding the 2nd term to the total so far.

Fill down to generate the sums to as many terms as you need.

Can you devise a formula for the sum to n terms using a, b and n?

13.2 Recurrence relations 2: $u_{n+1} = au_n$

In many contexts including banking interest, inflation, radioactive decay, cooling curves, the spread of disease, the growth of a population, for example, sequences are generated that can be described by a recurrence relation of the form $u_{n+1} = au_n$.

Example 1

A sequence is described by the recurrence relation $u_{n+1} = 3u_n,\ u_1 = 2$.

a List the first four terms in the sequence.

b Calculate: **i** $u_2 \div u_1$ **ii** $u_4 \div u_3$ **iii** $u_{n+1} \div u_n$.

c A related sequence is described by $v_{n+1} = 3v_n,\ v_1 = 4$.

i What is v_3 divided by u_3?

ii Explain your answer.

a $u_1 = 2$ (given); $u_2 = 3u_1 = 3 \times 2 = 6$ (from the given definition);
$u_3 = 3u_2 = 3 \times 6 = 18$; $u_4 = 3u_3 = 3 \times 18 = 54$.
The first four terms are 2, 6, 18, 54.

b **i** $u_2 \div u_1 = 6 \div 2 = 3$ **ii** $u_4 \div u_3 = 54 \div 18 = 3$ **iii** $u_{n+1} \div u_n = 3$

For this reason, in the recurrence relation $u_{n+1} = au_n$, a is often referred to as the **common ratio**.

c **i** The sequence starts 4, 12, 36, 108.
So $v_3 \div u_3 = 36 \div 18 = 2$.

ii In general, if the first term is p then the 3rd term will be $9p$; if the first term is q then the 3rd term will be $9q$.

The division is $9p \div 9q = \dfrac{p}{q}$.

Example 2

A sequence is described by the recurrence relation $u_{n+1} = au_n,\ u_1 = b$.
$u_2 = 15$ and $u_4 = 375$. What are the values of a and b?

Since $u_{n+1} = au_n$ and $u_1 = b$, the sequence starts $b, ab, a^2b, a^3b, \ldots$

$$u_2 = 15 \Rightarrow ab = 15 \qquad \ldots ①$$

$$u_4 = 375 \Rightarrow a^3b = 375 \qquad \ldots ②$$

Divide ② by ①: $\dfrac{a^3b}{ab} = \dfrac{375}{15} \Rightarrow a^2 = 25 \Rightarrow a = \pm 5$

Substituting into ①: If $a = 5$ then $b = 3$. If $a = -5$ then $b = -3$.

So there are two different recurrence relations that fit the description,
viz. $u_{n+1} = 5u_n$ and $u_1 = 3$ and $u_{n+1} = -5u_n$ and $u_1 = -3$.

Example 3

A virus hits the school. The number of people that have been hit by it n days after it started, u_n, is modelled by the recurrence relation $u_{n+1} = 1{\cdot}2u_n,\ u_1 = 10$.

a What is $u_2 \div u_1$?

b How many will be affected by day 10?

c The first few terms can be expressed as, $10, 1{\cdot}2 \times 10, 1{\cdot}2^2 \times 10, 1{\cdot}2^3 \times 10, \ldots$
How might you express the nth term?

d When will the number affected go above 300 people?

a 1·2

b $u_{10} = 10 \times 1{\cdot}2 \times 1{\cdot}2 \times 1{\cdot}2 \times 1{\cdot}2 \times 1{\cdot}2 \times 1{\cdot}2 \times 1{\cdot}2 \times 1{\cdot}2 \times 1{\cdot}2 = 10 \times 1{\cdot}2^9 = 52$ cases (to nearest person).

c $u_n = 10 \times 1{\cdot}2^{n-1}$

d $10 \times 1{\cdot}2^{n-1} = 300 \Rightarrow 1{\cdot}2^{n-1} = 30$ ... we'll need to use logarithms to solve this

Taking the log of both sides:

$$\ln(1{\cdot}2^{n-1}) = \ln 30$$

$$\Rightarrow (n - 1)\ln 1{\cdot}2 = \ln 30$$

$$\Rightarrow \quad n - 1 = \frac{\ln 30}{\ln 1{\cdot}2} = 18{\cdot}7$$

$$\Rightarrow \quad n = 19{\cdot}7$$

In the context, we are talking about day 20 before the number of cases gets to 300.

Again a **formula** for the nth term, rather than the **recurrence relation** is more useful.

Note that this type of recurrence relation leads to an exponential relationship between u_n and n, viz. $u_n = q \times p^{n-1}$, where p and q are constants.

Exercise 13.2

1 Find the first four terms of the following recurrence relations.

a $u_{n+1} = 3u_n,\ u_1 = 2$ **b** $u_{n+1} = 4u_n,\ u_1 = 1{\cdot}5$ **c** $u_{n+1} = \frac{1}{3}u_n,\ u_1 = 486$

d $u_{n+1} = \frac{1}{2}u_n,\ u_1 = 96$ **e** $u_{n+1} = 1{\cdot}1u_n,\ u_1 = 1$ **f** $u_{n+1} = \pi u_n,\ u_1 = \pi$

2 Each of these sequences has been generated using a recurrence relation of the type $u_{n+1} = au_n,\ u_1 = b$.

i Find the recurrence relation.

ii Express the nth term as a function of n.

iii Calculate the 10th term.

a 5, 15, 45, 135, ... **b** 7, 14, 28, 56, ... **c** 2, 8, 32, 128, ...

d 160, 80, 40, 20, ... **e** 6144, 1536, 384, 96, ... **f** 3, −3, 3, −3, ...

g 2, −6, 18, −54, ... **h** 10, 1, 0·1, 0·01, ... **i** a, ab, a^2b, a^3b ...

3 Sequences have been generated by a recurrence relation of the form $u_{n+1} = au_n,\ u_1 = b$. For each sequence a couple of terms have been given.
Use this information to find the values of a and b in each and state the recurrence relation.

a $u_2 = 10$ and $u_4 = 250$ **b** $u_3 = 16$ and $u_5 = 64$

c $u_2 = 18$ and $u_5 = 486$ **d** $u_3 = 5$ and $u_4 = 2{\cdot}5$

e $u_2 = 100$ and $u_4 = 1$ **f** $u_2 = 54$ and $u_4 = 19{\cdot}44$

4 Each exponential expression gives the nth term of a sequence.
Express each sequence in terms of a recurrence relation.

a $u_n = 3 \times 1{\cdot}6^n$ **b** $u_n = 2 \times 2{\cdot}8^n$ **c** $u_n = 1{\cdot}6 \times 0{\cdot}5^n$ **d** $u_n = b \times a^n$

5 Sequences generated by a recurrence relation of the form $u_{n+1} = au_n,\ u_1 = b$ have an nth term $u_n = b \times a^{n-1}$.

a Use this information in each case, to suggest to what the nth term is tending as n tends to infinity.

i $u_{n+1} = 4u_n,\ u_1 = 2$ **ii** $u_{n+1} = \frac{1}{2}u_n,\ u_1 = 100$ **iii** $u_{n+1} = 2u_n,\ u_1 = 1$

iv $u_{n+1} = \frac{3}{4}u_n,\ u_1 = 20$ **v** $u_{n+1} = 5u_n,\ u_1 = 3$ **vi** $u_{n+1} = 0{\cdot}1u_n, u_1 = 100\,000$

b For some of the sequences, the terms tend to infinity and for others they tend to zero. Can you identify what condition can be used to identify the two types?

6 Magpies have been introduced on to an island that originally had none. The growth in their population has been modelled by $M_{n+1} = 1{\cdot}05M_n,\ M_1 = 21$ where M_n is the number of magpies at the **end** of year n, and $n = 0$ at the moment of their introduction to the island.

a How many magpies were there initially? [$n = 0$]

b How many magpies will there be when $n = 11$?

c How long will it be before the population doubles to 40 birds?

7 Karen bought a car for £25 000. She is told that, because of depreciation, the value of her car can be modelled by $v_{n+1} = 0{\cdot}85v_n,\ v_0 = 25\,000$, where £$v_n$ is the value of the car at the end of the nth year of ownership.

a Calculate the value of her car at the end of each of the first 4 years (to the nearest pound).

b Calculate the value of her car after 10 years.

c When did the value first drop below £2000?

d The annual insurance premium on the car is $\frac{1}{100}$th of the value of the car.
Express the premium in year n as a recurrence relation.

e What will happen in the long run to the value of the car?

8 When Rory was born £10 000 was put into a saving scheme that he could use when he was 18 years' old.

The growth of the 'pot' of money was modelled by the recurrence relation $p_{n+1} = 1{\cdot}04p_n,\ p_0 = 10\,000$.

a What is the 'pot' worth on Rory's 5th birthday?

b What will it be worth when it matures on his 18th birthday?

c When will it first exceed £16 000?

9 Water lies on the flat roof of a public building. It will naturally evaporate according to the model $w_{n+1} = 0{\cdot}6w_n,\ w_0 = 10$, where w_n cubic metres are present at the end of day n.

a How much water is on the roof initially?

b How much is there after 9 days?

c When the amount of water goes below $0{\cdot}01\ \text{m}^3$, they consider the roof basically dry. After how many days will this happen?

d The longer water sits on the roof, the greater the chance of damage.
To increase the rate of evaporation, pebbles are scattered over the roof. This changes the model to $w_{n+1} = 0{\cdot}3w_n,\ w_0 = 10$.
Using the new model, how long will it be before the roof is effectively dry?

10 In number theory, the number 0·9 **recurring**, i.e. 0·99999999 ...., is **equal** to 1.

The argument goes like this:

$N = 0{\cdot}999999\ldots = 0{\cdot}9 + 0{\cdot}09999\ldots$

$\Rightarrow N = 0{\cdot}9 + \frac{1}{10}(0{\cdot}9999\ldots) = 0{\cdot}9 + \frac{1}{10}N$

$\Rightarrow \quad 10N = 9 + N$

$\Rightarrow \quad 9N = 9$

$\Rightarrow \quad N = 1$

a Use the same argument to show that 0·3 recurring is equal to $\frac{1}{3}$.

b Show that 0·1 recurring is a rational number.

c Adapt the argument to find a fraction equal to 0·1212121...

d The recurrence relation $u_{n+1} = 0{\cdot}1u_n,\ u_1 = 1$ will generate the series $S = 1 + 0{\cdot}1 + 0{\cdot}1^2 + 0{\cdot}1^3 + \ldots$

We can use an argument similar to the above to find an expression for S.

It starts: $S = 1 + 0{\cdot}1 + 0{\cdot}1^2 + 0{\cdot}1^3 + \ldots = 1 + 0{\cdot}1(1 + 0{\cdot}1 + 0{\cdot}1^2 + 0{\cdot}1^3\ldots)$.

Continue the argument to find a value for S.

Note that this argument only makes sense when the common ratio lies between 1 and -1. The terms you are adding get progressively smaller.

e Repeat the argument with $S = 1 + r + r^2 + r^3 + \ldots = 1 + r(1 + r + r^2 + r^3\ldots)$, where $-1 < r < 1$, to find an expression for $S = 1 + r + r^2 + r^3 + \ldots$ in terms of r.

f Check your formula works using $r = 0{\cdot}1, 0{\cdot}9$, and $\frac{1}{3}$.

13.3 Recurrence relations 3: $u_{n+1} = au_n + b$

In many situations, the context can involve both an addition and a multiplication as the next term is generated, i.e. the recurrence relation will take the form $u_{n+1} = au_n + b,\ u_1 = c$.

The nth term

With a recurrence relation of the type $u_{n+1} = u_n + b,\ u_1 = a$, we saw that $u_n = a + (n-1)b$.
With a recurrence relation of the type $u_{n+1} = bu_n,\ u_1 = a$, we saw that $u_n = ab^{n-1}$.
Finding the *n*th term of this third type of recurrence relation is beyond the scope of the course but knowing it will help explain a phenomenon that is part of the required course.

The limit when $-1 < a < 1$

If you did part **e** of Question 10 of Exercise 13.2, you will have discovered that:

$1 + r + r^2 + r^3 + r^4 + \ldots = \frac{1}{1-r},\ -1 < r < 1$ ... this condition is important

Consider the relation $u_{n+1} = au_n + b,\ u_1 = c$, when $-1 < a < 1$.

We generate the first few terms:

$c,\ ca + b,\ ca^2 + ba + b,\ ca^3 + ba^2 + ba + b,\ \ldots,\ ca^{n-1} + ba^{n-2} + ba^{n-3} + \ldots + ba^2 + ba + b$

So $u_n = ca^{n-1} + ba^{n-2} + ba^{n-3} + \ldots + ba^2 + ba + b$

$\Rightarrow u_n = ca^{n-1} + b(1 + a + a^2 + a^3 + \ldots + a^{n-3} + a^{n-2})$

However, if we were to continue the sum to infinity ...

$\Rightarrow u_\infty = ca^{n-1} + b(1 + a + a^2 + a^3 + \ldots)$

$\Rightarrow u_\infty = ca^{n-1} + \frac{b}{1-a}$

Since $-1 < a < 1$, as *n* tends to infinity, a^{n-1} will tend to zero, giving:

$$u_\infty = \frac{b}{1-a}.$$

As you continue the sequence generated by the recurrence relation

$u_{n+1} = au_n + b,\ u_1 = c,\ -1 < a < 1$, the terms will approach the limit of $u_\infty = \frac{b}{1-a}$.

Note that the limit is independent of the starting value *c*.

Example 1

A sequence is described by the recurrence relation $u_{n+1} = 0{\cdot}5u_n + 16,\ u_1 = 64$.

a List the first four terms in the sequence.

b **i** Explain why a limit exists for this sequence.

ii Find the limit.

a $u_1 = 64$ (given); $u_2 = 0{\cdot}5u_1 + 16 = 0{\cdot}5 \times 64 + 16 = 48$;
$u_3 = 0{\cdot}5u_2 + 16 = 0{\cdot}5 \times 48 + 16 = 40$; $u_4 = 0{\cdot}5u_3 + 16 = 0{\cdot}5 \times 40 + 16 = 36$.
The first four terms are: 64, 48, 40, 36.

b **i** Since the multiplier 0·5 lies between 1 and −1, a limit exists.

ii As *n* approaches infinity, u_{n+1} approaches u_n. Let *L* be the limit.
$u_{n+1} \to u_n$ and $u_{n+1} = 0{\cdot}5u_n + 16$ becomes $L = 0{\cdot}5L + 16$
$\Rightarrow 0{\cdot}5L = 16$
$\Rightarrow\ L = 32$
As *n* approaches infinity, u_n approaches the value 32.[2]

---

[2] If you were to use a spreadsheet with A1 containing 64, A2 containing =0.5*A1+16, a fill-down will show that by the 31st term, the computer's arithmetic can't tell the difference between u_{31} and the value 32.

- This method has the edge over using the formula. Although the formula is easy, it is also easily forgotten.
- The formula and the above method both provide 'answers' when a is not between 1 and -1, so it is essential that this condition is stated when declaring that there is a limit.

Example 2

A sequence is described by the recurrence relation $u_{n+1} = au_n + b,\ u_1 = 3$.
$u_2 = 16$ and $u_4 = 406$. What are the possible values of a and b?

Given $u_1 = 3 \Rightarrow u_2 = 3a + b$ ... thus $3a + b = 16$ ... ①
Given $u_2 = 16 \Rightarrow u_3 = 16a + b \Rightarrow u_4 = 16a^2 + ba + b$ ... thus $16a^2 + ba + b = 406$. ... ②
Substituting $16 - 3a$ for b in ②: $16a^2 + (16 - 3a)a + (16 - 3a) = 406$.
$\Rightarrow 16a^2 + 16a - 3a^2 + 16 - 3a = 406$
$\Rightarrow 13a^2 + 13a - 390 = 0$
$\Rightarrow (a - 5)(a + 6) = 0$
$\Rightarrow a = 5$ or -6
$\Rightarrow b = 1$ or 34
Thus the sequence might be 3, 16, 81, 406, 2031, ... or it could be 3, 16, -62, 406, -2402, ...

Example 3

In the country golf course, sheep keep down the grass in the rough.

Their eating habits over a week reduce the height at the start of the week by 80%, i.e. the height is reduced to 20% of what it was.

However, the grass grows 5 cm in a week.

a Let u_n represent the height of the grass in centimetres at the end of week n.
When the sheep were first allowed in, the grass was 20 cm high. [$u_0 = 20$.]
Form a recurrence relation of the form $u_{n+1} = au_n + b,\ u_1 = c$ to model the situation.

b Explain what will happen long-term if the sheep are allowed to graze.

a The grass will be reduced to 0·2 of its height.
So $u_{n+1} = 0{\cdot}2u_n + 5,\ u_0 = 20$.

b Since the multiplier, 0·2, lies between -1 and 1 a limit exists. Call it L cm.
$L = 0{\cdot}2L + 5 \Rightarrow 0{\cdot}8L = 5 \Rightarrow L = 6{\cdot}25$.
In the long-term, the grass in the rough will maintain a height of 6·25 cm.

Caveat

Care must be taken when interpreting recurrence relations in a context. The relation really only models 'snap-shots' of a situation ... where n is a whole number. In general, we cannot interpolate.
If we create a model to tell us the number of passengers on a bus as it leaves bus stop number n, the model might tell us there were 5 leaving stop 6 and 12 leaving stop 7 ... but it cannot be used to suggest a value for stop number 6·5!

Exercise 13.3

1 Find the first four terms of the following linear recurrence relations.

a $u_{n+1} = 2u_n - 10,\ u_1 = 6$
b $u_{n+1} = 3u_n + 1,\ u_1 = 5$
c $u_{n+1} = 4u_n - 6,\ u_1 = 2$
d $u_{n+1} = 0{\cdot}7u_n + 1,\ u_1 = 10$
e $u_{n+1} = 0{\cdot}9u_n + 8,\ u_1 = 300$
f $u_{n+1} = \frac{3}{4}u_n + 16,\ u_1 = 256$

2 Find the linear recurrence relation that would generate, as the first three terms,

a 5, 11, 23
b 7, 27, 87
c 4, 22, 112
d 64, 34, 19
e 360, 230, 152
f 40, 50, 58.

3 In each case, given the description of the recurrence relation and two specific terms, find the possible recurrence relations.

a $u_{n+1} = au_n + b,\ u_1 = 2, u_2 = 9, u_3 = 37$
b $u_{n+1} = au_n + b,\ u_1 = 3, u_2 = 17, u_4 = 437$
c $u_{n+1} = au_n + b,\ u_1 = 6, u_2 = 18, u_4 = 27$
d $u_{n+1} = au_n + 4,\ u_1 = 9, u_2 = 22, u_3 = 48$
e $u_{n+1} = au_n - 1,\ u_2 = 2, u_4 = 14$
f $u_{n+1} = 4u_n + b,\ u_2 = 6, u_3 = 78$
g $u_{n+1} = \frac{1}{2}u_n + b,\ u_2 = 8, u_4 = -1$
h $u_{n+1} = au_n + b,\ u_1 = 60, u_2 = 10, u_4 = -2$

4 In each case a limit exists. Find it.

a $u_{n+1} = \frac{1}{5}u_n - 2,\ u_1 = 3$
b $u_{n+1} = \frac{1}{3}u_n + 3,\ u_1 = 4$
c $u_{n+1} = \frac{3}{4}u_n + 1,\ u_1 = 1$
d $u_{n+1} = 0{\cdot}1u_n + 9,\ u_1 = 100$
e $u_{n+1} = 12 - \frac{1}{2}u_n,\ u_1 = -3$
f $u_{n+1} = 17 - 0{\cdot}7u_n,\ u_1 = c$

5 A satellite orbits the Earth. Using suitable units, its height at the completion of orbit n is given by $h_{n+1} = \dfrac{8h_n + 35}{10}$.
Initially it is 20 units up, but because of orbital decay it is losing height.

a What is its height at the end of the third orbit?
b If it goes below 17 units it will burn up in the atmosphere.
  i Does the sequence have a limit?
  ii Is it good or bad news for the satellite?

6 A thermostat in a room can be set to maintain a constant temperature.
The insulation in a room allows a certain fraction of the heat to be lost.
The thermostat kicks in regularly and boosts the temperature by 8 °C.
The situation can be modelled by $T_{n+1} = aT_n + 8$.

a What would the value of a have to be if the temperature is to settle at:
  i 20 °C ii 24 °C?
b Express a as a function of the desired temperature, L °C.

7 The nth term of a sequence is given by $u_n = 3^n - 1$.
It can be expressed as a recurrence relation of the form $u_{n+1} = au_n + b,\ u_1 = c$.

a Find the values of a, b and c.
b Repeat this question for the recurrence relation with nth term, $u_n = 4^n - 2$.
c It was thought that the sequence defined by $u_n = 3^n + 1$ could be expressed as a recurrence relation of the form $u_{n+1} = au_n + b,\ u_1 = c$.
  i Use the first three terms of the sequence to find a, b and c.
  ii Show that the fourth term of the sequence also matches the fourth term of the recurrence relation.
  iii Is this enough to prove that the nth term formula and the recurrence relation both generate the same sequence?

8 Money left in a special long-term account in the bank gets 4% interest per annum.
The saver is allowed to add to the account but not withdraw.
One particular saver deposits £10 000 initially and at the end of each year puts another £5000 into the account. Interest is calculated before his deposit is made.

a Write down a recurrence relation to model the amount in the account at the end of each year using £A_n as the amount in the bank at the end of the nth year.

b The bank use a formula $A_n = 135\,000 \times 1{\cdot}04^n - 125\,000$ to calculate the amount in the account at the end of year n. Compare this formula with your recurrence relation by calculating A_4 by both methods.

c Use the bank's formula to help you determine in which year the amount was over £100 000.

9 Graffiti in a section of the town marred 3000 m^2 of wall surface.
In a working day a man can clean 30% of the surface. The cost of this is £100 a day.
Unfortunately vandals can damage another 50 m^2 during the night.

a Let the amount of damage be u_n, where $u_0 = 3000$.
Write down a recurrence relation to model the amount of damage at the start of each working day if:
i 1 man is employed **ii** 2 men are employed **iii** 3 men are employed.

b What are the long-term prospects under each possibility?

c If a man is employed, at the cost of £100 a day, to be a night watchman, the overnight graffiti drops to 5 m^2. What are the long-term prospects now in each scenario?

d By considering the daily cost, which scenario do you think offers the best value for money?

10 As a moth flies past a light source, it tries to keep its flight path at a fixed acute angle of $\theta°$ to the source. At regular intervals it has to adjust its flight path to do this, correcting the angle and pulling 10 mm away from the light. After the nth adjustment, its distance from the source, d_n mm can be modelled by $d_{n+1} = d_n \sin\theta° + 10,\ d_0 = 2000$.

a Say why you know that for $0 < \theta < 90$, this recurrence relation has a limit.

b What is this limit when θ equals: **i** 15 **ii** 30 **iii** 60?

c Find a formula for the limit, L, as a function of θ.

d If it gets closer than 25 mm to the source, the moth will singe its wings.
What is the least value that θ can be set at and still keep the moth safe?

11 In a certain wood the grey squirrels are taking over from the red.
It is desired to take steps to reduce the grey population.

The grey population can be modelled by $G_{n+1} = 0{\cdot}9G_n + 45$,
the red population can be modelled by $R_{n+1} = 0{\cdot}8R_n + 30$,
where n is the year when the count is taken. Initially $G_0 = 500$ and $R_0 = 300$.

a If no steps are taken, what will be the long-term prospects for both populations?

b If the grey : red ratio gets to 3 : 1 then the models will break down and the reds will become extinct in the wood. Is this a possibility?

c A decision is taken to cull a fixed number, h, of the greys each year.
The model for the greys then becomes $G_{n+1} = 0{\cdot}9G_n + 45 - h$.
It is desired to keep this up until the reds outnumber the greys.
What is the minimum number of greys that need to be culled to ensure that the reds outnumber the greys by the 10th year?

12 The strength of the spreadsheet lies in its ability to perform number-crunching tasks. The following tasks are best explored using a spreadsheet.

a Definition: a **fixed point** in a recurring relation is where $u_{n+1} = u_n$.

i Check that each of the following meets this condition:

- $u_{n+1} = 3u_n - 12,$ when $u_1 = 6$
- $u_{n+1} = 0{\cdot}5u_n + 4,$ when $u_1 = 8.$

ii There is something fundamentally different about the nature of the two fixed points in **i**. By considering different starting values for each, see if you can detect the difference. One of the fixed points is called **stable** and the other **unstable**. Can you decide which is which?

iii Find the fixed point for each of these recurrence relations and state its nature:

- $u_{n+1} = 4u_n - 21$
- $u_{n+1} = 0{\cdot}6u_n + 5{\cdot}2.$

b Consider the recurrence relation $x_{n+1} = \frac{x_n}{2} + \frac{a}{2x_n}$ for different values of a.
In B1 type a value for a, e.g. 16
In A1 type a starting value for x, e.g. 1
In A2 type: =A1/2+$B$1/(2*A1) ... fill down to row 15.
Note that the relation has a stable fixed point.
What is the relation between the fixed point and the value of a?

c Isaac Newton explored ways of solving equations using recurrence relations. He discovered that when trying to solve an equation of the form $f(x) = 0$ the recurrence relation $x_{n+1} = x_n - \frac{f(x_n)}{f'(x_n)}$ often had a stable fixed point that was the solution to the equation.

i If $f(x) = x^3 + 2x^2 + 10$, calculate $f'(x)$ and hence show that Newton's relation in this case is $x_{n+1} = x_n - \frac{x_n^3 + 2x_n^2 + 10}{3x_n^2 + 4x_n}$.

ii Use a spreadsheet to find a solution to the equation $x^3 + 2x^2 + 10 = 0$.
[With a starting value in A1, say 1, in A2 type: =A1-(A1^3+2*A1^2+10)/(3*A1^2+4*A1) and fill down.]

iii Investigate the usefulness of this method for solving the equation $x^4 - x^3 - 200 = 0$.

Preparation for assessment

1 a The recurrence relation $u_{n+1} = \frac{3u_n - 4}{5}$ has a second term, $u_2 = 4$.

i Calculate u_3.

ii Find u_1.

iii Say why a limit exists and find it.

b In a recurrence relation of the form $u_{n+1} = au_n + b,\ u_1 = 50$ it is known that $u_2 = 30$ and $u_4 = 15$.
Find the possible values of a and b.

c For one value of a, the recurrence relation has a limit. Find it.

2 A cyclist went from Land's End to John O'Groats for charity.
Each day he covered 67 km.

a Form a recurrence relation to model his progress where d_n is the distance covered in kilometres by the end of day n of his journey.

b i Express the nth term of the sequence in terms of n.
ii How far had he travelled at the end of day 10?
iii The total distance is 1407 km.
How far has he still to travel by the end of day 15?

3 A genealogist traces back someone's pedigree.

The number of people in generation n can be modelled by $G_{n+1} = 2G_n,\ G_1 = 1$.

Generation 1 is the person whose roots are being traced.

a i List the first four terms in the sequence
ii Express the nth term in terms of n.

b In which generation will the number of people exceed 3000?

c The total number of people in the tree up to generation n is modelled by:
$T_{n+1} = 2T_n + 1,\ T_1 = 1$.
i List the first four terms in the sequence.
ii Express the nth term in terms of n.

d By which generation will the total number of people in the tree exceed 3000?

4 Janet takes out a loan of £5000. The bank adds 1% interest on to the debt every month, just before she makes a partial repayment of £500.

a Write out a recurrence relation to model the amount, £A_n, she owes after n months.
A_1 is set at 5000.

b How much does she owe after four repayments?

c The bank uses the nth term formula $A_n = 50\,000 - 45\,000 \times 1{\cdot}01^{n-1}$.
In which month will this go below zero?

5 Michael is in hospital. He is given an injection of 100 units of a medicine.

His body filters out 22% of the active ingredient every hour.

a Form a recurrence relation to model the amount of active ingredient, a_n units, in Michael's system n hours after the injection.

b The doctors know that the amount of active ingredient after n hours can be worked out by $a_n = 100 \times 0{\cdot}78^n$. What percentage of the active ingredient is still in his system after 6 hours?

c The doctors know that the medicine will become ineffective if the amount in Michael's system falls below 20 units. So every 6 hours they administer another 100-unit injection. Write down a recurrence relation to model d_n, the number of units in Michael's system immediately after the nth injection.

d The medicine does have a toxic effect and it is not considered sensible to let the amount in the system exceed 1·5 times the administered dosage.
Is the regime, as described, sensible?

Summary

1 $u_{n+1} = u_n + b,\ u_1 = c$ is a linear recurrence relation which generates a sequence whose terms differ by a fixed amount, b.

This constant b is called the **common difference** ... $u_{n+1} - u_n = b$.

2 Such a sequence as generated in **1** is often referred to as an arithmetic sequence. Its nth term is $u_n = c + (n - 1)b$.

3 $u_{n+1} = au_n,\ u_1 = c$ is a linear recurrence relation which generates a sequence whose terms increase by a fixed factor, a.

This constant a is called the **common ratio** ... $u_{n+1} \div u_n = a$.

4 A sequence generated by a linear recurrence relation with a common ratio is often referred to as a geometric sequence. Its nth term is $u_n = ca^{n-1}$.

5 When $-1 < a < 1$, the terms of a geometric sequence tend towards zero as n tends to infinity.

6 $u_{n+1} = au_n + b,\ u_1 = c$ is a linear recurrence relation.

7 When $-1 < a < 1$, the terms of the sequence tend towards a limit as n tends to infinity.

This limit, L, can be found by letting $u_{n+1} = u_n = L$ and solving $L = aL + b$ to get $L = \dfrac{b}{1 - a}$.

14 Preparation for course assessment

Before we start...

Expect to be given these facts and formulae in the external course assessment.

The rest you need to know.

Circle:

The equation $x^2 + y^2 + 2gx + 2fy + c = 0$ represents a circle centre $(-g, -f)$ and radius $\sqrt{g^2 + f^2 - c}$ ... as long as $g^2 + f^2 - c > 0$.

The equation $(x - a)^2 + (y - b)^2 = r^2$ represents a circle, centre (a, b) and radius r.

Scalar product:

$\mathbf{a}.\mathbf{b} = |\mathbf{a}||\mathbf{b}| \cos \theta$, where θ is the angle between $\mathbf{a}$ and $\mathbf{b}$.

or

$\mathbf{a}.\mathbf{b} = a_1b_1 + a_2b_2 + a_3b_3$, where $\mathbf{a} = \begin{pmatrix} a_1 \\ a_2 \\ a_3 \end{pmatrix}$ and $\mathbf{b} = \begin{pmatrix} b_1 \\ b_2 \\ b_3 \end{pmatrix}$.

Trigonometric formulae:

$$\sin(A \pm B) = \sin A \cos B \pm \cos A \sin B$$

$$\cos(A \pm B) = \cos A \cos B \mp \sin A \sin B$$

$$\sin 2A = 2 \sin A \cos A$$

$$\begin{aligned} \cos 2A &= \cos^2 A - \sin^2 A \\ &= 2\cos^2 A - 1 \\ &= 1 - 2\sin^2 A \end{aligned}$$

Standard derivatives:

$f(x)$	$f'(x)$
$\sin ax$	$a \cos ax$
$\cos ax$	$-a \sin ax$

Standard integrals:

$f(x)$	$\int f(x)\,dx$
$\sin ax$	$-\frac{1}{a}\cos ax + c$
$\cos ax$	$\frac{1}{a}\sin ax + c$

What you need to know

The SQA provides lots of information on the form and content of the course assessment.

The link to their Higher Mathematics page is http://www.sqa.org.uk/sqa/47910.html.

In its documentation the SQA lists 50 skills you must be able to exhibit in the course assessment.

The skills list is:

Algebraic and trigonometric skills

1. Factorise a cubic or quartic polynomial expression.
2. Simplify a numerical expression using the laws of logarithms and exponents.
3. Apply the addition or double angle formulae.
4. Apply trigonometric identities.
5. Convert $a \cos x + b \sin x$ to $k \cos(x \pm a)$ or $k \sin (x \pm a)$, $k > 0$.
6. Identify or sketch a function after a transformation.
7. Sketch $y = f'(x)$ given the graph of $y = f(x)$.
8. Sketch the inverse of a logarithmic or an exponential function.
9. Complete the square in a quadratic expression where the coefficient of x^2 is non-unitary.
10. Determine a composite function given $f(x)$ and $g(x)$, where $f(x)$, $g(x)$ can be trigonometric, logarithmic, exponential or algebraic functions.
11. Determine $f^{-1}(x)$ for linear functions.
12. Solve a cubic or quartic polynomial equation.
13. Given the nature of the roots of an equation, use the discriminant to find an unknown.
14. Solve a logarithmic or exponential equation.
15. Find the coordinates of the point(s) of the intersection of two curves (including straight lines).
16. Solve trigonometric equations in degrees or radians in a given interval (including those involving the wave function or trigonometric formulae or identities).

Geometric skills

17. Determine the resultant of vector pathways in three dimensions.
18. Work with collinearity.
19. Determine the coordinates of an internal division point of a line.
20. Use the scalar product.
21. Use unit vectors **i**, **j**, **k** as a basis.

Calculus skills

22. Differentiate an algebraic function that can be simplified to an expression in powers of x.
23. Differentiate $k \sin x$, $k \cos x$.
24. Differentiate a composite function using the chain rule.
25. Determine the equation of a tangent to a curve at a given point by differentiation.
26. Determine where a function is strictly increasing/decreasing.
27. Sketch the graph of an algebraic function by determining stationary points and intersections with the axes.
28. Integrate an algebraic function that can be simplified to an expression of powers of x.
29. Integrate functions of the form $f(x) = (x + b)^n$, $n \neq -1$.
30. Integrate functions of the form $f(x) = a \cos x$ and $f(x) = a \sin x$.
31. Integrate functions of the form $f(x) = (ax + b)^n$, $n \neq -1$.
32. Integrate functions of the form $f(x) = a \cos (bx + c)$ and $f(x) = a \sin (bx + c)$.
33. Solve differential equations of the form $\frac{dy}{dx} = f'(x)$.
34. Calculate definite integrals with limits that are integers, radians, surds or fractions.

35. Determine the optimal solution for a given problem.
36. Solve problems using rate of change.
37. Find the area between a curve and the x-axis.
38. Find the area between a straight line and a curve or between two curves.
39. Determine and use a function from a given rate of change and initial conditions.

Algebraic and geometric skills

40. Find the equation of a line parallel to, and a line perpendicular to, a given line.
41. Use $m = \tan\theta$ to calculate a gradient or angle.
42. Use properties of medians, altitudes and perpendicular bisectors in problems involving the equation of a line and intersection of lines.
43. Determine, and use, the equation of a circle.
44. Use properties of tangency in the solution of a problem.
45. Determine the intersection of circles or a line and a circle.
46. Determine a recurrence relation from given information.
47. Use a recurrence relation to calculate a required term.
48. Find and interpret the limit of a sequence, where it exists.

Reasoning skills

49. Interpret a situation where mathematics can be used and identify a strategy.
50. Explain a solution and/or relate it to context.

(Source: Scottish Qualifications Authority)

14.1 Mandatory skills

The first section of this chapter will act as a check on your grasp of the skills above.

If you can't do one of these questions, or a part of one of these questions, then you have identified a region for necessary revision.

Exercise 14.1

Check your algebraic and trigonometric skills

1 a Using synthetic division or otherwise, show that $(x - 1)$ is a factor of the expression: $2x^4 - x^3 - 14x^2 + 19x - 6$.

b Hence fully factorise: $2x^4 - x^3 - 14x^2 + 19x - 6$.

c Solve the equation: $2x^4 - x^3 - 14x^2 + 19x - 6 = 0$.

2 a Find the value of $\log_6 2 + 2\log_6 3 - \log_6 \frac{1}{2}$.

b Simplify $\dfrac{(2a^x)^2 \times a^x}{a^6}$ and then find its value when $x = 2$.

3 a A and B are acute angles. $\sin A = \frac{3}{5}$ and $\cos B = \frac{5}{13}$.
Calculate the **exact** value of: **i** $\sin 2A$ **ii** $\cos(A + B)$.

b If $\cos 2A = 0{\cdot}98$, with A acute, find the exact value of: **i** $\sin^2 A$ **ii** $\sin A$.

4 a Express $8\sin x° + 15\cos x°$ in the form $k\sin(x + a)°$.

b Hence state:

i the maximum value of $40 - 8\sin x° - 15\cos x°$

ii the value of x in the interval $0 \leqslant x \leqslant 360$ for which it is attained.

5 a The graph opposite shows $y = x^2$ after a translation.

i What is the equation of the transformed graph?

ii State the coordinates of the points marked A, B and C.

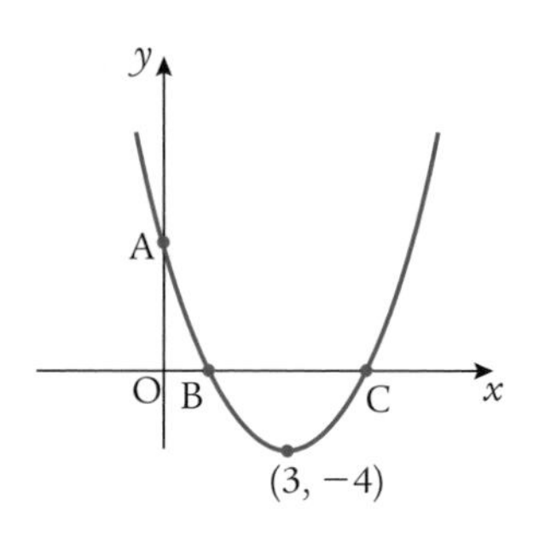

b The graph below shows $y = f(x)$ close to the origin.

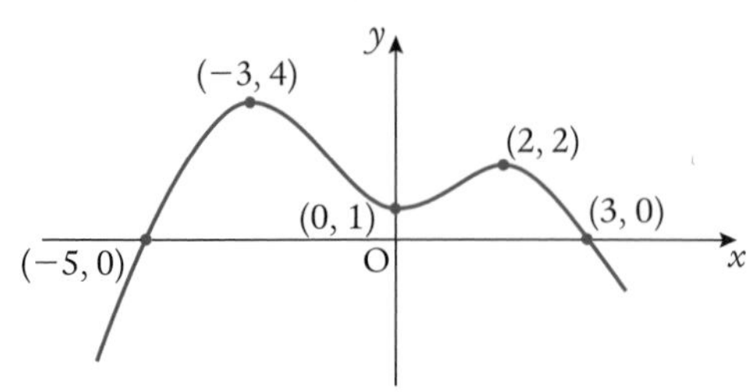

Make a sketch of: **i** $y = f(-x)$ **ii** $y = 3 + f(-x)$ **iii** $y = f'(x)$.

c The graph opposite is a sketch of $f(x) = 3\log_{10} x$.
Make a sketch of the inverse function, $f^{-1}(x)$, and state its equation.

6 a Express $2x^2 + 4x + 5$ in the form $a(x + b)^2 + c$.

b Hence find the maximum value of $\dfrac{6}{2x^2 + 4x + 5}$.

7 Functions f and g are defined by $f(x) = 1 - 3x$ and $g(x) = x^2 - 9$.

a Find: **i** $f(g(x))$ **ii** $g(f(x))$.

b For what values of x is $f(g(x)) = g(f(x))$?

c Find $f^{-1}(x)$.

8 For what value of k does the equation $2x^2 + 3x + k = 0$ have real distinct roots?

9 Solve, correct to 2 decimal places,

a $4 + 3 \times 1{\cdot}6^x = 6$

b $2 \ln x + 1 = 4$.

10 $y = 3x^2 + 2x - 1$ and $y = x^2 - 3x + 2$ are two parabolae.
Find the points at which they intersect.

11 a Solve $8 \sin x° + 15 \cos x° = 10$ in the interval $0 \leqslant x \leqslant 360$, to 1 decimal place.
[Refer to Question 4.]

b Solve $\sin 2x - \cos x = 0$ in the interval $0 \leqslant x \leqslant 2\pi$.

c Solve $\cos 2x + 3 \cos x - 1 = 0$ in the interval $0 \leqslant x \leqslant 2\pi$.

Check your geometric skills

12 A crystal is in the form of a hexahedron, PNQAZ.

The edges of the crystal represent vectors: $\overrightarrow{ZP} = \mathbf{u}$, $\overrightarrow{ZA} = \mathbf{v}$, $\overrightarrow{ZQ} = \mathbf{w}$.

B is the midpoint of PQ.

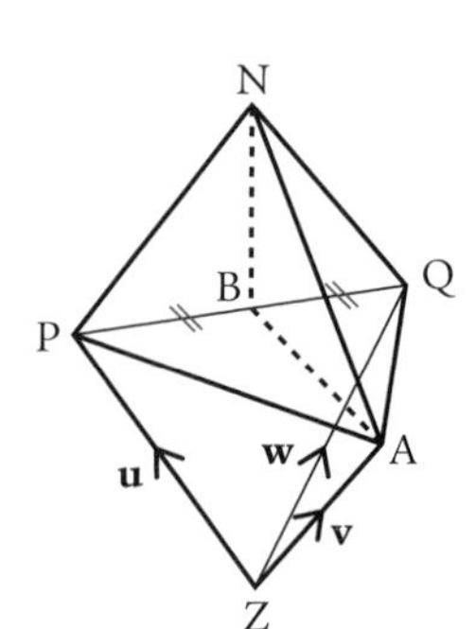

a Express, in terms of **u**, **v** and/or **w**,

i $\overrightarrow{PQ}$ **ii** $\overrightarrow{PB}$ **iii** $\overrightarrow{AB}$.

b $\overrightarrow{BN}$ and $\overrightarrow{ZA}$ are parallel and have the same magnitude.

i Express $\overrightarrow{AN}$ in terms of **u**, **v** and **w**.

ii Find the resultant of $\overrightarrow{PB} + \overrightarrow{AN}$.

13 Prove that the three points A(2, −3, 5), B(5, 2, 7) and C(11, 12, 11) are collinear.

14 Find the point, P, that divides the line joining A(2, −5, 3) to B(9, 9, −4) in the ratio 4 : 3.

15 Find the acute angle between the vectors **u** and **v**, where $\mathbf{u} = 3\mathbf{i} + 2\mathbf{j} - 3\mathbf{k}$ and $\mathbf{v} = 5\mathbf{i} - 7\mathbf{j} + \mathbf{k}$.

Check your calculus skills

16 Differentiate:

a $x^4 + x^2 + 1 + \frac{1}{x^2}$ b $x^3 + \frac{x+1}{\sqrt{x}}$ c $3\sin x + 4\cos x$ d $3\sin(4x - 1)$

e $\sqrt{x^3 + 2x + 1}$ f $\frac{3}{x^2 + 2x}$ g $(x + 4)(2x - 1)$.

17 Find the equation of the tangent to the curve $y = x^3 - 4x^2$ at the point where $x = 2$.

18 a Differentiate: $\frac{1}{3}x^3 + x^2 - 3x + 1$.

b Hence identify the values of x for which the function $f(x) = \frac{1}{3}x^3 + x^2 - 3x + 1$ is decreasing.

19 A function is defined by $f(x) = x^3 - 3x^2 - 144x - 140$.

a Given that it cuts the x-axis at $x = 14$, find its other x-intercepts.

b Where does it cut the y-axis?

c Find its stationary points and determine their nature.

d Sketch the curve.

20 Integrate:

a $8x^3 - \frac{1}{x^2}$ b $6\sqrt{x} - x + 4$ c $\frac{3 + x^2}{2\sqrt{x}}$.

21 Find:

a $\int (x + 3)^5 dx$ b $\int 3\cos x - 5\sin x\, dx$

c $\int \sqrt{4x - 5}\, dx$ d $\int 3\sin(3x + 1) + 4\cos(2x + 3) dx$.

22 A function $y = f(x)$ is such that $\frac{dy}{dx} = 3x^2 - 6x + 1$ and $f(1) = 3$. Find the function.

23 Evaluate:

a $\int_{-3}^{3} x^2 + 2x + 1 dx$ b $\int_0^{\frac{\pi}{2}} \cos\left(2x - \frac{\pi}{2}\right) dx$ c $\int_{\sqrt{2}}^{\sqrt{3}} -\frac{1}{x^3} dx$ d $\int_{\frac{1}{2}}^{\frac{3}{2}} x^2 + \frac{1}{x^2} dx$.

24 The cost of a job, £C, is calculated by $C(x) = 25x + \frac{200}{x}$, where x is the time in hours that you need the job completed by. For what number of hours will the job cost the least?

25 A particle is moving along the x-axis so that at time x seconds its displacement, s metres, can be calculated using $s(x) = 2x^3 + x^2 + 1$.

a Find an expression for the rate of change of s with respect to time, i.e. the velocity in metres per second.

b i How far is the particle from the origin when $x = 2$?

ii What is its velocity at this time?

c What is the acceleration of the particle in the third second?

26 Calculate the area trapped between the parabola $y = x^2 + x - 6$ and the x-axis.

27 The parabolae $y = 2x^2 - 7x + 18$ and $y = 3 + 11x - x^2$ intersect at two points.

a Find these two points.

b Calculate the area trapped between the two curves.

28 A sky-diver jumped from a stationary balloon. He was filmed falling through the air. The camera recorded the time in seconds elapsed from when it was first switched on. Its velocity in m s^{-1} is modelled by $v(t) = 40 + 10t$.
When the camera was started, the sky-diver's displacement from the balloon was 300 m.

a Express the displacement at time t seconds as a function of t.

b i How far from the balloon was he after 3 seconds?
ii How fast was he travelling?

Check your algebraic and geometric skills

29 ABCD is a rectangle. AB lies on the line $y = 3x + 1$.
C is the point (9, 8).

a Find the equation of the line BC.

b Find the equation of CD.

30 a What angle does the line with equation $3y = 4x + 1$ make with the x-axis?

b A line makes an angle of 60° with the x-axis and passes through the point (1, 5). What is the equation of the line?

31 The vertices of a triangle are A(0, 10), B(6, 0) and C(10, 12).

a Find the equation of the median through C.

b Find the equation of the altitude through A.

c Find the point of intersection of these two lines.

32 A(4, −4) and B(2, 10) are points on a circle.

a Find the perpendicular bisector of the chord AB.

b The centre of the circle has a y-coordinate of 4. Find the radius of the circle.

33 a Find the equation of the circle, centre (2, 5), passing through (8, −3).

b Say whether each of these points lies in, on or outside the circle:
i (11, 8)
ii (5, 15)
iii (8, 13).

c Find the equation of the tangent to the circle when the point of tangency is (−6, 11).

34 a Where does the line $y = 11 - 2x$ intersect the circle $(x - 1)^2 + (y - 4)^2 = 10$?

b i The circle $(x + 2)^2 + (y - 1)^2 = 40$ intersects $(x - 1)^2 + (y - 4)^2 = 10$ at two points. Show this is true by considering the lengths of the radii and the distance between centres.
ii Find the points of intersection.

35 A tank contains 26 litres of water. Because of the heat, it loses a certain percentage of its contents to evaporation every day.
A leaky inlet allows a constant number of litres back into the tank each day.
The situation can be modelled by a recurrence relation of the form $u_{n+1} = au_n + b,\ u_1 = 26$.
It was noted that $u_2 = 16$ and $u_3 = 10$.

a Determine the recurrence relation by finding a and b.

b Calculate u_4.

c i Explain why a limit exists for this sequence and find it.
ii Interpret the limit in the context.

14.2 Paper 1 questions

No calculator allowed

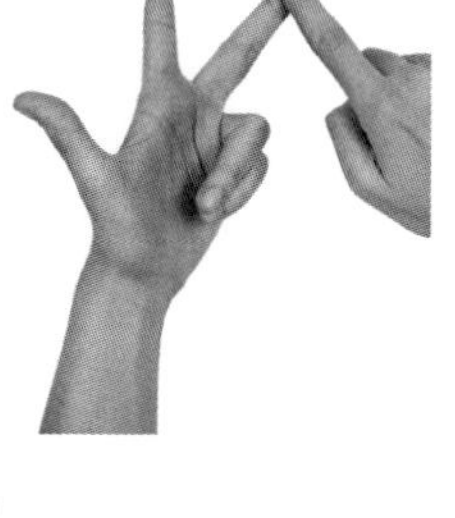

In Paper 1 you will be given questions that sample the syllabus content.

The balance of this sample will be algebra (30–45%), geometry (15–35%), calculus (15–40%), trigonometry (10–25%).

Sixty marks out of a total of 130 marks are allocated to this paper.

It is expected to take you 70 minutes.

The questions will be designed to check that you can, **without the aid of a calculator**, apply numerical, algebraic, geometric, trigonometric, calculus and reasoning skills specified in Section 14.1.

The SQA is trying to assess your understanding of underlying processes. Questions are selected for this paper where the use of a calculator may compromise this assessment.

Exercise 14.2

1 The Firth is busy with coastal trade. The harbour master keeps track of movements using a coordinate system. His station at M lies halfway between the port of Ardbeg, A(3, −4), and Balwearie, B(5, 4). A third port at Cairns, C(−6, 5), completes the Firth Triangle.

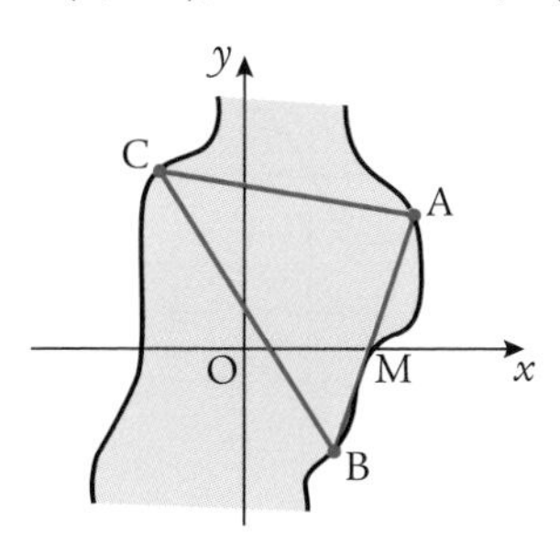

A ship leaves Balwearie at right angles to AC meeting it at T.

a What is the equation of the line which models the route of the ship from Balwearie to T? [3 marks]

b A second ship leaves Cairns and travels along the median of the triangle.
Find the equation representing this route. [3 marks]

c Find where the altitude from B intersects the median from C. [3 marks]

2 Two functions are defined by $f(x) = 3x + 1$ and $g(x) = x^2 + x + k$, where k is a constant.

a Express $g(f(x))$ as a quadratic in the form $ax^2 + bx + c$. [2 marks]

b i For what value of k does $g(f(x)) = 0$ have coincident roots? [2 marks]
ii State the root for this value of k. [1 mark]

3 A gardener is growing tomatoes. Each week he adds plant food to the grow-bag, and each week the plant absorbs 90% of it. The amount of food, F_n units, present in the bag at the end of week n can be modelled by: $F_{n+1} = \dfrac{9F_n + 50}{10}$, $F_0 = 150$.

a Evaluate F_2. [1 mark]

b i Explain why you know the process has a limit. [1 mark]
ii Evaluate the limit. [1 mark]

4 a Explain why the equation $x^2 + y^2 + 4x + 6y + 15 = 0$ does not represent a circle. [1 mark]

b In a computer model, a sphere is represented by a circle and a light source by a point. The circle has a centre (3, 4) and passes through the point (4, 1).

i What is the equation of the circle? [2 marks]

ii The line that separates the illuminated part from the dark part is modelled by $y = 5$. Where does it cut the circle? [1 mark]

iii The area in the dark is bordered by the tangents passing through these points. Find the equations of both tangents. [4 marks]

iv The light source is where the two tangents intersect. Find the coordinates of the light source. [1 mark]

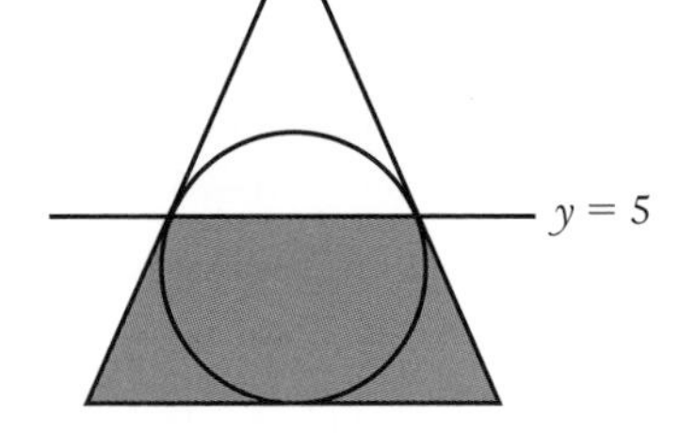

5 a Show that the parabola $y = 3x^2 + 6x + 5$ does not cut the x-axis. [2 marks]

b i Express $3x^2 + 6x + 5$ in the form $a(x + b)^2 + c$. [2 marks]

ii Hence say how close the parabola gets to the x-axis. [1 mark]

6 Two parts in a machine move independently.

The graph shows how their heights vary over time.

The equations of the curves are:

$h_1(x) = 2 - 3 \sin x$ and $h_2(x) = \cos 2x$,

where x is the number of seconds since observations began.

Solve the equation $2 - 3 \sin x = \cos 2x$ in the interval $0 \leqslant x \leqslant 2\pi$ to find when the parts are the same height. [4 marks]

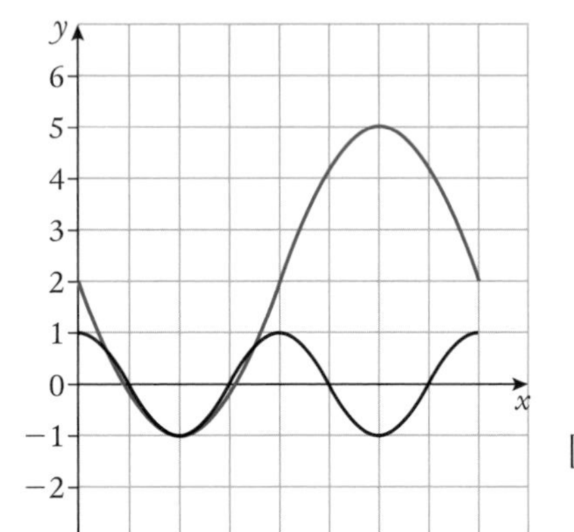

7 Take the log to the base 10 of 4 as 0·602 ... $\log_{10} 4 = 0{\cdot}602$.

a What is the value of:

i $\log_{10} 16$ ii $\log_{10} 2$ iii $\log_{10} \frac{1}{2}$? [3 marks]

b If $\log_{10} 3 = 0{\cdot}477$, calculate $\log_{10} 6$. [1 mark]

8 A structure is built from girders.

Two particular girders are joined at a point A(2, −4, −3).

One goes from A to the point B(8, 8, 1).

The other runs from A to the point C(4, 7, −13).

a Find, in component form, the vector represented by $\overrightarrow{AC}$. [1 mark]

b Find the magnitude of $\overrightarrow{AB}$. [1 mark]

c Find cos ABC. [2 marks]

d Estimate the size of angle ABC. [1 mark]

9 A function is defined by $f(x) = x^4 - 8x^2 - 9$.

a Find the coordinates of the x- and y-intercepts. [2 marks]

b Find the stationary points and determine their nature. [3 marks]

c Sketch the function $f(x) = x^4 - 8x^2 - 9$ close to the origin. [1 mark]

10 a Find where the cubic curve $y = x^3 - x^2 - 4x + 8 = 0$ intersects the parabola $y = 2 + 3x - x^2$. [5 marks]

b Calculate the area trapped between the two curves. [5 marks]

Total marks 60

14.3 Paper 2 questions

Sometimes the use of a calculator will expand the scope of a question giving more opportunities for the use of realistic applications and data and allow the candidate to exhibit their reasoning more clearly.

The balance of the topics sampled will complement that of Paper 1.

Seventy marks out of a total of 130 marks are allocated to this paper.

It is expected to take you 90 minutes.

Developing habits

You may use a calculator ... but make sure you practise on a calculator that you will be using in the exam. There is no point in developing skills and slick techniques on your mobile or computer tablet as these will not be allowed in the exam hall.

Exercise 14.3

1 AC and BD are the diagonals of trapezium ABCD.
E is the point where the diagonals intersect.
The points A, B and C are (3, 8), (10, 12) and (12, 2) respectively.

a E cuts AC in the ratio 1 : 2. Find the coordinates of E. [2 marks]

b Prove that BD is perpendicular to AC. [2 marks]

c Given that the x-coordinate of D is -2, find its y-coordinate. [2 marks]

2 A line passes through the point (4, 0) and makes an angle of 135° with the positive direction of the x-axis (see diagram below).

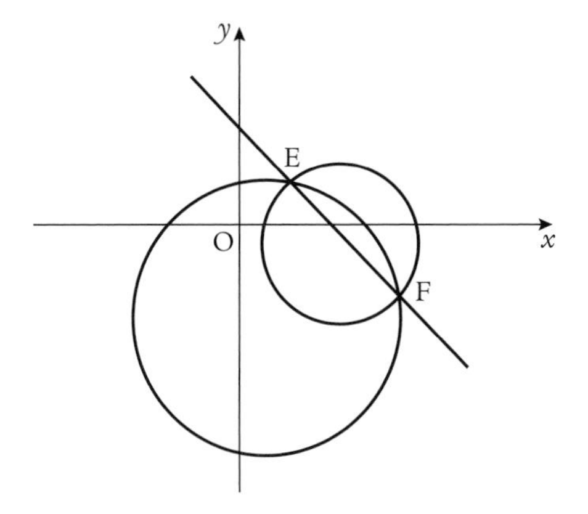

A circle, centre $(1, -5)$, passes through the point $(-1, 1)$.

a Find the points, E and F, where the line cuts the circle. [5 marks]

b Find the equation of the circle that has EF as a diameter. [2 marks]

3 *False summits.* You are climbing a hill, with your eye on the point you believe is the top.

Your eyeline is a tangent to the hill at this point.

When you get high enough, you see ... more hill!

Let the profile of the hill be modelled by the curve $y = x^3 + 7x^2 - 2x - 70$.

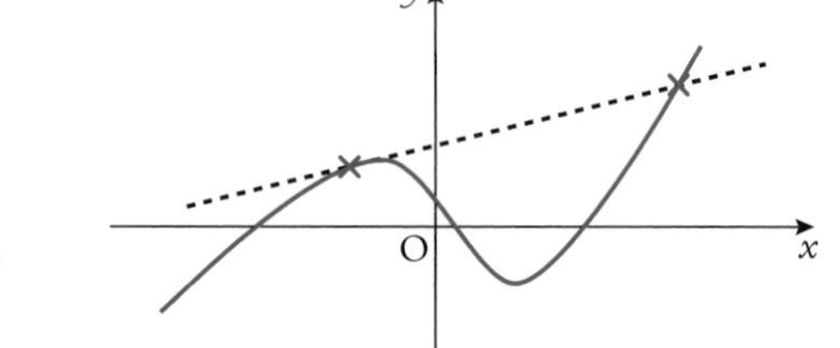

a Find the equation of the tangent when $x = -5$. [4 marks]

b Find the coordinates of the point where this tangent meets the curve again. [4 marks]

4 A clock face is being projected on to a wall by a computer.
The vertical velocity in centimetres per second of the tip of the second hand is modelled by

$v(t) = \cos\left(\frac{\pi t}{30} + \frac{\pi}{3}\right)$

where t is the time in seconds since observations began.

a Find the rate of change of this velocity with time when $t = 15$. [2 marks]

b What is the height, $s(t)$ cm, of the tip when $t = 10$, given that when $t = 15$, $s = 50$. [3 marks]

5 A capsule for delivering medicine must have a volume of $36\ \text{mm}^3$.

It is in the shape of a cylinder capped off at both ends by a hemisphere.

The cylinder has a radius of r mm and a height of h mm.

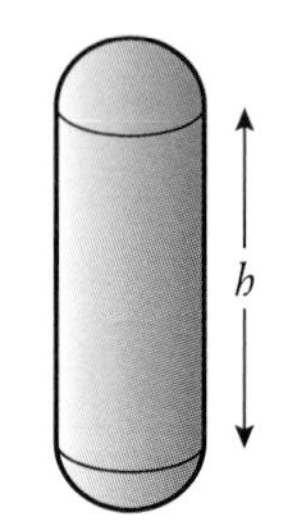

[Volume of a sphere $= \frac{4}{3}\pi r^3$; surface area of a sphere $= 4\pi r^2$.]

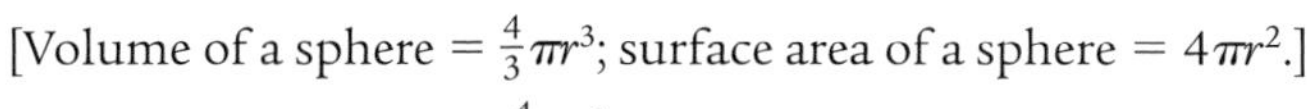

a Show that $h = \dfrac{36 - \frac{4}{3}\pi r^3}{\pi r^2}$. [1 mark]

b Hence show the surface area of the capsule is given by:
$A(r) = \frac{4}{3}\pi r^2 + \dfrac{72}{r}$. [1 mark]

c Find the value of r that minimises the surface area of the capsule. (Answers must be justified.) [5 marks]

6 A function is defined by $f(x) = \dfrac{k}{\sqrt{3x + 4}}$, $x > 1$ and k is a constant.

a $f'(7) = -6$. Calculate the value of k. [3 marks]

b Find the gradient to the curve $y = \dfrac{k}{\sqrt{3x + 5}}$ when $x = 32$. [2 marks]

7 A ladder leans against a wall at the regulation safe angle of $x°$ to the horizontal.
The regulation safe gradient for a ladder is 4.

4 units

$x°$

1 unit

a Find the exact value of: **i** $\sin x°$ **ii** $\cos x°$. [1 mark]

b Calculate the exact value of: **i** $\sin 2x°$ **ii** $\cos 2x°$. [3 marks]

c By appreciating that $\cos 3x° = \cos (2x + x)°$, find the exact value of $\cos 3x°$. [4 marks]

8 A botanist was studying the growth of a particular plant.

He believed its height on day x after potting out could be modelled by $h(x) = ae^{bx}$ for some constants a and b.

On day 1 after potting out it was 8 cm tall and on day 3 it was 12 cm. (Measurements are taken at the end of the day.)

a Find the values of a and b correct to 1 decimal place. [3 marks]

b What height was the plant: **i** when it was potted out **ii** after 10 days? [2 marks]

c When did the plant first exceed 1 metre in height?
[Trial and error will not be accepted as a valid strategy.] [3 marks]

9 The waters beside the shores at Gavrin are tidal. Their depth in centimetres can be modelled by: $d(x) = 28 \sin x + 45 \cos x + 60$
where x units is the time since the start of 1st January ... and a unit is $\frac{\pi}{6}$ hours.

a Express $28 \sin x + 45 \cos x$ in the form $k \cos(x + \theta)$, where $k > 0$ and $\theta > 0$. [4 marks]

b What is the difference between high tide and low tide? [1 mark]

c Find the first two times on 1st January that the depth was 20 cm. [3 marks]

10 The pyramid, ABCDE, and cube ABCDFGHI share a base.

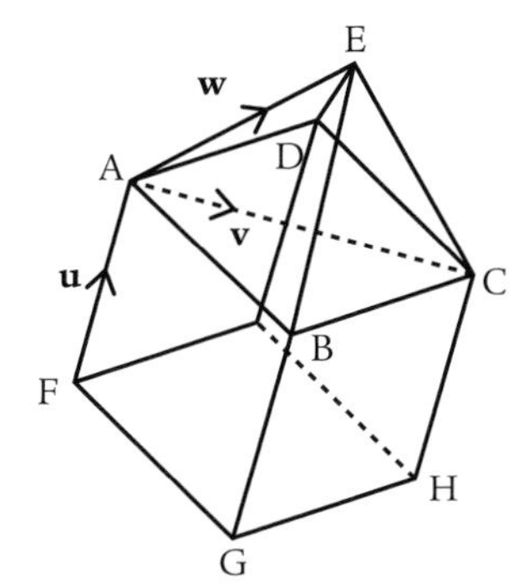

The sloping edges of the pyramid are equal in length to the diagonal of this base, i.e. ACE is an equilateral triangle.

The vectors **u**, **v**, and **w** are represented by $\overrightarrow{FA}$, $\overrightarrow{AC}$ and $\overrightarrow{AE}$ respectively.

a Express $\mathbf{u} \cdot (\mathbf{v} + \mathbf{w})$ in terms of $|\mathbf{u}|$, the magnitude of **u**. [3 marks]

b The position vector of C is $\mathbf{c} = -6\mathbf{i} - 4\mathbf{j}$ and the position vector of H is $\mathbf{h} = 5\mathbf{i} + 6\mathbf{j} + 2\mathbf{k}$ (where **i**, **j** and **k** are unit vectors parallel to the axes).
Find the position vector of the point that divides CH in the ratio 2 : 3. [3 marks]

c Find a unit vector parallel to **u**. [2 marks]

Total marks 70

Summary

1 Make sure you have checked the skills list for gaps in your knowledge.

2 Leave evidence on the exam paper that you have:
- properly interpreted the question
- selected appropriate strategies
- carried out the appropriate algorithm with accuracy
- communicated the answer/response in the context of the question.

Marks are awarded for such evidence.

[The marks associated with the questions above are for guidance only ... the SQA may well mark a similar question slightly differently.]

3 When sitting the exam, you **will** depend on the habits you have developed throughout the year.

Develop good habits in:
- laying out your work
- handling notation
- calculator use [don't spend the year using the calculator app on your phone ... you won't be allowed to use it in the exam].

Do this **throughout** the year. You won't develop these habits during the exam!

4 Practice makes perfect. A little and often is better that a lot the night before the exam.

5 Good luck.

Answers

1 Algebraic functions

What you need to know

1 a i 11 ii −9 iii 1 iv $4\pi - 1$
b i 1 ii $\frac{1}{4}$ iii $\frac{1}{3}$

2 a $1 \leqslant x \leqslant 52, x \in W$
b 7
c i 7 ii 7
d It is invariant, i.e. always 7

3 a Graph is a straight line
b i $2a + 1$ ii $2(a + 2) + 1 = 2a + 5$
c Real numbers

4 a 0
b i 9 ii $6 - 4 = 2$ iii $18 \times 12 = 216$

5 a Quadratic
b i −2 ii 0 iii −2
c 2, −1

6 a i 40 m ii 50 m iii 70 m
b 120

Exercise 1.1

1 a i
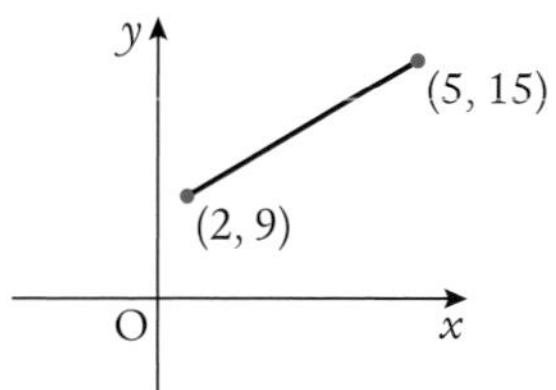

ii $y \geqslant 9$

b i
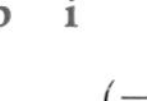
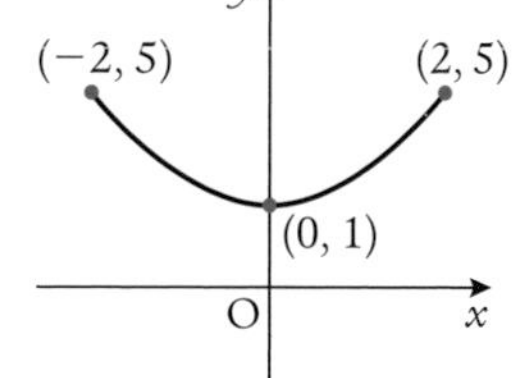

ii $1 \leqslant y \leqslant 5$

c i
ii $\frac{1}{5} \leqslant y \leqslant 1$

d i
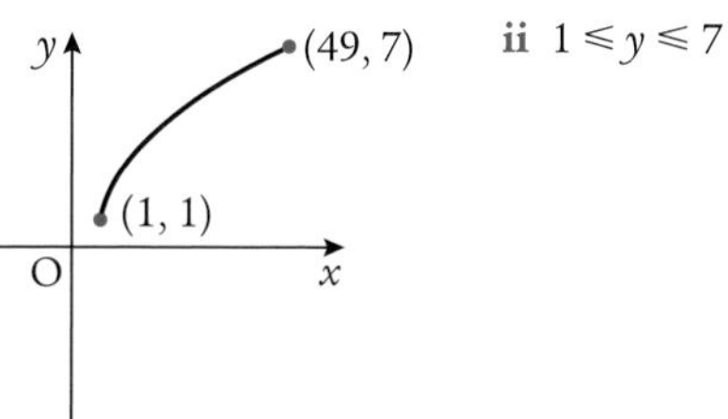

ii $1 \leqslant y \leqslant 7$

e i

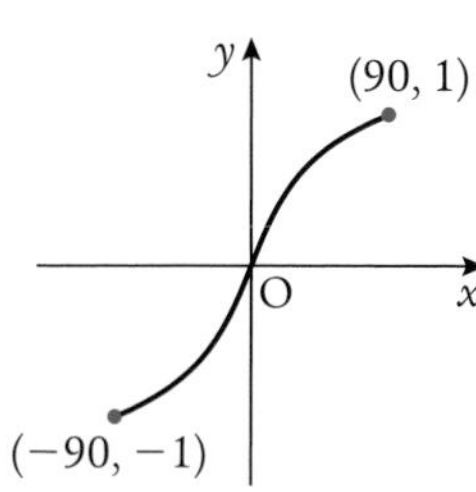

ii $-1 \leqslant y \leqslant 1$

2 a $R - \{-2\}$ b $R - \{1\}$ c $R - \{3\}$ d $R - \{1, 2\}$ e $R - \{1, 3\}$

3 a i $-1 \leqslant x \leqslant 3$ ii $-1 \leqslant y \leqslant 5$
b i $-2 \leqslant x \leqslant 5$ ii $-1 \leqslant y \leqslant 6$
c i $-4 \leqslant x \leqslant 4$ ii $0 \leqslant y \leqslant 4$

4 a $T(x) = \frac{45}{x}; 20 \leqslant x \leqslant 60$
b $0.75 \leqslant T \leqslant 2.25$
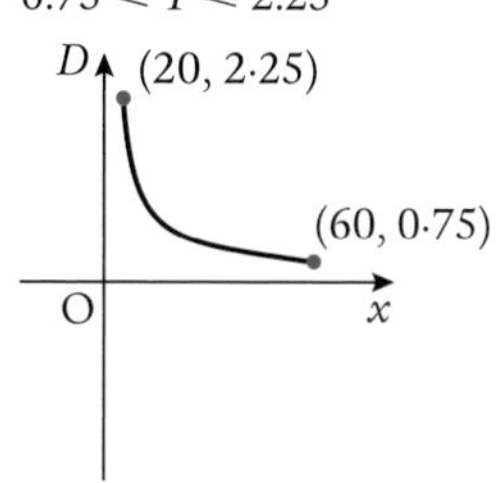

5 a $134 \leqslant H_{male} \leqslant 174$; $140 \leqslant H_{female} \leqslant 185$

b 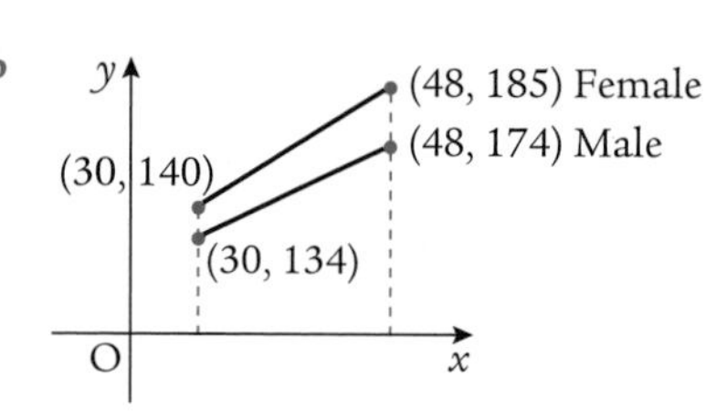

6 a i 20·25 ii 0
b $9 \leqslant D \leqslant 21{\cdot}16$

c 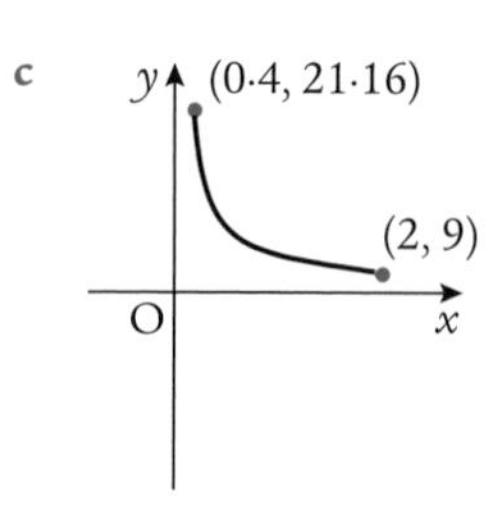

7 a i 2 ii 0·7 iii 0
b Error message
c Positive reals and zero ($R^+ + \{0\}$)
d So that $\sqrt{x}$ is unambiguous ... only has one meaning (the positive root is chosen for convenience)

8 a $1 \leqslant x - 2 \leqslant 5 \Rightarrow 3 \leqslant x \leqslant 7$
b $V(x) = 1\,.\,(x-2)(x-3) = x^2 - 5x + 6$
c $0 \leqslant V(x) \leqslant 20$
d Rising curve from (3, 0) to (7, 20)

9 a $13 \leqslant H \leqslant 18$
b 0·69 cm (to 2 d.p.)

10 90°

11 For example, $\frac{1}{x}(x \neq 0)$ $\sin^{-1}x(-1 \leqslant x \leqslant 1)$, $\tan x$, ($x \neq 90k$, where k is odd), $\sqrt{x}$, ($x \geqslant 0$) and others

Exercise 1.2

1 a i $2x - 5$ ii $2x - 1$
b i $7 - 15x$ ii $19 - 15x$
c i $8 - 35x$ ii $-4 - 35x$
d i $x^2 + 1$ ii $(x+1)^2$
e i $9x^2 - 6x + 2$ ii $3x^2 + 2$
f i $6x^2 - 18x + 4$ ii $36x^2 - 42x + 11$
g i $\frac{1}{x^2}$; $R - \{0\}$ ii $\frac{1}{x^2}$; $R - \{0\}$
h i $\sin(x+1)°$ ii $\sin x° + 1$
i i $\frac{1}{x} + 1$; $R - \{0\}$ ii $\frac{1}{x+1}$; $R - \{-1\}$
j i $\sqrt{\cos x°}$; $\cos x° \geqslant 0$ ii $\cos\sqrt{x°}$; $x \geqslant 0$
k i $x, x \geqslant 0$ ii x, R
l i x, R ii x, R

2 a i -5 ii -1
b i 7 ii 19
c i 8 ii -4
d i 1 ii 1
e i 2 ii 2
f i 4 ii 11
g i Undefined ii Undefined
h i sin 1° ii 1
i i Undefined ii 1
j i 1 ii 1
k i 0 ii 0
l i 0 ii 0

3 a $g(f(x)) = \frac{x}{100\,000}$, converts cm to km
b $g(f(x)) = \frac{5(x-32)}{9} - 273$, converts F to degrees absolute
c $g(f(x)) = 10x^2 + 20$, cost of laying a carpet of side x m

4 a $4x + 3$, domain and range are R
b $g(g(x)) = \frac{1}{\frac{1}{x}} = x$, but range and domain are $R - \{0\}$
c $\frac{1}{2x+1}$, $R - \{-\frac{1}{2}\}$; $R - \{0\}$
d $h(h(x)) = \frac{1}{2\left(\frac{1}{2x+1}\right) + 1} = \frac{2x+1}{2x+3}$ ... using intermediate step we find domain is $R - \{-0{\cdot}5, -1{\cdot}5\}$ and range is $R - \{1\}$

5 a i 33 ii 17 iii 9 iv 5
b i $\frac{x+1}{2}$ ii $\frac{x+3}{4}$ iii $\frac{x+7}{8}$ iv $\frac{x+15}{16}$
c nth term $= \frac{x + 2^n - 1}{2^n}$

6 a i x ii x iii x iv x v x
b All x
c All x

7 a i $(x-1)^2$ ii $x^2 - 1$
b $(x-1)^2 = x^2 - 1 \Rightarrow 2x = 2 \Rightarrow x = 1$

Exercise 1.3

1 a $\frac{x-5}{3}$ b $\frac{x}{4}$ c $\frac{1-x}{3}$ d $6-x$ e $\frac{2}{x}$

f $\frac{x}{x-2}$ g x^2-2 h $(1-x)^2$ i $\left(\frac{x-2}{x}\right)^2$

2 a $y = 2 - x$ b $f^{-1}(x) = 2 - x$, same as $f(x)$ c Perpendicular to $y = x$

3 a i Parallel line passing through (2, 0) ii Line passing through (−2, 0), (0, 1)

iii Line passing through (0, 1), (2, 0)

b i Half-parabola with vertex (0, 0) passing through (1, 1)

ii Half-parabola with vertex (0, 0) passing through (−1, 1)

iii Half-parabola with vertex (0, −1) passing through marked point on $y = x$

4 b only

5 a $-1 \leqslant f(x) \leqslant 1$

b In general each y has more than one image in the domain

c i Domain $0 \leqslant x \leqslant 180$, range $-1 \leqslant f(x) \leqslant 1$

ii One-to-one correspondence between domain and range

iii Domain $-1 \leqslant x \leqslant 1$ range $0 \leqslant f^{-1}(x) \leqslant 180$

iv

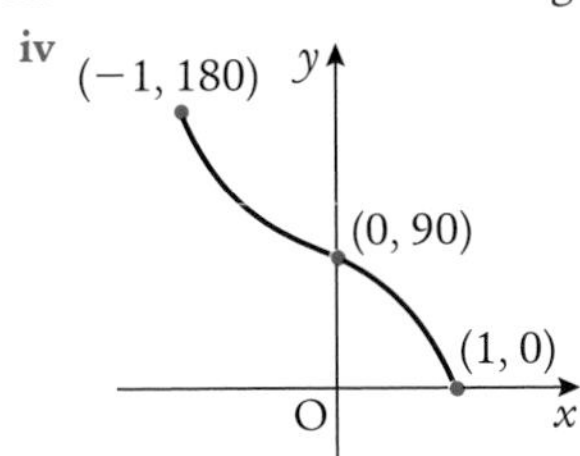

d $-90 \leqslant x \leqslant 90$

Exercise 1.4

1 a $(x+2)^2-7$ b $(x+5)^2-24$ c $(x-4)^2-12$ d $(x-1)^2-2$ e $\left(x+\frac{3}{2}\right)^2-\frac{5}{4}$

f $\left(x+\frac{1}{2}\right)^2-\frac{21}{4}$ g $\left(x-\frac{3}{2}\right)^2+\frac{15}{4}$ h $\left(x-\frac{1}{2}\right)^2-\frac{5}{4}$ i $\left(x+\frac{1}{4}\right)^2-\frac{13}{16}$ j $(x+0{\cdot}2)^2+0{\cdot}06$

2 a $(x-2)^2-6$ b $f_{\min} = -6$ at $x = 2$ c $x \geqslant 2$ d $f^{-1}(x) = \sqrt{x+6}+2, x \geqslant -6$

3 a i $(x+6)^2-37$ ii $f_{\min} = -37$ at $x = -6$ iii $x \geqslant -6$ iv $f^{-1}(x) = \sqrt{x+37}-6, x \geqslant -37$

b i $(x-4)^2-4$ ii $f_{\min} = -4$ at $x = 4$ iii $x \geqslant 4$ iv $f^{-1}(x) = \sqrt{x+4}+4, x \geqslant -4$

c i $(x-\frac{1}{2})^2+\frac{19}{4}$ ii $f_{\min} = \frac{19}{4}$ at $x = \frac{1}{2}$ iii $x \geqslant \frac{1}{2}$ iv $f^{-1}(x) = \sqrt{x-\frac{19}{4}}+\frac{1}{2}, x \geqslant \frac{19}{4}$

d i $(x-1)^2-2$ ii $f_{\min} = -2$ at $x = 1$ iii $x \geqslant 1$ iv $f^{-1}(x) = \sqrt{x+2}+1, x \geqslant -2$

e i $(x-\frac{7}{2})^2-\frac{17}{4}$ ii $f_{\min} = -\frac{17}{4}$ at $x = \frac{7}{2}$ iii $x \geqslant \frac{7}{2}$ iv $f^{-1}(x) = \sqrt{x+\frac{17}{4}}+\frac{7}{2}, x \geqslant -\frac{17}{4}$

f i $(x-\frac{1}{6})^2+\frac{29}{36}$ ii $f_{\min} = \frac{29}{36}$ at $x = \frac{1}{6}$ iii $x \geqslant \frac{1}{6}$ iv $f^{-1}(x) = \sqrt{x-\frac{29}{36}}+\frac{1}{6}, x \geqslant \frac{29}{36}$

4 a $2(x+3)^2-23$ b $3(x+3)^2-26$ c $4(x-1)^2-1$ d $5(x-1)^2-3$ e $2(x-\frac{3}{2})^2-\frac{5}{2}$ f $3(x-\frac{1}{3})^2+\frac{2}{3}$

g $-(x-1)^2+4$ h $-(x+3)^2+10$ i $-(x-\frac{3}{2})^2+\frac{25}{4}$

5 a $f(x) = 4(x+3)^2-39; f_{min} = -39$ at $x = -3$

b

y
O
x
(0, −3)
(−3, −39)

c Domain $x \geqslant -3$, range $f \geqslant -39$

d $f^{-1}(x) = \sqrt{\frac{x+39}{4}}-3, \frac{\sqrt{x+39}}{2}-3$

6 a $(x+1)^2+3$

b Maximum of a function will occur at the minimum of its reciprocal. The minimum of $(x+1)^2+3$ is 3 when $x = -1$; so maximum is $\frac{12}{3} = 4$.

Exercise 1.5

1 a $-3 \leq f(x) \leq 3$

b Sketches passing through the following points:

i $(-4, 12), (-3, 0), (0, -12), (2, -4), (5, -12)$ ... y-coordinates $\times 4$: stretch, domain $-4 \leq x \leq 5$, range $-12 \leq f(x) \leq 12$

ii $(-4, -3), (-3, 0), (0, 3), (2, 1), (5, 3)$ ... y-coordinates $\times -1$: reflect in x-axis, domain $-4 \leq x \leq 5$, range $-3 \leq f(x) \leq 3$

iii $(-4, -12), (-3, 0), (0, 12), (2, 4), (5, 12)$ ... y-coordinates $\times -4$: stretch and reflect in x-axis, domain $-4 \leq x \leq 5$, range $-12 \leq f(x) \leq 12$

iv $(4, 3), (3, 0), (0, -3), (-2, -1), (-5, -3)$ ... x-coordinates $\times -1$: reflect in y-axis, domain $-5 \leq x \leq 4$, range $-3 \leq f(x) \leq 3$

v $(-2, 3), (-1{\cdot}5, 0), (0, -3), (1, -1), (2{\cdot}5, -3)$ ... x-coordinates $\times\frac{1}{2}$: stretch by factor $\frac{1}{2}$, domain $-2 \leq x \leq 2.5$, range $-3 \leq f(x) \leq 3$

vi $(-4, -1), (-3, -4), (0, -7), (2, -5), (5, -7)$ ... y-translation ... down by 4, domain $-4 \leq x \leq 5$, range $-7 \leq f(x) \leq -1$

vii $(-4, 1), (-3, 4), (0, 7), (2, 5), (5, 7)$ ... y-coordinates $\times -1$ then add 4: reflect in x-axis then translate up 4, domain $-4 \leq x \leq 5$, range $1 \leq f(x) \leq 7$

viii $(-6, 4), (-5, 1), (-2, -2), (0, 0), (3, -2)$ ...translation left by 2 and up by 1, domain $-6 \leq x \leq 3$, range $-2 \leq f(x) \leq 4$

ix $(-4, 5), (-3, -4), (0, -13), (2, -7), (5, -13)$ ... y-coordinates $\times 3$ then subtract 4: stretch factor 3 then translate down 4, domain $-4 \leq x \leq 5$, range $-13 \leq f(x) \leq 5$

2 a Line segment joining (0, 1) and (4, 13), range $1 \leq F(x) \leq 13$

b A line joining

i (0, 2) and (4, 26), range $2 \leq 2F(x) \leq 26$

ii (0, 1) and $(\frac{4}{3}, 13)$, range $1 \leq F(3x) \leq 13$

iii $(0, -1)$ and $(4, -13)$, range $-13 \leq -F(x) \leq -1$

iv (0, 1) and $(-4, 13)$, range $1 \leq F(-x) \leq 13$

v (1, 3) and (5, 15), range $3 \leq F(x - 1) + 2 \leq 15$

3 a $(x - 3)^2 + 2$; min. value 2 when $x = 3$

b Sketch of parabola cutting y-axis at (0, 11) with min. turning point (TP) at (3, 2)

c i $4x^2 - 12x + 11$; sketch of parabola cutting y-axis at (0, 11) with min. TP at (1·5, 2)

ii $y = (2x + 1)^2 - 6(2x + 1) + 11 = 4x^2 - 8x + 6$; note: from all x-coordinates we subtract 1 then halve. The curve will be a parabola with y-intercept (0, 6) and min. TP at (1, 2).

iii y-intercept $(0, -8)$ and max. TP at (3, 1)

d i (0, 38), (2, 2) ii (0, 3), (0·5, 2) iii $(0, b^2 - 6b + 11), \left(\frac{3 - b}{a}, 2\right)$

4 a A downward sloping curve (called a hyperbola) passing through (1, 36), (2, 18), (3, 12), (4, 9)

b $h(x) = \frac{36}{x - 1}$, R − {1}:

i Add 1 to each x-coordinate: (2, 36), (3, 18), (4, 12), (5, 9)

ii y-coordinate $\times 2 + 1$: (1, 73), (2, 37), (3, 25), (4, 19); domain R − {0}

iii Halve x-coordinate, multiply y-coordinate by -1 then add 3: $(0{\cdot}5, -33), (1, -15), (1{\cdot}5, -9), (2, -6)$

5 a Range $0 \leq g(x) + 1 \leq 2$

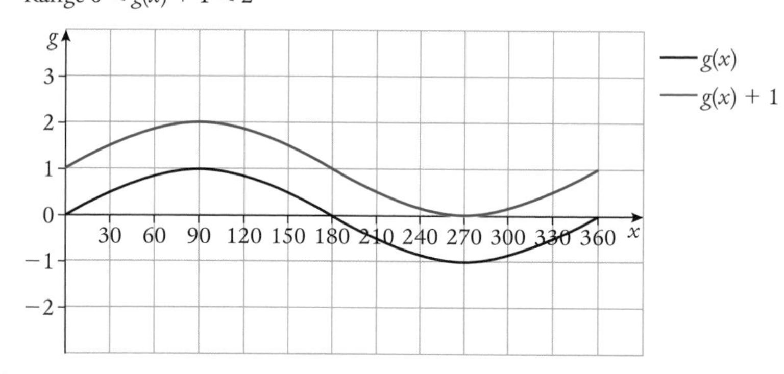

b Range $-1 \leqslant g(x-30) \leqslant 1$

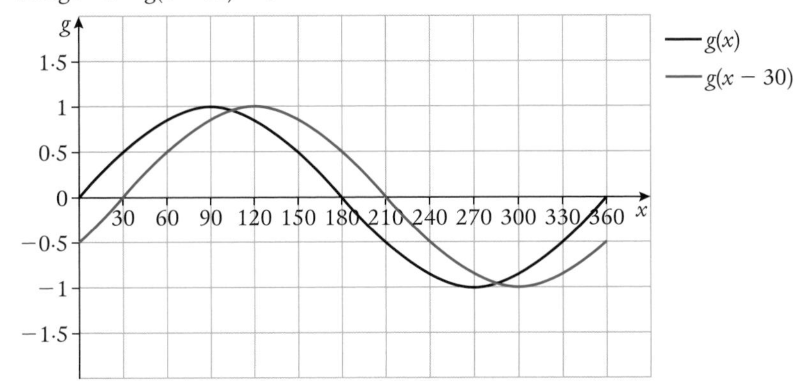

c Range $-1 \leqslant -g(x) \leqslant 1$

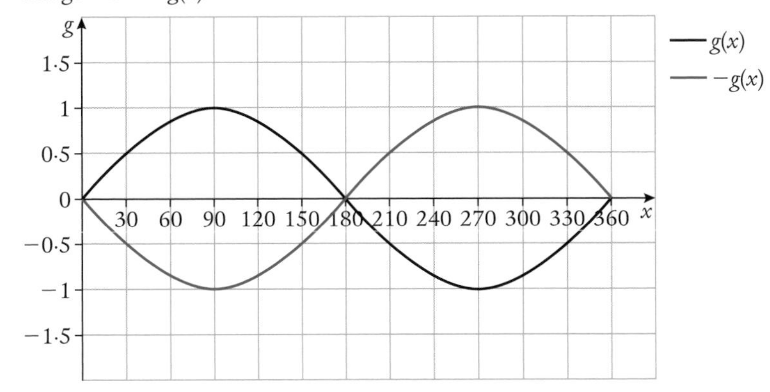

d Range $-1 \leqslant g(-x) \leqslant 1$

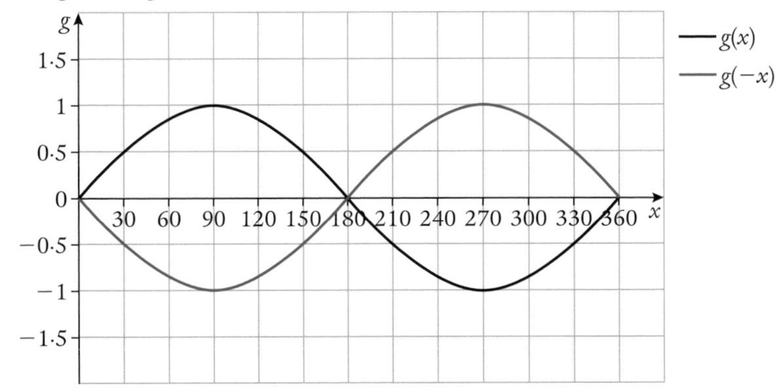

Note that $\sin(-x)° = -\sin x°$

e Range $-3 \leqslant 3g(x) \leqslant 3$

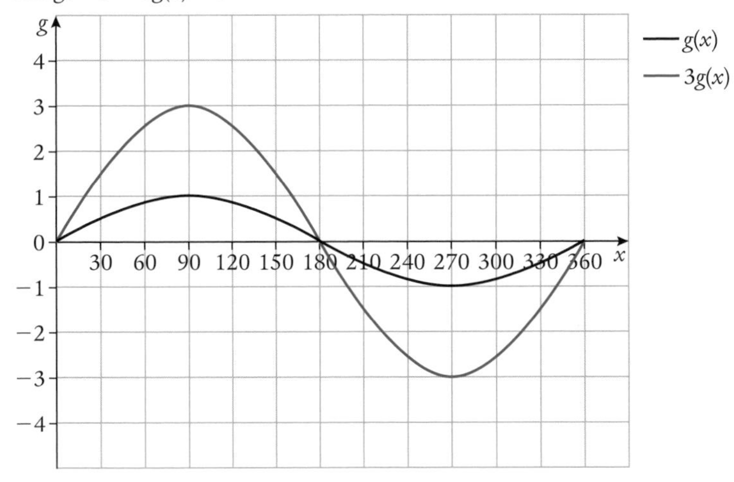

f Range $-1 \leq g(3x) \leq 1$

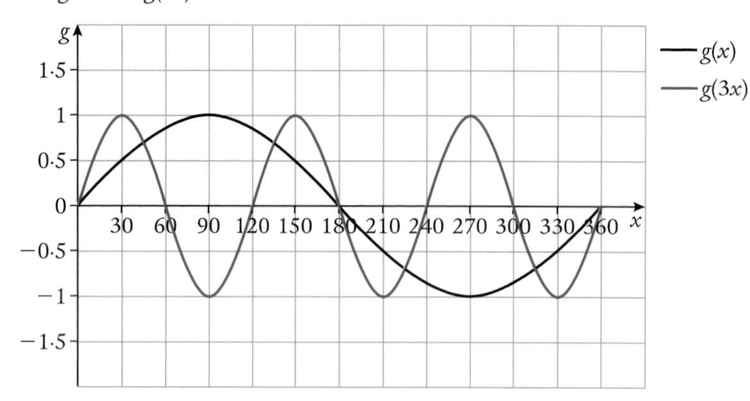

6 a iii ... reflection in y-axis

b i (1, 3), (2, 5), (3, 7), (4, 9)

ii (2, 5), (3, 7), (4, 9), (5, 11)

iii Mapping onto itself

iv $m = 4$, so for each x-step, y increases by 4. So one symmetry would be $4 + k(x - 1)$. There are infinitely many more.

Preparation for assessment

1 a i 13 ii 9 b i $1 \leq x \leq 10$ ii $5 \leq C(x) \leq 23$ c 8

2 a i 6 ii 18 iii $3 - 2 = 1$ b i $(x - 3)^2 + 2$ ii 2 c $2 \leq D \leq 18$

3 a i 2 ii 6 b $f^{-1}(x) = \dfrac{60 - x}{x}$ c 119

4 a i $(2x + 3)^2 + 1 = 4x^2 + 12x + 10$ ii $2(x^2 + 1) + 3 = 2x^2 + 5$

b $x = 0$

5 a $\sin(\sqrt{1 - x^2})$, 0.015 b i $\sqrt{1 - \sin^2 x}$ ii $\cos x$

6 a $h(x) = \dfrac{4}{x^2 + 2x + 2}$ b $(x + 1)^2 + 1$ c 4

7 Curve passing through:

a (0, 0), (1, −2), (4, 0) b (0, 2), (1, 0), (4, 2) c (−1, −1), (0, 1), (3, −1)

8 a $g(-2) = g(1) = 0$ so $g^{-1}(0)$ is ambiguous. (For $-1 < x < 4$, $f^{-1}(x)$ would have more than one possible image)

b i Reflect in y-axis (2, 0), (1, −1), (−1, 0), (−2, 4), (−3, 0) ii (2, 2), (1, 1), (−1, 2), (−2, 6), (−3, 2)

c $-1 \leq x \leq 2$

2 Trigonometry 1 – using radians

What you need to know

1 a i 0·5 ii −0·5 iii −0·5

b i −0·5 ii −0·5 iii 0·5

c i −1 ii 1 iii −1

2 a i 3 ii 180°

b Vertical translation of 1 unit

c Max. value = 4 occurs at 45° and 225°

d 170°, 280°, 350°

3 a Vertical translation of 2 units

b Reflection about the mid-line $y = 2$

c Horizontal translation of 30° to right

4 $y = \tan x°$

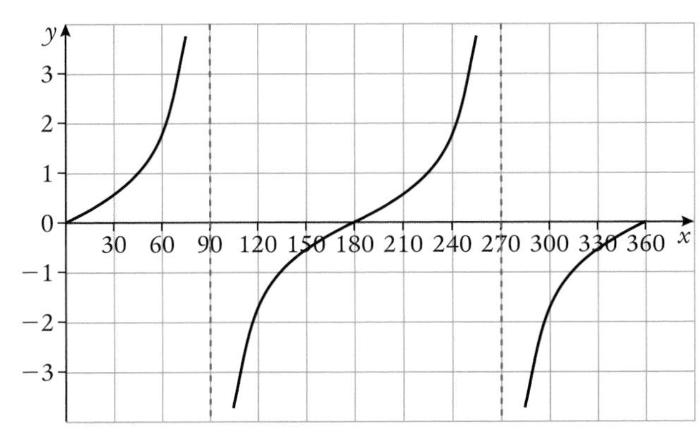

5 a $38{\cdot}7°, 141{\cdot}3°$ b $70{\cdot}5°, 289{\cdot}5°$
c $30°, 150°, 210°, 330°$ d $24{\cdot}7°, 95{\cdot}3°, 204{\cdot}7°, 275{\cdot}3°$

6 a $\sin^2 x° + \cos^2 x° = 1; \dfrac{\sin x}{\cos x} = \tan x$

Exercise 2.1

1 a i $\sin(\pi + x) = -\sin x$ ii $\cos(2\pi - x) = \cos x$ iii $\cos(\pi + x) = -\cos x$
iv $\tan(\pi - x) = -\tan x$ v $\sin(2\pi - x) = -\sin x$ vi $\tan(2\pi - x) = -\tan x$

b $\sin\left(\dfrac{\pi}{2} - x\right) = \cos x$

c The first quadrant lies between 0 and $\dfrac{\pi}{2}$. The second quadrant lies between $\dfrac{\pi}{2}$ and π.

The third quadrant lies between π and $\dfrac{3\pi}{2}$. The fourth quadrant lies between $\dfrac{3\pi}{2}$ and 2π.

2

	30°	45°	60°	90°
sin	$\frac{1}{2}$	$\frac{1}{\sqrt{2}}$	$\frac{\sqrt{3}}{2}$	1
cos	$\frac{\sqrt{3}}{2}$	$\frac{1}{\sqrt{2}}$	$\frac{1}{2}$	0
tan	$\frac{1}{\sqrt{3}}$	1	$\sqrt{3}$	Undefined

	$\frac{\pi}{6}$	$\frac{\pi}{4}$	$\frac{\pi}{3}$	$\frac{\pi}{2}$
sin	$\frac{1}{2}$	$\frac{1}{\sqrt{2}}$	$\frac{\sqrt{3}}{2}$	1
cos	$\frac{\sqrt{3}}{2}$	$\frac{1}{\sqrt{2}}$	$\frac{1}{2}$	0
tan	$\frac{1}{\sqrt{3}}$	1	$\sqrt{3}$	Undefined

3 a $-\dfrac{1}{\sqrt{2}}$ b $-\dfrac{1}{2}$ c $\sqrt{3}$ d $-\dfrac{1}{2}$ e $\dfrac{1}{\sqrt{2}}$
f $-\dfrac{1}{2}$ g $-\dfrac{1}{\sqrt{2}}$ h $-\sqrt{3}$ i $\sqrt{3}$ j $-\dfrac{\sqrt{3}}{2}$

4 a $-\dfrac{1}{\sqrt{2}}$ b $\dfrac{1}{2}$ c $-\sqrt{3}$ d $\dfrac{\sqrt{3}}{2}$ e $\dfrac{1}{\sqrt{2}}$
f -1 g $-\dfrac{1}{\sqrt{2}}$ h $-\sqrt{3}$ i Undefined j -1

5 a $\dfrac{\pi}{18}$ b $\dfrac{\pi}{5}$ c $\dfrac{10\pi}{9}$ d $\dfrac{5\pi}{3}$
e $\dfrac{5\pi}{6}$ f $\dfrac{5\pi}{36}$ g $\dfrac{19\pi}{180}$ h $\dfrac{23\pi}{12}$

6 a 1130 m b 1·40 radians c 1160 m d ii 30 m difference ii 4 significant figures

7 a $L = xr$ b $A = \dfrac{xr^2}{2}$ c $L = r$ cm d $A = r^2$

Exercise 2.2

1 **a** 2π

b Max. is 1 at $x = \frac{\pi}{2}$

c $\frac{5\pi}{2}$

d $-\frac{\pi}{2} < x < 0$ and $\pi < x < 2\pi$ and $3\pi < x < \frac{7\pi}{2}$

e **i**

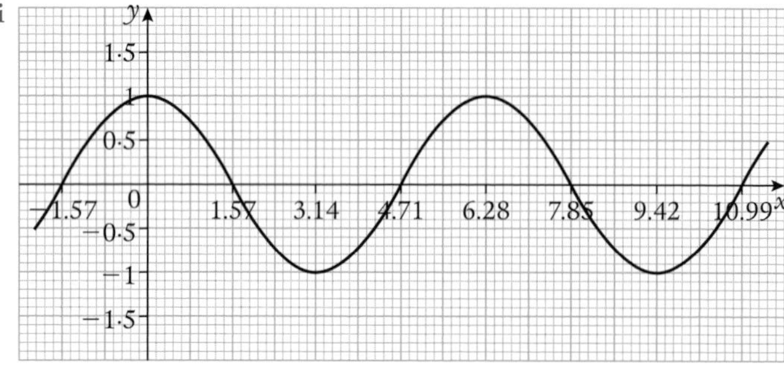

ii

2 Student's own investigation

3 **a, b**

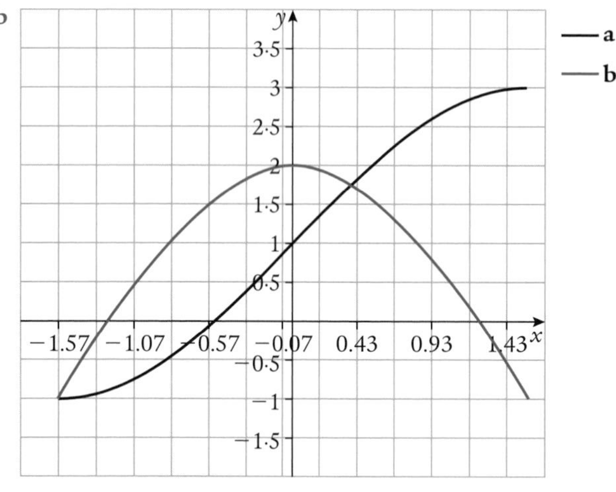

c, d, e, f

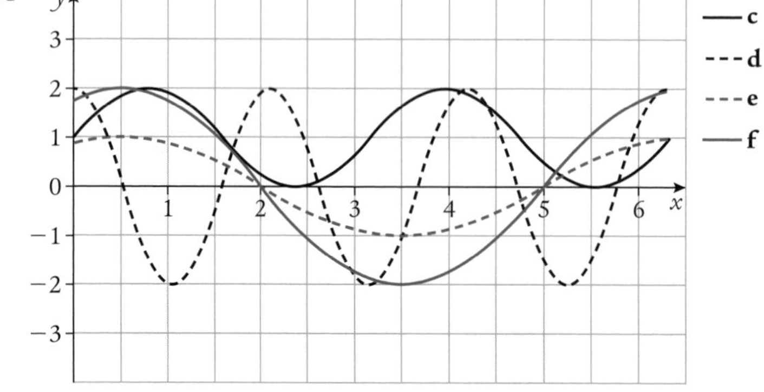

g, h, i

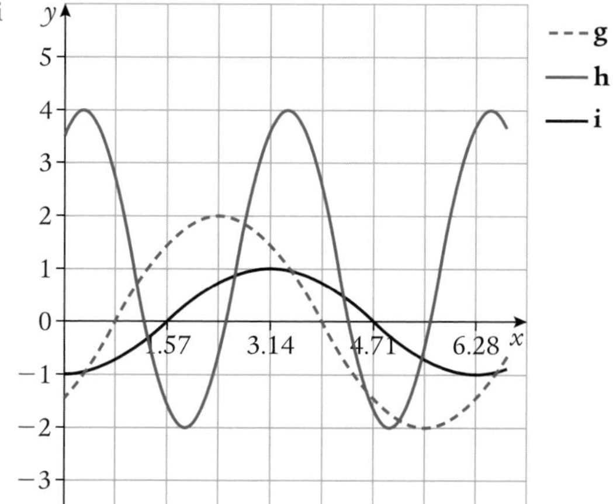

i is the black curve ... it is also $y = -\cos x$

j

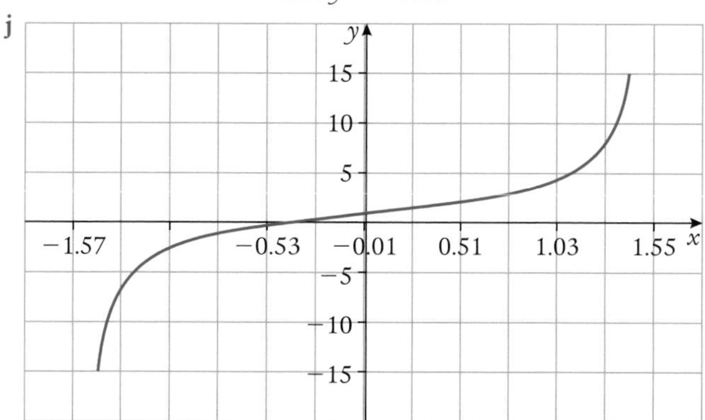

4 a $y = 2\cos 3x$ **b** $y = 3\sin 2x$ **c** $y = -\sin 2x$ **d** $y = 3\sin\dfrac{x}{2}$

5 a 50% **b i** 100% **ii** 50% **iii** 0%

c

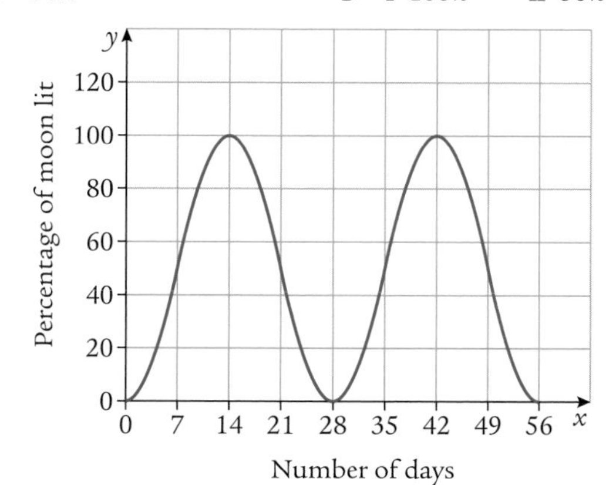

Exercise 2.3

1 a $\dfrac{\pi}{3}, \dfrac{2\pi}{3}$ **b** $\dfrac{\pi}{6}, \dfrac{11\pi}{6}$ **c** $\dfrac{\pi}{6}, \dfrac{5\pi}{6}$ **d** $\dfrac{\pi}{3}, \dfrac{4\pi}{3}$

e $\dfrac{\pi}{4}, \dfrac{3\pi}{4}$ **f** $\dfrac{3\pi}{4}, \dfrac{5\pi}{4}$ **g** $\dfrac{7\pi}{6}, \dfrac{11\pi}{6}$ **h** $\dfrac{2\pi}{3}, \dfrac{4\pi}{3}$

2 $\dfrac{\pi}{6}$

3 a 0·3398, 2·802 **b** $0, 2\pi$ **c** 0·1974, 3·339, 6·481, 9·622

d 1·231, 5·052 **e** 2·678, 5·820 **f** $\pi, 3\pi$

g −1·911, 1·911 **h** −2·214, 2·214, 4·069

4 a i 11·68 **ii** 11·82 **iii** 11·73 **b** 0·5236, 2·618

5 a $0, \frac{\pi}{6}, \pi, \frac{7\pi}{6}, 2\pi$ b $\frac{\pi}{9}, \frac{2\pi}{9}, \frac{7\pi}{9}, \frac{8\pi}{9}, \frac{13\pi}{9}, \frac{14\pi}{9}$ c $\frac{5\pi}{24}, \frac{13\pi}{24}, \frac{29\pi}{24}, \frac{37\pi}{24}$

d $\frac{\pi}{12}, \frac{7\pi}{12}, \frac{13\pi}{12}, \frac{19\pi}{12}$ e $\frac{\pi}{4}, \frac{3\pi}{4}, \frac{11\pi}{12}, \frac{17\pi}{12}, \frac{19\pi}{12}$ f 0·7058,1·865, 3·847, 5·007

6 a i 7 m ii 5.45 hrs, 17.45 hrs, or 5.27 a.m., 5.27 p.m.

b 11.27 a.m. c 1·06 hours, 9·84 hours, 13·06 hours, 21·84 hours or 1.03 a.m., 9.51 a.m., 1.03 p.m. and 9.51 p.m.

7 a $\frac{\pi}{2}, \frac{\pi}{6}, \frac{5\pi}{6}$ b 2·301, 3·983 c 0·4636, 3·605, 2·820, 5·961

d 3·394, 6·031, $\frac{3\pi}{2}$ e No solutions f 2·601, 5·743, 1·816, 4·957

8 0·340, 2·802, 3·665, 5·760, 6·623, 9·085, 9·948 (months)

Preparation for assessment

1 71·6°

2 0·1684

3 a $\frac{\sqrt{3}}{2}$ units^2 b $\sqrt{3}$

4 0·3398, 2·802

5 a 8·18 units b 0·6981 s, 1·396 s, 2·793 s, 3·491 s, 4·887 s, 5·585 s

6 a $0, \frac{\pi}{6}, \pi, \frac{7\pi}{6}, 2\pi$ b In the interval $0 < x < \frac{\pi}{6}$, the height is negative for $\pi < x < \frac{7\pi}{6}$

7 $\frac{3\pi}{2}$, 0·730, 2·41

3 Trigonometry 2 – compound angle formulae

What you need to know

1 a $(2\cos 135°, 2\sin 135°)$ b $(\cos 240°, \sin 240°)$

c $(3\cos 300°, 3\sin 300°)$ d $(3\cos(-60)°, 3\sin(-60)°)$

2 $\sin(30° + 60°) = \sin 90° = 1$; $\sin 30° + \sin 60° = \frac{1}{2} + \frac{\sqrt{3}}{2} \neq 1$

3 a $-\frac{\sqrt{3}}{2}$ b $\frac{\sqrt{3}}{2}$ c $-\frac{1}{2}$ d $\frac{1}{\sqrt{2}}$ e $-\frac{1}{\sqrt{2}}$

4 a $\sin x°$ b $\cos x°$ c $\cos x°$ d $\sin x°$ e $-\cos x°$

5 a 0·64 b $\frac{3}{5}$

6 PQ = 5

Exercise 3.1

1 a $\cos x \cos y - \sin x \sin y$ b $\cos p \cos q + \sin p \sin q$

c $\cos 2p \cos q - \sin 2p \sin q$ d $\cos 3a \cos 2b - \sin 3a \sin 2b$

2 a $\cos\left(\frac{\pi}{3} + \frac{\pi}{6}\right) = \cos\frac{\pi}{2} = 0$: $\cos\left(\frac{\pi}{3}\right)\cos\left(\frac{\pi}{6}\right) - \sin\left(\frac{\pi}{3}\right)\sin\left(\frac{\pi}{6}\right) = \left(\frac{1}{2}\right)\left(\frac{\sqrt{3}}{2}\right) - \left(\frac{\sqrt{3}}{2}\right)\left(\frac{1}{2}\right) = 0$

b $\cos(120 - 30)° = \cos 90° = 0$: $\cos 120° \cos 30° + \sin 120° \sin 30° = -\cos 60° \cos 30° + \sin 60° \sin 30° = 0$

c $\cos(180 - x)° = \cos 180° \cos x° + \sin 180° \sin x° = -\cos x° + 0 . \sin x° = -\cos x°$

3 a i $\cos 90° \cos x° - \sin 90° \sin x° = -\sin x°$ ii $\cos 90° \cos x° + \sin 90° \sin x° = \sin x°$

b i $\cos 180° \cos 90° - \sin 180° \sin 90° = 0$ ii $\cos 270° \cos x° - \sin 270° \sin x° = 0 - -1 . \sin x° = \sin x$

c $\cos 30° = \cos(10 + 20)° = \cos 10° \cos 20° - \sin 10° \sin 20°$

4 a i $\cos(45 + 30)° = \cos 45° \cos 30° - \sin 45° \sin 30° = \frac{\sqrt{3} - 1}{2\sqrt{2}}$ ii $\frac{1 - \sqrt{3}}{2\sqrt{2}}$ iii $\frac{\sqrt{3} + 1}{2\sqrt{2}}$

b i $\frac{\sqrt{3} - 1}{2\sqrt{2}}$ ii $\frac{\sqrt{3} + 1}{2\sqrt{2}}$

5 a $\frac{33}{65}$ **b** $\frac{323}{325}$ **c** $-\frac{155}{493}$

6 a $\frac{3\sqrt{3}}{2}\cos x° + \frac{3}{2}\sin x°$ **b** $2\cos x° - 2\sqrt{3}\sin x°$ **c** $-\sqrt{2}(\cos x° - \sin x°)$

d $\frac{5}{\sqrt{2}}\cos x° - \frac{5}{\sqrt{2}}\sin x°$ **e** $-\frac{\sqrt{3}}{2}\cos x + \frac{\sin x}{2}$ **f** $-\frac{3}{2}\cos x° - \frac{3\sqrt{3}}{2}\sin x°$

7 a $\cos(20 + 70)° = \cos 90° = 0$ **b i** -1 **ii** 1 **iii** -1

c i $\cos 111°$ **ii** $\cos 280°$ **iii** $-\cos 44°$

8 a 25 cm **b** $\frac{44}{125}$ **c** $\cos\angle ADC = -\frac{220}{625}$; $\cos\angle ADC = \cos(180 - \angle ABC) = -\cos\angle ABC$

9 a i 146 m **ii** 390 m **iii** $\frac{4233}{4745}$ **c** 27° to nearest degree

10 a i $2\cos A\cos B$ **ii** $2\cos 40°\cos 20°$ **b** $2\cos 42°\cos 25°$

Exercise 3.2

1 a $\sin 180°\cos A° - \cos 180°\sin A° = \sin A°$ **b** $\cos A$ **c i** $\frac{\sqrt{3}-1}{2\sqrt{2}}$ **ii** $\frac{\sqrt{3}+1}{2\sqrt{2}}$

2 a i $\frac{1+\sqrt{3}}{2\sqrt{2}}$ **ii** $\frac{1+\sqrt{3}}{2\sqrt{2}}$ **iii** $\frac{\sqrt{3}-1}{2\sqrt{2}}$ **b i** $\frac{1+\sqrt{3}}{2\sqrt{2}}$ **ii** $\frac{\sqrt{3}-1}{2\sqrt{2}}$

3 a $\sin x° - \sqrt{3}\cos x°$ **b** $\frac{\sqrt{3}}{2}(\sin x° + \cos x°)$ **c** $-\sqrt{2}(\sin x° + \cos x°)$

d $\frac{3}{2}(\sqrt{3}\sin x - \cos x)$ **e** $-\frac{1}{\sqrt{2}}(\sin x + \cos x)$ **f** $1 - \sqrt{3}\sin x + \cos x$

4 a $\sin 70°$ **b** $\sin 49°$ **c** $\sin 100°$ **d** $2\sin 35°$ **e** $2\sin 51°$ **f** $3\sin 49°$

5 a 75 cm **b i** $\frac{364}{725}$ **ii** $\frac{364}{725}$ **c** $\frac{21}{29}$

6 $\frac{63}{65}$

7 a i $\sqrt{10}$ **ii** $\sqrt{13}$ **b i** $\frac{3}{\sqrt{130}}$ **ii** $\frac{11}{\sqrt{130}}$ **iii** $\frac{3}{11}$

8 a i $\frac{19}{21}$ **ii** $-\frac{11}{21}$ **b i** $-\frac{4\sqrt{5}}{21}$ **ii** $\frac{8\sqrt{5}}{21}$ **c** $-\frac{19\sqrt{5}}{20}$

9 a $\frac{3\sqrt{15} + 7\sqrt{7}}{32}$ **b** 70·4°

10 a $2\sin A\cos B$ **b** A 30°, B 20° **c** $\sin 50° + \sin 10° = 2\sin 30°\cos 20° = \cos 20°$

d $\sqrt{3}\cos 10°$ **e** $\sqrt{\frac{3}{2}}$

11 a Since B is a constant in graph then so is $\sin B$. So it acts as a simple vertical translation.

b Diagram suggests $A = 0$ or 2π are solutions; a quick substitution verifies this.
Diagram suggests third point is where they are both zero simultaneously.
$\sin(A + B) = 0 \Rightarrow A + B = 0, \pi, 2\pi, \ldots$. In this example, $B = \frac{3\pi}{4} \Rightarrow A = -\frac{3\pi}{4}, \frac{\pi}{4}, \frac{5\pi}{4}, \ldots$
Checking these values into $\sin A + \sin B = 0$ gives us that this is true when $A = \frac{5\pi}{4}$ (in the case when $B = \frac{3\pi}{4}$).

Exercise 3.3

1 a i $-0·68$ **ii** $-0·28$ **iii** 1 **iv** -1 **b i** 0·5 **ii** 0·98 **iii** -1 **iv** 1

2 a $\frac{120}{169}$ **b** $-\frac{119}{169}$ **c** $\frac{24}{25}$ **d** $-\frac{7}{25}$ **e** $-\frac{3696}{4225}$ **f** $-\frac{837}{845}$

3 a i $\frac{20}{29}$ **ii** $\frac{840}{841}$ **iii** $-\frac{41}{841}$ **b i** $-\frac{68\,880}{707\,281}$ **ii** $-\frac{703\,919}{707\,281}$

4 a $(180 - 2x)°$ **b i** $\frac{336}{625}$ **ii** $-\frac{527}{625}$ **c i** $-\frac{354\,144}{390\,625}$ **ii** $\frac{164\,833}{390\,625}$

5 a $\cos x = \cos^2\frac{x}{2} - \sin^2\frac{x}{2}$ **b i** $\frac{2184}{7225}$ **ii** $-\frac{6887}{7225}$

6 a $y = 2\cos^2 x - \cos x - 1$
b $(2\cos x + 1)(\cos x - 1)$
c i $(\cos x + 2)(2\cos x - 1)$ ii $(\sin x + 1)(3 - 2\sin x)$ iii $(2\sin x + 1)(1 - \sin x)$
iv $(4\cos x + 3)(\cos x + 1)$ v $(-4\sin x + 1)(\sin x + 2)$ vi $(2\cos x - 1)(3\cos x + 2)$

7 a i $2\sin x\cos x + 2\sin x$ ii $2\sin x(\cos x + 1)$
b i $\sin x(6\cos x + 1)$ ii $\cos x(4\sin x - 1)$ iii $2\sin x(\cos x + 2)$ iv $\cos x(3 - 4\sin x)$

8 a $\sin 2x\cos x + \cos 2x\sin x$ b $3\sin x - 4\sin^3 x$ c $4\cos^3 x - 3\cos x$

9 a $\sin\frac{2\pi}{3}$ b $\cos\frac{\pi}{2}$ c $\cos\frac{\pi}{6}$ d $\cos\frac{\pi}{4}$ e $-\cos\frac{5\pi}{6}$ f $-\cos\frac{4\pi}{3} + 2$

10 a $\sqrt{2r^2 - 2r^2\cos 2x}$ b $2r\sin x$ c $2r^2 - 2r^2\cos 2x = 4r^2\sin^2 x \Rightarrow \cos 2x = 1 - 2\sin^2 x$

Exercise 3.4

1 a a b $-a$ c $x = \frac{\pi}{2} - b$; $\left(\frac{\pi}{2} - b, a\right)$

2 a $13\sin(x + 22{\cdot}6)°$; i 13 ii -13 iii 5 iv 22·6° left
b $25\sin(x - 16{\cdot}3)°$; i 25 ii -25 iii -7 vi 16·3° right
c $29\cos(x - 46{\cdot}4)°$; i 29 ii -29 iii 20 iv 46·4° right
d $2\cos(x + 30)°$; i 2 ii -2 iii $\sqrt{3}$ iv 30° left

3 a $5\sin(x - 307)°$ b $61\sin(x + 349{\cdot}6)°$ c $6{\cdot}5\cos(x + 346)°$ d $\sqrt{3}\cos(x - 215)°$

4 a $2\sin\left(x + \frac{\pi}{6}\right)$ b $2\sin\left(x - \frac{\pi}{3}\right)$ c $\sqrt{2}\cos\left(x + \frac{7\pi}{4}\right)$ d $5\cos(x - 2{\cdot}5)$

5 a i $17\sin(x + 61{\cdot}9)°$ ii $17\sin(x - 298{\cdot}0)°$ iii $17\cos(x - 28{\cdot}1)°$
b All cut y-axis at 15, amplitude 17, period 360, 1st max. (28·1, 17) using a spreadsheet all graphs are superimposed

6 a Max.: $3 + \sqrt{2}$; min.: $3 - \sqrt{2}$, 1st max. $\left(\frac{\pi}{4}, 3 + \sqrt{2}\right)$; y-intercept = 4 b Max.: 4; min.: 0, 1st max. (150, 4)

7 $25\cos(x - 0{\cdot}284) + 35$; max. = 60; min. = 10; $\min_y = \frac{1}{2}$, $\max_y = 3$

8 a $4\sin x + 6 = 5\cos x + 4$ b $\sqrt{41}\sin(x + 5{\cdot}39)$ c $x = 0{\cdot}896, 4{\cdot}038$

9 a 1·03 s, 4·172 s b $5\cos 1{\cdot}03 + 20 = 22{\cdot}6$ m; $5\cos 4{\cdot}172 + 20 = 17{\cdot}4$ m

10 Student's own work

Preparation for assessment

1 a $\frac{3\sqrt{3}}{2}\sin x° + \frac{3}{2}\cos x°$ b $2\cos x + 2\sqrt{3}\sin x$

2 a $\frac{67}{77}$ b $\frac{24\sqrt{10}}{77}$ c $\frac{31}{49}$ d $\frac{12\sqrt{10}}{49}$ e $\frac{744\sqrt{10}}{2401}$

3 a $\frac{2}{5\sqrt{5}}$ b $\frac{11}{5\sqrt{5}}$

4 a $\frac{\cos^2 A - \sin^2 A}{\sin A + \cos A} = \frac{(\cos A - \sin A)(\cos A + \sin A)}{\sin A + \cos A} = \cos A - \sin A$ b $\frac{3\pi}{4}, \frac{7\pi}{4}$

5 a $\sqrt{2}$ b 1

6 a $5\cos(x + 5{\cdot}356)$ b 1 c Yes; does not equal 0 ... for all x

4 Vectors

What you need to know

1 a

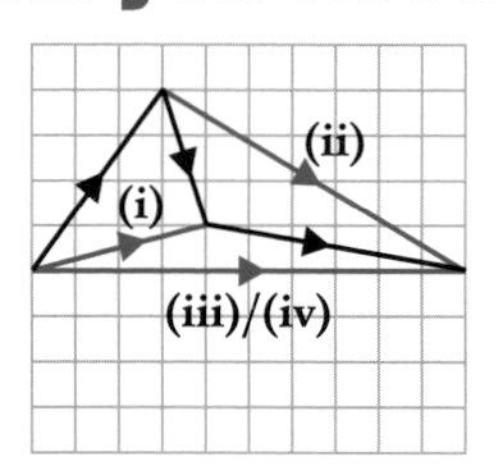

b All three expressions are equivalent

2 a i $\begin{pmatrix}7\\3\end{pmatrix}$ ii $\begin{pmatrix}-2\\-4\end{pmatrix}$ iii $\begin{pmatrix}0\\0\end{pmatrix}$

b i $\begin{pmatrix}-1\\-6\end{pmatrix}$ ii $\begin{pmatrix}3\\-2\end{pmatrix}$ iii $\begin{pmatrix}-5\\7\end{pmatrix}$ iv $\begin{pmatrix}2\\8\end{pmatrix}$

c i $\begin{pmatrix}2\\3\end{pmatrix}$ ii $\begin{pmatrix}-3\\3\end{pmatrix}$ iii $\begin{pmatrix}8\\-2\end{pmatrix}$

d i $\begin{pmatrix}4\\14\end{pmatrix}$ ii $\begin{pmatrix}4\\14\end{pmatrix}$ iii $\begin{pmatrix}10\\35\end{pmatrix}$ iv $\begin{pmatrix}-6\\-21\end{pmatrix}$

e i $\begin{pmatrix}4\\12\end{pmatrix}$ ii $\begin{pmatrix}6\\-15\end{pmatrix}$ iii $\begin{pmatrix}-4\\21\end{pmatrix}$ iv $\begin{pmatrix}11\\-11\end{pmatrix}$

3 a (0, 0, 0), (3, 0, 0), (3, 3, 0), (0, 3, 0), (0, 0, 3), (3, 0, 3), (3, 3, 3), (0, 3, 3)

b i $\begin{pmatrix}0\\2\\1\end{pmatrix}$ ii $\begin{pmatrix}2\\-3\\-1\end{pmatrix}$

c $\begin{pmatrix}0\\2\\1\end{pmatrix}+\begin{pmatrix}2\\-3\\-1\end{pmatrix}=\begin{pmatrix}2\\-1\\0\end{pmatrix}$

4 a $\begin{pmatrix}4\\-3\\3\end{pmatrix}$ **b** $\begin{pmatrix}7\\-6\\-1\end{pmatrix}$ **c** $\begin{pmatrix}9\\-8\\9\end{pmatrix}$ **d** $\begin{pmatrix}-7\\3\\14\end{pmatrix}$

5 $\mathbf{q}$

Exercise 4.1

1 a $\begin{pmatrix}1\\2\\7\end{pmatrix}$ **b** $\begin{pmatrix}-1\\0\\-3\end{pmatrix}$ **c** $\begin{pmatrix}-1\\-5\\4\end{pmatrix}$

2 a i $\begin{pmatrix}4\\-3\\-4\end{pmatrix}$ ii $\begin{pmatrix}-4\\-2\\-7\end{pmatrix}$ iii $\begin{pmatrix}4\\2\\7\end{pmatrix}$ iv $\begin{pmatrix}8\\-3\\-3\end{pmatrix}$ v $\begin{pmatrix}-8\\3\\3\end{pmatrix}$

b i $\begin{pmatrix}-x\\-y\\-z\end{pmatrix}$ ii $\begin{pmatrix}0\\0\\0\end{pmatrix}$

3 a 3 **b** 17 **c** 11 **d** 21

4 a i 15 ii 9 iii 7

b $|\overrightarrow{EF}| = |\overrightarrow{GH}| = 21; \begin{pmatrix}18\\9\\6\end{pmatrix} \neq \begin{pmatrix}8\\19\\4\end{pmatrix}$

5 $|\mathbf{a}| = 7, |\mathbf{b}| = 3, |\mathbf{a}+\mathbf{b}| = \sqrt{90} < 10$

6 AB and KL, CD and EF, GH and MN, IJ and OP

7 a 9 **b** $\overrightarrow{SP} = 2\overrightarrow{AM}$

8 a $\overrightarrow{BC} = 3\overrightarrow{AB}$; B common **b** $\overrightarrow{QR} = \frac{3}{2}\overrightarrow{PQ}$; Q common **c** $\overrightarrow{TV} = -\frac{5}{2}\overrightarrow{ST}$; T common **d** $\overrightarrow{XY} = \frac{1}{2}\overrightarrow{WX}$; X common

9 a $\overrightarrow{EP_1} \neq k\overrightarrow{EM}$ for any k **b** P_2 **c** $23 - 21 = 2$ units

Exercise 4.2

1 a i $\mathbf{v}$ ii $-\mathbf{u}$ iii $\mathbf{u}+\mathbf{v}$ iv $\mathbf{v}-\mathbf{u}$

b i $\overrightarrow{AB}+\overrightarrow{AD}$ ii $\overrightarrow{AD}-\overrightarrow{AB}$

c i $\mathbf{b}-\mathbf{a}$ ii $\mathbf{b}-\mathbf{a}=\mathbf{c}-\mathbf{d} \Rightarrow \mathbf{a}+\mathbf{c}=\mathbf{b}+\mathbf{d}$ iii $\frac{1}{2}(\mathbf{a}+\mathbf{c})$ iv $\frac{1}{2}(\mathbf{b}+\mathbf{d})$

v $\mathbf{m}=\frac{1}{2}(\mathbf{a}+\mathbf{c}), 2\mathbf{m}=\mathbf{a}+\mathbf{c}$

$\mathbf{n}=\frac{1}{2}(\mathbf{b}+\mathbf{d}), 2\mathbf{n}=\mathbf{b}+\mathbf{d}$

from **ii** $\mathbf{a}+\mathbf{c}=\mathbf{b}+\mathbf{d}, 2\mathbf{m}=2\mathbf{n}, \mathbf{m}=\mathbf{n}$

2 a i $-\mathbf{u}$ ii $-\mathbf{u}+\mathbf{v}$ iii $-\mathbf{u}+\mathbf{v}$ iv $\mathbf{v}$ v $2(-\mathbf{u}+\mathbf{v})$ vi $-2\mathbf{u}+\mathbf{v}$ vii $-\mathbf{u}+2\mathbf{v}$ viii $-2\mathbf{u}$

b $(-\mathbf{u}+2\mathbf{v})+(-\mathbf{u}-\mathbf{v})+(2\mathbf{u}-\mathbf{v})=\mathbf{0}$

3 a i $\mathbf{u}+\mathbf{v}$ ii $\mathbf{w}+\mathbf{v}$ iii $\mathbf{w}+\mathbf{v}+\mathbf{u}$ iv $\mathbf{u}+\mathbf{v}+\frac{1}{2}\mathbf{w}$ **b** $\mathbf{u}+\frac{1}{2}\mathbf{v}+\frac{1}{2}\mathbf{w}$ **c** $\frac{1}{2}\mathbf{u}+\frac{1}{2}\mathbf{v}$

4 a i $\mathbf{u}+\mathbf{v}$ ii $-\mathbf{w}-\mathbf{v}-\mathbf{u}$ iii $\mathbf{v}-\mathbf{w}$ iv $\frac{1}{2}(\mathbf{v}+\mathbf{w})$ **b** $\mathbf{u}+\frac{1}{2}(\mathbf{v}+\mathbf{w})$

c i $-\frac{1}{2}(\mathbf{v}+\mathbf{w})$ ii $-(\mathbf{v}+\mathbf{w})$; $MN=\frac{1}{2}DB$

5 a i $\mathbf{m}=\frac{1}{2}(\mathbf{a}+\mathbf{b}); \mathbf{n}=\frac{1}{2}(\mathbf{b}+\mathbf{c})$ ii $\overrightarrow{MN}=\mathbf{n}-\mathbf{m}=\frac{1}{2}(\mathbf{b}+\mathbf{c}-\mathbf{a}-\mathbf{b})=\frac{1}{2}(\mathbf{c}-\mathbf{a})$

iii $\overrightarrow{AC}=\mathbf{c}-\mathbf{a} \Rightarrow \overrightarrow{MN}=k\overrightarrow{AC} \Rightarrow MN \parallel AC$

iv $\mathbf{p}=\frac{1}{2}(\mathbf{a}+\mathbf{c}); \overrightarrow{MP}=\mathbf{p}-\mathbf{m}=\frac{1}{2}(\mathbf{a}+\mathbf{c}-\mathbf{a}-\mathbf{b})=\frac{1}{2}(\mathbf{c}-\mathbf{b}); \overrightarrow{BC}=\mathbf{c}-\mathbf{b} \Rightarrow \overrightarrow{MP}=k\overrightarrow{BC} \Rightarrow MP \parallel BC$

b i $\overrightarrow{AN}=\mathbf{n}-\mathbf{a}=\frac{1}{2}(\mathbf{b}+\mathbf{c})-\mathbf{a}$ ii $\overrightarrow{BP}=\frac{1}{2}(\mathbf{a}+\mathbf{c})-\mathbf{b}; \overrightarrow{CM}=\frac{1}{2}(\mathbf{a}+\mathbf{b})-\mathbf{c}$

iii $\frac{1}{2}(\mathbf{b}+\mathbf{c})-\mathbf{a}+\frac{1}{2}(\mathbf{a}+\mathbf{c})-\mathbf{b}+\frac{1}{2}(\mathbf{a}+\mathbf{b})-\mathbf{c}=0$

6 a $2\mathbf{u}+5\mathbf{v}=\mathbf{w}$. So $\mathbf{u}, \mathbf{v}, \mathbf{w}$ co-planar.

b $4x_p+2x_q=x_r$ and $4y_p+2y_q=y_r$ but $4z_p+2z_q \neq z_r$

Exercise 4.3

1 a (6, 12, 19) b (2, −1, 9) c (5, −4, 3) d (−2, −1, 3)
e (1, 3, 9) f (2, −3, 3) g (5·5, −0·5, 4)

2 a M(5, −2, 7) b N(6, 0, 2) c $\begin{pmatrix} 1 \\ 2 \\ -5 \end{pmatrix}$

d $\overrightarrow{QR} = \begin{pmatrix} 4 \\ 8 \\ -20 \end{pmatrix} = 4\begin{pmatrix} 1 \\ 2 \\ -5 \end{pmatrix} \Rightarrow QR \parallel MN$

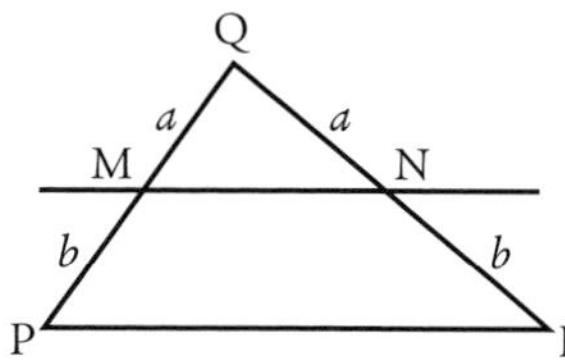

e $\mathbf{m} = \dfrac{a\mathbf{p} + b\mathbf{q}}{a+b}$; $\mathbf{n} = \dfrac{a\mathbf{r} + b\mathbf{q}}{a+b}$

$\overrightarrow{MN} = \dfrac{a\mathbf{r} + b\mathbf{q} - a\mathbf{p} - b\mathbf{q}}{a+b} = \dfrac{a(\mathbf{r} - \mathbf{p})}{a+b} = \dfrac{a}{a+b}\overrightarrow{PR}$

3 a P(5, 1, −2) b E($\frac{13}{3}$, 0, 0) c i Q(7, −2, 1) ii F($\frac{13}{3}$, 0, 0) iii E and F coincident

4 $\overrightarrow{AQ} = \mathbf{q} - \mathbf{a} = \frac{1}{2}(\mathbf{b} + \mathbf{c}) - \mathbf{a}$. Taking inspiration from Q3, find S the point that divides AQ in the ratio 2 : 1.
$\mathbf{s} = \dfrac{2.(\frac{1}{2}(\mathbf{b} + \mathbf{c})) + 1.\mathbf{a}}{2+1} = \frac{1}{3}(\mathbf{a} + \mathbf{b} + \mathbf{c})$. Similarly the point that divides CP and BR is also $\frac{1}{3}(\mathbf{a} + \mathbf{b} + \mathbf{c})$. Hence medians trisect each other at the point of concurrency $\mathbf{s} = \frac{1}{3}(\mathbf{a} + \mathbf{b} + \mathbf{c})$.

5 a AP : PB = 1 : 3 b AP : PB = 2 : 5 c i AP : PB = $(x - a_1) : (b_1 - x)$ ii $\mathbf{p} - \mathbf{a} : \mathbf{b} - \mathbf{p}$

6 a $\mathbf{t} = \dfrac{3\mathbf{r} + \mathbf{s}}{4}$ b $\mathbf{v} = \dfrac{4\mathbf{t} + 3\mathbf{p}}{7} = \dfrac{3\mathbf{r} + \mathbf{s} + 3\mathbf{p}}{7}$ c $\mathbf{q} = \mathbf{p} + \mathbf{r} - \mathbf{s}$

d i $\mathbf{u} = \dfrac{4\mathbf{s} + 3\mathbf{q}}{7}$ ii $\dfrac{3\mathbf{p} + 3\mathbf{r} + \mathbf{s}}{7}$; coincident ... i.e. same point

7 a B(38, 9, 15) b 38·8 units

8 a i $\mathbf{e} = \frac{1}{2}(\mathbf{a} + \mathbf{b})$ ii $\mathbf{f} = \frac{1}{2}(\mathbf{c} + \mathbf{d})$ b $\frac{1}{2}(\frac{1}{2}(\mathbf{a} + \mathbf{b}) + \frac{1}{2}(\mathbf{c} + \mathbf{d})) = \frac{1}{4}(\mathbf{a} + \mathbf{b} + \mathbf{c} + \mathbf{d})$
c Same result as b; the lines joining the midpoints of opposite sides of a tetrahedron bisect each other.

Exercise 4.4

1 a $3\mathbf{i} + 7\mathbf{j} + 9\mathbf{k}$ b $-\mathbf{i} + \mathbf{j} - \mathbf{k}$ c $\mathbf{i} - 2\mathbf{k}$ d $6\mathbf{k}$

2 a 19 b 9 c 11 d 39

3 a $28\mathbf{i} + 42\mathbf{j} - 14\mathbf{k}$ b $-8\mathbf{i} + 3\mathbf{j} + 21\mathbf{k}$ c $10\mathbf{i} + 19\mathbf{j} - 3\mathbf{k}$
d $3\mathbf{i} + 7\mathbf{j} - 5\mathbf{k}$ e $-24\mathbf{i} - 12\mathbf{j} + 24\mathbf{k}$ f $\sqrt{197} = 14·0$ (1 d.p.)

4 a $\frac{14}{15}\mathbf{i} - \frac{1}{3}\mathbf{j} + \frac{2}{15}\mathbf{k}$ b 8

5 a Scalar product = 3.(−2) + 5.4 + (−2).7 = −6 + 20 − 14 = 0 b $a = -2$

6 $8\mathbf{i} + 10\mathbf{j} + 8\mathbf{k}$

7 a 18 cos 35° = 14·7 (3 s.f.) b 16·8 cos 95° = −1·46 (3 s.f.) c 28 cos 40° = 21·4 (3 s.f.)

8 a 2.2.cos 60° = 2 b $2.\sqrt{3}$.cos 30° = 3 c 0 d 2.2.cos 120° = −2

9 a 0 b $2.2\sqrt{2}$.cos 45° = 4 c 0 d $2.2\sqrt{2}$.cos 45° = 4

10 a i 4 ii 7 iii 47 b 89·3° (3 s.f.)

11 a 44·0° b 76·5° c 52·7° (all to 1 d.p.)

12 a 8·11 m b 38·2°

13 $\angle A = 46·4°$, $\angle B = 60·7°$, $\angle C = 73·0°$

14 41·2°

Preparation for assessment

1 a $\begin{pmatrix} 2 \\ -10 \\ 4 \end{pmatrix}$ b $\sqrt{120} = 2\sqrt{30}$

2 a $\frac{3}{4}$ b -1 c $\begin{pmatrix} \frac{3}{13} \\ \frac{12}{13} \\ \frac{4}{13} \end{pmatrix}$

3 $\overrightarrow{AB} = \begin{pmatrix} 7-5 \\ 5-1 \\ -4-2 \end{pmatrix} = \begin{pmatrix} 2 \\ 4 \\ -6 \end{pmatrix}; \overrightarrow{AC} = \begin{pmatrix} 10-5 \\ 11-1 \\ -13-2 \end{pmatrix} = \begin{pmatrix} 5 \\ 10 \\ -15 \end{pmatrix} \Rightarrow \overrightarrow{AC} = \frac{5}{2}\overrightarrow{AB} = AC \parallel AB$

Since A is a common point, the three points (trees) are collinear

4 a $\mathbf{u} + \mathbf{v}$ b $\frac{1}{2}\mathbf{v}$ c $\mathbf{u} + \frac{1}{2}\mathbf{v}$ d $\frac{1}{2}\mathbf{v} - \mathbf{u}$ e $\frac{1}{2}\mathbf{v}$

5 It is easy to work out that $4\mathbf{u} - 3\mathbf{v} = \mathbf{w}$ ... so $\mathbf{u}$, $\mathbf{v}$, and $\mathbf{w}$ are coplanar

6 (5, 0, 2)

7 $\mathbf{a}.(\mathbf{b} + \mathbf{c}) = \mathbf{a}.\mathbf{b} + \mathbf{a}.\mathbf{c} = 2.2.\cos 60° + 2.2.\cos 30° = 2 + 2\sqrt{3}$

8 a A(1, 3, 3), B(0, 1, 1), C(3, 1, 3) b 49·7°

5 Quadratic theory
What you need to know

1 a $-4, 3$ b $-\frac{1}{2}, 2$

2 a $5, -7$ b $\frac{1}{3}, -\frac{3}{2}$

3 a $-3{\cdot}8, 1{\cdot}8$ b $-1{\cdot}7, 1{\cdot}4$

4 $a = -3, b = 2$

5 a

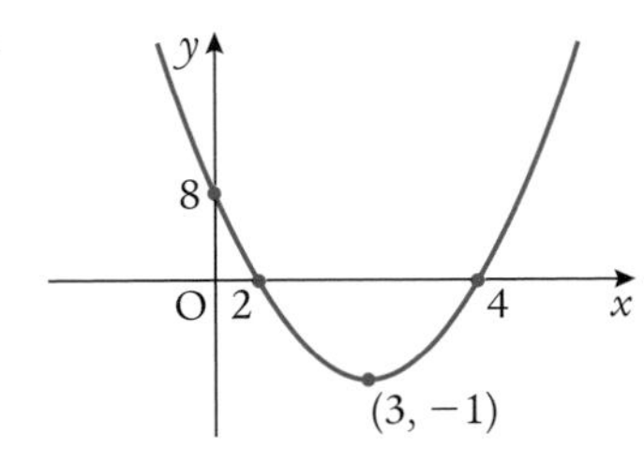

(−1, 16) y 15 −5 O 3 x

6 a $x = -2, y = 54$ b $x = 4, -8$

Exercise 5.1

1 a $x \leqslant 2$ or $x \geqslant 3$ b $-1 < x < 4$ c $-1 < x < 1$ d $x \leqslant -5$ or $x \geqslant 2$
e $-6 \leqslant x \leqslant -5$ f $x < -7$ or $x > 1$ g $x < -2$ or $x > 4$ h $x = -3$
i $-1 \leqslant x \leqslant \frac{1}{2}$ j $\frac{1}{2} < x < \frac{2}{3}$ k $-2 < x < -\frac{1}{3}$ l $x < -\frac{2}{5}$ or $x > \frac{1}{2}$
m $x < -\frac{3}{5}$ or $x > 0$ n $x \leqslant \frac{1}{7}$ or $x \geqslant 1$ o $x = -4$ p $-2 < x < 2$

2 a $-2 < x < 3$ b $x \leqslant -4$ or $x \geqslant 6$ c $x \leqslant -1$ or $x \geqslant 8$
d $1 < x < \frac{3}{2}$ e $-3 \leqslant x \leqslant \frac{7}{2}$ f $x < -\frac{4}{3}$ or $x > 4$

3 a $30 - x$ b $x(30 - x)$
c $x^2 - 30x + 200 > 0; x < 10$ or $x > 20$ d $x^2 - 17x + 60 < 0; 5 < x < 12$

4 a $(34 - x)$ m b $x(34 - x) \geqslant 240, 10 \leqslant x \leqslant 24$

5 a area of A $= (x - 4)(x + 1)$; area of B $= 2(x - 1)(x - 5)$
b $x > 5$
c $(x - 4)(x + 1) > 2(x - 1)(x - 5); 5 < x \leqslant 7$

6 $0 < x < 44$

Exercise 5.2

1 a Distinct roots b Coincident c No real
d Coincident e No real f Distinct roots
g Distinct roots h Coincident i No real

2 a $5x^2 + 7x - 1 = 0$; distinct b $3x^2 - 2x + 4 = 0$; no real c $x^2 + 8x + 5 = 0$; distinct
d $49x^2 - 14x + 1 = 0$; coincident e $2x^2 + x + 3 = 0$; no real f $25x^2 - 20x + 4 = 0$; coincident
g $2x^2 + x - 7 = 0$; distinct h $4x^2 - 2x - 3 = 0$; distinct i $x^2 + x + 1 = 0$; no real

3 a $b^2 + 12 > 0$ for all b b $k > -\frac{3}{4}$ c $m = \pm 6$

4 a i $20 - x$ ii $20x - x^2$ b $0 < A < 100$ c 99, 96, 91, 84, 75, 64, 51, 36, 19

5 a $\frac{x}{900}(400 - x) = k$ b $x^2 - 400x + 900k = 0$ c $k < 44\frac{4}{9}$
d i $44\frac{4}{9}$ m ii 200 m

6 a $m = 2, 6$ b $c = 3$

7 $k = 2, 10$

8 $0 \leqslant k < 2$ or $k > 6$

Exercise 5.3

1 a $\frac{4}{5}(x^2 - 4x - 5)$ b $\frac{2}{3}(x^2 - 9x + 14)$ c $-8(x^2 + 6x + 8)$ d $\frac{5}{16}x(x - 8)$

2 a i $-(a + b)$, i.e. $-$sum ii ab, the product b Any equation of form $y = k(x^2 - 7x + 6)$
c $y = -2x^2 + 32x - 96$

3 a $y = -\frac{3}{250}(x^2 - 120x + 1100)$ b $-13{\cdot}2$; 13·2 m below the road

4 a $h = -5t^2 + 40t - 60$ b $3 < t < 5$

Preparation for assessment

1 a $\frac{7}{2} \leqslant x \leqslant 5$ b $x < -3$ or $x > \frac{1}{5}$

2 a $x(20 - x)$ b $8 < x < 12$

3 a Real distinct b No real c Coincident

4 $n < -9$ or $n > 8$

5 $y = -\frac{1}{12}(x - 7)(x + 3)$ b $(2, \frac{25}{12})$; maximum

6 Polynomials

What you need to know

1 a $(x - 5)(x + 3)$ b $(2x - 3)(x + 4)$ c $(3x - 4)(2x - 5)$ d $2(2x + 1)(x - 5)$

2 $\frac{1}{6}(x^2 + x - 12)$

3 a $6x^2 + 7x - 3$ b $2x^3 + 6x^2 + 4x$ c $x^3 + 3x^2 - x - 3$ d $x^3 + 6x^2 + 11x + 6$

4 a $5(x + 1)(x + 2)$ b i 100 ii 150 iii 360 iv 0
c The factors $x + 1$ and $x + 2$ are consecutive numbers so one must be a multiple of 2. So the final number has a factor of 2.

5 $f(x)$, B, quadratic; $g(x)$, C, cuts x-axis in 3 places; $h(x)$, A, cuts x-axis in 4 places

Exercise 6.1

1 a i -8 ii 70 b i -16 ii 0 c i -4 ii 18 d i 0 ii -12
e i 12 ii -30 f i 0 ii 0 g i -1 ii -40 h i 0 ii -7
i i 0 ii -12 j i 1 ii 1

2 a i 4 ii −20 b i −7 ii 45 c i 0 ii 0
d i 9 ii −3 e i 35 ii $-26\frac{1}{4}$

3 a Both $f(x)$ and $g(x) = 4$ b $g(3) = 64, f(3) = 49; g(3) > f(3)$

4 a $24 - 6a$ b $a = 4$

5 a i 0 ii 0; volume reduces to zero at limits of $x = 1$ and $x = 3$. b $V(1{\cdot}5) = 6, V(2{\cdot}5) = 3; V(1{\cdot}5)$ is bigger

Exercise 6.2

1 a $x^2 + 5x + 4$ r 0 b $x^2 + 2x - 3$ r 1 c $x^2 - 9$ r 2 d $x^2 - 2x + 1$ r −2
e $x^2 + x - 6$ r 3 f $x^2 - 4x + 4$ r 0 g $4x^2 - 6x + 2$ r 3 h $2x^2 + 5x + 3$ r 4
i $3x^2 - 5x - 12$ r −2 j $2x^2 - 9x + 9$ r −3

2 a $x^2 - 2x + 1$ r −2 b $x^2 - 4x + 3$ r 0 c $x^2 + 6x + 9$ r 2
d $x^2 + 4x + 3$ r −1 e $3x^2 + 5x + 2$ r 1 f $5x^2 - 11x + 2$ r 0

3 a $x^2 - x - 6$ r 4 b $x^2 - 3x - 4$ r −6 c $2x^2 - x - 3$ r 0 d $3x^2 + 2x - 8$ r 5
e $2x^2 - 6x + 4$ r 4 f $5x^2 - 13x - 6$ r 10

4 a $x^2 + 4x + 3$ r 1 b i $x^2 + 5x + 4$ r 0 ii $x^2 - x - 12$ r 4 iii $x^2 + 4x + 4$ r 6

5 a $x^2 + 7x + 12 + \dfrac{4}{x+3}$ b $x^2 - 2x - 8 + \dfrac{5}{4x+1}$

6 a

$$\begin{array}{r|l} & ax^2 + (ae+b)x + (ae^2+be+c) \\ \hline x - e & ax^3 + bx^2 + cx + d \\ & \underline{ax^3 - eax^2} \\ & \quad (ae+b)x^2 + cx + d \\ & \quad \underline{(ae+b)x^2 - e(ae+b)x} \\ & \qquad (ae^2+be+c)x + d \\ & \qquad \underline{(ae^2+be+c)x - e(ae^2+be+c)} \\ & \qquad\qquad ae^3 + be^2 + ce + d \end{array}$$

b

$$\begin{array}{c|cccc} e & a & b & c & d \\ & & ae & ae^2+be & ae^3+be^2+ce \\ \hline & a & ae+b & ae^2+be+c & ae^3+be^2+ce+d \end{array}$$

7 a i $r = 0; x^2 + 7x + 12$ ii $r = 0; x^2 + 5x + 6$ iii $r = 0; x^2 - 2x - 8$
b $k = -18$ c $k = 24$

Exercise 6.3

1 a −3 b 5 c −10 d 0

2 a/b $(x+2)(x-5)(x+3)$

3 a $(x-3)(x-1)(x+5)$ b $(x-3)(x-2)(x+4)$ c $3(x-2)(x+1)(x+3)$
d $(x-3)(x^2+x+5)$ e $(x-5)(x+1)(x+4)$ f $(x-1)(x+4)(x+5)$
g $(x-2)(3x^2+2x+1)$ h $(5x-1)(x-1)(x+3)$ i $(x+2)(2x+1)(3x-1)$
j $(3x+2)(4x-1)(x+3)$ k $(x-1)(x+1)(x+4)(x-5)$ l $x(x+3)(x+2)(x-4)$
m $(x-1)(x-2)(x-3)(x-6)$ n $(x-2)(x+5)(x^2-x+2)$

4 a $f(x) = x^2 - 9 \Rightarrow f(3) = 0 \Rightarrow x - 3$ is a factor
b i $f(x) = x^2 - a^2 \Rightarrow f(a) = 0 \Rightarrow x - a$ is a factor; $x + a$ is other factor
ii $f(x) = x^3 - a^3 \Rightarrow f(a) = 0 \Rightarrow x - a$ is a factor; $x^2 + ax + a^2$ is other factor
iii $f(x) = x^4 - a^4 \Rightarrow f(a) = 0 \Rightarrow x - a$ is a factor; $x^3 + ax^2 + a^2x + a^3$ is other factor
c i $f(x) = x^n - a^n \Rightarrow f(a) = 0 \Rightarrow x - a$ is a factor;
ii $x^{n-1} + ax^{n-2} + a^2x^{n-3} + \ldots + a^{n-2}x + a^{n-1}$ is other factor

5 a 60 b $(x-6)(x-5)(x+2)$

6 a 13 b $(x-1)(x-5)(x+3)$

7 a $a = -15, b = 18$ b $(x-6)(x-1)(x+3)$

Exercise 6.4

1 a $-1, 1, 2$ b $1, -5, -6$ c $-1, 2, 3$
d $-1, 2, 4$ e $-1, -3, -6$ f $-1, -4, 5$
g $-\frac{1}{6}, -3, 5$ h $-\frac{1}{2}, 1$ (twice) i $-\frac{5}{4}, -3, -2$
j $-\frac{1}{3}, \frac{1}{2}, -1$ k $-\frac{2}{5}, -\frac{1}{2}, 1$ l $-\frac{3}{4}, \frac{2}{3}, -2$

2 Equation 1 with graph C: $x = -1$ and no other real roots, 1; equation 2 with graph B: $x = 1, 3, 6$; equation 3 with graph A: $x = -3, -3, 1$

3 a $-4, -2, -1, 1$ b $-2, 3, 4, 6$
c $-2, -2, \frac{1}{4}, 1$ d $-3, 1$, and no other real roots

4 a $(x-2)(x+3)(x+9) = x^3 + 10x^2 + 3x - 54 = V$ units^3
b i $2, -3, -9$
ii $x > 2$, otherwise negative edges
c i $x^3 + 10x^2 + 3x - 126 = 0$
ii Solutions: $x = -7, -6, 3$; practical solution $x = 3$

5 a $-1{\cdot}4, 1, 3{\cdot}4$ b $x = 1$; 6 hours after low tide

6 a $-2, 1, 7$ b Only $x = 1$ is in the domain

7 a $(x-3)$ is a factor twice b $x = \frac{5}{2}, y = 0$

8 $(1, 2)$ and $(\frac{4}{3}, \frac{29}{9})$

9 a $m = 10$ b $x = 2$ and $x = -5$

10 a $m = 11, n = 6$ b $x = \frac{1}{2}$

Exercise 6.5

1 a i $f(-2) = -13; f(-1) = 2 \Rightarrow$ root in range $-2 < x < -1$
ii $f(1) = 2; f(2) = -1 \Rightarrow$ root in range $1 < x < 2$
iii $f(2) = -1; f(3) = 2 \Rightarrow$ root in range $2 < x < 3$
b $x = 1{\cdot}54$ (2 d.p.) c $x = 2{\cdot}7$ (1 d.p.)

2 a Between 2 and 3 b Between 2·24 and 2·25 ... 2·2 (1 d.p.)

3 a £600
b i $D(0{\cdot}5) = 9{\cdot}875$ hundred, $D(0{\cdot}6) = 10{\cdot}896$ hundred; so between 0·5 and 0·6 takes the value 10 hundred
ii 0·51 year (2 d.p.)

4 a $\frac{1000}{3} = 10d^2 - \frac{d^3}{3}$, hence result b $f(6) = 136, f(7) = -127$ c 6·53 (to 2 d.p.)

Preparation for assessment

1 a i -6 ii 2 b i -1 ii 163

2 a $2x^2 + 3x + 8$ r 6 b $3x^2 - 4x + 4$ r -13 c $x^3 + x^2 + 5x + 5$ r 14

3 a $(x+3)(x+4)(x-1)$ b $(6x-1)(x-2)(x-2)$
c $2(x-1)(2x^2 + x + 3)$ d $(x-1)(x+6)(x-3)(x+2)$

4 a $x = 1, 6, -4$ b -4 (twice), $-2, 3$

5 a i Equation $5x + 2 = x^3 - 3x^2 - 4x - 3$ simplifies to $x^3 - 3x^2 - 9x - 5 = 0$. This has coincident roots at $x = -1$.
ii $(-1, -3)$
b $(5, 27)$
c i $f(4) = -3, f(4{\cdot}5) = 9{\cdot}375$ so root exists between 4 and 4·5
ii 4·14 (to 2 d.p.)

7 Differential calculus 1

What you need to know

1 a $(a-b)(a+b)$ b $(a-b)(a^2+ab+b^2)$ c $(a-b)(a^4+a^3b+a^2b^2+ab^3+b^4)$

2 a i $(x+h)^2+4(x+h)-3=x^2+2hx+h^2+4x+4h-3$ ii $x^3+3x^2h+3xh^2+h^3$

b i $\sin x\cos h+\cos x\sin h$ ii $\cos x\cos h-\sin x\sin h$

3 a 2 b $\dfrac{y-5}{x-1}$ c 3

4 a i (4, 20) ii (6, 40) b 10

c i 5 ii $(1+h)^2+4=5+2h+h^2$ d i $2+h$ ii 2

5 a i x^{-1} ii $2x^{-2}$ iii $4x^2$ iv $\frac{1}{2}x^{-3}$ v $\frac{2}{3}x$ vi $x^{\frac{1}{2}}$ vii $4x^{\frac{1}{3}}$ viii $x^{-\frac{1}{2}}$ ix $2x^{-\frac{1}{3}}$ x $x^{-\frac{1}{2}}$

b i x^3 ii $x^{\frac{7}{6}}$ iii $x^{\frac{5}{12}}$ iv x^{-1} v x vi $x^{\frac{1}{6}}$

c i $x^{-1}+x^{-\frac{1}{2}}$ ii $x^{-\frac{1}{2}}+x^{\frac{1}{2}}$ iii $x^{-\frac{1}{6}}+x^{\frac{1}{3}}$

6 a 1 hour b i 30 mph ii 50 mph

Exercise 7.1

1 a i $\lim\limits_{h\to 0}\dfrac{3(x+h)^2-3x^2}{h}=\lim\limits_{h\to 0}(6x+3h)=6x$

ii $\lim\limits_{h\to 0}\dfrac{(x+h)^3-x^3}{h}=\lim\limits_{h\to 0}(3x^2+3xh+h^2)=3x^2$

iii $\lim\limits_{h\to 0}\dfrac{2(x+h)^3-2x^3}{h}=\lim\limits_{h\to 0}(6x^2+6xh+2h^2)=6x^2$

iv $\lim\limits_{h\to 0}\dfrac{(x+h)^4-x^4}{h}=\lim\limits_{h\to 0}(4x^3+6x^2h+4xh^2+h^3)=4x^3$

v $\lim\limits_{h\to 0}\dfrac{3(x+h)^4-3x^4}{h}=\lim\limits_{h\to 0}(12x^3+18x^2h+12xh^2+3h^3)=12x^3$

b i $\lim\limits_{h\to 0}\dfrac{5(x+h)-5x}{h}=\lim\limits_{h\to 0}\left(\dfrac{5h}{h}\right)=5$ ii 3 iii -1 iv $\dfrac{f(x+h)-f(x)}{h}=\dfrac{3-3}{h}=0$, for all $h\neq 0$ v 0

2 a 10 b 30 c 75

3 $y=8x-6$

Exercise 7.2

1 a $6x$ b $35x^4$ c $-12x^3$ d $-1+10x$ e $-16x^{-3}$

f $-x^{-2}$ g $30x^{-6}$ h $-2-36x^{-5}$ i $-8x^{-3}$ j $-21x^{-4}$

k $-5x^{-6}$ l $-x^{-2}-6x^{-3}-15x^{-4}$ m $2x^{-\frac{1}{2}}$ n $\frac{15}{2}x^{\frac{1}{2}}$

o $-\frac{8}{7}x^{-\frac{5}{7}}$ p $x^{-\frac{1}{2}}+x^{-\frac{2}{3}}+x^{-\frac{3}{4}}$ q $\frac{3}{2}x^{-\frac{1}{2}}$ r $\frac{7}{3}x^{-\frac{2}{3}}$ s $-\frac{4}{3}x^{-\frac{1}{3}}$

t $\frac{1}{2}x^{-\frac{1}{2}}+\frac{1}{3}x^{-\frac{2}{3}}+\frac{1}{5}x^{-\frac{4}{5}}$

2 a -18 b -6 c 0 d 18

3 i y_1 ii y_2 iii y_2

4 a $6x+5$ b $x=-\frac{5}{6}$ c $y=5x-4$

5 a 24 b $x=-1, 2$ c $-1<x<2$ d $y=-12x$

6 a $5-27x^{-2}$ b i -22 ii $\frac{17}{4}$ c $x=\pm 3$ d $x=\pm 2{\cdot}32$ (to 2 d.p.)

7 a $-\frac{1}{2}x+8$ b $x=16$ c $y=7x-54$

Exercise 7.3

1 a $3x^2-2x^{-1}$; $6x+2x^{-2}$ b $3x^{-1}+x^{-2}$; $-3x^{-2}-2x^{-3}$

c $\frac{1}{3}x^{-2}-\frac{3}{2}x^{-1}$; $-\frac{2}{3}x^{-3}+\frac{3}{2}x^{-2}$ d $\frac{1}{5}x^{-4}+\frac{1}{7}x^{-3}+\frac{1}{9}x^{-1}$; $-\frac{4}{5}x^{-5}-\frac{3}{7}x^{-4}-\frac{1}{9}x^{-2}$

2 a $2x + 2$ **b** $-8 + 2x$ **c** $4x + 3$ **d** $\frac{3}{2}x^{\frac{1}{2}} + \frac{1}{2}x^{-\frac{1}{2}}$

3 a $2x - 2x^{-3}$ **b** $3x^{\frac{1}{2}} - \frac{1}{2}x^{-\frac{1}{2}}$ **c** $\frac{1}{3}x^{-\frac{2}{3}} + 3x^2 + 4x - 1$

4 a $-x^{-2}$ **b** $1 - x^{-2}$ **c** $-x^{-2} - 4x^{-3} - 3x^{-4}$ **d** $-\frac{3}{2}x^{-\frac{5}{2}} - \frac{1}{2}x^{-\frac{3}{2}}$

5 a $-\frac{3}{2}x^{-\frac{3}{2}}$ **b** $\frac{3}{2}x^{\frac{1}{2}} + \frac{3}{4}x^{-\frac{1}{2}}$ **c** $-\frac{1}{2}x^{-\frac{1}{2}} + \frac{3}{2}x^{\frac{1}{2}}$ **d** $-\frac{1}{2}x^{-\frac{3}{2}} - \frac{2}{3}x^{-\frac{5}{3}}$

6 a $-3x^{-2} - 2x^{-3}$ **b** $\frac{3}{2}x^{\frac{1}{2}} + 2x^{-\frac{1}{2}}$ **c** $4x^3 - 2x$ **d** $\frac{7}{3}x^{\frac{4}{3}} + \frac{5}{3}x^{-\frac{1}{6}}$

7 a $1 - x = 1^2 - \sqrt{x^2} = (1 - \sqrt{x})(1 + \sqrt{x})$

b **i** $f(x) = \dfrac{1 - x}{1 + \sqrt{x}} = \dfrac{(1 - \sqrt{x})(1 + \sqrt{x})}{1 + \sqrt{x}} = 1 - \sqrt{x}; f'(x) = -\frac{1}{2}x^{-\frac{1}{2}}$

ii $f(x) = \dfrac{1 - x}{1 + \sqrt{x}} = (1 + 2x) = (1 - \sqrt{x})(1 + 2x) = 1 + 2x - x^{\frac{1}{2}} - 2x^{\frac{3}{2}}; f'(x) = 2 - \frac{1}{2}x^{-\frac{1}{2}} - 3x^{\frac{1}{2}}$

iii $f'(x) = \frac{3}{2}x^{\frac{1}{2}} + \frac{3}{2}x^{-\frac{1}{2}} + 1$

8 a 26 **b** $6x^{\frac{1}{2}} - 4x^{-\frac{1}{2}} - 10x^{-\frac{3}{2}}$

c **i** $V'(1) = -8$ ... losing £800/year **ii** $V'(4) = 8\frac{3}{4}$ ... gaining £875/year

9 a $h = \dfrac{1000}{x^2}$ **b** **i** $k(x, h) = 2x^2 + 4xh$ **ii** $k(x) = 2x^2 + \dfrac{4000}{x}$

c **i** $k'(x) = 4x - \dfrac{4000}{x^2}$ **ii** $k'(5) = -140; k(5) = 850; y = -140x + 1550$ **iii** $x = 10$ **iv** $y = 600$

v It is a minimum turning point

Exercise 7.4

1 a **i** Increasing **ii** Decreasing **b** **i** Decreasing **ii** Increasing

c **i** Decreasing **ii** Stationary **d** **i** Increasing **ii** Increasing

e **i** Decreasing **ii** Decreasing

2 a $-2 < x < 5$ decreasing; $x < -2, x > 5$ increasing; $x = -2, 5$ stationary

b $-1 < x < 1$ decreasing; $x < -1, x > 1$ increasing; $x = -1, 1$ stationary

c $-1 < x < 2$ increasing; $x < -1, x > 2$ decreasing; $x = -1, 2$ stationary

3 a $f'(x) = 3x^2 + 6x + 3 = 3(x + 1)^2$ ... which is greater than or equal to zero for all x

b $f'(x) = 9x^2 - 6x + 1 = (3x - 1)^2 \geqslant 0$ ... so stationary at $x = \frac{1}{3}$, otherwise increasing

c **i** $f'(x) = 3x^2 + 4x + 2; b^2 - 4ac < 0$ ... so no real roots, so no stationary points

ii $f'(0) = 2$ ... so increasing at $x = 2$ ... so always increasing since to decrease it would need to pass through stationary point

d $f'(x) = -1 + 2x - x^2 = -(x - 1)^2$ ... so stationary at $x = 1$; decreasing otherwise, so $x_a < x_b \Rightarrow f(x_b) < f(x_a)$

4 a **i** $x = 2, -5$ **ii** $x = 2$, minimum; $x = -5$ maximum

x	→	−5	→	2	→
$x - 2$	−	−	−	0	+
$x + 5$	−	0	+	+	+
$f'(x)$	+	0	−	0	+
Tangent	/	—	\	—	/

b **i** $x = -2, -1, 1$ **ii** $x = -2$, minimum; $x = -1$ maximum; $x = 1$ minimum

x	→	−2	→	−1	→	1	→
$x - 1$	−	−	−	−	−	0	+
$x + 1$	−	−	−	0	+	+	+
$x + 2$	−	0	+	+	+	+	+
$f'(x)$	−	0	+	0	−	0	+
Tangent	\	—	/	—	\	—	/

c **i** $x = -1, 2, 2$ **ii** $x = -1$, minimum; $x = 2$ point of inflexion (rising)

d **i** $x = -2, 1, 3$ **ii** $x = -2$, maximum; $x = 1$ minimum; $x = 3$, maximum

5 a $x = -\frac{1}{3}, y = -\frac{28}{9}$; point of inflexion (rising)

b $x = -\frac{1}{3}, y = -\frac{49}{27}$; maximum; $x = 1, y = -3$ minimum

c $(1, \frac{29}{12})$ maximum; $(3, -\frac{17}{4})$ minimum; $(-3, -\frac{161}{4})$ minimum

d $(\frac{1}{4}, 3)$ minimum

Exercise 7.5

1 a

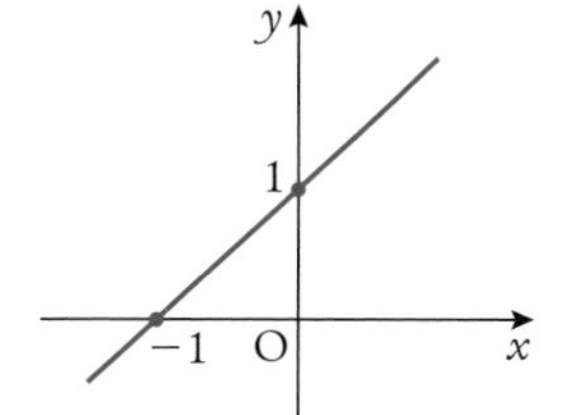

b

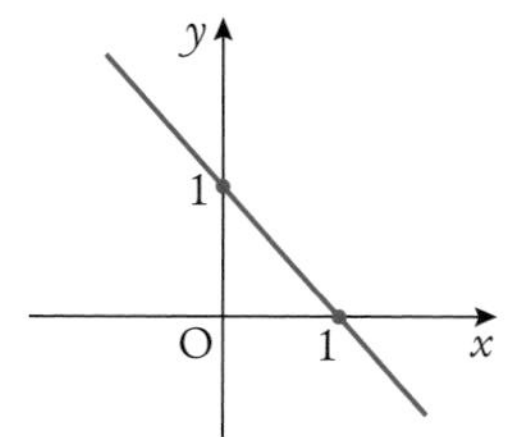

c

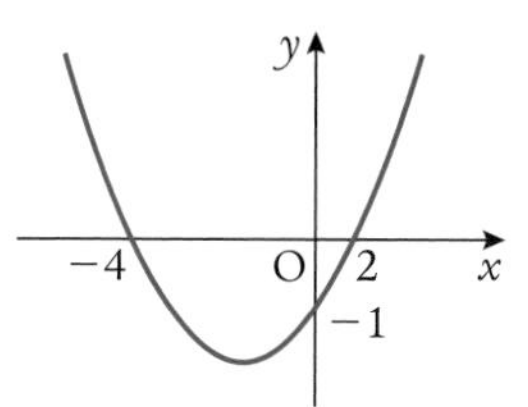

d

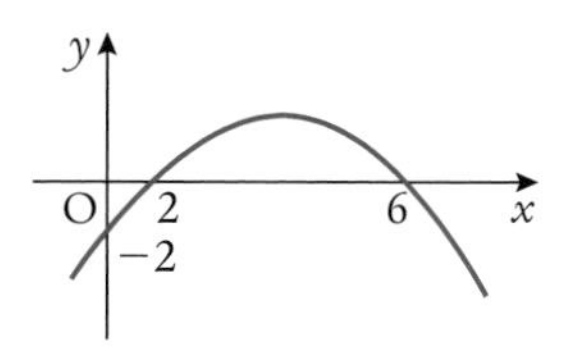

e

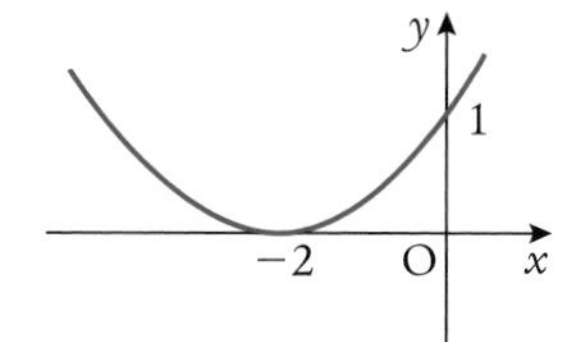

f

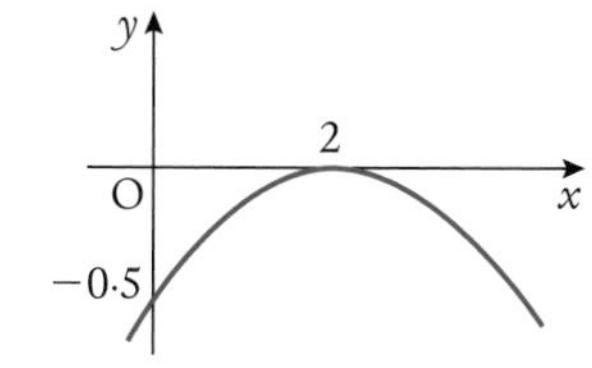

2 a

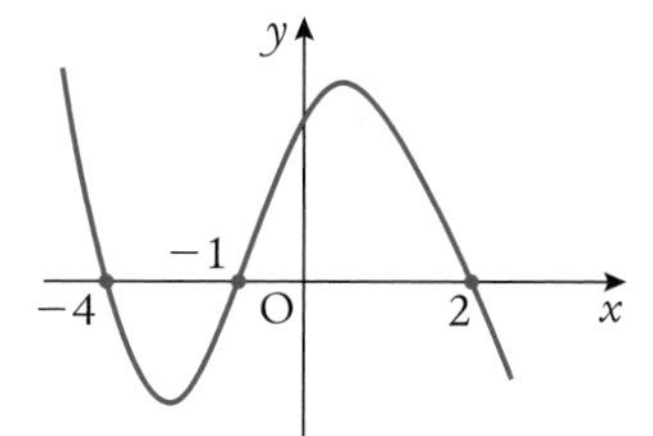

b

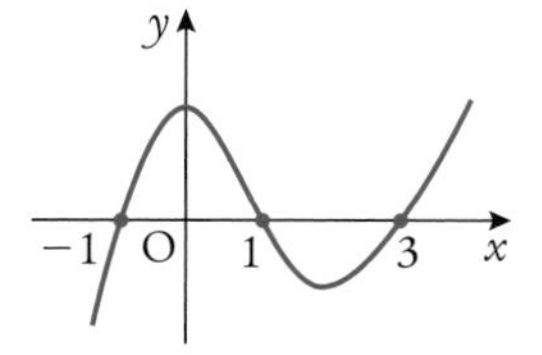

c

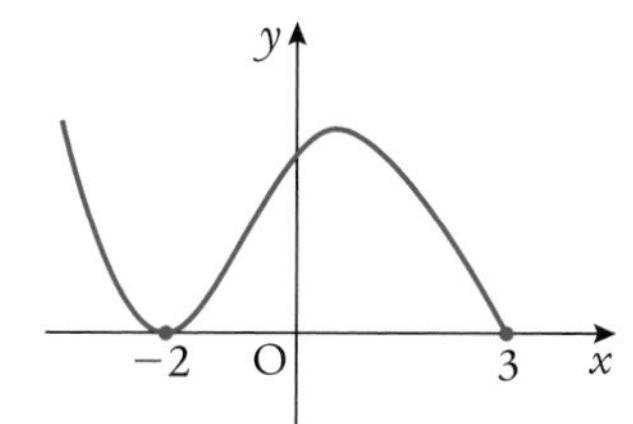

d

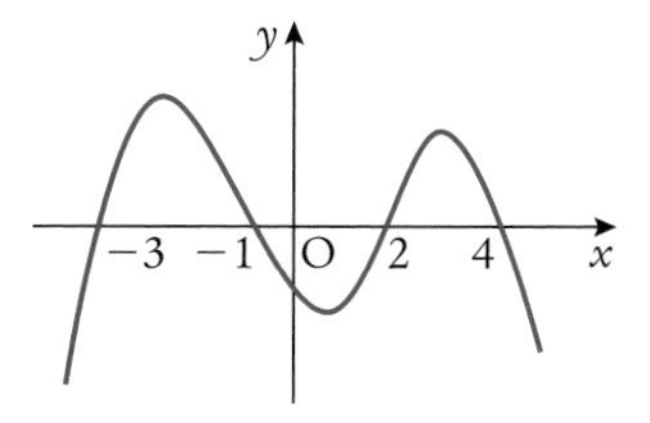

e

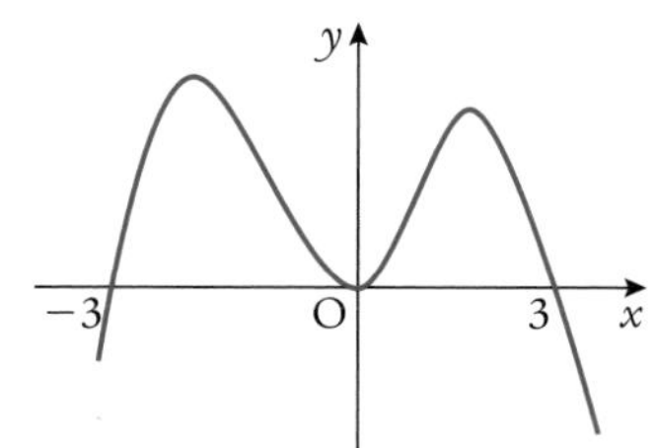

f

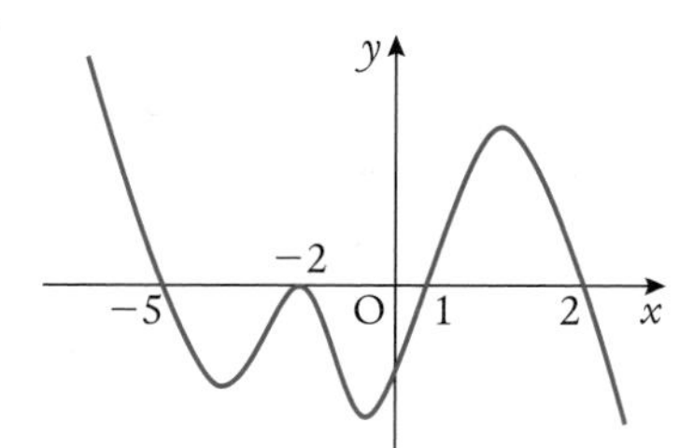

3

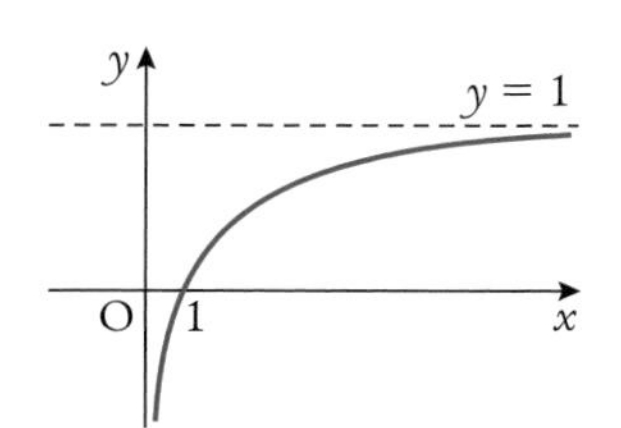

4 a Black curve

b Purple curve 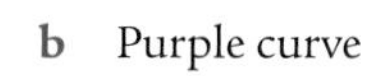

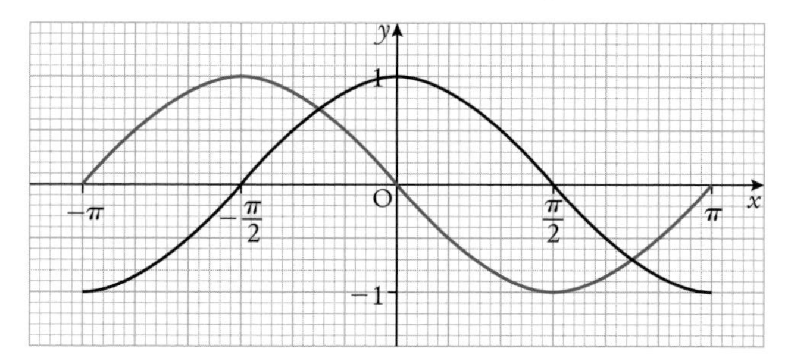

5 Student's own research. Though the derivative of $\sin x$ is $\cos x$ for x in radians, this is not the case when x is in degrees. One can hardly tell the difference between the derivative of $\sin x°$ and the x-axis in a scale that shows $y = \sin x°$ on same diagram.

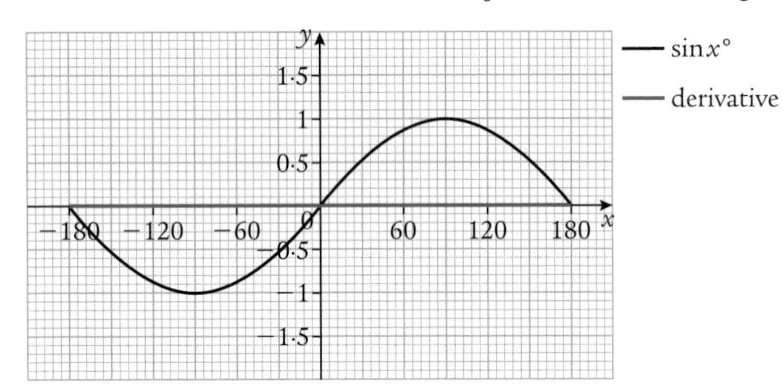

The derivative of $\sin x°$ is approximately $0{\cdot}01745 \cos x°$.

Exercise 7.6

1 **a** $3(2x+1)^2.2 = 6(2x+1)^2$ **b** $5(3x-2)^4.3 = 15(3x-2)^4$ **c** $4(x^2+x)^3(2x+1)$
d $-1.(x+1)^{-2}.1 = -(x+1)^{-2}$ **e** $-6(5-x)^5$ **f** $-18(1+6x)^{-4}$
g $\frac{x}{2} - 1$ **h** $5(2x^2+3x+1)^4(4x+3)$ **i** $-2(x^3+2x^2-2)^{-3}(3x^2+4x)$
j $3(4-3x)^{-2}$

2 **a** $(2x+1)^{-1}; -2(2x+1)^{-2}$ **b** $4(3x+2)^{-1}; -12(3x+2)^{-2}$
c $6(1-2x)^{-1}; 12(1-2x)^{-2}$ **d** $(3x-1)^{-2}; -6(3x-1)^{-3}$
e $(x^2+2x+1)^{-1}; -(x^2+2x+1)^{-2}(2x+2)$ **f** $(1-x)^{-3}; 3(1-x)^{-4}$
g $5(2x^2-3x+1)^{-1}; -5(2x^2-3x+1)^{-2}(4x-3)$ **h** $3x^{-1} + 2(x+1)^{-1}; -3x^2 - 2(x+1)^{-2}$
i $2(x+1)^{-2} + (x-1)^{-2}; -4(x+1)^{-3} - 2(x-1)^{-3}$ **j** $4(3x^3+2x^2+x)^{-1}; -4(3x^3+2x^2+x)^{-2}.(9x^2+4x+1)$

3 **a** $(2x+3)^{-\frac{1}{2}}$ **b** $\frac{5}{2}(1+5x)^{-\frac{1}{2}}$ **c** $x(3+x^2)^{-\frac{1}{2}}$ **d** $(3x+7)^{-\frac{2}{3}}$
e $\frac{1}{4}(x^2-3x+1)^{-\frac{3}{4}}(2x-3)$ **f** $-\frac{3}{2}(3x+1)^{-\frac{3}{2}}$ **g** $-4(4x-3)^{-\frac{4}{3}}$ **h** $-\frac{1}{2}x^{-\frac{3}{2}} - \frac{1}{2}(x+1)^{-\frac{3}{2}}$
i $(2x+1)^{-\frac{1}{2}} - (2x+1)^{-\frac{3}{2}}$ **j** $\frac{2x}{3}((x-1)(x+1))^{-\frac{2}{3}}$

4 **a** 36 **b** 225 **c** 576

5 **a** **i** -18 **ii** $-\frac{9}{4}$ **iii** $-\frac{2}{3}$ **b** For all $x > -1$, $(x+1)^{\frac{3}{2}} > 0 \Rightarrow \frac{-18}{(x+1)^{\frac{3}{2}}} < 0$

6 **a** **i** $\frac{-x}{\sqrt{25-x^2}}$ **b** **i** $\frac{3}{4}$ **ii** 0 **iii** $-\frac{4}{3}$ **c** $y = -\frac{3}{4}x + \frac{25}{4}$

d

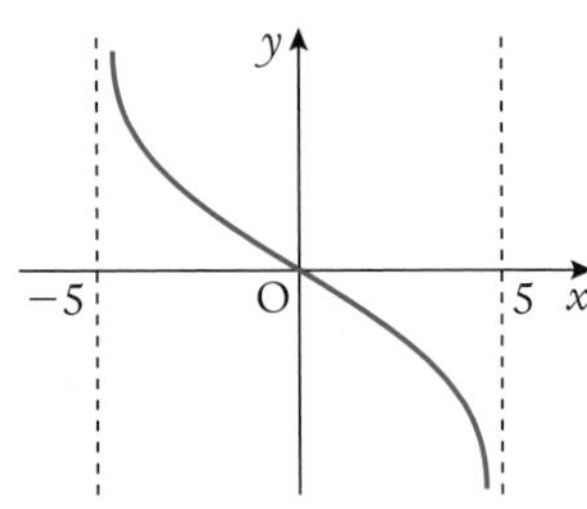

7 **a** $t'(x) = 20 - \frac{1620}{(2x+1)^2}$ **b** 12·8 **c** $y = -160x + 450$
d $x = 4$ **e** Minimum

Exercise 7.7

1 $$\frac{d \cos x}{dx} = \lim_{h \to 0} \frac{\cos(x+h) - \cos x}{h} = \lim_{h \to 0} \frac{\cos x \cos h - \sin x \sin h - \cos x}{h}$$

$$= \lim_{h \to 0} \frac{\cos x . 1 - \sin x . h - \cos x}{h} = \lim_{h \to 0} \frac{-\sin x . h}{h} = -\sin x$$

2 **a** $3 \cos x$ **b** $-2 \sin x$ **c** $-\sin x + \cos x$ **d** $-3 \sin x + 4 \cos x$

3 a $5\cos x - 12\sin x$ **b i** $\frac{5-12\sqrt{3}}{2}$ **ii** -12 **iii** -5 **c** $y = -12x + 6\pi + 5$

4 a $-\frac{1}{2}$ **b** $2\sqrt{3} + \frac{3}{2}$ **c** 1

5 a $3\cos 3x$ **b** $-5\sin 5x$ **c** $8\cos 4x$ **d** $5\sin(-x)$
e $2\cos(2x+1)$ **f** $-3\sin(3x-1)$ **g** $15\cos(1+5x)$ **h** $2\sin(3-x)$
i $-2\cos x\sin x$ **j** $1 - 2\sin x\cos x$ **k** $-2\sin x\cos x$ **l** $-4\sin x\cos x$
m $\frac{3\sin x}{\cos^2 x}$ **n** $\frac{-2\cos x}{\sin^2 x}$ **o** $\frac{2\sin x}{\cos^3 x}$ **p** $\frac{-3\cos x}{\sin^4 x} - \frac{2\cos x}{\sin^3 x}$
q $\frac{\cos x}{2\sqrt{\sin x}}$ **r** $-\sin x\cos(\cos x)$ **s** $\frac{\sin x}{2\sqrt{\cos^3 x}}$ **t** $\frac{\sin x - \cos x}{2\sqrt{(\sin x - \cos x)^3}}$

6 a $3\cos(3x-1)$ **b** $6\sin(3x-1)\cos(3x-1)$
c i $-4\cos(2x+3)\sin(2x+3)$ **ii** $-4(2x+3)\sin(2x+3)^2$ **iii** $\frac{\cos(2x-1)}{\sqrt{\sin(2x-1)}}$

7 a $\frac{d(\sin x°)}{dx} = \frac{d\left(\sin\frac{\pi}{180}x\right)}{dx} = \frac{\pi}{180}\cos\frac{\pi}{180}x = \frac{\pi}{180}\cos x°$
b $\frac{d(\cos x°)}{dx} = \frac{d\left(\cos\frac{\pi}{180}x\right)}{dx} = -\frac{\pi}{180}\sin\frac{\pi}{180}x = -\frac{\pi}{180}\sin x°$

8 a $\frac{-2\cos 2x}{\sin^2 2x}$ **b** $x = \frac{\pi}{4}$ **c** $f'\left(\frac{\pi}{5}\right) < 0; f'\left(\frac{\pi}{4}\right) = 0; f'\left(\frac{\pi}{3}\right) > 0$; minimum **d** 1

9 a $h'(t) = \frac{2\pi}{3}\cos\frac{\pi t}{3}$ **b i** $\frac{\pi}{3}$ m/s **ii** $-\frac{\pi}{3}$ m/s **c** 1·5 s

10 a $\frac{d}{dx}(x + \sin x) = 1 + \cos x$; since $-1 \leqslant \cos x \leqslant 1$, then $0 \leqslant 1 + \cos x \leqslant 2$ ... i.e. it is never negative, so function is never decreasing
b $\frac{d}{dx}(\cos x - x) = -\sin x - 1$; since $-1 \leqslant -\sin x \leqslant 1$, then $-2 \leqslant -\sin x - 1 \leqslant 0$ ... i.e. it is never positive, so function is never increasing

11 a i $\frac{d}{dx}2\sin x\cos x = \frac{d}{dx}\sin 2x = 2\cos 2x$ **ii** $\frac{d}{dx}\sin\frac{x}{2}\cos\frac{x}{2} = \frac{d}{dx}\frac{1}{2}\sin x = \frac{1}{2}\cos x$
b i $\frac{d}{dx}\cos 2x = -2\sin 2x$; $\frac{d}{dx}\cos^2 x - \sin^2 x = -4\sin x\cos x$
ii $-4\cos x\sin x = -2\sin 2x \Rightarrow \sin 2x = 2\sin x\cos x$ ... which is true **iii** All are equal to $-2\sin 2x$

Preparation for assessment

1 a $9x^2 + 4x + 1$ **b** $-\frac{1}{4}x^{-\frac{3}{2}} + \frac{3}{4}x^{-\frac{1}{2}}$ **c** $2x - 1$ **d** $1 + \frac{2}{3}x^{-\frac{5}{3}}$
e $2\cos x$ **f** $-5\sin x$ **g** $2\sin x$ **h** $2\cos x + 7\sin x$

2 a i -30 **ii** -48 **b** $x = 5$ or $x = -1$ **c** $-1 < x < 5$

3 a $k = 2$ **b** $y = 14x - 11$

4 a $\frac{1}{2}(x^2 + 2x - 1)^{-\frac{1}{2}}(2x + 2) = (x + 1)(x^2 + 2x - 1)^{-\frac{1}{2}}$ **b** $-(x - 3)^{-2}$ **c** $\frac{9}{4}(3x + 7)^{-\frac{1}{4}}$
d $10(2\sin x - 1)^4\cos x$ **e** $3\sin^2 x\cos x + 3\cos^2 x\sin x$ **f** $-6(\sin 2x)^{-2}\cos 2x$

5

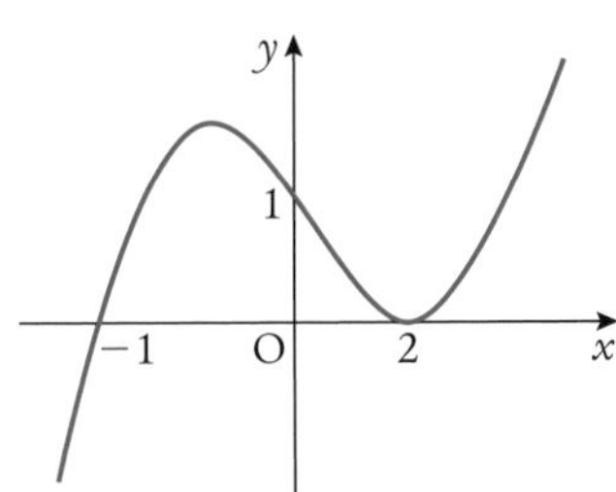

6 a $d'(x) = \frac{2\pi}{3}\cos\left(\frac{\pi}{6}x + 5\right)$ **b i** $d'(1) = 1·52$ (to 2 d.p.) **ii** $-0·59$ (to 2 d.p.)
c 5·45, 11·45, 17·45, 23·45 (all to 2 d.p.)

8 Trigonometry 3 – equations

Before we start...

The tethered goat: Goat has access to sector APBD + segment APE + segment PBF.

Working in radians, area of sector APBD $= \frac{x}{2\pi} \times \pi . L^2 = \frac{xL^2}{2}$.

$\angle ACP = \pi - x$, so area of sector ACPE $= \frac{\pi - x}{2\pi} \times \pi . 1^2 = \frac{\pi - x}{2}$.

Area of $\triangle ACP = \frac{1}{2} \times 1^2 \times \sin(\pi - x) = \frac{1}{2}\sin x$.

So segment APE = area of sector ACPE − area of $\triangle ACP = \frac{\pi - x}{2} - \frac{1}{2}\sin x$.

Area of segment PBF = area of segment APE $= \frac{\pi - x}{2} - \frac{1}{2}\sin x$.

i Thus area available to goat, $g(x, L) = \frac{xL^2}{2} + 2\left(\frac{\pi - x}{2} - \frac{1}{2}\sin x\right) = \frac{xL^2}{2} + \pi - x - \sin x$.

ii From the right-angled triangle PCG we get $\frac{L}{2} = \cos\frac{x}{2} \Rightarrow L = 2\cos\frac{x}{2}$.

$$g(x) = \frac{x\left(2\cos\frac{x}{2}\right)^2}{2} + \pi - x - \sin x = 2x\cos^2\frac{x}{2} + \pi - x - \sin x.$$

Using $\cos 2A = 2\cos^2 A - 1$, we get $2\cos^2 A = \cos 2A + 1$ ... $2\cos^2\frac{x}{2} = \cos 2 . \frac{x}{2} + 1 = \cos x + 1$.

Available to goat, $g(x) = x(\cos x + 1) + \pi - x - \sin x = x\cos x - \sin x + \pi$. The circular field has an area of $\pi . 1^2 = \pi$.

The goat has access to half of this. Thus, $x\cos x - \sin x + \pi = \frac{\pi}{2}$.

Thus $x\cos x - \sin x + \frac{\pi}{2} = 0$ or $\sin x - x\cos x - \frac{\pi}{2} = 0$.

iii Using spreadsheet method (see Section 8.4), $x = 1{\cdot}905695729$ radians (to 9 d.p.).
Thus $L = 2\cos(1{\cdot}905695729 \div 2) = 1{\cdot}158728473$ units (to 9 d.p.).

What you need to know

1 a 41·8°, 138·2° b 120°, 240°

2 127°, 413°, 487°

3 a 99·7°, 170·3°, 279·7°, 350·3° b 15°, 75°

4 30°, 150°, 54·7°, 305·3°

5 a i $(2a + 1)(a - 1)$ ii $(2\sin x + 1)(\sin x - 1)$ b 210°, 330°, 90°

Exercise 8.1

1 a 4·8°, 55·2°, 124·8°, 175·2°, 244·8°, 295·2° b 33·2°, 146·8°, 213·2°, 326·8°
c 0°, 40°, 120°, 160°, 240°, 280°, 360° d 17·6°, 72·4°, 107·6°, 162·4°
e 36·1°, 123·9°, 156·1°, 3·9°, −83·9°, −116·1° f 10°, 58°, 82°, 130°, 154°

2 a 0·464, 1·107, 3·605, 4·249 b 0·280, 1·814, 2·375, 3·908, 4·469, 6·003
c $\frac{\pi}{6}, \frac{11\pi}{18}, \frac{5\pi}{6}, \frac{23\pi}{18}, \frac{3\pi}{2}, \frac{35\pi}{18}$ d $\frac{\pi}{12}, \frac{\pi}{4}, \frac{13\pi}{12}, \frac{5\pi}{4}$
e $-\pi, -\frac{7\pi}{9}, -\frac{\pi}{3}, -\frac{\pi}{9}, \frac{\pi}{3}, \frac{5\pi}{9}, \pi$ f $\frac{3\pi}{10}, \frac{7\pi}{10}$

3 a 45°, 225°, 105°, 165°, 285°, 345°
b 36·5°, 83·5°, 156°, 204°, 276°, 324° and 20°, 100°, 140°, 220°, 260°, 340°
c 76·2°, 103·8° d 0°, 180°, 360° e −180°, −60°, 60°, 180° f −45°, 135°

4 a $\frac{\pi}{2}, \frac{7\pi}{6}, \frac{11\pi}{6}, \frac{\pi}{6}, \frac{5\pi}{6}, \frac{3\pi}{2}$ b $\frac{\pi}{6}, \frac{5\pi}{6}, \frac{7\pi}{6}, \frac{11\pi}{6}, 0, \pi, 2\pi$ c $\frac{\pi}{8}, \frac{5\pi}{8}$
d $-\frac{\pi}{3}, -\pi, \frac{\pi}{3}, \pi$ e $-\frac{5\pi}{6}, -\frac{4\pi}{6} = -\frac{2\pi}{3}, \frac{\pi}{6}, \frac{\pi}{3}$ f $\frac{\pi}{12}, \frac{7\pi}{12}, \frac{3\pi}{4}, \frac{5\pi}{4}, \frac{17\pi}{12}, \frac{23\pi}{12}$

5 a $\frac{\pi}{12}, \frac{5\pi}{12}, \frac{\pi}{4}$ b $\frac{13\pi}{12}, \frac{17\pi}{12}, \frac{5\pi}{4}$

6 a 2·02 m b 1·2 m c 8 a.m., 8 p.m.

Exercise 8.2

1 a $90°, 270°$ and $210°, 330°$ b $90°, 270°$ and $41.8°, 138.2°$ c $0°, 180°, 360°$

d $-180°, 0°, 180°$ and $-138.6°, 138.6°$ e $-90°, 90°$

f $0°, 180°, 360°$ g $0, 360°, 720°, 141.1°, 578.9°$ h $180°$ and $60°, 300°$

2 a $\frac{\pi}{2}, \frac{3\pi}{2}$ and $\frac{\pi}{4}, \frac{3\pi}{4}$, b $0, \pi, 2\pi, \frac{\pi}{6}, \frac{11\pi}{6}$ c $0, \pi, 2\pi$

d $-\frac{\pi}{2}, \frac{\pi}{2}$ e $\frac{\pi}{2}$ f $0, 2\pi, 4\pi$ and $\frac{3\pi}{2}, \frac{5\pi}{2}, \frac{11\pi}{2}, \frac{13\pi}{2}$

3 $90°, 270°$, and $14.5°, 165.5°$

4 a $0°, 360°$ and $60°, 300°$ b $19.5°, 160.5°$ and $210°, 330°$ c $48.2°, 311.8°$

d $90°$ e $60°, 300°$ f $263.6°, 360°$

5 a 0.201, 2.940 and $\frac{3\pi}{2}$ b $0, 2\pi$ and 0.841, 5.442 c 0.524, 2.618

d 1.318 e 7.330, 11.519 f 4.601

6 $\frac{\pi}{3}$ years, 400 rabbits; $\frac{5\pi}{3}$ years, 400 rabbits; 1.911 years, 233 rabbits; 4.373 years, 233 rabbits

7 a $128.2°, 231.8°$ b 1.53 units

8 a $-2 \sin 2x + \cos x$ b $\frac{\pi}{2}, \frac{3\pi}{2}$, 0.253, 2.889

c Minimum 1, maximum 4.125 (another TP has y-value 3)

Exercise 8.3

1 a i $0°, 90°, 360°$ ii $119.6°, 346.7°$ iii $98.0°, 205.9°$ iv $4.7°, 207.8°$ v $121.7°, 325.5°$ vi $40.2°, 252.4°$

b i 1.147, 3.566 ii $0, 4.428, 2\pi$ iii $0, 1.911, 2\pi$ iv $3.285, \frac{3\pi}{2}$

2 a $17 \sin (x - 5.793)$ b 0.294, 1.868

3 a $25 \cos (x + 286)°$ b $162°, 346°$ (to nearest degree)

4 a $\sqrt{5} \sin (x + 2.678)$ b $0, 4.069, 2\pi$

5 a $3\sqrt{2} \cos (2x - 160.5)°$ b $58°, 103°, 238°, 283°$

6 a Between 9.6 min and 170.4 min b 36.9 min, 110.6 min

7 a i $AD = 35 \sin x°$ ii $AE = 12 \cos x°$ b Perimeter $= 2(AD + AE) = 2(35 \sin x° + 12 \cos x°)$

c $x = 44.2°$

Exercise 8.4

1 a $\frac{dy}{dx} = 1 + \cos x$, since $-1 \leqslant \cos x \leqslant 1$ it follows $0 \leqslant 1 + \cos x \leqslant 2$; so $\frac{dy}{dx} \geqslant 0$ for all x; function is never decreasing

b $x = 0.51$ (to 2 d.p.) c 2.18 d SP at $\cos x = -1$, i.e. $x = \pm\pi$

2 a 0.86 b 3.43

3 a 0.70 b 0.49974 [to 5 d.p. for info only]

4 $x = -0.82(413...), 0.82(413...)$; $y = 0.6791... = 0.68$ (to 2 d.p.)

5 a 6.2832 b 0.3927 and 0.7854 c 4.7124 d 0.8893

Preparation for assessment

1 $70°, 140°, 250°, 320°$

2 $0, \frac{2\pi}{3}, \pi, \frac{5\pi}{3}, 2\pi$

3 $\frac{\pi}{6}, \frac{5\pi}{6}$

4 a $0, \pi, 2\pi, \frac{\pi}{6}, \frac{11\pi}{6}$ b 0, 120, 360

5 a Black ≡ lighthouse; purple ≡ beacon b 47·5 s, 132·5 s, 244·7 s, 295·3 s

6 a 126 m b 210·5 c 93, 328

7 a $13 \cos (x + 5{\cdot}888)$ b 1·088, 5·985

9 Logs and exponential functions

What you need to know

1 a 81 b $\frac{1}{2}$ c 2 d 1 e 3 f 9

2 a a^8 b a^2 c a^{15} d a^4 e $a^5 b^{-1}$

3 a i 10·5 ii 5·5 iii 8·3 iv 9·1 b i 2·3 ii 2·7 iii 5·8 iv 3·9
c i 8·4 ii 9·3 iii 7·6 iv 10·6 d i 3·1 ii 2·2 iii 1·5 iv 1·2

4 a i 2·00 ii 3·00 iii 4·00 iv 5·00 v 6·00
b i 2·4 ii 5·6 iii 7·1 iv 8·9 v 9·3

5 a i $1{\cdot}9 \times 2{\cdot}9 = 10^{0{\cdot}279} \times 10^{0{\cdot}462} = 10^{0{\cdot}741} = 5{\cdot}5$ ii $6{\cdot}4 \times 1{\cdot}3 = 10^{0{\cdot}806} \times 10^{0{\cdot}114} = 10^{0{\cdot}920} = 8{\cdot}3$
b i $8{\cdot}7 \div 3{\cdot}2 = 10^{0{\cdot}940} \div 10^{0{\cdot}505} = 10^{0{\cdot}435} = 2{\cdot}7$ ii $9{\cdot}3 \div 1{\cdot}6 = 10^{0{\cdot}968} \div 10^{0{\cdot}204} = 10^{0{\cdot}764} = 5{\cdot}8$
c i $2{\cdot}1^3 = (10^{0{\cdot}322})^3 = 10^{0{\cdot}966} = 9{\cdot}2$ ii $1{\cdot}5^5 = (10^{0{\cdot}176})^5 = 10^{0{\cdot}880} = 7{\cdot}6$
d i $\sqrt[3]{10{\cdot}6} = (10^{1{\cdot}025})^{\frac{1}{3}} = 10^{0{\cdot}342} = 2{\cdot}2$ ii $\sqrt[4]{5{\cdot}1} = (10^{0{\cdot}708})^{\frac{1}{4}} = 10^{0{\cdot}177} = 1{\cdot}5$

Exercise 9.1

1 a 0 b 1 c 2 d 3 e 4 f −1
g −2 h −3 i $\frac{1}{2}$ j $\frac{3}{2}$ k $\frac{2}{3}$

2 a $\log \frac{1}{3} = \log 3^{-1} = -\log 3$ b $\log 81 = \log 3^4 = 4 \log 3$
c i 4 ii $\log 4 \div \log 32 = \log 2^2 \div \log 2^5 = 2 \log 2 \div 5 \log 2 = \frac{2}{5}$

3 a $\log 3\,.\,2 = \log 6$ b $\log 4\,.\,6 = \log 24$ c $\log 2\,.\,3\,.\,7 = \log 42$ d $\log \frac{12}{6} = \log 2$
e $\log \frac{24}{8} = \log 3$ f $\log \frac{(5\,.\,6)}{15} = \log 2$ g $\log 7 + 0 = \log 7$ h $\log 7 - 0 = \log 7$
i $\log \left(\frac{1}{2}\,.\,\frac{2}{3}\right) = \log \frac{1}{3}$ j $\log \left(\frac{1}{3} \div \frac{2}{3}\right) = \log \frac{1}{2}$ k $\log \left(\frac{5}{1} \div \frac{5}{6}\right) = \log 6$ l $\log (0{\cdot}2 \times 5) = \log 1 = 0$

4 a $\log 3 + \log 2^2 = \log 3\,.\,2^2 = \log 12$ b $\log \left(\frac{16}{2^3}\right) = \log 2$ c $\log (5^2\,.\,4^3) = \log 1600$
d $\log(4^{\frac{1}{2}}\,.\,2) = \log 4$ e $\log(16^{\frac{1}{2}}\,.\,8^{\frac{1}{3}}) = \log 8$ f $\log(8^{\frac{2}{3}}\,.\,4) = \log 16$

5 a 1·748 b 2·748 c 3·748 d 4·748
e 1·940 f 2·544 g 3·279 h 4·908

6 a 2·2 b 2·8 c 5·5 d 6·6

Exercise 9.2

1 a 4 b 2 c −1 d $\frac{1}{2}$
e $\frac{1}{3}$ f −2 g −3

2 a $\log_{10} 10 = 1$ b $\log_4 16 = 2$ c $\log_9 729 = 3$ d $\log_6 36 = 2$
e $\frac{1}{2}$ f −1 g −2 h 2

3 a $x = \log_4 216 \Leftrightarrow 216 = 4^x$ b $m = \log_4 18 \Leftrightarrow 18 = 4^m$ c $n = \log_5 2 \Leftrightarrow 2 = 5^n$ d $p = \log_7 \left(\frac{1}{2}\right) \Leftrightarrow \frac{1}{2} = 7^p$

4 a $\log_3 x = 4 \Leftrightarrow x = 3^4 = 81$ b $\log_4 x = 2 \Leftrightarrow x = 4^2 = 16$
c $x = 2$ d $x = 5$

5 a i 0·2 ii 1·5 iii 2·5 iv 3·5 v 5.0
b $0{\cdot}67 \log T - 0{\cdot}8 = 2 \Rightarrow \log T = 4{\cdot}179 \Rightarrow T = 10^{4{\cdot}179} = 15\,104$ g = 15 kg (to 2 s.f.)

6 a i $\log_c A = \log_c B^x$ ii $\log_c A = x \log_c B \Rightarrow x = \frac{\log_c A}{\log_c B}$ b If $c = B \Rightarrow x = \frac{\log_B A}{\log_B B} = \log_B A$
c i $\log_7 12 = \frac{\log_{10} 12}{\log_{10} 7} = 1{\cdot}28$ ii 4·75 iii 0·721 iv 2·62 v 1

Exercise 9.3

1 Graphs of the form shown with the following points:

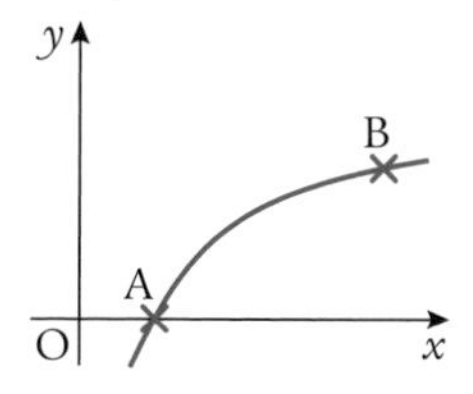

a $A(1, 0), B(3, 1)$ b $A(1, 0), B(5, 1)$ c $A(1, 0), B(6, 2)$
d $A(1, 0), B(10, 3)$ e $A(1, 0), B(3, 2)$ f $A(1, 0), B(5, 3)$
g $A(1, 0), B(6, -1)$ graph reflected in x-axis h $A(1, 0), B(10, \frac{1}{2})$

2 Graphs with curves passing through points:
a $(1, 0)\ (a, 2)$ b $(1, 0)\ (a, 3)$ c $(1, 0)\ (a, -1)$
d $(5, 0)\ (a + 4, 1)$ e $(1, 3)\ (a, 4)$ f $(1, 1)\ (a, 3)$

3 a 0 dB b i 10 dB ii 60 dB iii 85 dB c 632 000 mPa (to 3 s.f)

4 a $k = 50$ b 31% c $50 \log_{10} 2$ d 9th week

Exercise 9.4

1 a $y = 5^x$ b $y = 7^x$ c $y = e^x$ d $y = 5^{\frac{x}{2}}$
e $y = 4^{\frac{x}{3}}$ f $y = e^{\frac{x}{4}}$ g $y = 3^{x-2}$ h $y = 7^{x+1}$

2 a i 16 ii 2 iii $\frac{1}{4}$ b i 54·6 ii 2·00 iii 3·00

3 a $f(x) = e^{0.79x}$ b $f(x) = e^{3.47x}$ c $f(x) = e^{5.48x}$ d $f(x) = 5e^{1.39x}$

4 a 8 units b 9 units c 12·3 units

5 a i 181 litres ii 174 litres b i 157 litres ii 43 litres c $B = Ve^{-0.02x}$

6 a i 9·7 g ii 7·0 g b i $\frac{1}{2}A_0 = A_0(1 - r)^{10} \Rightarrow 1 - r = \sqrt[10]{\frac{1}{2}} \Rightarrow r = 0.067$ ii $A_t = A_0e^{-0.07t}$
c i $r = 0.00012$; $A_t = A_0 0.9999^t$; $A_t = A_0e^{-0.00012t}$ ii 13 400 years

7 a i 98·9 m ii 62·4 m iii 0 m b 34·4 m c $x = 60e^{-\frac{h}{90}}$

8 a $f(e^x) = 4e^x + 3$ b i e^{4x+3} ii $e^{4x+3} = e^{4x} . e^3 = 20e^{4x}$

9 a i e^{2x} ii e^{x^2} b $(e^x)^2$ and $(e^x)^x$ c 0 or 2

Exercise 9.5

1 a 0·847 b 0·811 c 1·19 d 2·83
e 1·79 f 2·10 g 1·91 h 1·77

2 a $x = 16\,384$ b $x = 1.22$ c $x = 1.40$

3 a $\frac{x + 8}{x - 8} = 3^2 \Rightarrow x = 10$ b $\frac{x^2}{x + 6} = 2^3 \Rightarrow x = 12$ or -4 but $\log_2 x$ not defined when $x = -4$

4 a $x = 1.11$ (to 3 s.f.) b $x = 1.71$ (to 3 s.f.)

5 $a = 4.66$ (to 3 s.f.), $b = 0.588$ (to 3 s.f.) ; $y = 4.66x^{0.588}$

6 a $13 = a . 5^b$ and $35 = a . 8^b$ b $a = 0.438, b = 2.11, f(x) = 0.438x^{2.11}$

7 a $0.24 = a . 0.39^b$ and $1.88 = a . 1.52^b$ b $a = 1.0, b = 1.5$
c $y = x^{1.5}$ [Kepler gave it as $y^2 = x^3$]

8 $a = 4.8, b = 0.46$ (to 2 s.f.)

9 a $\log 5 = b \log 2 + \log a$; $\log 106 = b \log 874 + \log a$ b $a = 3.5, b = 0.50$ c Distance = 58·6 km

10 $a = 1.84, b = 1.28$ (to 3 s.f.); $y = 1.84 \times 1.28^x$

11 a $a = 1000$, $b = 1.03$ (to 3 s.f.); $y = 1000 \times 1.03^x$ b £1344 (to nearest £)

12 $a = 100, b = \sqrt{10}$

13 a $\log h = k \log 0.6 + \log 500$ b 9th bounce

Exercise 9.6

1 a $\log y = b\log x + \log a$

b $\log y = x\log b + \log a$

c $\log y = 4\log x + \log 3$

d $\log y = x\log 4 + \log 3$

2 $y = 1000\,x^2$

3 $y = e^{\frac{3}{4}}x^{\frac{5}{4}} \approx 2{\cdot}12x^{\frac{5}{4}}$

4 $y = 10^{-4} \times 10^{4x}$

5 $a = 9, b = 3, y = 9 \times 3^x$

6 a Table to 3 d.p. although graph drawing will be to 1 d.p.

log (time)	0·000	0·301	0·477	0·602	0·699	0·778
log (drop)	0·699	1·301	1·653	1·903	2·097	2·255

b Straight line passing through (0, 0·7) and (0·7, 2·1)

c $y = 10^{0{\cdot}7}x^2$

7 a Student's own check

b $a = 50, b = -0{\cdot}15$ (to 2 s.f.)

c 11 years

Preparation for assessment

1 $x = \log_2 \frac{y}{3}$

2 $A = 5000e^{\frac{y}{33}}$

3 a

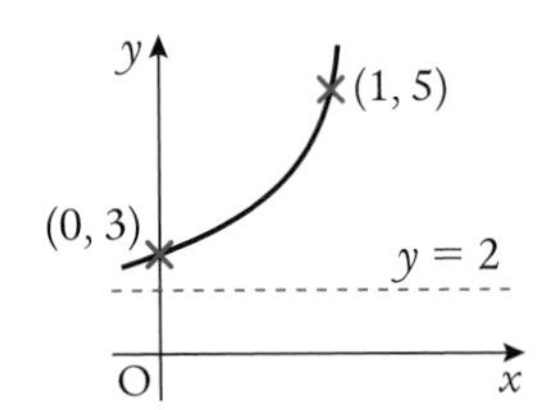

b $y = 3^x + 2$

4 $\log_5\left(\frac{2 \times 10^2}{\sqrt{64}}\right) = 2$

5 a $x = 0$

b $x = 3$

6 a $4e^x - 1$

b i e^{4x-1} ii $0{\cdot}37e^{4x}$

7 $a = 3, b = \frac{1}{2}$

8 $a = \frac{1}{4}, b = 2$

9 a Argument leading to $y = ax^b \Rightarrow \log y = b\log x + \log a$, concluding that this is of the form $Y = mX + c$

b $a = 3{\cdot}3$ and $b = 2{\cdot}2$ (to 2 s.f.) Answers will vary – discuss.

c 523

10 a Show that $\log P = t\log b + \log a$

b Show data linear by graph or finding 'constant' gradient of 0·02 approx.

c $a = 4000, b = 1{\cdot}05$

d Year 14

10 Differential calculus 2 – applications

What you need to know

1 a $8x^3 + 2x$

b i -10 ii 0 iii 10

c Factorising gives $2x(4x^2 + 1)$. The bracketed term is always positive, so the sign of $f'(x)$ will be the same as the sign of $2x$. Hence conclusion.

d i From c, $f(x)$ increases when $x > 0$ ii $f(x)$ decreases when $x < 0$ iii $f(x)$ stationary at $x = 0$

e
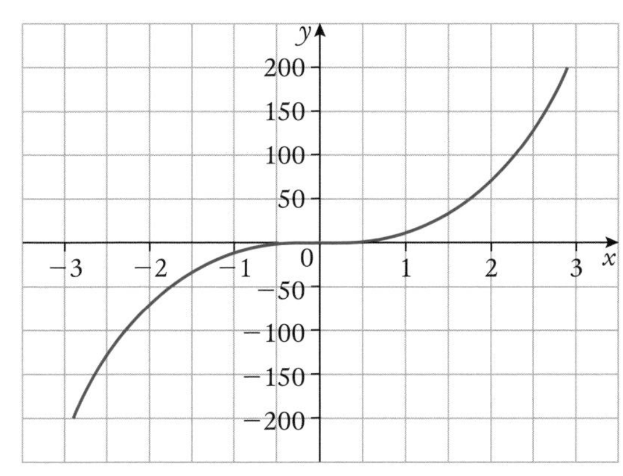

2 a $2\cos\left(2x+\frac{\pi}{3}\right)$ b i -1 ii 1 c $\frac{\pi}{12}, \frac{7\pi}{12}, \frac{13\pi}{12}, \frac{19\pi}{12}$ d Max., min., max., min.

3 a $3x^2 - 6x - 9$ b $(3, -26)$, min.; $(-1, 6)$, max. c $-1 < x < 3$

4 a $V = \pi x^2 h \Rightarrow h = \frac{250}{\pi x^2}$ b $A = \frac{500}{x}$ c $f(x) = 2\pi x^2 + \frac{500}{x}$ d $f'(x) = 4\pi x - \frac{500}{x^2}$

e (3·41, 220); for values of x less than the stationary point, the gradient is negative; for values of x greater than the stationary point, the gradient is positive; e.g.

x	$f'(x)$
3·41392	−0·0000119
$\sqrt[3]{\frac{500}{4\pi}}$	0
3·41393	0·000365067

thus the stationary point is a minimum.

Exercise 10.1

1 a $\frac{d\$}{dx} = 1{\cdot}6$ b i 1·6 ii 16 c i $f(x) = ax + b \Rightarrow f'(x) = a$ (a constant) ii 0

2 a $\frac{dA}{dt} = 3$ b i 3 ii 3 c i 103 ii 3 iii 3%

3 a 20 b $f'(x) = 0{\cdot}8x + 4 \Rightarrow$ 4 pheasants/year
c i 50 pheasants ii 8 pheasants/year
d Model predicts 100 pheasants, got 120 pheasants; model predicts 12 pheasants/year, get 18 pheasants/year. Population is doing better than model predicts.

4 a 20 °C b i 64 °C ii 2·2 °Cs^{-1} c 0·55 °Cs^{-1}

5 a 3 cms^{-1} b 8π cm^2cm^{-1} c i Substitute $3t$ for r in πr^2 ii $18\pi t$ cm/s iii 90π cm^2/sec
d $\frac{dA}{dt} = 18\pi t; \frac{dA}{dr} = 2\pi r; \frac{dr}{dt} = 3 \Rightarrow \frac{dA}{dt} = 6\pi r = 6\pi . 3t = 18\pi t$

6 a $P'(t) = \frac{10\pi}{3}\sin\left(\frac{\pi}{15}t\right)$ b $\frac{5\pi}{\sqrt{3}} = 9{\cdot}07\%$/day c i 0%/day ii −9·07%/day
d i Rate greater than zero ... waxing Moon ii Rate less than zero ... waning Moon

7 a i 38 people per day ii 160 people per day b 369 people per day

8 a $P(R) = R^{\frac{3}{2}}$ b i $\frac{dP}{dR} = \frac{3}{2}R^{\frac{1}{2}}$ ii 7·5 years per AU
c $R(P) = P^{\frac{2}{3}}$ d $\frac{dR}{dP} = \frac{2}{3}P^{-\frac{1}{3}}; \frac{1}{3}$ AU per year

Exercise 10.2

1 a $v(t) = -3 + 4t$ b i 1 m s^{-1} ii 9 m s^{-1} c $a(t) = 4$; it is constant

2 a $s(3) - s(2) = 147$ m b $v(t) = 2 + 20t + 15t^2$ c i $v(1) = 37$ m s^{-1} ii $v(2) = 102$ m s^{-1}
d $a(t) = 20 + 30t$ e 80 m s^{-2}

3 **a** $v(t) = 24 - 10t; a(t) = -10$
b $s(0) = 5$ m
c 2·4 seconds after being struck
d $-10\ \text{m s}^{-2}$
e Strikes ground at $s(t) = 0$, giving $t = 5$ in the context: $v(5) = -26\ \text{m s}^{-1}$; $a(5) = -10\ \text{m s}^{-2}$

4 **a** By Pythagoras' theorem: $s^2 = t^2 + 100^2$, hence result
b $v(t) = \frac{1}{2}.2t.(t^2 + 10\,000)^{-\frac{1}{2}} = \dfrac{t}{\sqrt{t^2 + 10\,000}}$
c $0{\cdot}51\ \text{m s}^{-1}$
d $0{\cdot}001\ \text{m s}^{-2}$

5 **a** $v(t) = 3.\frac{\pi}{6}.\cos\left(\frac{\pi t}{6}\right) = \frac{\pi}{2}\cos\left(\frac{\pi t}{6}\right)$
b $\frac{\pi}{4}$ cm s^{-1}
c $a(t) = -\frac{\pi}{6}.\frac{\pi}{2}.\sin\left(\frac{\pi t}{6}\right) = -\frac{\pi^2}{12}\sin\left(\frac{\pi t}{6}\right)$
d $-\frac{\pi^2}{24}$ cm/s^2

6 **a** $v(t) = s'(t) = u + at$
b $v(0) = u$
c **i** $a(t) = v'(t) = a$ **ii** It is constant

Exercise 10.3

1 **a** y-intercept $(0, -700)$; x-intercepts $(-14, 0)$, $(-5, 0)$, $(10, 0)$; max. TP $(-10, 400)$, min. TP $(4, -972)$; profile: Figure 1
b y-intercept $(0, 245)$; x-intercepts $(-5, 0)$, $(7, 0)$; max. TP $(-1, 256)$, min. TP $(7, 0)$; profile: Figure 1
c y-intercept $(0, -3848)$; x-intercepts $(-37, 0)$, $(-13, 0)$, $(8, 0)$; max. TP $(-27, 4900)$, min. TP $(-1, -3888)$; profile: Figure 1
d y-intercept $(0, 2)$; x-intercepts $(-2, 0)$, $(1, 0)$; max. TP $(-1, 4)$, min, TP $(1, 0)$; profile: Figure 1
e y-intercept $(0, 98)$; x-intercepts $(-7, 0)$, $(2, 0)$; max. TP $(-1, 108)$, min. TP $(-7, 0)$; profile: Figure 2
f y-intercept $(0, 238)$; x-intercepts $(-7, 0)$, $(2, 0)$, $(17, 0)$; max. TP $(-3, 400)$, min. TP $(11, -972)$; profile: Figure 1

Figure 1

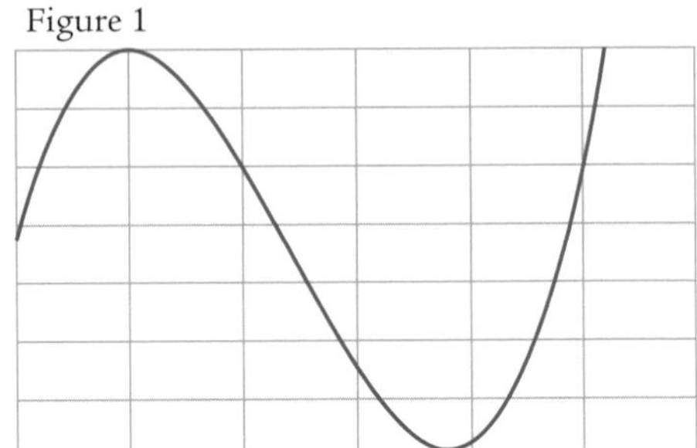

Figure 2

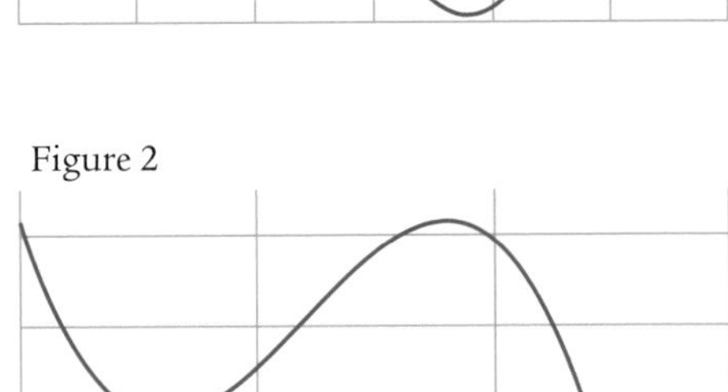

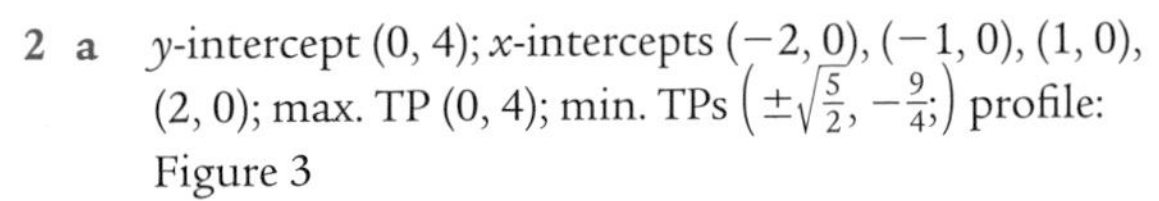

2 **a** y-intercept $(0, 4)$; x-intercepts $(-2, 0)$, $(-1, 0)$, $(1, 0)$, $(2, 0)$; max. TP $(0, 4)$; min. TPs $\left(\pm\sqrt{\frac{5}{2}}, -\frac{9}{4};\right)$ profile: Figure 3
b y-intercept $(0, 49)$; x-intercepts $(-7, 0)$, $(-1, 0)$, $(1, 0)$, $(7, 0)$; max. TP $(0, 49)$, min. TPs $(-5, -576)$, $(5, -576)$; profile: Figure 3
c y-intercept $(0, 0)$; x-intercepts $(0, 0)$, $(4, 0)$; max. TP $(3, 81)$, min PI $(0, 0)$; profile: Figure 4

Figure 3

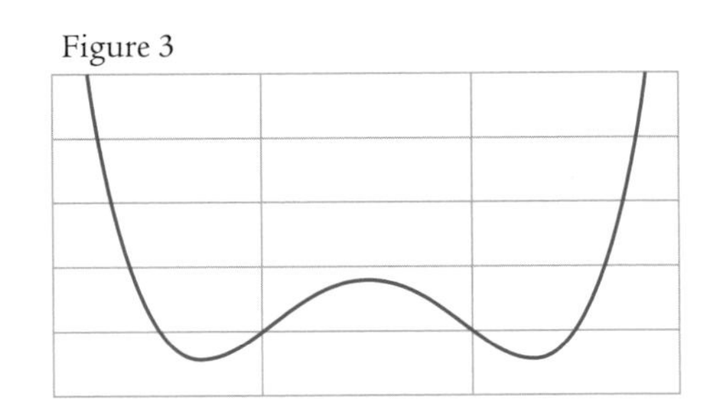

3 **a** $(x + 3)(x - 1)(x^2 + 2x + 3)$
b $(0, -9)$ min., $(-2, -9)$ min., $(-1, -8)$ max.
c Profile: Figure 3;

Figure 4

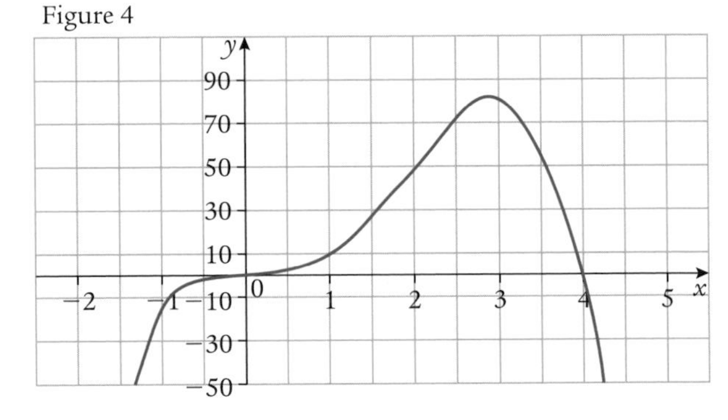

4 a (5, 100); min. **b** **i** $C(1) = 260$ **ii** $C(10) = 125$
c **i** Decreasing **ii** Increasing **d** Figure 5

Figure 5

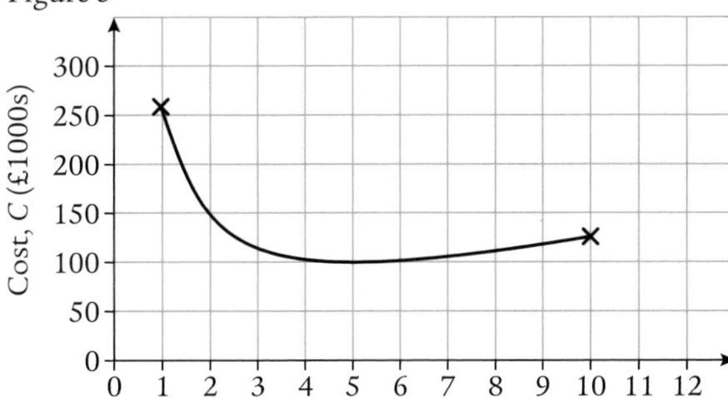

Exercise 10.4

1 a Min. −13; max. 19 **b** Min. −9; max. 3·25 **c** Min. −51; max. 1
d Min. 5; max. 9 **e** Min. −189; max. 64 **f** Min. −16; max. 1
g Min. −12; max. 20 **h** Min. 32; max. 257 **i** Min. 0·5; max. 1
j Min. −2; max. 4

2 a $60 - x$ **b** $60x - x^2$ **c** $x = 30$; maximum

x	→	30	→
$60 - x$	+	0	−
	↗	—	↘

3 a $\dfrac{49}{x}$ **b** $x + \dfrac{49}{x}$ **c** $x = 7$; minimum

x	→	7	→
$1 - \dfrac{49}{x^2}$	−	0	+
	↘	—	↗

4 a $(25 - x)$ m **b** $A(x) = (25x - x^2)\ \text{m}^2$ **c** $x = 12{\cdot}5$; maximum

x	→	12·5	→
$25 - 2x$	+	0	−
	↗	—	↘

5 a $(200 - 2x)$ m
b $A(x) = x(200 - 2x) = 200x - 2x^2$; $x = 50$ maximises area; length = 100 m, breadth = 50 m
c Length = breadth = 100 m

6 a $V(x) = 4x^3 - 66x^2 + 216x$
b $x = 9$ min.; $x = 2$ max.
c $x = 2$ maximises volume at 200 cm^3
d $x = \frac{5}{3}$ maximises volume

x	→	2	→	9	→
$x - 2$	−	0	+	+	+
$x - 9$	−	−	−	0	+
$V'(x)$	+	0	−	0	+
	↗	—	↘	—	↗

7 a **i** $\sqrt{1 + x^2}$ **ii** $T = \dfrac{D}{S} = \dfrac{\sqrt{1 + x^2}}{4}$ **iii** $10 - x$ **iv** $\dfrac{10 - x}{5}$ **v** $\dfrac{\sqrt{1 + x^2}}{4} + \dfrac{10 - x}{5}$
b $x = \frac{4}{3}$ **c** $\frac{129}{60}$ h = 129 min

8 a $h = \frac{1000}{x^2}$

b **i** Top = bottom = x^2; each face = $x . \frac{1000}{x^2} = \frac{1000}{x}$; $s(x) = 2$ tops $+ 4$ faces $= 2x^2 + \frac{4000}{x}$.

ii $x = 10$ minimises surface area; so box is cube of edge 10 cm

9 a Area of kite is half product of diagonals; so diagonals are x cm and $\frac{10\,000}{x}$ cm.

b $L(x) = x + \frac{10\,000}{x}$ **c** $x = 100$ minimises length; minimum length = 200 cm

10 a $h = \frac{500}{\pi x^2}$ **b** $A(x) = 2\pi x^2 + \frac{1000}{x}$

c $x = \sqrt[3]{\frac{1000}{4\pi}}$ minimises surface area: this is a radius of about 4·3 cm

11 a 4·5 m

b **i** Hypotenuse $= 9 - x$; by Pythagoras' theorem other side is $\sqrt{(9-x)^2 - x^2} = \sqrt{81 - 18x}$;
$A = \frac{1}{2}bh = \frac{1}{2}x\sqrt{81 - 18x} = \frac{1}{2}\sqrt{81x^2 - 18x^3}$. **ii** 3

12 a $36 - \frac{x^2}{12}$ **b** $A(x) = x\left(36 - \frac{x^2}{12}\right)$ **c** 12 **d** 12 m by 24 m

13 a $C'(x) = 50 - \frac{180\,000}{x^2}$ **b** 60 tonnes **c** £6000

Preparation for assessment

1 a $V = 2t$ **b** $V = \frac{4}{3}\pi r^3$

c **i** 4π litres/cm **ii** 100π litres/cm **d** $r = \sqrt[3]{\frac{3V}{4\pi}}$

e **i** 0·21 cm/litre **ii** 0·07 cm/litre **f** $\frac{1}{2\pi r^2}$ cm/minute **g** 0·0044 cm/minute

2 a $v(t) = 3t^2 + 10t$ **b** **i** 1 s **ii** 2 s **c** $a(t) = 6t + 10$

d $34\ \text{m s}^{-2}$ **e** 10 s **f** $v(0) = 0\ \text{m s}^{-1}$; $a(0) = 10\ \text{m s}^{-2}$

3 a −2 cm **b** $\frac{\pi}{2}$ s, $\frac{3\pi}{2}$ s **c** $v(t) = 2\cos 2t - 2\sin t$ **d** $2\ \text{cm s}^{-1}$

e $a\left(\frac{3\pi}{2}\right) = -4\sin 3\pi - 2\cos\frac{3\pi}{2} = 0$; $v\left(\frac{3\pi}{2}\right) = 2\cos 3\pi - 2\sin\frac{3\pi}{2} = 0$; $s\left(\frac{3\pi}{2}\right) = \sin 3\pi + 2\cos\frac{3\pi}{2} = 0$

4

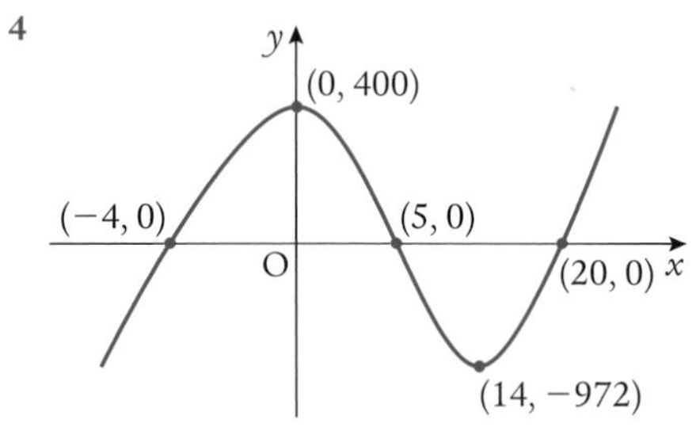

5 a $(2x + x\theta)$ m **b** $\theta = \frac{16}{x} - 2$ **c** **i** $A = \frac{x^2\theta}{2} \Rightarrow A(x) = \frac{x^2}{2}.\left(\frac{16}{x} - 2\right) = (8x - x^2)\ \text{m}^2$ **ii** 4

11 Integral calculus

What you need to know

1 a $9x^2 + 8x - 5$ **b** **i** $\frac{1}{2}x^{-\frac{1}{2}}$ **ii** $-x^{-2}$ **iii** $-3x^{-4}$

c $-\frac{1}{2}x^{-\frac{3}{2}} + \frac{3}{2}x^{\frac{1}{2}}$ **d** **i** $\cos x$ **ii** $-\sin x$

2 a **i** $6(3x + 1)$ **ii** $-6(3x + 1)^{-3}$ **iii** $\frac{3}{2}(3x + 1)^{-\frac{1}{2}}$

b **i** $3\cos 3x$ **ii** $3\cos(3x - 4)$ **iii** $-5\sin(5x - 1)$ **iv** $5\sin(1 - 5x)$

3 90 miles

4 a $\sqrt{24}$ **b** $\frac{5+\sqrt{24}}{2}$ **c** $\frac{\sqrt{24}+\sqrt{21}}{2}, \frac{\sqrt{21}+4}{2}$, 3·5, 1·5
d 18·98 (to 2 d.p.) **e** 19·63 (to 2 d.p.)

Exercise 11.1

1 a **i** $4x + c$ **ii** $x^2 + c$ **iii** $2x^3 + c$ **iv** $\frac{2}{5}x^5 + c$ **v** $\frac{1}{2}x^6 + c$
b **i** $x^2 + x + c$ **ii** $\frac{7}{2}x^2 - 2x + c$ **iii** $x^3 + x^2 - x + c$ **iv** $\frac{1}{4}x^4 + \frac{4}{3}x^3 + \frac{1}{2}x^2 - 5x + c$ **v** $x - 2x^2 + c$
vi $x - 2x^2 - x^3 + c$ **vii** $3x^3 - x^4 - 3x + c$ **viii** $-x + \frac{1}{2}x^2 + c$

2 a **i** $-2\cos x + c$ **ii** $3\sin x + c$ **iii** $x - 3\cos x + c$ **iv** $2x + 5\cos x + c$
v $\frac{1}{3}x^3 - \sin x + c$ **vi** $x\cos\frac{\pi}{3} + c$ **vii** $\frac{1}{2}x^2 + x\sin\frac{\pi}{2} + c$ **viii** $\sin x - \cos x + c$
ix $2\sin x + 5\cos x + c$
b **i** $\frac{2}{3}x^{\frac{3}{2}} + c$ **ii** $-3x^{-1} + c$ **iii** $3x^{\frac{5}{3}} + c$ **iv** $\frac{3}{2}x^{\frac{2}{3}} + c$
v $\frac{6}{5}x^{\frac{5}{2}} - \frac{4}{3}x^{\frac{3}{2}} + c$ **vi** $-4x^{-1} - 8x^{\frac{7}{4}} + c$ **vii** $\frac{2}{3}x^{\frac{3}{2}} + c$ **viii** $6x^{\frac{4}{3}} + c$
ix $x + \frac{2}{5}x^{\frac{5}{2}} + c$ **x** $2x^{\frac{1}{2}} + c$ **xi** $\frac{2}{3}x^{\frac{3}{2}} + 6x^{\frac{1}{2}} + c$ **xii** $-2x^{-\frac{1}{2}} + c$

3 a **i** $\frac{1}{2}x^2 + c$ **ii** $x + c$ **iii** $\frac{1}{3}x^3 + \frac{3}{2}x^2 + 2x + c$ **iv** $\frac{2}{3}x^{\frac{3}{2}} + \frac{3}{4}x^{\frac{4}{3}} + c$
v $-x^{-1} + \frac{1}{2}x^{-2} + c$ **vi** $x^3 + x^2 + c$ **vii** $2x^3 + \frac{1}{2}x^2 - x + c$
b **i** $-\frac{x^{-2}}{2} - x^{-1} + c$ **ii** $\frac{2}{5}x^{\frac{5}{2}} - 2x^{\frac{1}{2}} + c$ **iii** $x - 6x^{-\frac{1}{2}} - 2x^{-1} + c$ **iv** $x + 6x^{\frac{1}{2}} + c$
c **i** $x + 2x^{\frac{1}{2}} + c$ **ii** $x - 2x^{\frac{1}{2}} + c$

4 a $y = 2x^3 + x + c$ **b** $y = 3x^4 + 3x^3 + x + c$ **c** $y = 7x + c$
d $y = c$ **e** $y = \frac{8}{3}x^{\frac{3}{2}} + c$ **f** $y = x - \frac{2}{3}x^3 - \frac{3}{4}x^4 + c$
g $y = x + \frac{x^2}{2} + \frac{x^3}{6} + \frac{x^4}{24} + \frac{x^5}{120} + c$ **h** $y = 2x - 3\cos x + c$ **i** $y = x - \cos x - 5\sin x + c$
j $y = 3\sin x - 4\cos x + \pi x + c$

5 a $y = \frac{2}{3}x^3 - x + 4$ **b** $y = 2x^{\frac{3}{2}} + 3$ **c** $y = 2x^3 + 3x^2 + 1$
d $y = 3x + 4$ **e** $y = 10$ **f** $y = 1 + x + \frac{x^2}{2} + \frac{x^3}{6} + \frac{x^4}{24} + \frac{x^5}{120}$
g $y = 2x^{\frac{1}{2}} + \frac{2}{3}x^{\frac{3}{2}} + \frac{2}{3}$ **h** $y = -3\cos x + 4$

6 a $y = x^2 + x - 11$ **b** $y = -3\cos x + x + 5$

7 a $v(t) = 10t + 5$ **b** $15\ \text{m s}^{-1}$ **c** $s(t) = 5t^2 + 5t$ **d** **i** 2 s **ii** $25\ \text{m s}^{-1}$

8 a $v(t) = \frac{3}{2}t^2$ **b** $6\ \text{m s}^{-1}$ **c** $s(t) = \frac{t^3}{2} + 2$ **d** **i** 64·5 m **ii** $37{\cdot}5\ \text{m s}^{-1}$

9 a $v(t) = 5\sin t$, $5\ \text{m s}^{-1}$ **b** $s(t) = -5\cos t + 5$ **c** **i** 10 m **ii** 10 m

10 a $P = 0{\cdot}35x^2 + 400$ **b** 1275 **c** By about 7 people

11 a $F = 0{\cdot}45x^2 - 2{\cdot}5x + 5$ **b** Population much less than at start **c** Year 10

12 a $C(t) = 3(t + t^2 + 0{\cdot}4t^3) + 20$ **b** 88 cases **c** 68 cases

13 a $w(x) = 2\sqrt{x} + 0{\cdot}5$ **b** 6·5 lb **c** 2 lb **d** 20·5 lb

Exercise 11.2

1 a $\frac{1}{18}(3x+1)^6 + c$ **b** $\frac{1}{8}(2x-1)^4 + c$ **c** $-\frac{1}{10}(3-2x)^5 + c$ **d** $-\frac{1}{4}(1-x)^4 + c$
e $-\frac{2}{3}(1-\frac{1}{2}x)^3 + c$ **f** $-\frac{1}{2}(2x+3)^{-1} + c$ **g** $(1-x)^{-1} + c$ **h** $-(\frac{1}{2}x-2)^{-2} + c$
i $-3(\frac{1}{3}x+7)^{-1} + c$ **j** $\frac{1}{6}(4-2x)^{-3} + c$ **k** $\frac{2}{15}(5x+3)^{\frac{3}{2}} + c$ **l** $\frac{1}{3}(6x-1)^{\frac{1}{2}} + c$
m $-\frac{3}{2}(1-x)^{\frac{2}{3}} + c$ **n** $\frac{2}{7}(2x+1)^{\frac{7}{4}} + c$ **o** $\frac{2}{3}(2-3x)^{-\frac{1}{2}} + c$ **p** $\frac{2}{21}(7x)^{\frac{3}{2}} + c$
q $-\frac{2}{3}(1-x)^{\frac{3}{2}} + c$ **r** $\frac{1}{3}(2x+5)^{\frac{3}{2}} + c$ **s** $\frac{3}{16}(4x-1)^{\frac{4}{3}} + c$ **t** $\frac{3}{5}(x+1)^{\frac{5}{3}} + c$

2 a $\frac{3}{28}(7x)^{\frac{4}{3}} + c$ **b** $2(x+1)^{\frac{1}{2}} + c$ **c** $3(2x+1)^{\frac{1}{2}} + c$ **d** $x + 2(1-2x)^{\frac{1}{2}} + c$

3 **a** $\frac{1}{3}\sin(3x+4)+c$ **b** $-\frac{1}{2}\cos(2x-1)+c$ **c** $-\sin(5-2x)+c$ **d** $2\cos(2-3x)+c$
e $x+\sin(1-x)+c$ **f** $3x-\cos(5-x)+c$ **g** $\frac{3}{2}\sin 2x+c$ **h** $-\frac{1}{9}\cos(3x-1)+c$
i $-\frac{5}{7}\cos 7x+c$ **j** $\frac{1}{2}\sin 2x+\frac{1}{5}\cos 5x+c$ **k** $x+3\cos 7x+2\sin 4x+c$ **l** $\frac{x^3}{3}-\frac{1}{3}\cos 3x+c$

4 **a** $\cos^2 x=\frac{1}{2}\cos 2x+\frac{1}{2}$ **b** $\frac{1}{4}\sin 2x+\frac{1}{2}x+c$ **c** $\sin^2 x=\frac{1}{2}-\frac{1}{2}\cos 2x; \frac{1}{2}x-\frac{1}{4}\sin 2x+c$
d **i** $\frac{1}{2}\sin 2x+c$ **ii** $x+c$ **e** $\frac{1}{2}\sin x+\frac{1}{2}x+c$ **f** $-\frac{1}{4}\cos 2x+c$

5 **a** $v(t)=\frac{3}{\pi}\sin\left(\frac{\pi t}{6}\right)$ **b** $\frac{3}{\pi}\,\text{m s}^{-1}$
c $s(t)=-\frac{18}{\pi^2}\cos\left(\frac{\pi t}{6}\right)+1+\frac{18}{\pi^2}$ **d** **i** $\frac{9}{\pi^2}+1=1\cdot9\,\text{m}$ (to 2 s.f.) **ii** $\frac{1}{4}\text{m s}^{-2}$

6 **a** $-\frac{180}{\pi}\cos x°+c$ **b** $\frac{180}{\pi}\sin x°+c$

7 **a** $-\frac{1}{8}(1-2x)^4+\frac{1}{4}$ **b** $-3\cos 2x+2$

8 **a** $C(x)=294(2x+8)^{\frac{1}{3}}+12$ **b** £3721 **c** £1·23/booklet

9 **a** Increasing at 5·33 minutes per day **b** $L=740-315\cos\left(\frac{2\pi t}{365}\right)$
c 787 min **d** 1055 min

Exercise 11.3

1 **a** 45 **b** −14 **c** 6 **d** 48 **e** 511
f 21 **g** 8 **h** $17\frac{1}{3}$ **i** 5 **j** $-\frac{5}{6}$
k 1 **l** 10 **m** 1 **n** 0·293 **o** 0
p 0·890 **q** $\frac{1}{2}$ **r** 0·190 (to 3 d.p.) **s** 0 **t** 0

2 **a** 12 **b** 35 **c** 3·4 **d** $10\frac{2}{3}$ **e** $-20\frac{5}{6}$
f $20\frac{5}{6}$ **g** $1\frac{1}{12}$ **h** $165\frac{1}{3}$ **i** 1

3 **a** 13 **b** $11\frac{5}{6}$ **c** 12

4 **a** $(-5, 0), (3, 0)$ **b** $85\frac{1}{3}$ unit² **c** **i** $(-\frac{1}{2}, 0), (4, 0)$; 30·375 unit² **ii** $(-7, 0), (3, 0)$; $166\frac{2}{3}$ unit²

5 **a**

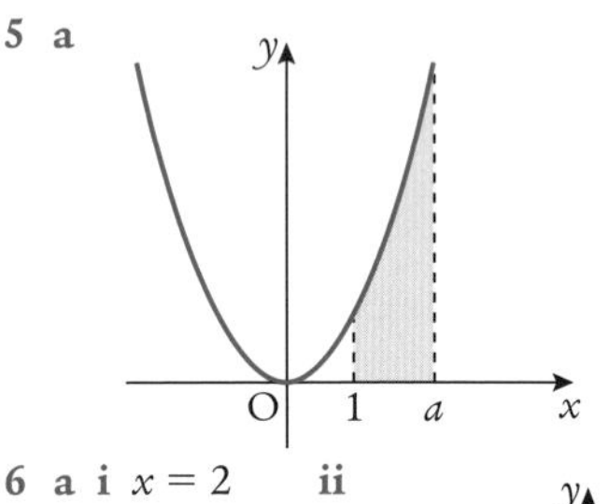

b 5

6 **a** **i** $x=2$ **ii**

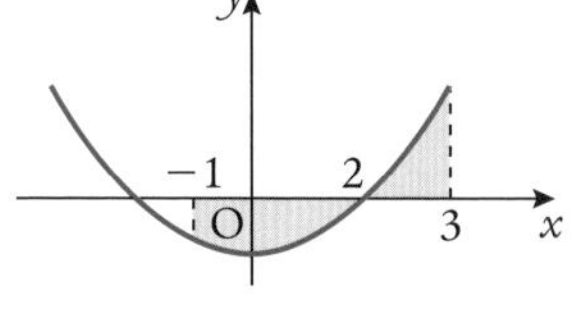

iii $11\frac{1}{3}$

b **i** $x=3$ **ii**

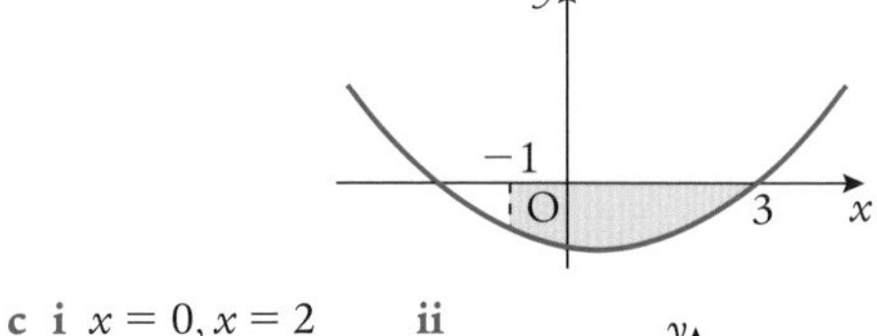
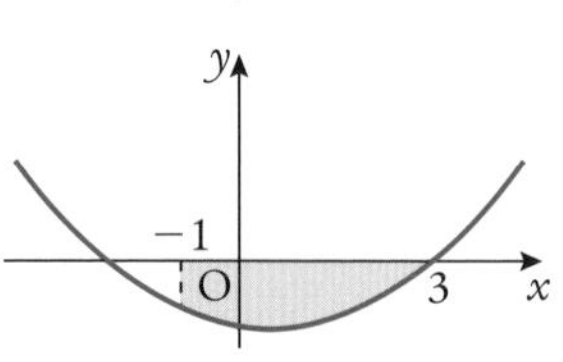

iii $18\frac{2}{3}$

c **i** $x=0, x=2$ **ii**

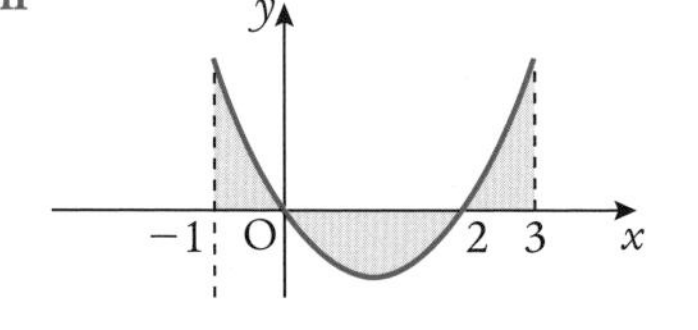

iii 4

7 a 9 b $18\frac{1}{6}$ c 3·75 d $\frac{2}{\pi}$

8 740 minutes

9 a i $v(t) = 2t$ ii 84 m b i $v(t) = \frac{3t^2}{2} + 4$ ii 492 m

Exercise 11.4

1 a $(-4, 8), (2, 8)$ b 36 unit^2 c $10\frac{2}{3}$ unit^2

2 a $(3, 9), (-2, 4)$ b $20\frac{5}{6}$ unit^2 c $57\frac{1}{6}$ unit^2

3 a $(-3, 1), \left(\frac{1}{2}, \frac{11}{4}\right)$ b 14·3 unit^2 (to 3 s.f.) c 91·5 unit^2 (to 3 s.f.)

4 a $(-3, -11), (-1, -1), (0, 1)$ b 3·08 unit^2 (to 3 s.f.) c 210 unit^2 (to 3 s.f.)

5 a $(1, 0), (-2, -9), (-1, 0)$ b $3\frac{1}{12}$ unit^2 c $78\frac{1}{12}$ unit^2

6 a $\frac{\pi}{4}, \frac{5\pi}{4}$ b $2\sqrt{2}$ unit^2

7 a $(-6, 0), (6, 0)$ b 28 unit^2

8 a

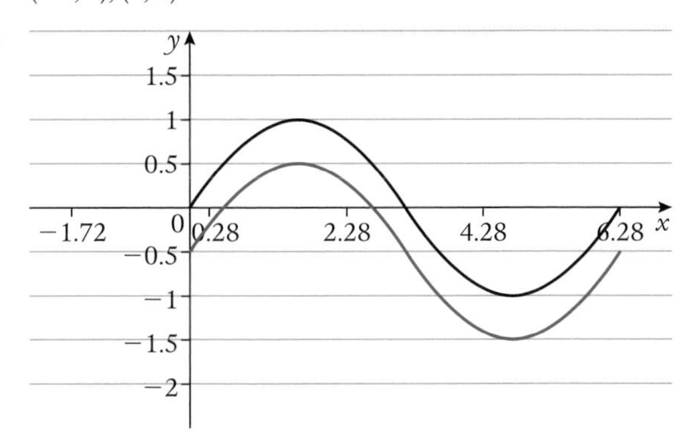

b $\int_0^{2\pi} \sin x - (\sin x - 0{\cdot}5)\,dx = \int_0^{2\pi} 0{\cdot}5\,dx$ c π unit^2

d i $y = f(x) - 5$ ii $\int_0^{100} f(x) - (f(x) - 5)\,dx = \int_0^{100} 5\,dx$ iii 500 m^2

Preparation for assessment

1 $2x^4 - \frac{1}{2x^2} + \frac{3}{5}x^{\frac{5}{3}} + x + c$

2 a $\frac{x^2}{2} + 2\sqrt{x} - 4$ b $\frac{1}{2} - \frac{1}{2}\cos\left(2x + \frac{\pi}{3}\right)$

3 a $\frac{2}{7}x^{\frac{7}{2}} + 2x^{\frac{1}{2}} + c$ b $\frac{(3x + 4)^{\frac{5}{3}}}{5} + c$ c $-\frac{3}{4}\sin(5 - 4x) + c$

4 a $a(t) = 4t$ b $s(t) = \frac{2}{3}t^3 + t + 1$ c i 22 m ii 19 m s^{-1} iii 12 m s^{-2}

5 a $x = 0, x = 120$ b 1440 unit^2

6 a Substituting $x = 1$ into the cubic gives $y = 5$; into the parabola also gives $y = 5$. So $(1, 5)$ is a point common to both. The curves intersect at $(1, 5)$. They also intersect at $(2, 7)$ and $(-3, -23)$.

b 32·75 unit^2

12 Lines and circles

What you need to know

1 a $\frac{3}{4}$ b $-\frac{1}{4}$ c 0 d undefined

2 a $m = 3, c = -1$ b $m = 0, c = 3$ c $m = 3, c = 7$
d m undefined, c undefined e $m = -\frac{3}{4}, c = 2$

3 a $y - 5 = 2(x - 1)$ b $y - 3 = -\frac{1}{3}(x - 3)$

4 (0, 0), (1, 1)

5 (6, 4)

6 a i 13 units ii $\sqrt{(x_2 - x_1)^2 + (y_2 - y_1)^2}$ b i 25 ii $x^2 + y^2$

7 a $(-2, 1)$ b $(-4, 3)$ c $(-y, x)$

Exercise 12.1

1 a $m_{AB} = \frac{1}{2} = m_{CD}$ b PQ $\parallel$ SR($m_{PQ} = \frac{2}{3} = m_{SR}$); PS $\parallel$ QR($m_{PS} = -\frac{2}{5} = m_{QR}$)
c i for both: $m = 3$ ii for both: $m = -4$ d i gradients different ... not parallel ii $d = 1$

2 a $m_1 = -\frac{1}{2} = m_2 = 2 \Rightarrow m_1 m_2 = -1$ b $-\frac{1}{3}$ c $\frac{1}{5}$
d i $m_{EG} = \frac{1}{3}$; $m_{FH} = -3$ ii (1, 2) for both iii diagonals bisect at right angles so it is a rhombus

3 $m_{DG} = \frac{47 - 17}{58 - 34} = \frac{5}{4}$; $m_{GF} = \frac{47 - 22}{58 - 38} = \frac{5}{4} \Rightarrow$ DG $\parallel$ GF. Since G is a common point, points are collinear.

4 a 36·9° b 36·9° c 90° d 22·6°

5 a (2, 1) b $x + 2y - 4 = 0$ c $y = 2x - 3$

6 a $\frac{2}{3}$ b $3x + 2y = 19$

7 a $y - 11 = \frac{1}{2}(x + 3)$ or $y = \frac{1}{2}(x + 25)$ b i A: south of ii B: on iii C: north of

8 a $m = -\frac{b}{a}$; $y - b = -\frac{b}{a}(x - 0)$; hence result b $\frac{x}{b} - \frac{y}{a} = -1$

Exercise 12.2

1 a P(−2, 2), Q(5, 3), R(3, −2) b i $x + 4y - 6 = 0$ ii $y = -3x + 7$ c S(2, 1)
d i $2x - 3y - 1 = 0$ ii (2, 1) satisfies the equation

2 a i $y - 2 = \frac{1}{3}(x - 4)$ ii $y - 10 = -3(x - 8)$ b (10, 4) c $y - 5 = -\frac{1}{2}(x - 13)$
d (9, 7) e $y - 2 = 1(x - 4)$; the coordinates of P satisfy this equation

3 a i $y - 1 = -\frac{1}{2}(x - 3)$ ii $y - 8 = -3(x - 4)$ b (7, −1)

4 a i 14·0° ii 58·0° b 44·0° c $y = 0{\cdot}73x$

5 a $y + 6 = \frac{1}{4}(x + 6)$ b Q(2, −4) c $\sqrt{68}$

6 $2\sqrt{10}$

7 a $y = x + 1$ b $y = -3x + 25$ c (6, 7)
d $\sqrt{10}$ e $y = -3x + 25$; same as altitude

Exercise 12.3

1 a (0, 0), $r = 1$ b (0, 0), $r = 12$ c (0, 0), $r = 14$ d (0, 0), $r = 5$
e (0, 0), $r = \frac{1}{3}$ f (0, 0), $r = 10$ g (0, 0), $r = 6$ h (0, 0), $r = 12$

2 a i $8^2 \leqslant x^2 + y^2 < 16^2$ ii $16^2 \leqslant x^2 + y^2 < 24^2$ iii $24^2 \leqslant x^2 + y^2 < 32^2$
b i Red ii Gold iii Blue iv Black v Red c i $x^2 + y^2 = 256$ ii $x^2 + y^2 = 1024$

3 a $x^2 + y^2 = 625$; $r = 25$ b $x^2 + y^2 = 2809$; $r = 53$ c $x^2 + y^2 = 3721$; $r = 61$ d $x^2 + y^2 = 676$; $r = 26$

4 $x^2 + y^2 = 5476$; $d = 74$ [by Pythagoras, radius is 37]

5 a $x^2 + y^2 = 100$
b If $x^2 + y^2 < 100$ then spots are not stars. All stars are behind Moon, i.e. **i**, **iv**, **vi** are spots.

6 a (−3, −6), (−6, 3) b (2, −6), (−3·6, 5·2)
c (5, 3), (3, −5) d (1, 3), (−3, 1)
e (2, 6), (−6, 2) f (2, 4), (−4, 2)

Exercise 12.4

1 a $(5, 3); 10$ b $(3, -2); 12$ c $(-1, 1); 4$
d $(-2, -5); 2\sqrt{2}$ e $(-6, 7); 4\sqrt{2}$ f $(0, -1); 3\sqrt{3}$
g $(-7, 0); 5\sqrt{2}$ h $(-1{\cdot}5, 3{\cdot}5); 0{\cdot}7$ i $(0, 8); \frac{1}{\sqrt{2}}$

2 a $(x-5)^2+(y-4)^2=9$ b $(x-1)^2+(y-1)^2=100$ c $(x+2)^2+(y-1)^2=25$
d $(x+1)^2+(y+3)^2=10$ e $x^2+(y-4)^2=2$ f $(x-5)^2+y^2=1$

3 a $(x-1)^2+(y-4)^2=25$ b $(x-5)^2+(y-3)^2=169$ c $(x+3)^2+(y-6)^2=289$
d $(x+1)^2+(y+1)^2=841$ e $(x-2)^2+y^2=58$ f $x^2+(y+3)^2=85$
g $(x-1)^2+(y+2)^2=41$ h $(x-6)^2+(y+9)^2=136$

4 a i $(7, 6)$ ii $\sqrt{29}$ b $(x-7)^2+(y-6)^2=29$ c $(x-7)^2+(y-6)^2=36$

5 a $m_{AB}=\frac{3}{4}; m_{BC}=-\frac{4}{3} \Rightarrow AB \perp BC$
b i Midpoint of AC, i.e. $(10, -13)$ ii 25 iii $(x-10)^2+(y+13)^2=625$

6 a $(-2, -3)$ b $(x-1)^2+(y-1)^2=25$
c i $y-4=\frac{1}{7}(x+3)$ ii $(1, 5)$ is above the line AB but $(1-1)^2+(5-1)^2<25$, so it's inside circle

7 a i $\sqrt{34}$ ii $(x-1)^2+(y-2)^2=34$ b $(4, 7), (-4, -1)$ c $(10, 14), (-7, -3)$

8 a $(x-4)^2+(y+1)^2=16$ b C: $(x-4)^2+(y-1)^2=64$; D: $(x-4)^2+(y-1)^2=16$

9 a i ± 6, but only 1st quadrant ... $y_b=6$ ii $(x-8)^2+(y-6)^2=25$ iii $y_c=9$ or 3
b i $y_b=8$ ii $(x-6)^2+(y-8)^2=25$ iii $y_c=13$ or 3
c i $y_b=\sqrt{96}$ ii $(x-2)^2+(y-\sqrt{96})^2=25$ iii $y_c=\sqrt{96}\pm\sqrt{21}$

10 a $(8-3)^2+(4+8)^2=25+144=169$ ... point lies on circle b Centre $(3, -8)$; $P'=(-2, -20)$
c i $(x+3)^3+(y-\frac{28}{3})^2=\frac{169}{9}$ ii $(3, 5)$

Exercise 12.5

1 a $(-5, -1)$, 5 units b $(-2, 9)$, 10 units c $(20, 6)$, 13 units
d $(7, 18)$, 17 units e $(7, 12)$, 20 units f $(29, 9)$, 25 units
g $(1, 3)$, $\sqrt{14}$ units h $(-4, 0{\cdot}5)$, $\sqrt{11{\cdot}25}$ units i $(2{\cdot}5, -1{\cdot}5)$, $\sqrt{17{\cdot}5}$ units
j $(-0{\cdot}5, -0{\cdot}5)$, $\sqrt{1{\cdot}5}$ units k $(3, 0)$, $\sqrt{3}$ units l $(0, -1)$, 2 units

2 a $x^2+y^2-4x-6y-23=0$ b $x^2+y^2+2x-10y-74=0$ c $x^2+y^2-14x+4y-28=0$
d $x^2+y^2-2y=0$ e $x^2+y^2-8x-48=0$ f $x^2+y^2+10x-12y+59=0$
g $x^2+y^2-2x-2y-1=0$ h $x^2+y^2+16x+2y+24=0$

3 a $k<4{\cdot}25$ b $k<-4$ or $k>4$ c $k<-\sqrt{3}$ or $k>\sqrt{3}$ d $k<0$ or $k>2$

4 a i -2 ii -8 b 2 units

5 a $x^2+y^2-4x-2y-4=0$ b $x^2+y^2-16x-8y-64=0$

6 a Touches at $x=4$
b i $x^2+y^2+8x-10y+16=0$ ii Intersect on axis of reflection, i.e. where $x=0$ at $y=2$ and $y=8$.
c i $(0, 2), (7, 9)$ ii $(0, 8), (7, 1)$ d $C(3, 5)$, so all lie on line $y=5$

7 a $(6, 1)$ b i $y=-\frac{1}{2}x+5$ ii $x=2$
c $(2, 4)$ d $x^2+y^2-4x-8y-5=0$
e i $x^2+y^2-4x-8y+15=0$
ii Radius $=\sqrt{5}$; distance from centre to AB $=3$, so AB is not a tangent to the circle, hence it is not the incircle.

8 a $(4, 8)$, 6 units b $(4, 2)$ c $x^2+y^2+8x-8y+28=0$

9 a $(3, 5)$, 7 units b $(x-11)^2+(y-13)^2=36$
c i Cell 1 ii Both iii Cell 2 iv Neither

Exercise 12.6

1 a $(2, 7)$ b $(2, -4)$ c $(-1, 5)$ d $(-2, 1)$
e $(4, 5)$ f $(-6, 7)$ g $(-4, 2)$

2 a $y = 2x + 3$ b $y = 1 - 3x$ c $y = 4x - 3$ d $y = 2 - 5x$
e $y = \frac{1}{2}x - 2$ f $y = \frac{1}{3}x + 1$ g $y = 2 - \frac{1}{3}x$ h $y = \frac{2}{3}x - 1$

3 a $A(6, 7), r_A = \sqrt{26} \approx 5{\cdot}1$ units; $B(-2, -1), r_B = \sqrt{58} \approx 7{\cdot}6$ units b $8\sqrt{2} \approx 11{\cdot}3$ units
c i $r_A + r_B \approx 12{\cdot}7$ units ii $r_A - r_B \approx 2{\cdot}5$ units
d Since $2{\cdot}5 < 11{\cdot}3 < 12{\cdot}7$, the two circles intersect at two places. e (5, 2) and (1, 6)

4 a (3, 3) and (−1, 7) b (−2, 1) and (2, 3) c (2, 1) and (4, 3) d (5, 2) and (3, −2)

5 a i $x^2 + m^2x^2 - 4x - 8mx + 4 = 0$ ii $(1 + m^2)x^2 - (4 + 8m)x + 4 = 0$
b i $\Delta = 64m + 48m^2$ ii $m = 0$ or $m = -\frac{4}{3}$ iii $y = 0$ or $y = -\frac{4}{3}x$ c $y = \frac{1}{3}x$ or $y = -3x$

6 a i $x^2 + (x + c)^2 + 2x + 6(x + c) + 8 = 0$ ii $2x^2 + (2c + 8)x + (c^2 + 6c + 8) = 0$
b i $\Delta = -4c^2 - 16c$ ii $-4c^2 - 16c = 0 \Rightarrow c = 0$ or -4
iii When $c = 0$, equation is $y = x$ and point of tangency is $(-2, -2)$; $c = -4$, equation is $y = x - 4$ and point of tangency is $(0, -4)$.
c a i $x^2 + (x + c)^2 + 2x + 6(x + c) + 2 = 0$ a ii $2x^2 + (2c + 8)x + (c^2 + 6c + 2) = 0$
b i $\Delta = -4c^2 - 16c + 48$ b ii $-4c^2 - 16c + 48 = 0 \Rightarrow c = -6$ or 2
b iii When $c = 2$, equation is $y = x + 2$ and point of tangency is $(-3, -1)$; $c = -6$, equation is $y = x - 6$ and point of tangency is $(1, -5)$.

7 $c = 10$ or -10

8 a (0, 0), radius 10 units; (3, 4), radius 5 units b 5 units
c $d = r_1 - r_2$, so smaller circle touches larger from inside
d Point of contact lies on line joining centres, viz. $y = \frac{4}{3}x$; substituting into $x^2 + y^2 = 100$ gives (6, 8) as the point of contact.

9 a (4, −5), (−1, 5) b $5\sqrt{5}$ units

10 a (2, 5), 5 units b (−1, 1) and (2, 10)
c Closest point $\left(\frac{1}{2}, \frac{11}{2}\right)$ (where line and perpendicular through centre meet), which is $\frac{\sqrt{10}}{2}$ units from centre.

Preparation for assessment

1 $y = \sqrt{3}x + 5$

2 45°

3 $m_{AC} = -\frac{3}{2} = m_{BC}$ and C is a common point

4 $m_{AB} = -\frac{2}{3}$; $m_{AC} = \frac{3}{2}$; $m_{BC} = \frac{4}{7} \Rightarrow m_{AB} \times m_{AC} = -1 \Rightarrow AB \perp AC$, so triangle right-angled

5 a $y + 2 = -\frac{1}{3}(x - 13)$ b $y = \frac{1}{2}x - 1$ c (4, 1)

6 a i $y = 3$; $y = -2x + 15$ ii (6, 3) iii $(x - 6)^2 + (y - 3)^2 = 65$
b The point is equidistant from all 3 villages.
c i $y = -\frac{2}{3}x + \frac{8}{3}$ ii Closest at (4, 0), distance $\sqrt{13}$ units

7 a Use discriminant as test b $m_{radius} = -\frac{1}{m_{tangent}} = \frac{3}{4}$. It passes through (0, 0), so equation is $y = \frac{3}{4}x$.

8 a (10, 4), (2, 8) b $y = \frac{1}{3}x + \frac{22}{3}$; $y = -3x + 34$ c (8, 10)

9 a $x^2 + y^2 - 10x - 24y = 0$ [line joining intercepts is a diameter ... angle in semicircle]
b i (0, 0) and (17, 17) ii $y = -\frac{5}{12}x$ and $y - 17 = -\frac{12}{5}(x - 17)$

10 a i Substitute, form quadratic, check discriminant = 0; hence coincident roots and tangency. ii (8, 7)
b $(x - 6)^2 + (y - 6)^2 = 5$

11 (7, −1) and (5, 5)

13 Recurrence relations

What you need to know

1 a i 11 ii −1 iii −4 iv −4 b i 8 ii 10 iii 12 iv 1

2 a i 2 ii 3 iii 4 iv 5 b Produces a sequence of whole numbers starting at 2.

3 a 3, 5, 7, 9, 11
b i 1, 3, 5, 7, ... the odd numbers ii 2, 4, 6, 8, ... the even numbers starting at 2
iii 1, 3, 6, 10, ... the triangular numbers

4 a $3x^2$ b 0 c 0

5 Lengths of radials to anchor points form the sequence $\sqrt{1}, \sqrt{2}, \sqrt{3}, \sqrt{4}, \ldots u_n = \sqrt{n}$.

Exercise 13.1

1 a i 6, 10, 14, 18 ii $u_n = 4n + 2$ iii 402
b i 1, −2, −5, −8 ii $u_n = -3n + 4$ iii −296
c i −4, −2, 0, 2 ii $u_n = 2n - 6$ iii 194
d i −2, −7, −12, −17 ii $u_n = -5n + 3$ iii −497
e i 0·5, 2, 3·5, 5 ii $u_n = 1{\cdot}5n - 1$ iii 149
f i $a, a + b, a + 2b, a + 3b$ ii $u_n = a + (n - 1)b$ iii $a + 99b$

2 a $a = 2, b = 5$ b $a = -4, b = 8$ c $a = 7, b = -2$
d $a = -1, b = -4$ e $a = \frac{1}{2}, b = \frac{1}{2}$ f $a = \frac{1}{2}, b = \frac{1}{3}$
g $a = x, b = x$ h $a = x, b = x - 1$ i $a = 1 - x, b = 2 + x$

3 a $u_1 = 6, u_{n+1} = u_n + 2$ b $u_1 = 4, u_{n+1} = u_n + 5$ c $u_1 = 2, u_{n+1} = u_n - 1$
d $u_1 = 6\frac{1}{2}, u_{n+1} = u_n + \frac{1}{2}$ e $u_1 = 2, u_{n+1} = u_n - 3$ f $u_1 = 5\frac{2}{3}, u_{n+1} = u_n - \frac{1}{3}$
g $u_1 = 3 + b, u_{n+1} = u_n + 3$ h $u_1 = a + b, u_{n+1} = u_n + a$

4 a $a = 1, b = 3; u_{n+1} = u_n + 3, u_1 = 1$ b $a = -2, b = 4; u_{n+1} = u_n + 4, u_1 = -2$
c $a = 3, b = -2; u_{n+1} = u_n - 2, u_1 = 3$ d $a = \frac{1}{4}, b = \frac{3}{4}; u_{n+1} = u_n + \frac{3}{4}, u_1 = \frac{1}{4}$
e $a = x, b = x + 1; u_{n+1} = u_n + x + 1, u_1 = x$ f $a = x + 1, b = x; u_{n+1} = u_n + x, u_1 = x + 1$

5 a 875 m^3 b 35 m^3 c i 525 m^3 ii 350 m^3
d i $u_{n+1} = u_n + 35, u_1 = 35$ ii 15 trips e 25 trips

6 a 6 assistants: by 90 people; 10 assistants: by 150 people
b 15 people/hour
c Student check on 6-assistant case; 10-assistant case: $Q_n = q - 150n$
d Look for $Q_{10} = 0$: (6 assistants) $q = 900$; (10 assistants) $q = 1500$

7 a i nth odd number = $2n - 1$; so 99 is the 50th odd number; sum = $(1 + 99) \times 50 \div 2 = 2500$
ii nth term = $5n$; so 100 is the 20th number; sum = $(5 + 100) \times 20 \div 2 = 1050$
iii Sum = $\frac{n}{2}[2a + (n - 1)b]$.
b Student's own investigation leading to confirmation that $S_n = \frac{n}{2}[2a + (n - 1)b]$.

Exercise 13.2

1 a 2, 6, 18, 54 b 1·5, 6, 24, 96 c 486, 162, 54, 18
d 96, 48, 24, 12 e 1, 1·1, 1·21, 1·331 f π, π^2, π^3, π^4

2 a i $u_{n+1} = 3u_n, u_1 = 5$ ii $u_n = 5 \times 3^{n-1}$ iii 98 415
b i $u_{n+1} = 2u_n, u_1 = 7$ ii $u_n = 7 \times 2^{n-1}$ iii 3584
c i $u_{n+1} = 4u_n, u_1 = 2$ ii $u_n = 2 \times 4^{n-1}$ iii 524 288
d i $u_{n+1} = \frac{1}{2}u_n, u_1 = 160$ ii $u_n = 160 \times \frac{1}{2}^{n-1}$ iii 0·3125
e i $u_{n+1} = \frac{1}{4}u_n, u_1 = 6144$ ii $u_n = 6144 \times \frac{1}{4}^{n-1}$ iii 0·0234375
f i $u_{n+1} = -u_n, u_1 = 3$ ii $u_n = 3 \times (-1)^{n-1}$ iii −3
g i $u_{n+1} = -3u_n, u_1 = 2$ ii $u_n = 2 \times (-3)^{n-1}$ iii −39 366
h i $u_{n+1} = \frac{1}{10}u_n, u_1 = 10$ ii $u_n = 10 \times \frac{1}{10}^{n-1}$ iii 0·00000001
i i $u_{n+1} = bu_n, u_1 = a$ ii $u_n = a \times b^{n-1}$ iii ab^9

3 a $a = \pm5, b = \pm2; u_{n+1} = \pm5u_n, u_1 = \pm2$
b $a = \pm2, b = 4; u_{n+1} = \pm2u_n, u_1 = 4$
c $a = 3, b = 6; u_{n+1} = 3u_n, u_1 = 6$
d $a = 0{\cdot}5, b = 20; u_{n+1} = 0{\cdot}5u_n, u_1 = 20$
e $a = \pm\frac{1}{10}, b = \pm1000; u_{n+1} = \pm\frac{1}{10}u_n, u_1 = \pm1000$
f $a = \pm0{\cdot}6, b = \pm90; u_{n+1} = \pm0{\cdot}6u_n, u_1 = \pm90$

4 a $u_{n+1} = 1{\cdot}6u_n, u_1 = 4{\cdot}8$
b $u_{n+1} = 2{\cdot}8u_n, u_1 = 5{\cdot}6$
c $u_{n+1} = 0{\cdot}5u_n, u_1 = 0{\cdot}8$
d $u_{n+1} = au_n, u_1 = ab$

5 a i ∞ ii 0 iii ∞ iv 0 v ∞ vi 0
b If, for a common ratio of r, $-1 < r < 1$, it tends to zero; if $r > 1$ or $r < 1$, it tends to infinity; if $r = 1$, term is constant; if $r = -1$, term has constant magnitude but the sign alternates.

6 a $M_1 = M_0 \times 1{\cdot}05 \Rightarrow M_0 = 21 \div 1{\cdot}05 = 20$ magpies
b $u_n = 20 \times 1{\cdot}05^n \Rightarrow u_{11} = 34$ magpies, to the nearest bird
c By 15th year

7 a £21 250, £18 063, £15 353, £13 050
b £4922 (to nearest pound)
c Year 16
d $P_{n+1} = \frac{1}{100}v_{n+1} = \frac{1}{100}v_n\,.\,0{\cdot}85 = P_n\,.\,0{\cdot}85$
e $v_n \to 0$ as $n \to \infty$

8 a £12 167 (to nearest pound) **b** £20 258 (to nearest pound) **c** by his 12th birthday

9 a 10 m^3
b 0·10 m^3 (to 2 d.p.)
c 14th day is 1st time below 0·01 m^3
d 6th day

10 a $N = 0{\cdot}\dot{3} = 0{\cdot}3 + \frac{1}{10}(0{\cdot}\dot{3}) = 0{\cdot}3 + \frac{1}{10}N \Rightarrow \frac{9}{10}N = 0{\cdot}3 \Rightarrow N = \frac{1}{3}$
[Note $0{\cdot}\dot{3}$ is notation for 0·3 recurring, the dots signifying the start and finish of the repeating period, so, for example, $0{\cdot}\dot{2}1\dot{3} = 0{\cdot}21321321...$]
b $N = 0{\cdot}\dot{1} = 0{\cdot}1 + \frac{1}{10}(0{\cdot}\dot{1}) = 0{\cdot}1 + \frac{1}{10}N \Rightarrow \frac{9}{10}N = 0{\cdot}1 \Rightarrow N = \frac{1}{9}$, which is rational
c $N = 0{\cdot}\dot{1}\dot{2} = 0{\cdot}12 + \frac{1}{100}(0{\cdot}\dot{1}\dot{2}) = 0{\cdot}12 + \frac{1}{100}N \Rightarrow \frac{99}{100}N = 0{\cdot}12 \Rightarrow N = \frac{12}{99}$
d $S = 1 + 0{\cdot}1S \Rightarrow 0{\cdot}9S = 1 \Rightarrow S = \frac{10}{9}$
e $S = \dfrac{1}{1-r}$
f Student's own check

Exercise 13.3

1 a $6, 2, -6, -22$ **b** 5, 16, 49, 148 **c** 2, 2, 2, 2
d 10, 8, 6·6, 5·62 **e** 300, 278, 258·2, 240·38 **f** 256, 208, 172, 145

2 a $u_{n+1} = 2u_n + 1, u_1 = 5$ **b** $u_{n+1} = 3u_n + 6, u_1 = 7$ **c** $u_{n+1} = 5u_n + 2, u_1 = 4$
d $u_{n+1} = 0{\cdot}5u_n + 2, u_1 = 64$ **e** $u_{n+1} = 0{\cdot}6u_n + 14, u_1 = 360$ **f** $u_{n+1} = 0{\cdot}8u_n + 18, u_1 = 40$

3 a $u_{n+1} = 4u_n + 1, u_1 = 2$
b $u_{n+1} = 5u_n + 2, u_1 = 3$ or $u_{n+1} = -6u_n + 35, u_1 = 3$
c $u_{n+1} = 0{\cdot}5u_n + 15, u_1 = 6$ or $u_{n+1} = -1{\cdot}5u_n + 27, u_1 = 6$
d $u_{n+1} = 2u_n + 4, u_1 = 9$
e $u_{n+1} = 3u_n - 1, u_1 = 1$ or $u_{n+1} = -\frac{5}{2}u_n - 1, u_1 = -\frac{6}{5}$
f $u_{n+1} = 4u_n + 54, u_1 = -12$
g $u_{n+1} = \frac{1}{2}u_n - 2, u_1 = 20$
h $u_{n+1} = 0{\cdot}2u_n - 2, u_1 = 60$ or $u_{n+1} = -1{\cdot}2u_n + 82, u_1 = 60$

4 a $-2{\cdot}5$ **b** 4·5 **c** 4 **d** 10 **e** 8 **f** 10

5 a 18·78 units **b** i Yes, 17·5 ii Good news 17·5 > 17

6 a i $\frac{3}{5}$ ii $\frac{2}{3}$ **b** $a = \dfrac{L-8}{L}$

7 a $a = 3, b = 2, c = 2$ **b** $a = 4, b = 6, c = 2$
c i $a = 3, b = -2, c = 4$
ii The r.r. generates 82 as u_4, which is indeed $3^4 + 1$.
iii No, special cases, no matter how many, don't constitute a proof.

8 a $A_{n+1} = A_n \times 1{\cdot}04 + 5000$
b Both the formula and the recurrence relation give 32930·9056 as A_4.
c $135\,000 \times 1{\cdot}04^n - 125\,000 = 100\,000 \Rightarrow 1{\cdot}04^n = 1{\cdot}667 \Rightarrow n = \dfrac{\ln(1{\cdot}667)}{\ln(1{\cdot}04)} = 13$ to the nearest year

9 a i $u_{n+1} = 0{\cdot}7u_n + 50, u_0 = 3000$ ii $u_{n+1} = 0{\cdot}4u_n + 50, u_0 = 3000$ iii $u_{n+1} = 0{\cdot}1u_n + 50, u_0 = 3000$
b Limits for each are i $166\frac{2}{3}\,\text{m}^2$ ii $83\frac{1}{3}\,\text{m}^2$ iii $55\frac{5}{9}\,\text{m}^2$
c Night watchman affects only the added constant; 50 becomes 5 and limits become
i $16\frac{2}{3}\,\text{m}^2$ ii $8\frac{1}{3}\,\text{m}^2$ iii $5\frac{5}{9}\,\text{m}^2$
d Employing the night watchman has a significant effect.
Going from 1 man + the night watchman to 2 men + the night watchman halves the graffiti. Going to 3 men and the night watchman has a less dramatic effect. Employing 2 men + a night watchman would seem the most effective and financially efficient way to tackle the graffiti problem.

10 a If $0 < \theta < 90$, then $-1 < \sin\theta < 1$. Since $\sin\theta$ is the multiplier in the linear recurrence relation then a limit exists.
b To 1 d.p., i 13·5 ii 20·0 iii 74·6
c Where a limit, L, exists, $L = \dfrac{10}{1 - \sin\theta°}$
d To 1 decimal place, if it is less than 36·9 then limit will be less than the safe distance of 25 mm.

11 a Greys stable at 450; reds at 150
b $G:R = 450:150 = 3:1$ ... extinction possible
c $h = 47$; at this point $G_{10} = 161$ and $R_{10} = 166$

12 a i Check 6 and 8 are fixed points
ii The second fixed point, 8, is stable. Use a starting point away from 8 and the sequence heads back to 8; the first fixed point, 6, is unstable. Use a starting point away from 6 and the sequence 'runs away' from 6.
iii Fixed points in order are 7 (unstable) and 13 (stable). The stable fixed points occur when $-1 < x < 1$.
b Fixed point is always $\sqrt{a}$. This has been used in the past as an algorithm for finding the square root of a number.
c i Student's proof
ii After term 14, the computer output suggests a fixed point of −3·064695385 ... Substituting this into the equation confirms it as a root of the equation.
iii Newton's method gives the recurrence relation $x_{n+1} = \dfrac{x_n - x_n^4 - x_n^3 - 200}{4x_n^3 - 3x_n^2}$. Putting 1 in A1 and typing into
A2: =A1-(A1^4-A1^3-200)/(4*A1^3-3*A1^2) ... and filling down, will yield the solution $x = 4{\cdot}0379$. (to 5 s.f.)

Preparation for assessment

1 a i $u_3 = \frac{8}{5}$
ii $u_1 = 8$
iii A limit exists because the recurrence relation can be written as $u_{n+1} = \frac{3}{5}u_n - \frac{4}{5}$ and we can see the multiplier in the linear recurrence relation lies between 1 and −1, i.e. $-1 < \frac{3}{5} < 1$; limit = −2
b $a = 0{\cdot}5, b = 5$ or $a = -1{\cdot}5, b = 105$ **c** Limit = 10

2 a $d_{n+1} = d_n + 67, d_1 = 67$ **b** i $67n$ ii 670 km iii 402 km

3 a i 1, 2, 4, 8 ii $G_n = 2^{n-1}$ **b** By 13th generation **c** i 1, 3, 7, 15 ii $T_n = 2^n - 1$
d By 12th generation

4 a $A_{n+1} = 1{\cdot}01A_n - 500, A_1 = 5000$ **b** £3172·82 (to nearest penny) **c** $n = 11{\cdot}6$, i.e. 12th month

5 a $a_{n+1} = 0{\cdot}78a_n, a_0 = 100$ **b** 22·5% **c** $d_{n+1} = 0{\cdot}225d_n + 100$
d Limit exists and is 129 units. Since this is only 1·29 times the administered dose, it is considered sensible.

14 Preparation for course assessment

Exercise 14.1

1 a Check that remainder is zero so that $(x - 1)$ is a factor. **b** $(x - 1)(x - 2)(x + 3)(2x - 1)$
c $x = 1, 2, -3, \frac{1}{2}$

2 a 2 **b** $4a^{2x+x-6} = 4a^0 = 4$

3 a i $\frac{24}{25}$ ii $-\frac{16}{65}$ **b** i 0·01 ii 0·1

4 a $17\sin(x + 62)°$ **b** i 57 ii 208

5 a i $y = x^2 - 6x + 5$ ii A(0, 5), B(1, 0), C(5, 0)

b i
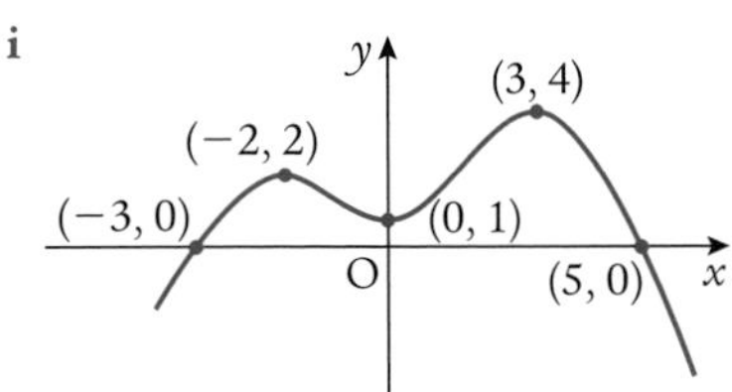

ii
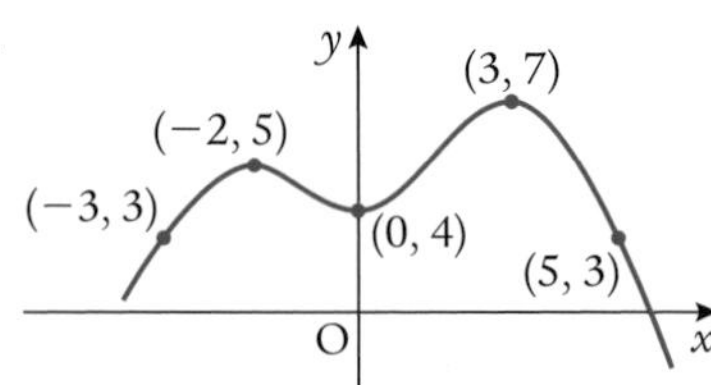

iii
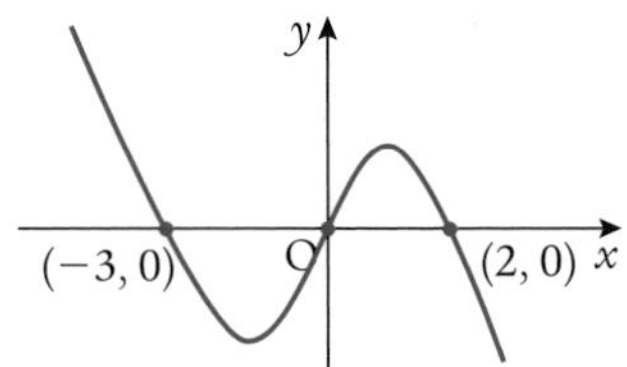

c $f^{-1}(x) = 10^{\frac{x}{3}}$
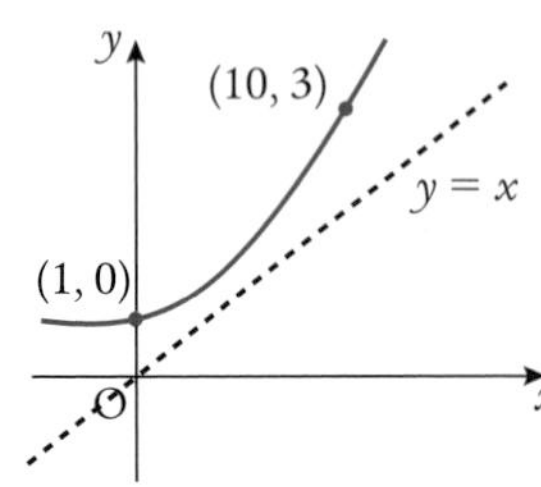

6 a $2(x + 1)^2 + 3$ **b** 2

7 a i $28 - 3x^2$ ii $9x^2 - 6x - 8$ **b** $x = -\frac{3}{2}$ or $x = 2$ **c** $f^{-1}(x) = \dfrac{1 - x}{3}$

8 $k < \frac{9}{8}$

9 a $-0{\cdot}86$ **b** $4{\cdot}48$

10 $\left(\frac{1}{2}, \frac{3}{4}\right), (-3, 20)$

11 a $82{\cdot}0°, 334{\cdot}0°$ **b** $\dfrac{\pi}{6}, \dfrac{\pi}{2}, \dfrac{5\pi}{6}$ or $\dfrac{3\pi}{2}$ **c** $\dfrac{\pi}{3}, \dfrac{5\pi}{3}$

12 a i $\mathbf{w} - \mathbf{u}$ ii $\frac{1}{2}(\mathbf{w} - \mathbf{u})$ iii $-\mathbf{v} + \mathbf{u} + \frac{1}{2}(\mathbf{w} - \mathbf{u})$

b i $\overrightarrow{BN} = \mathbf{v}; \overrightarrow{AN} = \overrightarrow{AB} + \overrightarrow{BN} = \frac{1}{2}(\mathbf{u} + \mathbf{w})$ ii $\mathbf{w}$

13 $\overrightarrow{AB} = \begin{pmatrix} 3 \\ 5 \\ 2 \end{pmatrix}; \overrightarrow{AC} = \begin{pmatrix} 3 \\ 5 \\ 2 \end{pmatrix}; \Rightarrow \overrightarrow{AB} \parallel \overrightarrow{AC}$ ***and*** A a common point, so A, B, C collinear

14 $(6, 3, -1)$

15 $\cos\theta = \dfrac{-2}{\sqrt{22}\sqrt{75}} \Rightarrow \theta = 92{\cdot}8 \Rightarrow$ acute angle is $87{\cdot}2°$

16 a $4x^3 + 2x - 2x^{-3}$ **b** $3x^2 + \frac{1}{2}x^{-\frac{1}{2}} - \frac{1}{2}x^{-\frac{3}{2}}$ **c** $3\cos x - 4\sin x$ **d** $12\cos(4x - 1)$
e $\frac{1}{2}(x^3 + 2x + 1)^{-\frac{1}{2}}(3x^2 + 2)$ **f** $-3(x^2 + 2x)^{-2}(2x + 2)$ **g** $4x + 7$

17 $y = -4x$

18 a $x^2 + 2x - 3$ **b** $f'(x) < 0$ when $-3 < x < 1$

19 a $(-10, 0), (-1, 0), (14, 0)$ **b** $(0, -140)$ **c** $(8, -972)$ min., $(-6, 400)$ max.

d
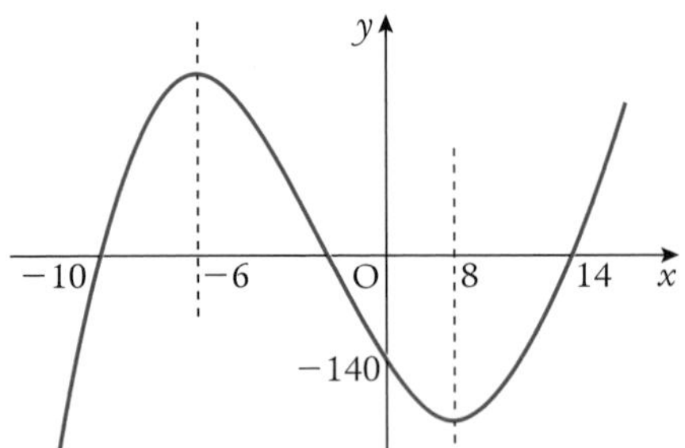

20 a $2x^4 + \dfrac{1}{x} + c$ **b** $4x^{\frac{3}{2}} - \dfrac{x^2}{2} + 4x + c$ **c** $3x^{\frac{1}{2}} - \frac{1}{5}x^{\frac{5}{2}} + c$

21 a $\frac{1}{6}(x + 3)^6 + c$ **b** $3\sin x + 5\cos x + c$
c $\frac{1}{6}(4x - 5)^{\frac{3}{2}} + c$ **d** $-\cos(3x + 1) + 2\sin(2x + 3) + c$

22 $y = x^3 - 3x^2 + x + 4$

23 a 24 **b** 1 **c** $-\frac{1}{12}$ **d** $2\frac{5}{12}$

24 $x = \sqrt{8}$ hours

25 a $v(x) = 6x^2 + 2x$ **b** **i** 21 m **ii** $28\ \text{m s}^{-1}$ **c** $a(x) = 12x + 2$; $a(3) = 38\ \text{m s}^{-2}$

26 $20\frac{5}{6}$ unit^2

27 a (5, 33) and (1, 13) **b** 32 unit^2

28 a $s(t) = 40t + 5t^2 + 300$ **b** **i** 465 m **ii** $70\ \text{m s}^{-1}$

29 a $y = -\frac{1}{3}x + 11$ **b** $y = 3x - 19$

30 a 53·1° **b** $y - 5 = \sqrt{3}(x - 1)$

31 a $y = x + 2$ **b** $y = -\frac{1}{3}x + 10$ **c** (6, 8)

32 a $y - 3 = \frac{1}{7}(x - 3)$ **b** 10

33 a $(x - 2)^2 + (y - 5)^2 = 100$ **b** **i** Inside **ii** Outside **iii** On
c $y - 11 = \frac{4}{3}(x + 6)$

34 a (4, 3) and (2, 7) **b** **i** $r_1 + r_2 \approx 9{\cdot}5$; $|r_1 - r_2| \approx 3{\cdot}2$; $d \approx 4{\cdot}2$: $|r_1 - r_2| < d < r_1 + r_2 \Rightarrow$ intersects twice
b **ii** (4, 3) and (0, 7)

35 a $a = 0{\cdot}6, b = 0{\cdot}4$; $u_{n+1} = 0{\cdot}6u_n + 0{\cdot}4$, $u_1 = 26$ **b** $u_4 = 6{\cdot}4$
c **i** $-1 < 0{\cdot}6 < 1$ **ii** Limit is 1. Over time the contents of the tank will stabilise at 1 litre.

Exercise 14.2

1 a $y = x - 1$ **b** $y = -\frac{1}{2}x + 2$ **c** (2, 1)

2 a $9x^2 + 9x + (2 + k)$ **b** **i** $k = \frac{1}{4}$ **ii** $-\frac{1}{2}$

3 a $F_2 = 131$ **b** **i** $F_{n+1} = 0{\cdot}9F_n + 5$; $-1 < 0{\cdot}9 < 1 \Rightarrow$ limit exists **ii** 50 units

4 a $g^2 + f^2 - c < 0$
b **i** $(x - 3)^2 + (y - 4)^2 = 10$ **ii** (0, 5) and (6, 5) **iii** $y = 3x + 5$; $y = -3x + 23$ **iv** (3, 14)

5 a $\Delta = 6^2 - 4.3.5 = 36 - 60 < 0 \Rightarrow$ no real roots
b **i** $3(x + 1)^2 + 2$ **ii** Min. value when $x = -1$, viz. $y = 2$ units

6 $x = \frac{\pi}{6}, \frac{5\pi}{6}, \frac{\pi}{2}$

7 a **i** 1·204 **ii** 0·301 **iii** −0·301 **b** 0·778

8 a $\overrightarrow{AC} = \begin{pmatrix} 2 \\ 11 \\ -10 \end{pmatrix}$ **b** $|\overrightarrow{AB}| = 14$ **c** $\cos ABC = \frac{104}{210} \approx 0{\cdot}50$ **d** $\angle ABC \approx 60°$

9 a Cuts y-axis at $(0, -9)$, cuts x-axis at $(-3, 0)$ and $(3, 0)$ **b** Derivative = 0 when $x = -2, 0, 2$

x	→	−2	→	0	→	2	→
$4x$	−	−	−	0	+	+	+
$x - 2$	−	−	−	−	−	0	+
$x + 2$	−	0	+	+	+	+	+
$\frac{dy}{dx}$	−	0	+	0	−	0	+
Slope	\	—	/	—	\	—	/
Nature		Min		Max		Min	

Stationary points:
minimum at (−2, −25)
maximum at (0, −9)
minimum at (2, −25)

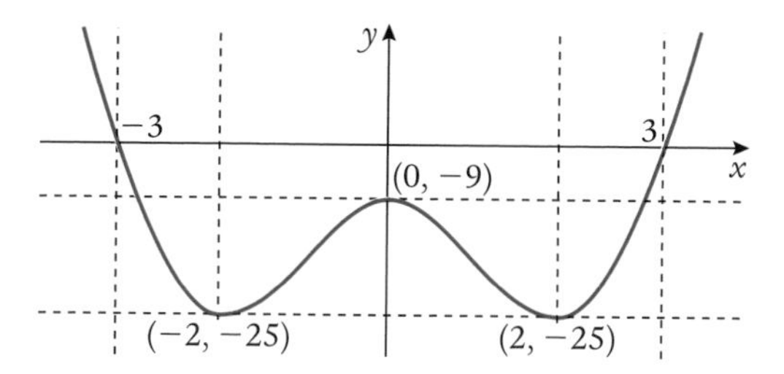

10 a Intersect at $x = -3, x = 1, x = 2$
b 32·75 unit^2

Exercise 14.3

1 a E(6, 6)

b $m_{BD} = m_{BE} = \frac{12-6}{10-6} = \frac{3}{2}$; $m_{AC} = \frac{2-8}{12-3} = -\frac{2}{3}$; $m_{BD} \cdot m_{AC} = -1 \Rightarrow BD \perp AC$

c $y_D = -6$

2 a E(3, 1), F(7, −3) **b** $(x-5)^2 + (y+1)^2 = 8$

3 a $y = 3x + 5$ **b** (3, 14)

4 a $-\frac{\pi}{60}\,\text{cm s}^{-2} \approx -0.052\,\text{cm s}^{-2}$ **b** $\left(\frac{15\sqrt{3}}{\pi} + 50 - \frac{15}{\pi}\right)\text{cm} \approx 53.5\,\text{cm}$

5 a $V = \frac{4}{3}\pi r^3 + \pi r^2 h$. Hence result. **b** $A = 4\pi r^2 + 2\pi rh$. Hence result.

c $r_{min} = \frac{3}{\sqrt[3]{\pi}} \approx 2.05$ mm [Remember to justify the nature of the stationary point.]

6 a $k = 500$ **b** $-\frac{3}{4}$

7 a **i** $\frac{4}{\sqrt{17}}$ **ii** $\frac{1}{\sqrt{17}}$ **b** **i** $\frac{8}{17}$ **ii** $-\frac{15}{17}$ **c** $-\frac{47}{17\sqrt{17}}$

8 a $a = 6.5, b = 0.2$ **b** **i** 6·5 cm **ii** 48 cm **c** day 14

9 a $53\cos(x + 5.727)$ [you should be working in radians] **b** 106 cm

c 2·98 time units and 4·41 time units [5 h 42 min and 8 h 26 min]

10 a $-|\mathbf{u}|^2\sqrt{\frac{3}{2}}$ **b** $-\frac{8}{5}\mathbf{i} - \frac{4}{5}\mathbf{k}$ **c** $\frac{11}{15}\mathbf{i} + \frac{10}{15}\mathbf{j} + \frac{2}{15}\mathbf{k}$